The compound eye
and vision of insects

The compound eye and vision of insects

EDITED BY
G. A. HORRIDGE

CLARENDON PRESS · OXFORD
1975

Oxford University Press, Ely House, London W.1

GLASGOW   NEW YORK   TORONTO   MELBOURNE   WELLINGTON
CAPE TOWN   IBADAN   NAIROBI   DAR ES SALAAM   LUSAKA   ADDIS ABABA
DELHI   BOMBAY   CALCUTTA   MADRAS   KARACHI   LAHORE   DACCA
KUALA LUMPUR   SINGAPORE   HONG KONG   TOKYO

ISBN 0 19 857375 8

© OXFORD UNIVERSITY PRESS 1975

PRINTED IN GREAT BRITAIN BY
WILLIAM CLOWES & SONS LTD.,
LONDON, COLCHESTER, AND BECCLES

Preface

THIS symposium originated when fourteen of the contributors presented their recent findings on the eye and vision of insects to the International Congress of Entomology held in Canberra, Australia in August 1972. Of the remainder of the contributors, some were unable to appear at the meeting but the majority responded to an invitation to add a contribution to the volume. Bearing in mind that those who responded were a small fraction of the experts who were invited, and that the visitors to the Congress were self-selected, it is fortunate that the collection of papers has turned out to cover many aspects of the anatomy, physiology, optics, and visual behaviour related to compound eyes. Even so the editor is aware of the deficiencies of the volume. In symposium volumes the subject is usually covered unevenly, according to the varied experience and application of the authors. Contributors often write about their own research rather than provide the general survey which most readers expect. Some topics, hopefully few, have not been dealt with at all, but where this has unfortunately happened it is because the most appropriate experts have declined on the grounds that they have written accounts that are available elsewhere.

One of the features of this volume is that it is published in English although this is not the native tongue of a number of the contributors. In these cases the script has been revised by the editor and changes have been referred back to the authors. While this process has removed the obvious infelicities, the characteristic styles of the originals are to some extent retained. The editor thanks these authors for their meek acceptance of his somewhat high-handed revision of their texts in that this acceptance accelerated the process and reduced the work. In return these authors expressed their thanks, which have not been repeated in each case throughout the book.

The compound eye and the part it plays in insect behaviour has been the subject of intensive research in the last decade. The number of references and their dates bears witness to this. The recent interest in the topic has grown partly because the techniques of electron microscopy, microelectrode recording, optics of light guides, and progressively more sophisticated analysis of behaviour have all recently made feasible a wide range of observations. As a result of this recent advance to a new plateau of understanding, the main types of insect eyes have now been described and the steps towards the understanding of the mechanisms can be outlined. Most of the problems of the optics of the retina, transduction by the receptors, neuron anatomy of the optic lobes and characterization of optic lobe neurons by their responses can now be outlined as problems although we are still far from complete solutions.

Oxford   G. A. H.
June 1973

Contributors

F. Baumann
Department of Physiology, Ecole de Médecine 1211 Genève 4, Switzerland

P. Carricaburu
Museum National d'Histoire Naturelle, 57, rue Cuvier, 75005 Paris, France

T. Collett
School of Biological Sciences, University of Sussex, Brighton, Sussex BN1 9QG

C. Taddei Ferretti
Laboratorio di Cibernetica, Via Toiano 2, 80072 Arco Felice, Napoli, Italy

L. J. Goodman
Department of Zoology, Queen Mary College, Mile End Road, London E.1.

F. G. Gribakin
Sechenov Institute, Academy of Sciences, Leningrad, U.S.S.R.

G. A. Horridge
Department of Neurobiology, Research School of Biology, Box 475, Canberra, Australia

W. Kaiser
Zoologisches Institut, Technischen Hochschule, Schnittspahnstrasse 3, 61 Darmstadt Germany

J. Kien
Department of Neurobiology, Research School of Biology, Box 475, Canberra, Australia

A. J. King
School of Biological Sciences, University of Sussex, Brighton, Sussex BN1 9QG

S. B. Laughlin
Department of Neurobiology, Research School of Biology, Box 475, Canberra, Australia

M. Land
School of Biological Sciences, University of Sussex, Brighton, Sussex BN1 9QG

G. A. Mazokhin-Porshnyakov
Institute for Information Transmission Problems, Academy of Science, Awiamotornaja Str. 8-a, Moscow E 24, U.S.S.R.

S. Meggitt
Department of Applied Mathematics, School of General Studies, Australian National University, Box 4, Canberra, Australia

R. Menzel
Zoologisches Institut, Technischen Hochschule, Schnittspahnstrasse 3, 61 Darmstadt, Germany

V. B. Meyer-Rochow
Department of Neurobiology, Research School of Biology, Box 475, Canberra, Australia

K. Mimura
Nagasaki University, Faculty of Liberal Arts, Nagasaki 852, Japan

W. J. Mueller
Masonic Medical Research Laboratory, Utica, N.Y., U.S.A.

R. B. Northrop
Bioengineering Laboratory, Box U.157, University of Connecticut, Storrs, Conn. 06268, U.S.A.

J. Palka
Departments of Zoology and Electrical Engineering, University of Washington, Seattle, Washington 98195, U.S.A.

H. F. Paulus
Biologisches Institüt der Universitat, 78 Freiburg, Katharinestr. 20, Germany

R. B. Pinter
Departments of Zoology and Electrical Engineering, University of Washington, Seattle, Washington 98195, U.S.A.

A. W. Snyder
Department of Applied Mathematics, Research School of Physical Sciences, Australian National University, Box 4, Canberra, Australia

A. Fernandez Perez de Talens
Laboratorio di Cibernetica, via Toiano 2, 80072 Arco Felice, Napoli, Italy

B. Walcott
Department of Anatomy, University of New York, Stonybrooke, Long Island, N.Y., U.S.A.

R. WEHNER
Zoologisch-vergl. Anatomisches, Institut der Universitat, Zurich 8006, Kunstlergasse 16, Switzerland

V. J. WULFF
Masonic Medical Research Laboratory, Utica, N.Y., U.S.A.

Contents

PART III
OPTICS

xvi   *Contents*

PART V
BEHAVIOURAL ANALYSIS

Part I
Receptor Anatomy

1. The compound eyes of apterygote insects*

H. F. PAULUS

AMONG the primitive wingless insects Protura, Diplura, Collembola and Thysanura, only the two latter have eyes, and true facetted compound eyes occur only in Archaeognatha (Machilidae), while the others have unconsolidated lateral eyes consisting of groups of ommatidia. The ommatidium in these forms is interesting as an indication of the precursors and evolution of the units of the pterygote compound eye.

1.1. Atypical compound eyes of Collembola

The Collembola usually have two groups of up to eight relatively large circular ommatidia, in size and arrangement characteristic of the species. They are frequently reduced or even absent, as in *Onychiurus*. These groups are recognized as large pigmented spots in the upper part of the head, directly behind the antennae, or pressed between the antennae, as in *Sminthuridae* (Fig. 1.1). A section through the group shows that it is surrounded by an extensive pigment layer which is composed of hypodermal cells with long slender pigment-filled proximal projections. Pigment cells also extend among the ommatidia, where they can be called secondary or accessory pigment cells. Each ommatidium is further isolated optically by pigment grains of the retinula cells.

From the work of Willem (1897) it was established that the grouped ommatidia of Collembola are not ocelli as found in larvae of holometabolic insects, but genuine eucone ommatidia as characterized by Grenacher (1880). As an exception in the genera *Anurida* and *Anurophorus*, Willem (1900) described the eyes as 'larval' ocelli. Subsequently, Hesse (1901) published a scheme for collembolan eyes, which now requires revision as a result of the electron microscope studies of Barra (1971*a,b*) and Paulus (1970*a,b*; 1972*a,b*; 1973*a,b*). This recent work shows the structure to be very diverse, as briefly summarized below for the two main types of Collembola.

* Research supported by 'Deutsche Forschungsgemeinschaft.'

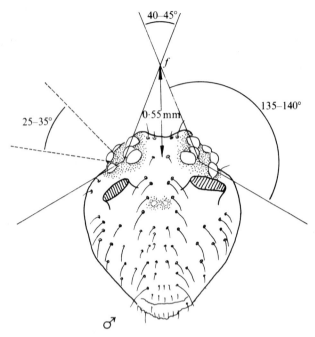

FIG. 1.1. Front view of the head of *Sminthurides aquaticus*, with antennae removed, showing the positions of the ommatidia and the presumed fields of view of one unit and of both groups, with overlap.

Abbreviations used in all illustrations

A	axon	KK	crystalline cone
C	cornea	N	nucleus
Co	corneagenous cell	R	retinula cell
EP	extracellular pigment	VS	palisade
HP, HPZ	primary or principal pigment cell		
NP	accessory or secondary cell		
S, SZ	cone or Semper cell		

1.1.1. *Lateral eyes of arthropleonean Collembola*

These are mostly unspecialized negatively-phototactic animals with strongly reduced eyes. A common characteristic is that the crystalline cone is a round homogeneous ball (pear shaped in *Tomocerus*). On account of its unique extracellular position, I term this cone 'ecteucone' to distinguish the type from the intracellular or 'enteucone' cone of other insects. Adjacent to the four 'Semper' cells forming the cone are two other cells which (except in *Entomobrya*) contain no pigment grains or other obvious cytoplasmic components, and, therefore, resemble the corneagenous cells of Crustacea. A constant number of eight retinula cells contribute to the rhabdom in very diverse ways in different species, and even in a single species. The rhabdom of poduromorph arthropleonids consists of two tiers, while in the other group, the entomobryomorphs, the rhabdom forms a single layer. The function of a tiered rhabdom, which is found also in dragonflies, some butterflies, e.g. *Pieris*, and many beetles, is discussed by Gribakin and also by Snyder (Chapter 9).

1.1.1.1. *Poduromorpha*

(1) *The dioptric apparatus.* The cornea is thin but strongly convex on the outer surface. The cornea is biconvex only in *Anurida maritima*, but in this species the crystalline cone is lacking so that a corneal lens of short focal length is more essential. The cornea has the layered structure typical of the general cuticle and usually bears the complicated honeycomb pattern found all over many collembolans, called 'grains primaires' by Massoud (1969) and described by Paulus (1971). A transition to the enlarged corneal nipples of some nocturnal insects has been found only in *Tomocerus* (Barra 1971*a,b*; Paulus 1972*a*), which has weak and bent thorn-shaped corneal extensions 80–150 nm long. True corneal nipples are found only in Symphypleona, and, in contrast, some Poduromorpha (*Podura*, *Anurida* and *Hypogastrura*) have a completely smooth cornea.

The four components of the extracellular ecteucone arise each as a separate intracellular inclusion of one of the four cone cells, and only later fuse to form a sphere. Remnants of membrane persist in the cone of *Hypogastrura* (Barra 1971*a,b*). The four cells of the adult cone remain associated with it, with nuclei situated laterally. Against the membrane of the cone are numerous mitochondria which are active presumably during the numerous ecdyses which involve continuous enlargement of the cone. According to Barra the cone contains proteins rich in sulphide and disulphide groups. Each cone cell sends a process, densely packed with microtubules, between the retinula cells in the rhabdom region. As in several other groups of insects these cone cell roots enlarge and contain pigment in the basal region of the eye.

(2) *Rhabdoms of Poduromorpha.* In all species examined in this group the rhabdom has two layers (Thibaud 1967*a,b*; Paulus 1972*a,b*), but details are available only for two species. *Podura aquatica* has eight ommatidia whereas *Neanura sp.* has only two or three. Both these species have eight retinula cells in two layers of four, but only six of the eight cells form rhabdomeres. In the dorsal ommatidia of these species there are only four rhabdomeres. The four distal retinula cells are pressed against the cone (Fig. 1.2). In *Neanura* two of the distal cells have microvilli at right angles to each other and form a fused rhabdom, whereas in *Podura* the rhabdomeres of cells 3 and 4 are separated by a process from the proximal retinula cell 5 (Fig. 1.2).

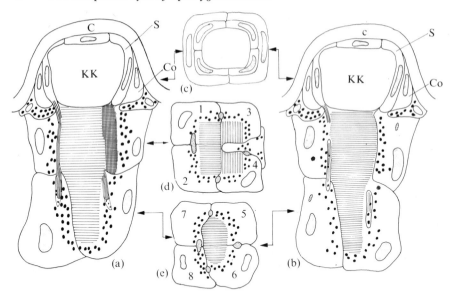

FIG. 1.2. The two types of ommatidia of the double eye (see Fig. 1.6) of *Podura aquatica.* (a) Ventral ommatidium in LS. (b) One of the two dorsal ommatidia in LS. The transverse sections are at the levels indicated.

Two of the rhabdomeres of the proximal layer are displaced somewhat laterally beneath a second group (cells 1/2 or 3/4) of the distal rhabdomeres in which case their microvilli are always parallel to those overlying them (contrary to expectations deduced by Gribakin (Chapter 9) and by Snyder (Chapter 11)). The more lateral rhabdomeres, found only in the distal layer, are always oriented at right angles to the others. All eight retinula cells, even those without a rhabdomere, have an axon.

In *Podura aquatica* the rhabdom of the two most dorsal ommatidia, A and B, has only four rhabdomeres whereas the other ommatidia have six. Both types have eight retinula cells in two layers of four, but only cells 1/2 and 5/6 develop rhabdomeres with a single orientation of microvilli (Fig. 1.2b). We have, therefore, a simple form of the double eye, with different structure (and presumably function) in dorsal and ventral parts. From the illustration of it (Barra 1971), the eye of *Hypogastrura* is apparently similar to *Neanura*, and *Anurida maritima* has a similar pattern (Paulus, unpublished observations).

1.1.1.2. *Lateral eyes of Entomobryomorpha.* The rhabdom in this group is always of one layer (according to present knowledge) although small areas of overlap may occur. The great variability of the structure is perhaps a sign of the primitive nature of the eye, or of its insignificance in natural selection, and is certainly in contrast to the constancy of rhabdom form within most insect orders.

Rhabdom patterns can be distinguished according to the following considerations of symmetry (Paulus 1972b):

(a) completely irregular, as in *Orchesella* (Fig. 1.3a).

(b) approximately radial symmetry of rhabdomeres, as in *Tomocerus* (Fig. 1.3b) and in primary eyes (see later) of *Entomobrya muscorum* (Fig. 1.3).

(c) bilateral symmetry, as in 'accessory eyes' of *Entomobrya muscorum*, and in 'primary eyes' of another *Entomobrya sp*.

In *Orchesella* eight retinula cells have their microvilli at right angles to the light path but arranged almost indiscriminately beneath the cone (Fig. 1.3a). In *Tomocerus* the intra-specific variability is still great but some symmetry is evident, with seven rhabdomeres and one basal retinula cell with few microvilli. Four of the rhabdomeres have microvilli mutually at right angles in an outer ring, and extend distally around the cone. The central three rhabdomeres have microvilli inclined *c*. 120° to each other (Fig. 1.3b), but in a variable position in relation to the outer ring. For the most part these rhabdomeres are pressed laterally against the outer ring with a space for the retinula cytoplasm in the centre, or they may be displaced proximally and be fused together. For this reason, the regularity noted in a previous work (Paulus 1972a) could not be confirmed.

The remarkable eye of *Entomobrya muscorum* has eight ommatidia, of

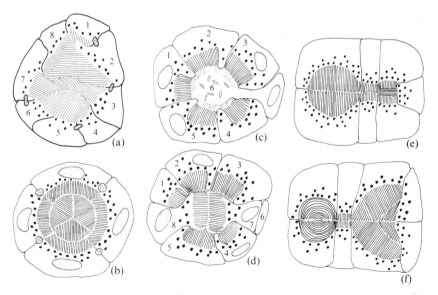

FIG. 1.3. The arrangement of the rhabdom in transverse section in Collembola. (a) *Orchesella*. (b) *Tomocerus sp*. (c) *Entomobrya muscorum* primary eyes, distal rhabdom level. (d) Ditto, of proximal level. (e, f) *Allacma fusca*.

which two (so-called accessory eyes) have a smaller lens than the others (primary eyes). On account of the difference in size and rhabdom structure it is justifiable to speak here of a double eye. In this species alone among Arthropleona the two corneagenous cells contain large pigment grains, and are in fact justifiably called primary pigment cells (Figs. 1.4a and 1.10f).

(1) *The open rhabdom of the primary eyes.* Five retinula cells have irregularly interrupted microvilli borders around a sixth central cell (Fig. 1.3c) along the whole longitudinal axis. The sixth cell has an eccentric nucleus and a small irregularly oriented rhabdomere. Its central cytoplasm is often packed densely with mitochondria. Peripheral to cell 6 lie cells 7 and 8, each with a large rhabdomere arranged as in Fig. 3d.

(2) *The fused rhabdom of the secondary eyes.* At first glance the secondary eyes appear quite different from the primary ones, but the rhabdomere pattern in cross-section is derived by distortion of the latter into a bilaterally symmetrical plan (Fig. 1.4c,d). The arrangement and layers

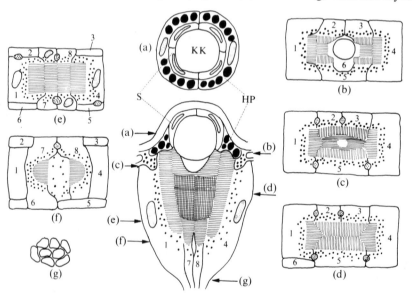

FIG. 1.4. The secondary or accessory ommatidium of *Entomobrya muscorum* with transverse sections at the levels indicated in the longitudinal section.

of cells remains unchanged. At its distal end, however, the rhabdom extends around the cone forming a characteristic pattern (Fig. 1.4b). Cell 6 forms a border of microvilli along two opposite sides of the cell, and this is larger and more regular than in the primary eyes. Microvilli of cells 1 and 4 run in the same direction: those of cells 2, 3 and 5 are at right angles to them. Although separated distally, these rhabdomeres meet proximally to form a remarkably elongated rhabdom (Fig. 1.4c); further proximally cell 6 disappears and 2, 3 and 5 meet in the axis. Unlike the situation in the primary eyes, the rhabdomeres of 2, 3 and 5 end where the more proximal rhabdomeres of 7 and 8 begin, forming a tiered rhabdom (Fig. 1.4). An interesting feature is that the microvilli of corresponding cells in the two secondary eyes are parallel. It certainly appears as if the dorsal part of this 'double eye' has a distinct function.

1.1.2. *The lateral eyes of Symphypleona*

This group of highly organized Collembola perform better than the others in behavioural tests (Mayer 1957). The eye is less reduced than in arthropleonen Collembola and there is a smaller divergent angle between ommatidia. In fact the ommatidium of *Sminthurus* has a structure typical of ptergote insects. Only one species, *Dicyrtomina ornata*, has been described fully, but interspecific differences are probably less than in the previous group (Paulus 1973*a*,*b*). From the limited data available on the cone structure, *Allacma fusca*, *Sminthurides aquaticus* (Paulus 1972*a*) and *Dicyrtoma atra* (Barra 1971*b*) are similar.

1.1.2.1 *Dioptric apparatus.* The cornea is strongly convex and, except in *Sminthurides*, slightly thickened. The outer surface bears nipples, which are of various forms in different species. In *Allacma* they are no more than flat humps 40–50 nm high and less near the edge of the lens. In *Sminthurides* the nipples are wide at the base with a stub-like apex, hexagonal in section, and 150–180 nm high. Nipples, similar to those described in some winged insects (Bernhard *et al.* 1970), are 150–190 nm high in *Dicyrtoma* and *Dicyrtomina*.

The crystalline cone, with four segments, is egg-shaped and formed intracellularly within the four cone cells (*enteuconically*) (Fig. 1.5b). Numerous mitochondria lie along the edge of the cone and four cone cell processes extend to the base of the ommatidium. Species of this group, investigated so far, have two typical primary pigment cells, which touch the cuticle or cornea solely by a narrow process, and which are densely filled with black pigment grains larger than those of the other pigment-bearing cells. In *Allacma fusca*, processes from both of these cells to the base of the rhabdom contain microtubules, and, as well as a supporting function, may contribute to pigment movement during adaptation.

1.1.2.2. *The structure of the rhabdom.* In *Allacma fusca* only random cross-sections are available, so that reconstruction is impossible (Paulus 1972*a*). However, the ommatidium consists of eight retinula cells with rhabdomeres in a bilaterally symmetrical arrangement (Fig. 1.3).

The rhabdom of *Dicyrtomina ornata* consists of five retinula cells with

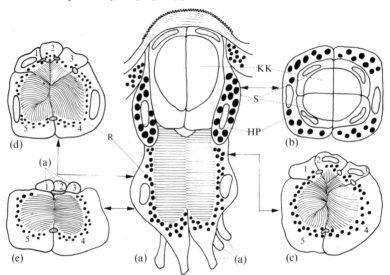

F IG. 1.5. Ommatidium of *Dicyrtomina ornata* with transverse sections at the levels indicated.

rhabdomeres. Three rudimentary cells lie somewhere within the complex, but have no axons, in contrast to retinula cells without rhabdomeres in other Collembola (e.g. *Podura*). That there are eight retinula cells in *Dicrytomina* is inferred because the eye consists of eight ommatidia and contains 64 candidate retinula cells. Cells without a rhabdomere usually have a rudimentary cell body. The voluminous rhabdom consists mainly of two large rhabdomeres with microvilli bent into an S-shape, rather than running parallel to each other as in most cases (Fig. 1.5). A 'tucking-in' of the microvilli surface has occurred in cell 5, as previously illustrated for *Allacma* (Fig. 1.3), and described by Horridge (1969) for the beetle *Photuris*. Lateral to the two cells 4 and 5 in the distal area are three additional considerably smaller cells, 1, 2, and 3, with parallel microvilli at an angle of 45° to each other in the different cells. These laterally adjacent cells give the cylindrical rhabdom a bilateral symmetry. Cell 2 is situated directly distally, so that only four rhabdomeres are present along the longitudinal axis, where the microvilli of cells 1 and 3 meet at 90°. In the proximal zone rhabdomeres 1 and 3 disappear also (Fig. 1.5).

1.1.3. *The orientation of the rhabdoms in the group of ommatidia*

Most Collembola exhibit a species-specific arrangement of ommatidia which are isolated at a relatively great distance from each other and loosely gathered in a group. On account of their large admittance angle (30–60°) and angles of divergence, the small number of visual elements have a large total field of view (Fig. 1.1), but their resolution is necessarily poor and they are presumably capable of only limited pattern recognition. In behavioural tests this has been demonstrated as an appreciation of bold contours (Mayer 1957; Schaller 1969). Orientation according to the plane of polarization has not yet been observed, although the rhabdom structure appears appropriate (Paulus 1972*a*). In the compound eyes of many pterygote insects the orientation of the individual retinula cells and of their microvilli runs parallel across the whole eye. Among Collembola, however, this applies only to *Podura aquatica*, if one projects the rhabdom cross-sections upon a plane (Fig. 1.6).

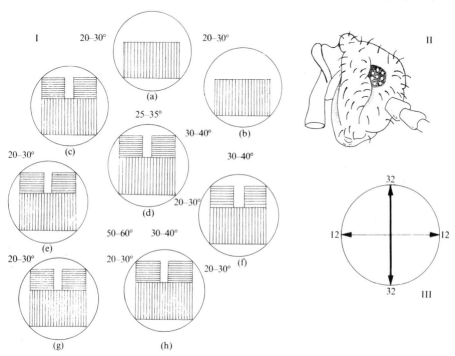

FIG. 1.6. (I) Arrangement and directions of microvilli of the rhabdomeres of *Podura aquatica*. The angles of divergence between neighbouring ommatidia are shown in the appropriate positions. (II) The natural position of the head, with eye oriented as in (I). (III) The axes of the microvilli, with 32 vertical and 12 horizontal rhabdomeres.

Other species show neither exact regularity nor a constant distribution of angles. Two cases which have been analyzed more thoroughly must be elaborated, since it is possible that their arrangement of rhabdoms bears some relation to function.

In *Entomobrya sp.* the six ommatidia (A)–(F) of the whole eye have a characteristic distribution and arrangement. Projection upon the plane of section showed that the rhabdoms of ommatidium (A) are parallel to corresponding ones in (C) and (D) and those in (B) are parallel to corresponding ones of (E) and (F). The extremely elongated bilateral form assists the analysis. In *Entomobrya muscorum* the rhabdoms of the primary eyes apparently have no regular arrangement, but those of the accessory eyes are always parallel to each other. This rhabdom is also remarkable in extending far distally on two sides of the cone so that the admittance angle to light is enlarged in this plane.

In *Dicyrtomina ornata* (Fig. 1.7) the axes of symmetry through the cross-

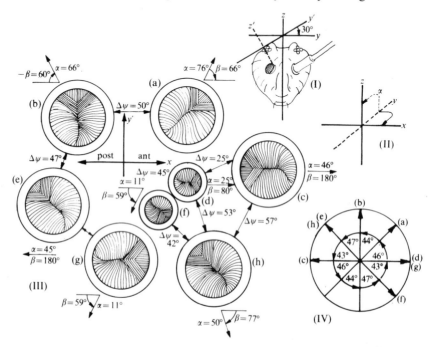

FIG. 1.7. Location of ommatidia on the head, and directions of microvilli in *Dicyrtomina ornata*. (I) Frontal view of the head with left antenna removed, showing the axes used in (II) and (III) as follows: y, right left axis; z, dorsoventral axis; y', the y-plane tilted by 30° to provide the best plane of projection of the sections; z', plane at right angles to y'. The x-axis is caudo-cranial and at right angles to the paper. (II) α is the angle between the ommatidial axis in the y'-plane and the z'-axis; β is the corresponding angle with the x-axis. Values of α and β are shown in (III). (III) The eight rhabdoms, with mutual inclinations $\Delta\psi$. The rhabdoms are projected upon the y'-plane. (IV) When all microvilli directions are plotted it is clear that they lie at 45° to each other.

sections of the rhabdoms are rotated by 45° between ommatidia. These arrangements cannot be related to function at present.

1.2. Presumed photoreceptors in Protura and Diplura

These groups of insects have no compound eyes or single ommatidia comparable to those of Collembola and all of them are small blind forms. A sense organ called a pseudoculus on each side of the head of Protura has been regarded as a rudimentary eye (Berlese 1909), but recent electron microscope work has shown that it is a chemoreceptor, probably a hygroreceptor (Bedini and Tongiorgi 1971; Haupt 1972), and probably homologous to the postantennal organ of collembolans or the Tömösvary organ of myriapods. In *Campodea*, George (1963) described sensory cells in the ventrolateral part of the head as photoreceptors. The position, serial arrangement, and histological structure of these cells indicate that they may be the rudiments of a compound eye, but this requires confirmation by the electron microscope.

1.3. The compound eye of Thysanura

1.3.1 *Zygentoma*

As in Collembola, the sugar-mites possess a reduced compound eye, which consists of 12 ommatidia in *Lepisma, Thermobia*, and *Ctenolepisma*. *Tricholepidion gertschii*, a living representative of the family Lepidotrichidae, otherwise known only from fossil material, has an eye of 40–50 ommatidia. The eyes of Thysanura are apparently reduced compound eyes; their structure was examined by Hesse (1901), Hanström (1940), and Elofsson (1970). An electron-microscope study by Brandenburg (1960) is available for *Lepisma saccharina*. The dioptric apparatus of *Thermobia domestica* was investigated by Paulus (1972a). Furthermore, I have a few electron microphotographs of the ommatidium of *Ctenolepisma* (Meyer-Rochow, unpublished observations), which agree, in general, with those of *Thermobia*.

1.3.1.1. *The dioptric apparatus and its cells.*   The thick biconvex cornea is covered superficially with flat nipples *c.* 50 nm tall. The eye is acone in *Lepisma*, eucone in *Thermobia* and *Ctenolepisma* and supposedly ecteucone in *Tricholepidion* (Elofsson 1970) although the latter requires confirmation by electron microscopy. The cone is always formed as four closely-fitting segments by the four surrounding cells, and is usually spherical.

While there is no doubt that, in *Thermobia, Ctenolepisma*, and, particularly, in *Tricholepidion*, in addition to the four cone cells, there are two other large cells containing characteristically large pigment grains and pigment cells which are unequivocally primary, Brandenburg (1960) and, in particular, Elofsson (1970) describe four primary pigment cells in addition to two corneagenic cells. This would imply a marked deviation from the usual structure of the ommatidium. But I believe that an incorrect interpretation of the number and origin of cell spaces has been made. With independent confirmation from Meyer-Rochow on *Ctenolepisma*, I find in *Thermobia* a situation which explains the difficulty in *Lepisma*. Large pigment grains are often found outside the two primary pigment cells, and Brandenburg and Elofsson regarded the space which the large grains occupy as distinct cells. On the basis of our pictures, however, one can clearly see that these large

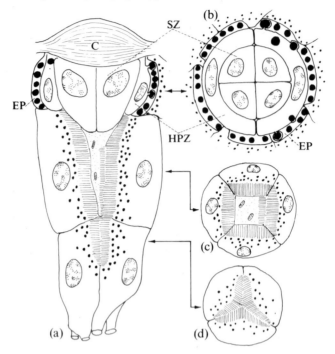

FIG. 1.8. *Lepisma saccharina* in longitudinal section with transverse sections at the levels indicated. Note the extruded pigment grains derived from the primary pigment cells.

grains arise within the primary pigment cells, but migrate from them, carrying an enveloping membrane with them. Each individual grain is evaginated and tied off then assumes a position that is extracellular although enclosed in neighbouring cells (Fig. 1.8b). It is not known why pigment displacement of this kind takes place. Similar processes were observed by Fuge (1967) in the eye of *Drosophila*. I believe that in *Lepisma* the same interpretation would yield the standard complement of cells, unless more cell nuclei can be found.

In all species investigated the rhabdom is formed by seven retinula cells, which lie one on top of the other, as in poduromorphic Collembola. Four rhabdomere borders are in the distal layer, three in the proximal layer. Distally they are arranged in a square with a four-sided space in the centre, and extend around the cone laterally. In the proximal layer the three microvilli borders form a rhabdom with the shape of a three-pointed star (Fig. 1.8d). In the middle, between the four rhabdomeres of the distal zone, both Brandenburg (1960) and Meyer-Rochow found cytoplasm apparently of an eighth retinula cell with a very narrow irregular border of microvilli on one side only in *Lepisma* but on all sides in *Ctenolepisma*. Neither author observed the corresponding nucleus, so perhaps we are dealing with an equivalent to cell 6 in *Entomobrya*, with a nucleus located far from the rest of the cell.

1.3.2. *The compound eye of Archaeognatha*

The Machilidae are the only apterygotes with fully-developed compound eyes which are large and meet on the dorsal midline of the head. General accounts are available from Oudemans (1887), Hesse (1901) and Hanström (1940), a comment on the dioptric system from Paulus (1972*a*) and on the rhabdom from Meyer-Rochow (1972). A more detailed analysis has been conducted independently by Paulus on *Machilis* and by Horridge on *Machilis* and *Allomachilis frogatti*.

1.3.2.1. *The structure of the dioptric system.* The slightly-thickened cornea is biconvex with inner curvatures less pronounced than the outer, which is covered with little nipples, *c*. 150 nm high. Directly beneath the lens, two large cells, enclosing the cone, have nuclei situated (at least in part) between the cornea and the crystalline cone, and cytoplasm extending some distance proximally (Fig. 1.9a,b). Around the tip of the cone these cells contain pigment grains which are only slightly larger than those in retinula cells (Fig. 1.9c). This was not recognized in earlier work, where they were as corneagenous cells by the position of the nuclei, but they have all the features of primary pigment cells. The crystalline cone, which tapers sharply, consists of four segments formed by four cone cells with nuclei arranged in four quadrants between the crystalline cone and the lens. This dioptric system is surrounded by numerous secondary pigment cells and hair cells, which form the numerous thin bristles between the lenses.

1.3.2.2. *The structure of the rhabdom.* At the continuation of the cone tip is an elongated rhabdom formed by seven retinula cells, arranged as in a typical apposition fused-rhabdom eye. Along most of their length they form a cylindrical almost radially-symmetrical rhabdom (Fig. 1.9e). In *Machilis* a previously-undescribed eighth small retinula cell lies directly below the crystalline cone, with a few microvilli between the cone tip and the main rhabdom (Fig. 1.9d). In *Allomachilis frogatti* Meyer-Rochow (1972) also found an eighth retinula cell, but at the base of the rhabdom. In the proximal region the rhabdom is symmetrical because the seven rhabdomeres form four segments with their microvilli at right angles to each other (Fig. 1.9f).

The structure of the machilid rhabdom is remarkable in two respects. First, an unusually large and extensive palisade of vacuoles (or Schaltzone in previous work) surrounds the rhabdom. In several insects it has now been shown that this is a region of lower refractive index surrounding the rhabdom (Horridge and Barnard (1965) on the locust; Butler (1973*b*) on the cockroach) and its function in improving the properties of the rhabdom as a light guide have been analysed by Snyder (Chapter 11). In some insects the palisade is a feature of the dark-adapted eye and disappears in the light, but in *Machilis* it is fully developed in the eye even when light-adapted to normal diffuse daylight, and therefore is hardly likely to be a part of a light–dark adaptation

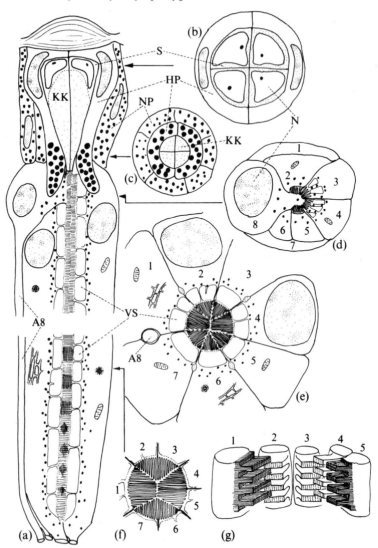

FIG. 1.9. *Machilis sp.* (not to scale), transverse sections at the levels indicated. Accessory pigment cells are omitted in (b). Note rhabdomere 8 in (d), the bilateral symmetry in (f), and the crustacean pattern of the proximal rhabdom as detailed in (g).

mechanism as in the cockroach (Snyder and Horridge 1972). In eyes of *Machilis* exposed to sunlight for an hour, however, Horridge found that the pigment grains migrate radially inwards towards the rhabdom until they lie in the spokes of the palisade and immediately around the rhabdom. Even this treatment does not reduce the palisade but the pigment must now absorb some of the light energy conducted along the rhabdom.

In the second place, while the distal half of the rhabdom exhibits only uniform microvilli in longitudinal section through the rhabdom, in the proximal region there is a periodic change in the direction of the rhabdomeres as in crustacean rhabdoms. Each retinula cell has a periodically-interrupted border of microvilli in longitudinal section, and the space which thus arises is filled by the microvilli of another retinula cell. Therefore, incident light traverses a rhabdom which consists of a stack of plates with crossed microvilli (Fig. 1.9g). Similar rhabdoms occur in a few Coleoptera and Lepidoptera (Butler *et al.* 1970; Meyer-Rochow 1972), as well as in most crustaceans.

1.4. The apterygote ommatidium and the phylogeny of the compound eye

A full discussion of this topic would involve many problems of the phylogeny of arthropods or at least of the Mandibulata, requiring a lengthy treatise, whereas at present we are concerned only with the relevance of the eye structure. Let us consider first the number of ommatidia. Hesse (1901) believed that compound eyes arose by progressive aggregation of more ommatidia, and he proposed an evolutionary series from Collembola with eight to Lepismatidae with twelve and so on. Numerous details as described in this paper suggest, however, that the eyes of Collembolans are reduced structures, and similarly Hanström (1940) considered that *Lepisma* has a reduced number of ommatidia. It is probable that the ancestors of apterygote insects already had compound eyes.

Let us now consider the form of unit structure. As support for an evolutionary series, Hesse (1901) showed that one can start with the cup-shaped ocellus of myriapods, and he related this to the folded-in and layered stemmata as found in the larva of *Dytiscus*, lepidopteran caterpillars and larvae of other holometabolic insects, and finally to the single-layered retina of higher insects. One question that has assumed some importance is the cellular constitution of the ommatidium inherited by primitive insects and perhaps also by crustaceans. In other words, the unit when the compound eye was first appearing may have had the following features:

(1) A cornea formed by two unpigmented cells, the corneagenous cells.
(2) A crystalline cone formed by four cells, the Semper or cone cells.
(3) Eight retinula cells.
(4) A varying number of pigment cells which isolate the individual ommatidia optically.

These cell numbers are exactly those found in the ommatidia of crustaceans and insects. In insects, the two corneagenous cells became the two primary pigment cells. However, during development they participate in the secretion of the cornea and produce subsequently the large pigment grains characteristic of this type of cell. This step can be demonstrated very nicely in apterygote insects. In *Machilis* the two cells still lie between the cornea and

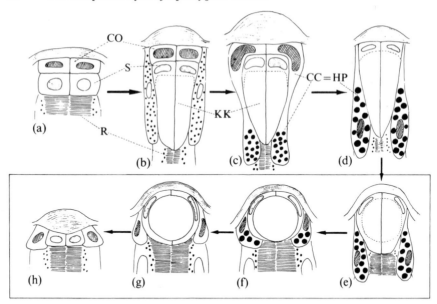

FIG. 1.10. Suggested homologies of corneagenous and primary pigment cells, from Crustacean to Insecta in (a) to (d), and the progressive reduction seen in Collembola (e)–(h). (a) Suggested protomandibulate ommatidium with 2 corneagenous and 4 cone cells, indicated differently by the drawing of the nucleus. (b) Typical crustacean. (c) *Machilis*, with 2 corneagenous cells developed also as pigment cells. (d) Pterygote insect of eucone type. The primary pigment cells are connected tenuously to the cornea in the imago. (e) Symphyleonid collemboian (Sminthuridae) similar to (d). (f) *Entomobrya*, retaining primary pigment cells. (g) Arthropleonid collembolan, with reduced unpigmented cells. (h) *Anurida maritima* with reduced eye but retaining the same cells.

the lens (Fig. 1.10c). In higher insects both cells are pushed further and further proximally so that they surround the cone, although they may remain connected to the cornea by a narrow tubular process (Fig. 1.10d). In the Collembola one can suggest progressive reduction beginning with the Symphyleonids and ending with *Anurida maritima* (Fig. 1.10e–h). In crustaceans, on the one hand, the two corneagenous cells function solely to produce the cornea (Fig. 1.10b) whereas in insects they assume an additional function as pigment cells.

Unfortunately the myriapod eye does not fit into the above scheme. Since myriapods are closely related to insects on the basis of numerous other synapomorphic characters (Hennig 1966), their eyes must be considered modified secondarily, without the constant number of cells. The larvae of several other primitive orders of insects possess typical compound eyes. Furthermore, parallel cases can be found among holometabolic imagines, which have eyes comparable to the myriapod ocellus, for example, such an eye was described by Wachmann (1972*a*) in *Stylops* (Strepsiptera). The ocellus or prosommatidium of fleas demonstrates how such a structure can arise (Wenk 1953; Wachmann 1972*b*). Fusion of the ommatidia produces a rhabdom contributed by many retinula cells beneath a single large lens, which arises from many corneagenic cells. In the Dorylinae ants a corresponding pathway of reduction occurs (Werringloer 1932). A peculiar case is presented by *Stylops* in which a compound eye is reconstructed secondarily from pseudo-ocelli which were present, and not from ommatidial units which have disappeared. This is an example of the irreversibility of evolution (Dollo's Law). I believe that the diverse forms of the myriapod eye can be explained similarly as derived rather than primitive, so that the compound eyes of the Mandibulata can be traced back to a single rather well-defined ancestral compound eye, and the components of the ommatidia are homologous structures. A more detailed discussion will be found in Paulus (1974).

2. Anatomical changes during light-adaptation in insect compound eyes

B. WALCOTT

INSECT eyes can adapt to light, as can the eyes of other animals. In any one state of adaptation an insect visual receptor can detect changes in light intensity over a range of c. 3 log units and its most efficient range is only c. 1–1·5 log units. Since the environmental intensity range between moonlight and sunlight is about 8 log units, adaptation mechanisms have been evolved to alter the effective range in which the receptors operate. These mechanisms can operate at several levels.

(1) Attenuation can occur before the light reaches the receptors, or when the light has reached the receptors but before it is absorbed by the visual pigment.
(2) The receptors can be affected by bleaching of their photopigment and by membrane changes associated with the transduction process.
(3) There can be adaptation in the integrative processes in the optic neuropiles.

This paper will be concerned only with those mechanisms of adaptation that affect the light before the receptors and in the receptors before the transduction process occurs. These mechanisms are varied and involve movement of dense pigment granules, membrane vacuoles, and whole groups of cells. Studies of the morphological adaptation are essential to the analysis of optical mechanisms in compound eyes. On the basis of these anatomical changes in light–dark adaptation, insect compound eyes can be divided into four major groups (see Table 2.1). Group (I) are apposition-type compound eyes which show radial changes in the position of screening pigment granules and in the position and extent of a vacuolar 'palisade'. Groups (II) and (III) are the 'clear zone' or superposition type of eye. These two groups are distinguishable by the presence or absence of cellular movement as well as longitudinal pigment movements. Group (IV) consists of the acone eyes which show extensive movements of cone cells and rhabdoms (Fig. 2.1).

TABLE 2.1

Types of insect compound eyes

(I) Eyes with palisade formation and radial pigment movement

Locusta	Horridge and Barnard 1965; Tunstall and Horridge 1967
Periplaneta	Butler and Horridge 1972
Musca and Calliphora	Kirschfeld and Franceschini 1969; Seitz 1970
Aeschna	Horridge 1969
Formica	Menzel and Lange 1971; Menzel 1972
Apis	Kolb and Autrum 1972

(II) Eyes with reported longitudinal pigment movements only

Bombycoidea	Tuurala 1954
Saturniidae	Yagi and Koyama 1963
Lampyridae	Horridge 1969
Sphingidae	Höglund 1966
Elateridae	Horridge 1971

(III) Eyes with movement of retinula cell bodies and iris pigment

Pyraloidea	Tuurala 1954
Gelechioidea	Tuurala 1954
Tineoidea	Tuurala 1954
Notodontidae	Yagi and Koyama 1963
Noctuidae (most)	Yagi and Koyama 1963
Scarabaeoidea	Horridge and Giddings 1971
Carabidae	Horridge and Giddings 1971
Dytiscidae	Walcott 1969
Hydrophiloidea	Horridge and Giddings 1971
Megaloptera	Walcott and Horridge 1971

(IV) Eyes with movement of cone, tract, and rhabdom

Notonecta	Lüdtke 1953; Eckert 1968
Belostomatidae	Walcott 1969; 1971
Forficula (Dermaptera)	Eckert 1968
Culex	Sato, Kato and Toriumi 1957
Chironomidae	Tuurala 1963
Tipulidae	Sotavalta, Tuurala, and Oura 1962; Eckert 1968
Anoplognathus	Meyer-Rochow 1972
Tenebrio	Eckert 1968
Staphylinidae	Meyer-Rochow 1972
Rhodnius	Müller 1970

2.1. Eyes with palisade formation and radial pigment movements

This type of adaptation mechanism has so far been found only in apposition compound eyes. In these eyes, the rhabdom of the retinula cells abuts the crystalline cone and extends centrally through most of the width of the retina. The fused rhabdom, or in open rhabdom eyes the separate rhabdomeres are 1–5 μm in diameter, and 50–500 μm long. Light–dark adaptation is associated with movements of vesicular structures and with radial screening pigment movement around the rhabdom.

In the dark-adapted state, the rhabdom is surrounded by sacks of endoplasmic reticulum, the palisade. The contents of the sacks of the palisade

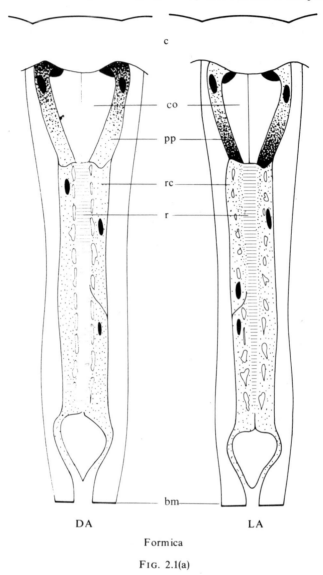

co

pp

rc

r

bm

DA   LA

Formica

FIG. 2.1(a)

always appear less dense than the surrounding cytoplasm of the retinula cells when viewed under light or with an electron microscope. The vesicles of the palisade are usually large and elongate, occupying most of the space of the retinula cell from the rhabdom outwards up to $\frac{1}{3}$ of the width of the cell. The pigments granules are displaced radially and occur between the vesicles of the palisade and in the peripheral cell cytoplasm.

In the light-adapted state, the palisade is reduced in extent and displaced radially. Some of the retinula cell pigment granules now lie close against the rhabdom.

These changes however, do not appear uniform along the length of the retinula cells. In *Aeschna* (Horridge 1969), *Periplaneta* (Butler 1971) and *Formica* (Menzel and Lange 1971; Menzel 1972) the palisade moves away from the rhabdom and is reduced on light-adaptation to a greater extent on the distal half of the ommatidium. In *Formica*, the palisade in the proximal

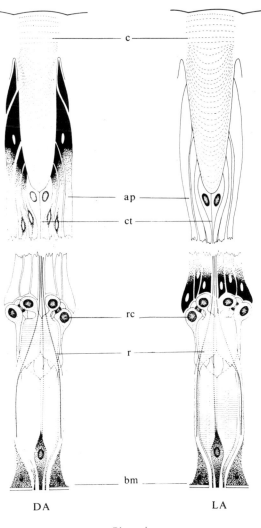

DA   LA

Photuris

FIG. 2.1(b)

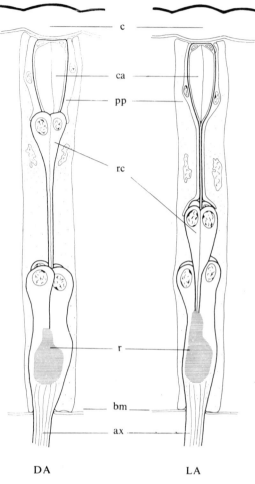

DA   LA

Archichauliodes

FIG. 2.1(c)

half moves only slightly and so on light-adaptation there is a wider and more
dense ring of cytoplasm surrounding the rhabdom. Further, most of the
retinula cell pigment granules in *Formica* occur in the distal half of the cell,
and most of the pigment granule movement occurs here. It is interesting to
note that not all the pigment granules move on adaptation (Menzel, personal
communication). The granules in the peripheral third of the cell do not move
at all, and it is only the granules nearer to the rhabdom that move.

In *Locusta*, pigment movements have not been found on adaptation
although there are extensive palisade changes and movements of mito-
chrondia (Horridge and Barnard 1965). In *Musca*, the pigment granules move

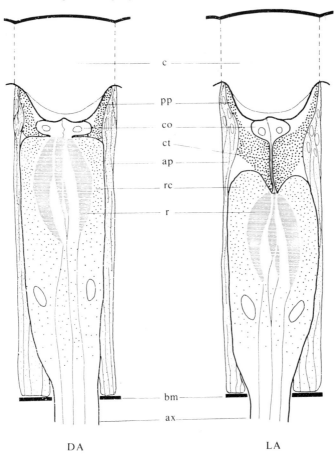

Lethocerus

FIG. 2.1(d)

FIG. 2.1. Drawings of the ommatidial structure in dark (left) and light (right) adapted compound eyes representing each type of adaptation mechanism as outlined in Table 2.1. Key: C. cornea; Co, cone cells; ct, crystalline tract; pp, primary pigment cells; ap, accessory pigment cells; rc, retinula cells; p, palisade; r, rhabdom; bm, basement membrane; ax, axons. (a) Ommatidia of *Formica*. There are radial movements of the palisade and pigment granules adjacent to the distal half of rhabdom on adaptation. (b) Ommatidia of *Photuris*. There are longitudinal pigment movements on adaptation. (c) Ommatidia of *Archichauliodes*. There are movements of the four distal retinula cell bodies, and the crystalline tract, but no movement of the rhabdom on adaptation. (d) Ommatidia of *Lethocerus*. There are movements of retinula cells, rhabdom, crystalline tract and primary pigment cells on adaptation.

towards the rhabdomeres on light adaptation (Kirschfeld and Franceschini 1969), but no palisade has been clearly demonstrated. Recent freeze-etch studies (Seitz 1970) show small vesicles (*c.* 0·2 μm in diameter) which appear near the rhabdom on dark-adaptation. However, they are very different from the vesicles of the palisade in other eyes.

The rates of these anatomical changes have been studied in only a few cases. The pigment granule movements in *Musca* on light-adaptation are very rapid and occur in only a few seconds (Franceschini and Kirschfeld 1969). The palisade in *Locusta* requires *c*. 15 min of dark-adaptation to develop fully (Tunstall and Horridge 1969). In cockroach, where both palisade and pigment granule changes have been observed, anatomical light-adaptation requires *c*. 10 min of exposure to light (Butler 1971).

Experiments on *Musca* (Kirschfeld and Franceschini 1969), *Periplaneta* (Butler 1971) and *Formica* (Menzel 1972) all indicate that these anatomical adaptation changes are induced by the individual retinula cells and that there is little or no interaction between retinula cells. In *Periplaneta* and *Formica* selective adaptation by light of different colours produces different anatomical adaptation states of retinula cells in the same ommatidium. *Periplaneta* ommatidia were shown by this technique to contain a particular pattern with two u.v.-sensitive cells and the other six of the green-sensitive type. Intracellular recordings from the cockroach had already demonstrated the two different cells in one ommatidium (Mote and Goldsmith 1971). All these results show that the light falling on an individual retinula cell causes the morphological adaptation changes of that cell alone. Further, the chromatic adaptation studies show that the visual pigment of the cell is likely to be involved in the mechanisms which cause the adaptation changes although the action spectrum for adaptation is known in no case. It is possible, therefore, that the trigger for the adaptational changes is the change in membrane potential.

The anatomical changes regulate, in several ways, the amount of light that is absorbed by the visual pigment in the rhabdom. The rhabdom is of a higher refractive index than the surround and so behaves as a light guide (see Snyder, Chapter 9). However, pigment granules near the rhabdom in the light-adapted state are able to absorb a fraction of the energy travelling along the rhabdom, and also the refractive index around the rhabdom is increased by the pigment so that light escapes to be absorbed by the pigment grains in the retinula cell. However, in the dark-adapted state, the pigment moves away and in some eyes a palisade of lower refractive index lies around the rhabdom, so that light is better retained within the rhabdom. In *Musca*, optical studies have clearly shown a correlation between pigment position and the colour and amount of light transmitted by the rhabdom as predicted (Franceschini and Kirschfeld 1969). However, in *Locusta*, intracellular recordings show a rapid rise in the sensitivity of units by almost 2 log units on dark-adaptation which occurs in *c*. 1 min. But, the palisade has changed only slightly in one minute and requires *c*. 15 min to fully develop (Tunstall and Horridge 1967). The slight palisade changes observed in 1 min may be sufficient to alter the sensitivity, or more significant but unobserved changes may occur at the very distal portion of the rhabdom. An exact study employing the new light-guide equations is required.

In some apposition eyes, not only the sensitivity of the retinula cells changes on adaptation, but also the field of view of the units. In *Locusta* (Tunstall and Horridge 1967) and *Periplaneta* (Butler and Horridge 1972), the acceptance angle approximately doubles in size on dark-adaptation, in *Locusta* from 3·4 to 5·6° and in *Periplaneta* from 2·4 to 6·7°. This increase in visual field will effectively increase the sensitivity to a large source by increasing the angle within which light can be absorbed by the photopigment. In *Musca*, however, intracellular recordings show no change in the 2·5° acceptance angle on adaptation (Scholes 1969). The optical changes of the rhabdom can make themselves felt only when they operate over a range which is smaller than, or similar to, the aperture of the dioptric system, even if there is a control of the refractive index differential between rhabdom and surround. Changes in field size on adaptation will not necessarily be found, therefore, in worker bee where the dark-adapted angle is only 2·7° (Laughlin and Horridge 1971) and in *Aeschna* where the dark-adapted angle is only *c.* 2° (Horridge 1969*c*).

2.2. Eyes with longitudinal pigment movements only

This group contains mainly lepidopteran compound eyes, but also includes the fireflies. Most of the animals in this group are nocturnal and are not active in the presence of bright light. Light–dark adaptation is associated anatomically with longitudinal movements of pigment granules in the retinula and pigment cells, and there are no changes in the relative positions of cells in the retina. This group of eyes are clear zone eyes in that there is a crystalline tract consisting of extensions of either four cone cells or approximately eight retinula cells, which run from the tip of the cone to the rhabdom below. They are 'fixed crystalline tract eyes' in contrast to group III which are 'movable crystalline tract eyes' (Horridge 1971).

The compound eye of the firefly, *Photuris*, has been studied anatomically in detail (Horridge 1969). In each ommatidium the cornea forms a long (80 μm) cone-like projection. The tip of the corneal cone is surrounded by the four cone cells which form a long ($\leqslant$ 100 μm), thin (1–4 μm) crystalline tract to the retinula cells below. This thread is surrounded by a pair of primary pigment cells which are, in turn, surrounded by many accessory pigment cells. These pigment cells contain many dense pigment granules that migrate on dark–light adaptation.

In the dark-adapted state, the pigment grains are concentrated mainly between the corneal cones, with a few scattered grains in the distal part of the clear zone. On light-adaptation the pigment moves centrally in both types of pigment cell, until there is a scattering of grains between the cones and in the top half of the clear zone. The central region of the clear zone, particularly within 25 μm of the retinula cell layer is densely packed with pigment grains. The crystalline tract is surrounded by pigment with many granules closely applied to the tract membrane.

From the anatomy and from optical experiments on eye slices, Horridge (1969) stated the hypothesis that in the light-adapted state all the light reaching a receptor comes from its own facet and travels by wave-guide modes down the crystalline tract of that facet. Only light rays on, or very near, the optical axis of a facet will be accepted. No other light path can be imagined, because rays not on axis will be absorbed by the dense pigment between the corneal cones. In the dark-adapted state, however, an additional light path is created. Rays on axis will still be transmitted by the crystalline tract. However, light at a large angle to the ommatidial axis can now be accepted by a receptor belonging to different ommatidium. Therefore, light can reach a receptor from a large patch of facets. The effect is to increase the aperture of the optical system, thus increasing sensitivity. This effect would be functionally useful for a firefly detecting the faint flash from other fireflies, and for orientation to a light of low intensity.

Døving and Miller (1969) studied optically a number of insects with this type of eye, including *Photuris* (but they do not specifically mention skipper butterflies). Using dark-adapted fresh eye slices, they observed an intensity difference of approximately 1 log unit between the light in the crystalline tract and that in the surround. Therefore, they concluded that light passing down the crystalline tract was effective at the receptors, even in the dark-adapted state. However, not all clear-zone eyes have a light-guide mechanism. even when light-adapted, for recent experiments on Australian Skipper butterflies demonstrated the formation of a superposition image and no trace of functional crystalline tracts (Horridge, Giddings, and Stange 1972). By moving a small light guide or a slit 60 μm wide of light across the eye, it was shown that light does in fact reach a receptor unit from a number of different facets. Skipper butterflies, however, are unique in that their clear-zone eyes show virtually no pigment movement or other anatomical adaptation changes on exposure to light. They are permanently dark adapted.

The retinae of the Sphingid moths, *Manduca sexta* and *Deilephila elpenor* have been studied extensively in a different way (see Hoglund 1966b). Using the electroretinogram (ERG), Hoglund (1966a) showed a variation in sensitivity of retinula cells of 3 log units depending on the adaptation state. Recent intracellular recordings have shown a range of 1–1·9 log units for a single photoreceptor (Hoglund and Struwe 1971). Clearly, therefore, the movements of pigment granules alters the sensitivity of the retinula cells by affecting the amount of light reaching them. However, from these experiments, one cannot ascertain whether, in the dark-adapted state, light enters by many facets to reach one receptor. It may be that, when dark-adapted, more light crosses between ommatidia in the clear zone, and that, therefore, more light reaches a receptor, so increasing its sensitivity.

Therefore, one can postulate that in the light adapted state, all light for a given receptor enters by its facet and travels down the crystalline tract of its ommatidium. Such a system would provide good acuity and a reduction in the amount of light, particularly with the pigment grains close to the tract. In the dark-adapted state, however, in addition to the tract light path, light can cross between ommatidia in the clear zone, due to the retraction of pigment from around the cone tip. Therefore, a receptor could receive light from more than its own facet. This would effectively increase the sensitivity or efficiency of photon capture by the eye, but presumably would reduce the acuity of the eye, in so far as these rays are not focused by a special super-position of the cornea and cone.

Unfortunately, in the 'clear-zone' eyes of this type, as well as group (III), nothing is known about the control, mechanisms or rates of pigment move-ment. As this is clearly important for the optics (see Horridge, Chapter 11) it should be studied.

2.3. Eyes with movement of retinula cell bodies and iris pigment

This group of eyes, as group (II), are all of the clear-zone type, i.e. they contain, at least in the dark-adapted state, a clear zone of pigment-free cells between the cone and the rhabdom. The most notable and distinguishing feature of group (III) eyes is the movement of some of the retinula cell bodies on light–dark adaptation. A typical example is the retina of the beetle, *Dytiscus* (Walcott 1969; Horridge and Giddings 1971).

In the light-adapted state, the four cone cells taper and form a crystalline tract which extends to the retinula cells some 120 μm below. This tract is only 1–2 μm in diameter, and contains many microtubules. A pair of primary pigment cells surround the cone and tract, stopping at the top of the retinula cell bodies. The pigment grains of these pigment cells are numerous and surround closely the tract; many actually touch the tract membrane. The retinula cell bodies lie distal to the rhabdom and one cell contains a small, but well developed, rhabdom at this level. The other seven distal cells contribute rhabdomeres to the large cross-shaped rhabdom which lies proximal to the clear zone. The two rhabdomere positions show that there are two functional systems in parallel, and changes on adaptation affect these differently.

In the dark-adapted state the crystalline tract disappears. During dark adaptation, four of the distal retinula cell bodies move distally and the tract first becomes shorter and thicker, then, in the extreme case, the tract disap-pears and four distal cell bodies, one with a distal rhabdom, surround the tip of the cone. The other four retinula cell bodies do not move and so the space between them and the cone is now bridged by a column of four retinula cell processes. During this change the primary pigment cells contract between the cone cells so that a real clear zone extends from the tips of the cone cells to the main rhabdom layer.

Some of the eyes in this category differ from *Dytiscus* in detail. The Megaloptera, for example, have very few pigment grains in their primary pigment cells (Walcott and Horridge 1971) and, therefore, the crystalline tract is not immediately surrounded by light-absorbing granules. Many of the retinae in this group also do not have a distal rhabdomere, which has been found only in Hydrophilidae, Dytiscidae, Gyrinidae and Carabidae (Horridge 1971). However, distal rhabdomere or not, the general mechanism for adaptation seems to be the same.

Unfortunately, very little is known about the physiology of the retinae in this group. Intracellular recordings from units in the *Dytiscus* retina have shown that light can enter from a number of different facets and affect one receptor (Horridge, Walcott, and Ioannides 1971). However, this experiment was always carried out on dark-adapted eyes and no recordings have been made from light-adapted eyes when the structure is quite different. Optical experiments on fresh slices of this type of compound eye have yielded some new information although they are far from complete (see Horridge 1971). From our present knowledge, it seems that in the dark-adapted eyes of this group a number of facets admit light to a receptor. In effect, this increases the absolute sensitivity of the unit while at the same time reducing its acuity. In the light-adapted eye, however, the crystalline tract must act as a light guide as no other light path has been discovered. The pigment granules of the primary pigment cells which lie close to the tract would act as a variable attenuator of transmission down this tract. This action and the narrow aperture at the tip of the cone would account for the reduced sensitivity, which is *c*. 1/1000 of that of the dark-adapted eye. Further, the extent of the primary pigment cells would limit the lateral spread of light between ommatidia. Therefore, one would expect receptors to receive light only from their own facet when light adapted. This action must control sensitivity and the light-adapted eye may have much smaller fields of view than when dark adapted, in so far as the light crossing between ommatidia is unfocused. The articles by Horridge (Chapter 12) and by Meyer-Rochow (Chapter 12) discuss other aspects of the eyes of these beetles.

2.4. Eyes with movement of cone, tract and rhabdom

Compound eyes in this group are exclusively acone eyes. That is, the four cone cells are 'watery' in appearance and do not have the rigid structure of the crystalline cones. Light-dark adaptation in these eyes is associated with extensive changes in shape and movement of the cone cells, the 'crystalline' tract they form, and the rhabdom of the retinula cells.

The compound eye of the Hemipteran, *Notonecta*, is typical of this group

(Lüdtke 1953; Eckert 1968). In the light-adapted state the four cone cells form a cone-shaped structure with its broad end against the cornea and its tapered end against the distal end of the rhabdom. The smaller rhabdom end of this tract is less than one quarter the diameter of the cone end. Surrounding the cone are a pair of primary pigment cells whose numerous pigment grains closely approach the cone cell membranes. These pigment cells run from the cornea to the distal end of the retinula cells. The retinula cells contain many pigment granules and these are densely packed around the distal portion of the rhabdom. The whole complex of primary pigment cells and retinula cells is surrounded by many accessory pigment cells which contain pigment granules evenly distributed throughout their length from cornea to basement membrane.

In the dark-adapted state the anatomy is different. Now, the primary pigment cells are shorter and are displaced laterally, and the pigment granules are displaced laterally within the cells away from the ommatidial axis. The rhabdom occurs at the very top of the retinula cells and indents into the cone cells. Thus the rhabdom moves from a position 25 μm from the cornea to 10 μm (Lüdtke 1953), a movement of 15 μm. I have observed a greater movement with the rhabdom coming to within 2–5 μm of the cornea. Thus, the cone cells are pushed in at the middle and now lie around the outside of the retinula cells. The pigment grains of the retinula cells are distributed more centrally, and not around the distal portion of the rhabdom.

These anatomical changes are typical of the Chironomidae, *Notonecta*, *Corixa*, *Forficula*, and *Culex*. However, the changes in the Belastomatidae, Tipulidae, *Anoplognathus* and *Tenebrio* differ in detail although not in principle. In the Belastomatidae, for example, *Lethocerus*, the four cone cells, in a light-adapted eye, are oval in longitudinal section and form a long (30 μm) thin (5 μm) tract to the rhabdom below (Walcott 1971a). This tract is closely surrounded by the primary pigment cells, many of the pigment grains of which touch the tract membranes. In the dark-adapted state the rhabdom moves *c*. 35 μm closer to the cornea, the cone cells are flattened between the two, and the tract is not present.

These changes necessarily alter the optical pathway to the rhabdom. In the light-adapted *Lethocerus* eye, the tract has an aperture of only 5 μm, and is closely surrounded by the dense pigment granules of the primary pigment cells. In the dark-adapted state, the long tract is gone and the aperture is now *c*. 20 μm in diameter. Therefore, one would expect changes in the sensitivity and field of view of the retinula cells.

Intracellular recordings have been made from units in the eye of *Lethocerus* and have been correlated with the state of adaptation (Walcott 1971b). Dark-adapted units are at least 1000 times more sensitive than the light-adapted units. This difference in sensitivity is related to the distance between the cornea and rhabdom. In addition to the sensitivity change, there is also a change in the angular sensitivity. Light-adapted units have an acceptance angle of 3·5°, which increases when dark-adapted to 9·0°. This change probably does not represent the possible extremes.

The time required to accomplish these cell movements is relatively long. *Lethocerus* requires 40 min in sunlight to become fully light adapted, with no observed movement in the first 10 min (Walcott 1971*b*). *Notonecta* and *Corixa* require *c.* 60 min for full anatomical light adaptation (Eckert 1968). Histological sections of *Lethocerus* retina often show a great variation in adaptation among ommatidia (Walcott 1971*b*), and therefore the ommatidia are likely to control their adaptation state independently. Stimulation of single facets of dark-adapted eyes with light guides shows that little if any light enters a receptor except through its own facet (J. Kien, unpublished observation).

2.5. Discussion

From the available data, one can conclude that insects have a variety of mechanisms to regulate the sensitivity of their visual system. An ommatidium with a narrow acceptance angle for greater acuity must, in general, sacrifice light-gathering ability. A receptor unit with a wide visual field can capture light from a greater area and, therefore, be more sensitive to a large source but thereby must necessarily sacrifice acuity. Insects obviously try to compromise in different ways between these two extremes depending on the environmental conditions and eye structure. In general, fast day flying insects like flies, dragonflies, bees and butterflies can regulate their sensitivity over a small range but have narrow (2°) acceptance angles under all conditions. Thus they have acute vision but do not have extreme sensitivity at very low light levels. Insects active at dusk or at night, such as locusts, *Lethocerus* or *Periplaneta* can change their acceptance angles. In the day, they are relatively narrow (3–4°) while at night the angles are larger, which provides greater sensitivity but reduces the acuity to small objects.

In the clear zone eyes (groups (II) and (III)), very few unit recordings have been made and none of these have been concerned with adaptation. It is not known for these eyes whether the acceptance angles change, for example, or whether in the light-adapted state light can enter by only the facet of the receptor. In the eyes that show movement of their retinula cell bodies (group (III)) we do not know whether the retinula cell columns across the clear zone act significantly as light guides or whether light crossing the clear zone freely is all important.

In all the various anatomical mechanisms for adaptation, the presence or absence of light is the cause of the reaction. In the case of apposition eyes (group (I)), each retinula cell seems to be an independent unit, and here possibly the degree of the depolarizations of the membrane of a unit determines its own adaptation state. However, in the acone eyes (group (IV) the retinula cells must move together and act as a unit. However, in *Lethocerus* at least, the receptors within an ommatidium having microvilli of the rhabdomeres oriented in different directions do not appear to be coupled electrically as significant polarized-light sensitivity can be recorded (Walcott 1971*b*). Also in these eyes, the retinula cells are not the only cells that move. The cone cells show the greatest changes and yet do not appear to contain any specialized structures for light detection. What then causes them to move? Are they coupled electrically to the retinula cells?

The mechanisms involved in effecting movements in insect eyes are unknown. Pigment granules within cells move as do whole groups of cells. There are no apparent morphological structures associated with pigment granules. Yet in the fly, for example, granules move several μm radially in the retinula cells in a predictable direction. Perhaps they are affected directly by an ionic field which is altered by depolarization of the cell. In the case of cell movements, there are sub-cellular structures which could affect some of the movements. In *Lethocerus*, for example, microtubules occur in great numbers in all the cells that move *except* the retinula cells (Walcott 1971*a*). In the cone cells and primary pigment cells, the microtubules are highly orientated in the directions of movement. Similar orientations are seen in the cone cells of a Megalopteran (Walcott and Horridge 1971) and *Notonecta* (Walcott, unpublished observations). Anatomically, therefore, the microtubules could be the 'motor' that causes the cell movements, while the retinula cells are pulled along 'passively'. However, there is no experimental evidence to support this hypothesis.

Part II
Receptor Physiology

3. The origin of the receptor potential of the lateral eye of Limulus

V. J. WULFF and W. J. MUELLER

THE origin of the light-evoked transretinal potential change, the retinal action potential, in the lateral eye of the horseshoe crab, *Limulus polyphemus*, has concerned visual physiologists since Hartline first described the event (Hartline 1928; Hartline and Graham 1932). Hartline, Wagner, and MacNichol (1952) demonstrated that ommatidia, surgically isolated but still attached to the cornea, generated a light-evoked potential change, the ommatidial action potential, which resembled the transretinal action potential of the entire eye. Thus, the retinal action potential of the lateral eye is composed of the discrete contributions of each illuminated ommatidium.

However, ommatidia of the *Limulus* lateral eye consist of a complex assembly of sense cells whose individual contributions to the ommatidial action potential were not resolved for some time. Demoll (1914) was the first to show that each ommatidium consists of a variable number (7–15) of radially-arranged retinula cells (Fig. 3.1) whose central ends envelop the thin distal process of a cell located more peripherally, the eccentric cell. The central region of the retinula cell membrane and that of the eccentric cell dendrite is modified to form a structure called the rhabdom (Demoll 1914; Miller 1959; Lasansky 1967; Fahrenbach 1969). This modification consists of numerous cylindrical projections of the membrane (the microvilli), packed in tight array (Miller 1957; Bass and Moore 1970), which abut those of neighbouring cells and form junctions (Fig. 3.1). Both retinula cells and the eccentric cell give rise to axons which leave the proximal end of the ommatidium as a bundle, but those of retinula cell origin are thinner.

Hartline and Graham (1932) first demonstrated that spike potentials could be recorded from small bundles of fibre dissected from the lateral eye optic nerve and the spike discharge, which is of constant magnitude and rather uniform frequency under constant illumination, was attributed to a single axon in the monitored bundle. Subsequently Hartline et al. (1952) showed that a similar unit discharge of spike potentials could be recorded from the nerve bundle leaving the ommatidium and, on occasion, could be

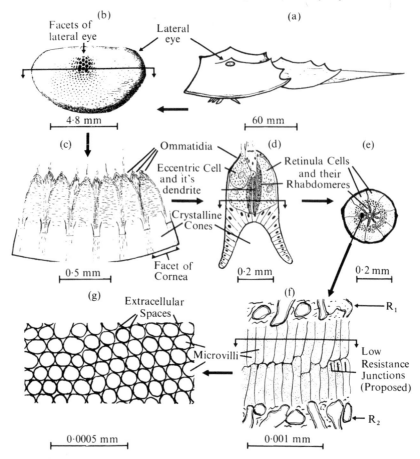

FIG. 3.1. A diagrammatic presentation of the gross and fine structure of the lateral eye of *Limulus polyphemus*: (a) lateral view of *Limulus*; (b) high lateral view of the lateral eye; (c) downward view of lower portion of lateral eye after cutting in the horizontal plane as indicated in (b); (d) longitudinal section through a single ommatidium; (e) cross-section of an ommatidium cut in the plane indicated in (d); (f) fine structure of a portion of the rhabdomeres of two adjacent retinula cells, R_1 and R_2; (g) tangential view of a portion of the rhabdomere of a retinula cell cut in the plane indicated in (f). These diagrams were adapted from the following sources: Miller 1957; Kennedy and Wulff 1960; Ratliff, Hartline, and Lange 1966; Lasansky 1967; Fahrenbach 1969; Bass and Moore 1970. By courtesy of Pergamon Press.

recorded superimposed on the ommatidial action potential. In the same communication Hartline and his colleagues demonstrated that a saline-filled glass micropipette inserted blindly into an ommatidium occasionally found a cell which generated a slow positive response to illumination on which relatively large positive spike potentials were superimposed. These spikes were synchronous with those recorded from the nerve bundle. Subsequently MacNichol and Wagner (MacNichol 1956) penetrated, under visual control, exposed eccentric cells at the cut edge of the lateral eye, demonstrating that the eccentric cell generated a slow light-evoked response with relatively large superimposed spikes. These observations, subsequently confirmed by others (Waterman and Wiersma 1954; Tomita 1957; Fuortes 1958; Tomita, Kikuchi, and Tanaka 1960; Behrens and Wulff 1965), led to the concept, still followed today, which stated all-or-nothing spike potentials are generated in eccentric cell axons but not in retinula cell axons.

The role of the retinula cell in the production of the light-evoked retinal action potential remained obscure for several years (see MacNichol 1956). Waterman and Wiersma (1954) first suggested that the light-evoked electrical response, the receptor potential, originated in retinula cells and was responsible for the potential changes recorded from eccentric cells and their axons. Similar conclusions were offered by Tomita and his co-workers (1960), Borsellino, Fuortes, and Smith (1965), and Stieve (1965). Tomita (1956) demonstrated that light-evoked slow potential changes without superimposed spikes or with rudimentary superimposed spikes could be recorded from cells within ommatidia of the lateral eye of the Japanese horseshoe crab. Behrens and Wulff (1965) and Kikuchi, Inuhma, and Tachi (1965) demonstrated unequivocally, using intracellular dye injection, that retinula cells generated light-evoked responses on which rudimentary spikes may, or may not, be superimposed and Behrens and Wulff (1967) demonstrated that such responses could be obtained in the absence of a functional eccentric cell within the same ommatidium. Smith and Baumann (1969) in a definitive paper (see earlier studies Borsellino, Fuortes and Smith 1965); Smith, Baumann and Fuortes (1965) demonstrated that the cells within each ommatidium of the *Limulus* lateral eye are coupled electrically by virtue of low-resistance junctions between neighbouring retinula cells and between retinula cells and the dendrite of the eccentric cell. The properties of these junctions is such that the light-evoked receptor potential (depolarization of retinular cells) causes depolarization of the eccentric cell which, in turn, results in the initiation of spike potentials which are propagated along the eccentric cell axon.

The low-resistance junctions responsible for the electrical coupling of cells within ommatidia are believed to exist where microvilli from neighbouring cells meet (Miller 1957 and Fig. 3.1). At these junctions, the outer leaflets of the cell membranes fuse to form a five-layered structure with a relatively thick middle layer (Lasansky 1967; Fahrenbach 1969); these junctions are believed to be the low-resistance 'tight junctions' between adjacent cells. Thus, the cells within ommatidia of the lateral eye of *Limulus* are coupled electrically to form a functional unit in which, under the influence of light or other agents, the response of retinula cells is co-ordinated rather effectively to produce depolarization of the eccentric cell.

Although it is now known that retinular cells in ommatidia of the lateral

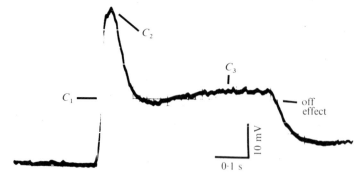

C_2

C_3

C_1 —

— off effect

10 mV

0·1 s

FIG. 3.2. An intracellularly-recorded receptor potential from a retinula cell in an ommatidium of the lateral eye of *Limulus* elicited by a 0·5 s light flash.

eye of *Limulus* produce receptor potentials when illuminated, the intracellular site(s) of origin of this potential is not yet established firmly. The retinula cell receptor potential (Fig. 3.2) consists of (1) an initial non-propagated 'spike', (2) a transient component which peaks and then subsides to a plateau when illumination is prolonged (Hartline and Graham 1932; Hartline *et al.* 1952; Tomita 1956; Kikuchi and Tazawa 1960; Benolken and Russel 1966; Yeandle 1967; Smith and Baumann 1969). For convenience, the components of the receptor potential will be referred to as C_1, C_2, and C_3, respectively (Fig. 3.2). Component C_3 terminates in an off-effect shortly after illumination ceases. Tomita (1956) demonstrated that non-rhabdomeric membrane of retinula cells (Figs. 1 and 9) is involved actively in the production of the receptor potential. This was achieved using a double-barrelled electrode assembly in a pencil configuration, enabling the inner electrode to protrude and impale cells with the outer electrode surrounding the site of impalement. The light-evoked potentials consisted of components C_2 and C_3 (C_1 cannot be distinguished in the published records). The intracellular potential changes evoked by light always have positive polarity but the polarity of the simultaneously-recorded extracellular potential change is variable; C_2 is always negative but C_3 is either positive or negative. Inspection of extra-cellularly-recorded light-evoked potentials probably of retinula cell origin (Tomita 1956, Fig. 9a and b) clearly show a negative C_2 and a positive C_3. These oscillograms resemble that of Yeandle (1967) recorded with a micro-pipette partially withdrawn from a retinular cell, consisting of a positive C_1, a negative C_2, and a positive C_3. Yeandle (1967) concluded that C_2 originated in a different region of the cell to either C_1 or C_3. The study by Tomita (1956) shows that superficial non-rhabdomeric membrane (Figs. 1 and 9) of retinula cells is certainly involved in the production of C_2.

This communication is addressed to the hypothesis that C_1 and, perhaps, C_3 of the receptor potential of retinula cells in the *Limulus* lateral eye originate in the rhabdomeric membrane deep within the retinula cell. This hypothesis derives from observations in addition to those described above. It has been shown (Wulff and Mendez 1970; Wulff 1971a) that hyperpolarization of retinula cells produced by injecting extrinsic current into the monitored cell increased the latent period of the receptor potential and Smith and Baumann (1969) concluded that the effect was produced by inward current flowing through the rhabdomeric 'tight junctions'. Wasserman (1970) corroborated the above observation independently and, in addition, demonstrated that hyperpolarizing currents strong enough to evoke a hyperpolarizing response also result in an attenuation of the response to light. Smith and Baumann (1969) demonstrated that a retinula cell exhibiting a hyperpolarizing response is effectively 'uncoupled' from its nearest neighbours, i.e. the low resistance 'tight junctions' between neighbouring retinula cells transiently assume a higher resistance. It is possible that this change in the rhabdomeric membrane system may have been responsible for the attenuation of receptor potential observed by Wasserman (1970). This is not incompatible with the localized inactivation of excitable membrane proposed by Wasserman.

Observations on photoreceptors of invertebrates other than *Limulus* also lend support to the hypothesis presented above. Lasansky and Fuortes (1969) demonstrated that the microvillar membrane of the leech photoreceptor is the site of an inward light-evoked current and conclude that the receptor potential of this photoreceptor originates in this structure; and the demonstration that visual pigments are located in the rhabdomeres of retinula cells in the eye of the blowfly (Langer and Thorell 1966) indicates that the first step in the visual process—the absorption of radiant energy—occurs in the rhabdomeres of sense cells in the compound eye (see also Goldsmith, Dizon and Fernandez 1969; Hays and Goldsmith 1969).

3.1. Methods

All experiments are performed on excised lateral eyes of the horseshoe crab, *Limulus polyphemus* (Behrens and Wulff 1965), clamped in a lucite chamber containing sea water (*c*. 40 ml). Selected ommatidia are penetrated from above with one or two independently-manipulated micropipettes filled with 2·7 M KCl using the technique of Tasaki *et al.* (1968) with resistances between 20 and 40 MΩ. The microelectrodes are connected to the input terminals of neutralized input capacity preamplifiers via short leads with silver–silver chloride ends inserted into the barrel of the pipettes. The outputs of the preamplifiers are connected to a 502 Tektronix cathode ray oscilloscope, the display of which is recorded on film using a Grass kymograph camera. In all double impalement experiments one channel of the recording system is used in the single-ended mode; the second channel is used in the differential mode, using a third micropipette immersed in the bath and a third preamplifier. Criteria used to determine double impalement of the same retinular cell are described (Wulff 1971a). A thick silver–silver chloride electrode is immersed in the bath and connected to ground through a 100 kΩ resistor.

The extrinsic current to depolarize impaled cells passes through a 400 MΩ resistance between the current source and the micropipette and is recorded as the potential drop across the 100 kΩ resistor. A manually-operated switch connects the current source to one micropipette and grounds the input of the single-ended preamplifier. A second switch in the input circuit of the preamplifiers permits passage of current through either of two impaling micropipettes and permits the measurement of the membrane potentials of doubly-impaled sense cells. The circuit between the current source and the impaling micropipette is either closed manually or by an electronic switch activated by a pulse generator. When activated, the electronic switch remains closed for 0·8 s and the time of occurrence of this current pulse may be varied relative to a pulse which triggers both the Grass Photo-stimulator, which produced a brief (10 μs) flash of light, and the traces of the oscilloscope.

Experiments are also performed with singly-impaled retinula cells in which the receptor potential and the output of a differentiating circuit are simultaneously recorded as a function of vertical position of the impaling micropipette (Fig. 3.8). The receptor potential is recorded in the differential mode and the same signal served as the input to the differentiator. Micropipettes are advanced or withdrawn in steps of 5 or 10 μm using a manually-operated microdrive. Relatively high resistance micropipettes (40–80 mΩ) with a gentle taper are employed in these experiments. All electrodes are inspected microscopically prior to use, cleaned if dirty and many are coated to the tip with silicone to facilitate free movement of the pipette.

Experiments involving the depletion and replenishment of Na^+ in the medium bathing the eye are performed with singly and doubly impaled retinula cells. The original solution is withdrawn by suction and the new solution enters the chamber via a baffle which prevents disturbance of the impalement. Exchange of the solution required *c*. 10 s. The osmotic pressure of sea water is 887 mosmol and that of the choline chloride artificial sea water is 867 mosmol.

All experiments are performed at room temperature (21–24 °C); during any one experiment the temperature fluctuated by 1 °C or less.

The ability of retinula cells in excised lateral eye preparations to withstand impalement is rather variable. Many doubly-impaled retinula cells fail to recover from the insult. Of those that survive, only those experiments in which the cell has a resting potential of 40 mV or more and a typical (Behrens and Wulff 1965) and stable receptor potential of 30 mV or more for at least 1 h are recorded. Some cells which withstand double penetration often exhibit reproducible current–voltage relationships only after 1 h or more after impalement. Part of the data reported here are selected from results obtained from sixty-seven doubly-impaled retinula cells.

3.2. Results and discussion

3.2.1. *The reversal of the receptor potential*

When a retinula cell is depolarized by the injection of extrinsic current intense enough to make the cell interior sufficiently electrically positive with respect to the external environment, the light-initiated receptor potential is of reversed polarity (Kikuchi, Naito, and Tanaka 1962; Smith and Baumann 1969; Wulff 1971a). However, the potential of reversal is not the same for all components of the receptor potential, as illustrated in Fig. 3.3. The receptor potential elicited by a 10 μs light flash with no current injection (lowest trace in Fig. 3.3a) exhibits positive components C_1 and C_2. At 6 mV above zero membrane potential, component C_2 is of reversed polarity but C_1 persists as a positive deflection. The response sequence of Fig. 3.3b, obtained from the same cell depolarized to a greater extent, shows the disappearance of C_1 at

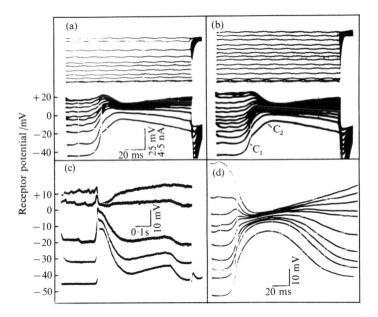

FIG. 3.3. Composite oscillograms of receptor potentials recorded from doubly-impaled retinula cells showing the effect of injected depolarizing currents of progressively increasing strength. (a) and (b) each have two parts: the upper traces represent the voltage drop across a 100 KΩ resistor produced by constant current pulses of 0·8 s duration which begin before the onset of each trace and terminate just before the end of each tree; the lower set of traces in (a) and (b) and the traces in (d) represent receptor potentials elicited at 10 s intervals by 10 μs light flashes coincident with the onset of each trace. The control responses are at the bottom of each frame and, with increasing strength of depolarizing current, the baseline is displaced upward. (d) shows a series of receptor potentials elicited by light flashes of 0·5 s duration occurring at 10 s intervals superimposed on current pulses of 0·8 s duration which begin on the onset of the sweep and terminate as shown by the second oscillogram from the bottom. The control response is at the bottom and with increasing strength of depolarizing current the baseline is displaced upward. These responses were arranged manually.

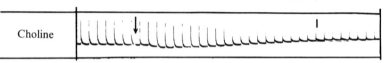

FIG. 3.4. Intracellularly recorded receptor potentials elicited by a 10 μs light flash at 10 s intervals before and after the application of artificial sea water in which choline replaced Na^+. Two rinses were applied at the time indicated by the arrow. Note the subsequent hyper- and de-polarization of the cell and the decline in response magnitude. The vertical calibration represents 20 mV [millivolts].

+25 mV, to give in an apparently longer latent period. The latency shortens when a cell is depolarized beyond the reversal potential of C_1 (Fig. 3.3d), which suggests that C_1 has reversed polarity. Similar results are obtained with sixty-seven doubly-impaled retinula cells.

The reversal potential of C_3 also lies on the positive side of zero membrane potential, as shown by the receptor potentials in Fig. 3.3c, elicited by 0·5 s light flashes superimposed on 0·8 s pulses of depolarizing current with increasing magnitude. The upper two oscillograms show clearly that C_1 and C_3 are still positive deflections whereas C_2 is quite negative. The reversal potential for C_2 was +4 mV and for C_1 +16 mV; C_3 reverses approximately at the same potential as C_1.

The fact that the receptor potential reverses polarity on the positive side of zero membrane potential (i.e. cell interior positive) together with the observation that depletion of Na^+ in the external environment reduces

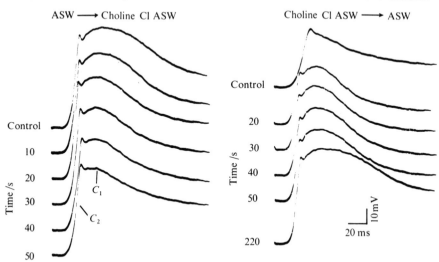

FIG. 3.5. Receptor potentials recorded intracellularly showing the effect of Na^+ depletion (left column) and replenishment (right column) on components C_1 and C_2. The time in seconds to the left of the records is the elapsed time from the instant of evacuation of the chamber prior to refilling with the test solution. The left column shows the effect of one application of choline chloride artificial sea water (ASW); the right column shows the effect of two applications of artificial sea water.

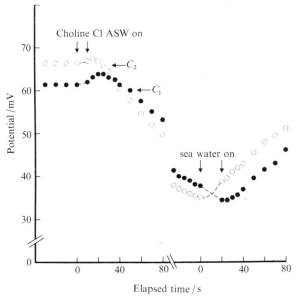

FIG. 3.6. Graphs illustrating the changes in magnitude of components C_1 and C_2 with time after depletion of the external $[Na^+]$ (left) and replenishment of the external $[Na^+]$ (right). The vertical arrows mark the beginning and end of the fluid exchange in the chamber containing the lateral eye preparation; choline chloride artificial sea water was applied only once (left) and artificial sea water (ASW) (right) was applied twice. It is estimated that *c*. 10% of the original solution remained after evacuation of the chamber.

markedly the receptor potential magnitude (Figs. 3.4, 3.5, and 3.6; Kikuchi *et al.* 1962) indicates that the influx of Na^+ is largely responsible for the production of the retinula cell receptor potential in the *Limulus* lateral eye. Similar conclusions are reached for the *Limulus* ventral photoreceptor (Millecchia and Mauro 1969*a*,*b*), the barnacle ocellus (Brown, Meech, Koike, and Hagiwara 1969; Brown, Hagiwara, Koike, and Meech 1970), the compound eye of the hermit crab (Stieve 1964), the compound eye of the drone (Fulpius and Baumann 1969) and the compound eye of the crayfish (Stieve and Wirth 1971). The fact that components C_1 and C_3 of the *Limulus* lateral eye retinula cell receptor potential reverse polarity at a higher transmembrane potential than component C_2 indicates that the membrane mechanisms responsible for their production are different.

3.2.2 *The effect of Na$^+$ depletion*
When normal sea water (435 mM with respect to NaCl), bathing the lateral eye preparation, is replaced with artificial sea water in which the NaCl is replaced by choline chloride, retinula cells first hyperpolarize then depolarize, and the receptor potentials exhibit initially a transient increase in magnitude followed by progressive reduction (Figs. 3.4 and 3.5). Upon

return to normal NaCl concentration retinula cells slowly hyperpolarize and the response magnitudes return approximately to the previous control values. When the immediate effects of Na^+ depletion and replenishment on components C_1 and C_2 of the receptor potential are examined (Figs. 3.5 and 3.6) it is evident that the magnitude of C_2 is affected more rapidly than the magnitude of C_1. These results are obtained only when superficially-located retinular cells within the upper row of exposed ommatidia are impaled and they suggest that the site of origin of C_1 is affected less rapidly by the altered external environment than is the site of origin of C_2. To reverse the argument, if C_1 originates in the rhabdomeric membrane of the retinula cell which lies near the centre of the ommatidium and C_2 originates in non-rhabdomeric membrane, then the observed results are expected. The possibility that the two sites exhibit different sensitivities to the external $[Na^+]$ is unlikely, for eventually the rate of decline after depletion and the rate of increase after replenishment of Na^+ is similar for both components (Fig. 3.6).

However, although the magnitude of C_1 is affected more slowly than that of C_2, its rate of rise is affected just as rapidly after Na^+ depletion or replenishment (Fig. 3.4, right-hand column, 20 s response). This observation is confirmed in ten experiments with superficial impalement of superficially-located retinular cells subjected to Na^+ depletion and replenishment while both the receptor potential and its differential were recorded. During these experiments it is observed that the onset of change in the rate of rise of C_1 can be delayed by driving the micropipette deeper into the retinular cell. A tentative explanation for this rapid effect of altered external $[Na^+]$ on the rate of rise of C_1 is that the conductivity of the cytoplasm in the vicinity of the tip of the voltage-sensing superficially-located micropipette changes rapidly when external Na^+ is either depleted or replenished; a reduction in conductivity would decrease the rate of rise of the initial transient (C_1) and an increase in conductivity would increase the rate of rise. This problem is still under investigation.

3.2.3. *The configuration of the receptor potential*

The wave forms of the receptor potential of the *Limulus* lateral eye retinula cell recorded from the same cell via two impaling micropipettes are usually so similar that the cell has been considered isopotential, even with respect to voltage transients (Smith and Baumann 1969; Wulff 1971a). However, variations in wave form of receptor potentials recorded via two impaling micropipettes have been reported (Behrens and Wulff 1965). In sixty-eight experiments performed recently in this Laboratory with doubly-impaled retinula cells, differences in the rate of rise of component C_1 of the receptor potential were observed in ten experiments. This rather rare occurrence is perhaps not surprising, for micropipettes are purposely not driven deep into retinula cells to minimize cell damage.

These observations of differences in waveform of the receptor potential generated the idea that the configuration may vary with the location of the microelectrode tip within the retinula cell. Accordingly, in six experiments with doubly-impaled retinula cells, one electrode (usually that which had not been used for current injection) is advanced purposely in steps of 5 or 10 μm until the impalement was lost. Receptor potentials were recorded after each advance of the micropipette. Changes in configuration were observed in two experiments, and the results of one experiment are illustrated in Fig. 3.7. In

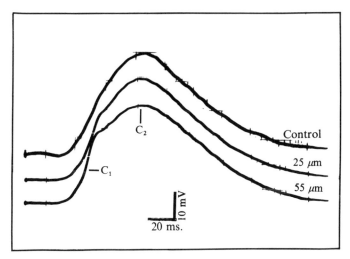

FIG. 3.7. Photographically arranged oscillograms of receptor potentials recorded by a micropipette driven progressively further into an impaled retinular cell. The potentials were elicited by a 10 μs light flash coincident with the onset of each trace. The control response was recorded with the pipette near the outside (upper) border (see Fig. 4.8) of a retinula cell; the remaining responses were recorded after the electrode had been advanced by 25 and 55 μm.

the other experiments, the receptor potentials deteriorate progressively with each advance of the micropipette until the impalement is lost.

To confirm these observations eighty-five experiments are performed on singly-impaled retinular cells in which the micropipette is advanced manually in steps of 5 or 10 μm, and, after each advance, the receptor potential and its differential are simultaneously recorded. Despite the precautions, clear evidence of a change in rate of rise of C_1 in the absence of a change in magnitude of the receptor potential is obtained in only ten experiments. The results of two experiments are presented in Fig. 3.8. These results show that the rate of rise of the first component of the receptor potential increases with the depth of the impaling microelectrode and that, at the maximum depth, the latency and time to peak (C_1) are both shorter.

While these experiments were in progress, it became apparent that a number of conditions must be satisfied in order to demonstrate changes in

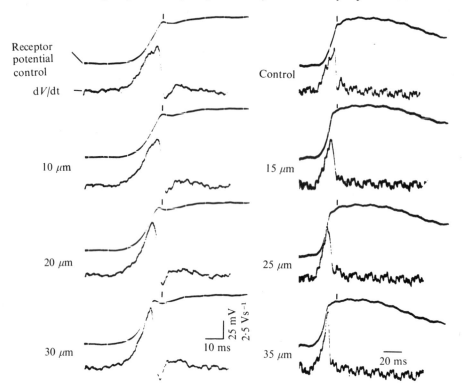

Fɪɢ. 3.8. Oscillograms illustrating the effect of advancing the recording micropipette into two superficially located retinula cells on the receptor potential (upper trace) and its differential (lower trace). The responses were elicited by a 10 μs light pulse coincident with the beginning of each trace. The vertical bars above the receptor potentials mark the peak of component C_1 in the control responses. The records are arranged manually.

receptor-potential configuration with electrode position. First, proper alignment of the axis of the micropipette and the cell must exist. Since all impalements were blind because lateral eye ommatidia are covered by a pigmented epithelium, proper alignment of electrode and cell was a matter of chance. Secondly, the micropipette must move freely into and out of the cell after the initial penetration which was usually achieved by gentle tapping of the microdrive. It was apparent that, in many experiments, movement of the micropipette gave movement of the tissue. Thirdly, displacement of the micropipette must produce negligible injury to the cell, i.e. response magnitude and membrane potential must remain constant. Many experiments fail to meet this condition. In view of these considerations, the relatively low number of experiments showing configurational changes is not surprising.

In all experiments in which the impalement persists after advancement of the micropipette, the pipette is withdrawn in step-wise fashion and, in some, the electrode is advanced again. Those experiments which yield an increase in rate of rise of C_1 upon first advancing the pipette show a reduction in rate of rise of C_1 upon withdrawal but the latter changes are always less pronounced.

3.3. Conclusion

The results of this investigation demonstrate that: (1) C_2 of the *Limulus* lateral eye receptor potential reverses its polarity at a lower transmembrane potential (inside positive) than either C_1 or C_3; (2) sudden depletion or replenishment of the external $[Na^+]$ affects C_2 more rapidly than C_1; and (3) the rate of rise of C_1 increases and the latent period and time to peak of C_1 decreases as the monitoring micropipette is driven deeper into retinula cells located superficially.

It has been demonstrated that illumination of the lateral eye of *Limulus* produces an increase in conductivity of the retinula cell membrane synchronous with the receptor potential (Tomita 1956; Fuortes 1958, 1959; Rushton 1959; Borsellino *et al.* 1965; Stieve 1965; Smith and Baumann 1969). The facts that the $[Na^+]$ in the fluid bathing the excised lateral eye exerts a pronounced effect on the magnitude of the receptor potential and that reversal of the receptor potential occurs only when the cell interior is electrically positive with respect to the outside indicate that influx of Na^+ is chiefly responsible for the production of the receptor potential. Thus, the absorption of radiant energy by photosensitive pigment presumably located in the rhabdom of *Limulus* lateral eye ommatidia results, after the elapse of some time, in the opening of ionic channels permitting the influx of Na^+ (and perhaps other cations), which produce the transient voltage changes collectively called the receptor potential.

The contention that the influx of Na^+ is chiefly responsible for the genesis of the receptor potential has not been without challenge. Smith, Stell, Brown, Freeman, and Murray (1968) advanced the hypothesis that a light-sensitive electrogenic sodium pump was responsible for the production of the receptor potential in the ventral eye of *Limulus*. This hypothesis was based upon the following: (1) that depletion of Na^+ and K^+ or increasing Ca^{2+} in the external environment, lowering the temperature to 2 °C or exposing the preparation to ouabain (1 mM) in sea water depolarized ventral photoreceptor cells by 10–20 mV; (2) that the temperature dependence of the membrane potential exceeded that expected from the Nernst relation and that intracellular injection of Na^+ hyperpolarized the cell; and (3) that 1 mM ouabain irreversibly abolished the receptor potential. Millecchia and Mauro (1969b) criticized the electrogenic-pump hypothesis on the following grounds: (1) the maximum e.m.f. of the pump must be 2–3 V to supply the maximum light-induced current (200–300 nA) which they measured in *Limulus* ventral photoreceptor cells under voltage-clamp conditions; (2) the e.m.f. of the electrogenic pump must vary with the membrane potential and must reverse in sign when the photoreceptor is 15 mV inside positive (i.e. above the reversal potential of the receptor potential); (3) the electrogenic sodium pump must be insensitive to light and must not vary in time at the reversal potential but must vary with light intensity and with time at all other voltages. They conclude that the light-sensitive electrogenic pump proposed by Smith *et al.* (1968) is unrealistic and that the receptor potential of the *Limulus* ventral eye is adequately explained by a light-modulated conductance. The results of the investigations of Brown *et al.* (1970) on the barnacle photoreceptor using the voltage-clamp technique also led to a rejection of the electrogenic sodium-pump hypothesis. Stieve, Bollman-Fisher, and Braun (1971) investigated the effects of ouabain and DNP on the receptor potential of the isolated crayfish retina, and observed (i) the loss of excitability in the presence of both drugs is hastened by increasing the frequency and duration of light stimulation; and (2) the effects of DNP and ouabain on the wave form of the receptor potential differ. They concluded that their results contradict the hypothesis that the receptor potential is caused by a light-modulated electrogenic sodium pump. Brown *et al.* (1970) report that ouabain at high concentrations decreases the receptor-potential of the barnacle eye and attribute this effect to a reduced Na^+ concentration gradient; at a concentration of 10 μM ouabain has no effect on the receptor potential but abolishes the hyper-polarization following illumination.

Post-illumination hyperpolarization of photoreceptor cells is described for the *Limulus* lateral eye (Benolken 1961; Kiluchi *et al.* 1962), the eye of the dragonfly (Naka 1961), the eye of the barnacle (Brown *et al.* 1970; Koike, Brown, and Hagiwara 1971) and the ventral photoreceptor of *Limulus* (Brown and Lisman 1972). In the barnacle, Koike *et al.* (1971) demonstrate that post-illumination hyperpolarization is not associated with an increase in conductance of the photoreceptor membrane. Intracellular injection of Na^+ also gives an hyperpolarization of the *Limulus* ventral photoreceptor and the barnacle eye (Brown and Lisman 1972; Koike *et al.* 1971). The hyper-polarization induced by light or injection of Na^+ is abolished in the presence of 10 μM ouabain (Koike *et al.* 1971), 5 μM strophanthidin (Brown and Lisman 1972), or by bathing the preparation in K^+-free sea water. These investigators conclude that the post-illumination hyperpolarization is pro-duced by an electrogenic sodium pump which was activated by Na^+ influx during illumination. Brown and Lisman (1972) also conclude that an electro-genic Na^+ pump contributes directly to the dark-adapted receptor cell membrane potential but does not generate directly the receptor potential.

Lisman and Brown (1972) demonstrate that intracellular injection of Ca^{2+} produces a reversible decrease in the magnitude of the response of *Limulus* ventral photoreceptor cells. Intracellular injection of Na^+ a similar reversible decline in magnitude when cells are bathed in media containing $CaCl_2$ at concentrations of 10 mM or more but not when the medium was deficient in $CaCl_2$ (0·1 mM). These investigators conclude that an increase in the internal $[Na^+]$ gives an increase in the internal $[Ca^{2+}]$ and that the latter produces the reduction in the response to light.

In summary, the available evidence argues against the hypothesis that the receptor potential is generated by a light-modulated electrogenic pump; rather, the accumulated evidence from the investigation of a number of invertebrate photoreceptors supports the hypothesis that illumination results in an increase in conductance of the photoreceptor membrane and that the influx of Na^+ is responsible chiefly for the production of the generator potential (Tomita 1956; Fuortes 1958, 1959; Rushton 1959; Benolken 1961; Kikuchi *et al.* 1962; Stieve 1964, 1965; Smith, Stell, and Brown 1968; Fulpius and Baumann 1969; Millecchia and Mauro 1969*b*; Brown *et al.* 1969, 1970; Stieve and Wirth 1971; Stieve, Bollmann-Fisher, and Braun 1971; Koike, Brown, and Hagiwara 1971; Brown and Lisman 1972).

The results of the present investigation suggest: (1) the membrane mechanism responsible for the generation of component C_2 is different from that responsible for the generation of C_1 and C_3; (2) the site of generation of C_1 is affected more slowly by changes in the external $[Na^+]$ than is the site of generation of C_2; and (3) that components C_1 and, perhaps, C_3 originate in the microvillar membrane system (the rhabdomere) of retinular cells and that component C_2 originates chiefly or entirely in non-rhabdomeric membrane (see also Tomita 1956 and Yeandle 1967). This model is summarized in Fig. 3.9.

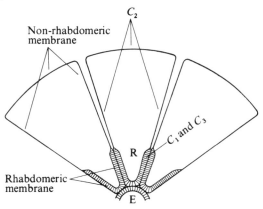

FIG. 3.9. A diagrammatic cross-section of part of an ommatidium of the lateral eye of *Limulus* showing three of the radially-arranged retinula cells and a portion of the eccentric cell dendrite. The cross-hatched regions represent the microvillar membrane which forms the rhabdome of the ommatidium. It is proposed that components C_1 and, perhaps, C_3 originate in the microvillar rhabdomeric membrane and that C_2 originates in non-rhabdomeric membrane. For more detailed morphological data please refer to Miller (1957), Fahrenbach (1969) or Wulff (1971*a*).

Evidence other than that presented and cited above bears upon the previous conclusion. Dowling (1968) suggested that large spontaneous potential fluctuations from *Limulus* lateral eye retinula cells are seen most readily when the recording micropipette tip is close to the rhabdom region. The effects of Ca^{2+} depletion or enrichment, $MnCl_2$, cyclic AMP, and Na^+ depletion on C_1 and C_2 are consistently different (Wulff 1971*b* and unpublished observations). Depletion of external $[Na^+]$ has a particularly strong effect on component C_2 (Fig. 3.5) so that, after equilibration, the only measurable component is C_1, suggesting that C_2 may be more dependent on external Na^+ than C_1.

The findings and inferences reported here indicate that retinula cells in the lateral eye of *Limulus* have specialized sub-cellular mechanisms which make unique contributions to the light-evoked receptor potential.

ACKNOWLEDGEMENT

We wish to express our appreciation to Mrs. Jane Fahy for her loyal and considerable assistance during the conduct of this research. This study was supported in part by grant number 5 R01 EY00236 from the National Eye Institute of the National Institutes of Health, Public Health Service, Department of Health, Education, and Welfare.

4. Electrophysiological properties of the honey bee retina

F. BAUMANN

THE retina of the honey bee (*Apis mellifera*), and especially that of the honey bee drone is a convenient preparation for the study of visual receptors with electrophysiological techniques. Honey bees have colour vision (von Frisch 1914; Daumer 1956; Autrum and von Zwehl 1964) and discriminate between different polarization planes of light (von Frisch 1949).

The bee retina is well separated from more central structures, which is typical for insect eyes. Also, apart from the most distal, corneal regions where cells of the dioptric apparatus are found, the retina is composed of only two types of cells. These are the retinula cells and the pigment cells (Figs. 4.1 and 4.2). It is the retinula cells which are the photoreceptor cells. They contain the photopigment and respond to light with an electrical signal called the receptor potential. Within the retina, the retinula cells are grouped together to form ommatidia, in which either nine (in the drone; Perrelet and Baumann 1969a) or eight or nine (in the worker bee; Phillips 1905; Varela and Porter 1969) cells are seen in transverse section as a circular flower-like structure (Fig. 4.1). The pigment cells (between twenty-five and thirty per ommatidium) surround the receptor cells and contain a dark screening pigment, the function of which is probably the optical isolation of neighbouring ommatidia (Goldsmith and Philpott 1962; Varela and Porter 1969). In the drone, both retinula and pigment cells are large (*ca.* 400 μm long and 10 μm in diameter) and are easily impaled with microelectrodes which can be kept inside a cell for periods of time frequently reaching one hour. The cells of the worker bee retina are smaller and more difficult to impale.

Some investigators have worked on eyes of living bees fixed and immobilized with wax or collophonium (Goldsmith 1963; Autrum and von Zwehl 1964); others have used isolated heads either intact (Shaw 1969) or separated into two halves by a cut which exposed the retinula and pigment cells (Naka and Eguchi 1962; Baumann 1968). The cut heads were maintained in a chamber filled with humid air (Naka and Eguchi 1962) or perfused with a

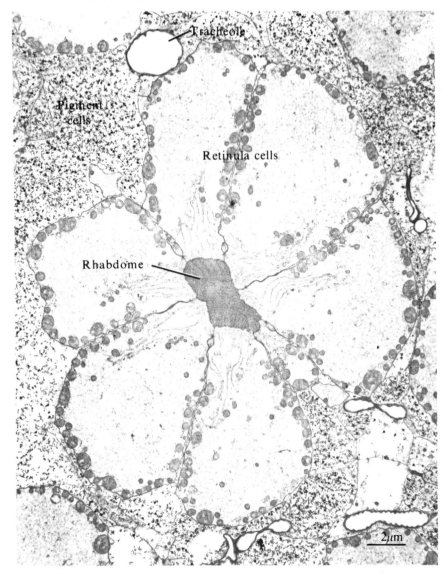

FIG. 4.1. Electron micrograph illustrating a cross-section of a drone ommatidium in the proximal part of the retina. One sees clearly that the ommatidium is formed by six large retinula cells, each of which contributes microvilli to the central rhabdom. Other retinula cells are small at this level and contribute few or no microvilli to the rhabdom. The retinula cells are surrounded closely by pigment cells (so-called accessory pigment cells which sheath each ommatidium from the base of the corneal facet to the basement membrane). These cells are rich in glycogen (small black dots) and wedge deeply in between retinula cells. Tracheoles are interspersed between pigment cells. Magnification *ca.* 6000×.

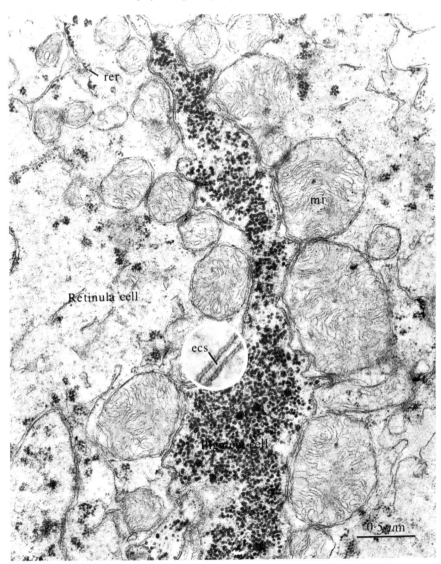

FIG. 4.2. High magnification electron micrograph illustrating the topographical relationships between the retinula and the pigment cells (accessory pigment cell in the drone). In this picture one sees parts of two retinula cells in cross-section. The cytoplasm of the retinula cells contains many profiles of rough endoplasmic reticulum (rer) and numerous mitochondria (mi) which are situated at the periphery of the cell. The accessory pigment cell wedged in between the retinula cells is characterized by its cytoplasm packed with beta particles of glycogen. Between these two types of cell there is a fairly regular extracellular space, measuring *ca*. 15 nm in width (circular insert). Magnification *ca*. 38 000 × ; circular insert *ca*. 150 000 × .

physiological solution of known composition (Baumann 1968). Despite the variety of experimental procedures, similar results have been obtained by all investigators who have worked with honey bee retinas.

This article reviews electrophysiological experiments on the honey bee retina, briefly summarizing experiments on the resting potential, the receptor potential and the spike potential which have been published previously, and dealing in greater detail with the process of visual adaptation. Using published and unpublished data, an attempt will be made to formulate and illustrate a hypothesis on the mechanism of visual adaptation. Finally, recent experiments on pigment cells of the drone retina will be described. These cells are influenced by the activity of the retinula cells and simultaneous recording from retinula and pigment cells has contributed to an understanding of certain unexplained properties of retinula cells. These experiments may also be the beginning of an attempt to consider the retina of the bee as a complete organ and not as a collection of a large number of individual cells.

4.1. Resting potential

Retinula cells of the honey bee drone have an average resting potential of -60 mV. When the ionic composition of the medium surrounding the cells is changed (Fulpius and Baumann 1969), the resting potential depends strongly on the external potassium concentration but is only slightly affected when the external sodium concentration is altered. These observations suggest that the resting potential is determined, mainly by the relative permeabilities of the cell membrane to potassium and sodium ions. Modifying the external chloride concentration gives rise only to transient potential changes, implying that chloride ions are distributed passively across the retinula cell membrane and play no role in the determination of the membrane potential in steady-state conditions.

Recently, it has been suggested that, in *Limulus*, part of the resting potential of retinula cells in steady-state conditions is due to the activity of an electrogenic Na–K pump, i.e. a mechanism in which active sodium extrusion exceeds active potassium uptake (Brown and Lisman 1972). Such a mechanism produces a net transfer of electrical charge across the membrane, which tends to make the inside of the cell more negative. Blocking the pump in these cells leads to a rapid decrease in membrane potential. In the drone, experimental conditions such as anoxia (Baumann and Mauro 1973), low temperature (Duruz and Baumann 1968), replacement of sodium in the bathing solution by lithium (Fulpius and Baumann 1969), or administration of ouabain (unpublished observation), conditions which are known to inhibit the activity of electrogenic Na–K pumps, all lead to a rapid loss in membrane potential of stimulated cells. However, this finding cannot be considered as definite evidence for participation of an electrogenic pump in determining the steady-state resting potential. These experimental conditions are expected to inhibit all Na–K pumping by the retinula cells and recent experiments have shown that the rapid depolarization is probably not due at all to suppression of an electrogenic pump, but is the consequence of an uncompensated leak of potassium from the retinula cell, which accumulates in the narrow extracellular space between the retinula and pigment cells (see later).

There is evidence, however, that retinula cells of the honey bee drone have an electrogenic pump, which is responsible for the transient hyperpolarization when the intracellular sodium concentration is increased, either by iontophoretic injection of sodium (unpublished observation) or by stimulation with light (Baumann and Hadjilazaro 1971). It remains an open question whether this electrogenic pump makes any contribution to the membrane potential of the resting cell.

4.2. Receptor potential

The response to light of the honey bee retinula cell consists of a slow depolarization with two phases, a transient phase, the transient, and a steady phase, the plateau (Fig. 4.3). A spike potential is regularly observed in the drone on the rising phase of the receptor potential as soon as its amplitude reaches a threshold value (Naka and Eguchi 1962; Baumann 1968). The amplitudes of the transient and of the plateau increase with an increase in light intensity. The transient saturates at zero membrane potential. Strong light thus abolishes temporarily the potential difference between the inside and the outside of the cell. In a normal bathing solution the amplitude of the plateau saturates at levels always below zero potential.

The cause of the drone receptor potential is a light-evoked increase of the membrane conductance and an inflow of current mainly carried by sodium ions. In dark-adapted cells, the effective membrane resistance is *ca.* 3 MΩ, while at the peak of the light response it drops to < 1 MΩ (Baumann and Hadjilazaro 1972). Replacement of the sodium in the bathing medium by larger cations such as Tris or choline decreases the amplitude of the receptor potential (Fulpius and Baumann 1969). The effects of an increase in the potassium concentration on the receptor potential are probably the consequence of depolarization of the cell membrane (Fulpius and Baumann 1969). Replacement of external chloride by impermeant anions increases the amplitude of the receptor potential, indicating that chloride anions normally shunt the membrane potential of the retinula cell (Baumann and Hadjilazaro 1972).

Changes in the external calcium concentration have been found to affect the amplitude and the shape of the receptor potential. A decrease in calcium concentration in the bathing medium increases the amplitude of the whole response to a weak stimulus and increases the amplitude of the plateau in

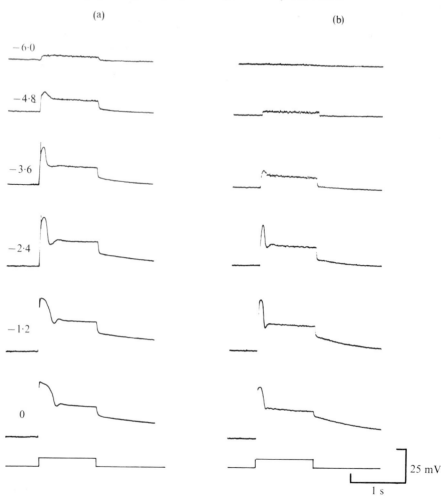

FIG. 4.3. Responses of drone (a) and worker bee (b) retinula cells to light of increasing intensity. Light intensity is expressed in log units. The three components of the response; spike, transient and plateau, are best seen in (a) with intensity −3·6. In this and other figures the bottom trace represents the light flash.

response to a strong light (Fig. 4.4); increasing the external calcium concentration has opposite effects (Fulpius and Baumann 1969). Similar changes in response to alteration of the external calcium level have been found for retinula cells of other invertebrate eyes. In the ventral photoreceptor of *Limulus*, it has been shown that the calcium concentration influences the light-induced sodium current (Millecchia and Mauro 1969).

(a)

(b)

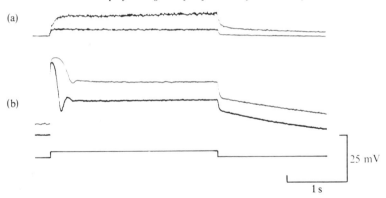

25 mV

1 s

FIG. 4.4. Increase in the amplitude of the receptor potential due to a decrease of the calcium concentration in the bathing medium from 1·8 to 0 mM. The two superimposed responses in (a) are responses to a weak flash; in (b) responses to a strong flash. The shift of the baseline with [Ca] = 0 observed in the response to the strong flash may be attributed to the slow repolarisation of the cell in this bathing medium, which exceeded the duration of the interval between flashes. Time calibration 1 s.

4.3. Spike potential

A striking feature of the drone retinula cells is the large spike superimposed on the rising phase of the receptor potential. Except in response to a weak flash, when depolarization barely reaches the threshold for the spike, it is an all-or-nothing phenomenon (Fig. 4.5). In most cells, the spike overshoots

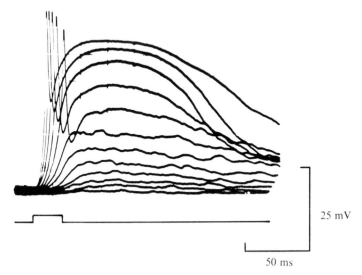

25 mV

50 ms

FIG. 4.5. Responses of a cell to 25 ms flashes. Starting with a dim light the preparation was stimulated with twelve flashes with an interval of 6·3 s between each flash. For each flash the light intensity was doubled.

zero potential, and is followed by a hyperpolarizing afterpotential which depresses the receptor potential. Occasionally, the initial spike is followed by a second spike of smaller amplitude. A single spike can also be evoked when a depolarizing current pulse is passed through the intracellular micropipette (Baumann 1968). The cause of the single spike is a regenerative inflow of sodium ions into the retinula cell. This interpretation is based on the finding that the spike is abolished in a sodium-free solution (Fulpius and Baumann 1969) and that it can be inhibited reversibly by Tetrodotoxin (Baumann 1968). When recorded extracellularly, the polarity of the spike potential is not reversed (Fig. 4.6). This has been taken to imply that the spike originates in

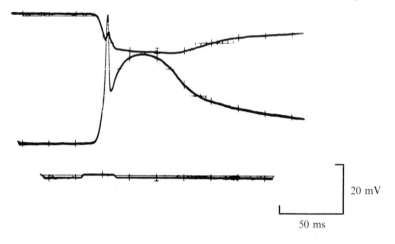

20 mV

50 ms

FIG. 4.6. Intracellular and extracellular recording of a response to a flash of light. A response was first recorded intracellularly (middle trace), the micropipette was then withdrawn from the cell and a second response recorded extracellularly (upper trace). Note that the receptor potential reverses polarity, whereas the spike potential does not. The preparation was stimulated throughout the experiment at a constant rate with flashes of 25 ms duration.

the retinula cell axon and passively invades the soma (Naka and Eguchi 1962). However, as mentioned by Shaw (1969), this does not exclude active invasion of the receptor cell soma (or even an origin for the spike in this part of the cell). It has been shown that active invasion of closed-end terminals gives rise to diphasic spikes, the main component of which is positive (Katz and Miledi 1965).

A puzzling phenomenon is the repetitive spiking which is seen sometimes during the plateau phase of the receptor potential. Based on this observation, it was originally thought that drone retinula cells transform light into a receptor potential and the receptor potential into trains of spikes (Naka and Eguchi 1962). Subsequently, Autrum and von Zwehl (1964), Baumann (1968) and Shaw (1969) pointed out that repetitive firing is relatively rare, occurring only in very special experimental conditions, and its absence in normal circumstances cannot be explained as an experimental artifact (see also Laughlin (Chapter 15)).

In drones, repetitive spikes are never found in response to short pulses of light applied to a previously dark-adapted preparation; at most, two spikes occur during the rising phase of the receptor potential. Repetitive spikes are recorded in cells illuminated with long flashes of light, some time after the onset of the flash (Fig. 4.7). The most likely origin of this repetitive firing is the small potential fluctuations which light evokes in retinula cells. Fig. 4.7

20 mV

0·5 s

FIG. 4.7. Response to a strong long-lasting illumination. Note the occurrence of repetitive firing a long time after the onset of light. Besides large spikes, two small spikes are seen during the plateau phase of the receptor potentials. These may have their origin in a different retinula cell of the same ommatidium since the retinula cells are known to be coupled electrically (Shaw 1969).

shows that during the long-lasting illumination these fluctuations increase progressively in amplitude, finally reaching a size sufficient to trigger a spike. The progressive increase in size of the fluctuations probably arises from the slow increase in membrane resistance which is associated with prolonged illumination (Baumann and Hadjilazaro 1972). Another experimental condition in which the fluctuations are increased in size is replacement of chloride in the bathing medium by large impermeant anions (unpublished observation). This produces an increase in membrane resistance and, in this case, repetitive firing is frequently found from the beginning of the receptor potential.

Repetitive firing is not evoked if the retinula cell is depolarized in the absence of potential fluctuations by, for example, applying depolarizing current, or increasing extracellular potassium, in the absence of light. This behaviour of retinula cells is very different from that of spike-producing cells such as the eccentric cell of *Limulus* lateral eye (Fuortes 1959; Fuortes and Mantegazzini 1962), or stretch receptor cells in crustaceans (Eyzaguirre and Kuffler 1955), where repetitive spiking occurs when the cell is depolarized in any way. For these receptor cells, it is well established that the frequency of firing is a function of the level of depolarization and that the repetitive spikes are the expression of receptor excitation propagated to more central structures.

On the other hand, the behaviour of the drone retinula cell has a certain resemblance to that of the retinula cells of *Limulus*, for in the latter, a single spike is also found superimposed on the rising phase of the receptor potential and small spikes are sometimes triggered by the potential fluctuations (quantum bumps) which light produces in these cells as well as in the drone. Whether the spike potential of the drone and *Limulus* retinula cells propagate to more central structures, and whether they have any functional significance, are questions which have not yet been answered.

4.4. Visual adaptation

Light and dark adaptation in the honey bee have been studied by intracellular recordings from retinula cells (Naka and Kishida 1969; Baumann 1968), or by recording the electroretinogram (Goldsmith 1963; Seibt 1967) as well as in behaviour experiments (Seibt 1967). All these experiments indicate that light and dark adaptation in the bee (as in many other invertebrates) are rapid processes. The responses evoked by flashes of light superimposed upon a uniform background illumination reach a steady level (light adaptation) in less than one second after the onset of the background light. Dark adaptation, as seen in the increase of the amplitude of the receptor potential after the termination of a background light, is also rapid. Recovery after an adapting light is slower when either the duration or the intensity of the adapting light are increased (Goldsmith 1963).

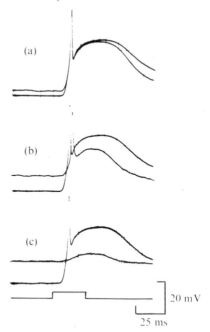

(a)

(b)

(c)

20 mV

25 ms

FIG. 4.8. Response to a bright flash recorded during regular stimulation (one flash every 6·3 s). Each recording shows a response in the dark superimposed on the response seen in the presence of a steady background light of three different intensities, $-2·4$ in (a), $-1·2$ in (b), and 0 in (c). The upward shift of the baseline is due to the steady background light.

It is well known (Fuortes and Hodgkin 1964) that light adaptation occurs as an aftereffect of illumination and also as a result of the presence of a background of light. Recorded with intracellular microelectrodes (Fig. 4.8), light adaptation produced by applying to a preparation a steady background light is characterized by an increase of the rate of repolarization and by a decrease of the duration and the amplitude of the receptor potential. Weak backgrounds mainly affect the rate of repolarization of the receptor potential and its duration; stronger backgrounds are required to depress its amplitude. The amplitude of the receptor potential after a strong adapting light recovers very rapidly (Fig. 4.9). Figure 4.9 also shows that if a preparation is left in darkness for a long time, there is a further decrease in the rate of depolarization and repolarization and an increase of the duration of the receptor potential. That is, recovery of these parameters of the receptor potential is much slower than recovery of the amplitude.

Experiments on visual adaptation in vertebrates suggest that light and dark adaptation has at least two origins, both in the receptor cell. One is photochemical and related to the bleaching and resynthesis of the photo-pigment, the other, sometimes called 'receptor process', is independent of changes in the pigment concentration (Dowling 1963; Dowling and Ripps 1972). The nature of this second process is so far unknown.

Examining intracellularly recorded responses of the drone retinula cells, Naka and Kishida (1969) come to the conclusion that the 'receptor process' is the major cause of visual adaptation, and that changes in the photochemical part of the visual process are of little importance. They also conclude that the 'receptor process' is related to the changes taking place in the cell when the amplitude of the receptor potential evoked by a long-lasting flash drops

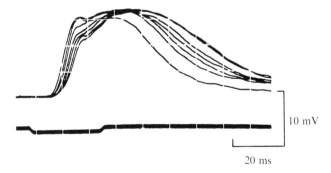

10 mV

20 ms

FIG. 4.9. Superimposed recordings of responses to a flash applied at different time intervals: immediately after and 1, 3, 7, 24, 30, and 47 min after the termination of a strong adapting light. The response recorded immediately after the termination of the background light shows a slightly diminished amplitude and an increased rate of depolarization and of repolarization when compared to responses obtained at later times. Note that the amplitude of the response recovers much faster than its duration.

from the initial transient to the plateau of lower amplitude. The passage from the transient phase to the plateau phase of the receptor potential is due at least partially to changes in the intracellular ion concentration resulting from flash or constant illumination of the visual cell (Lisman and Brown 1972).

The hypothesis of Naka and Kishida, which, surprisingly, has been almost unnoticed, is based on the finding that the linear regions of the *V* versus log *I* curves of a light-adapted cell (describing the amplitude, ΔV, of responses to flashes in the presence of background illumination; see Fig. 4.10) can be made

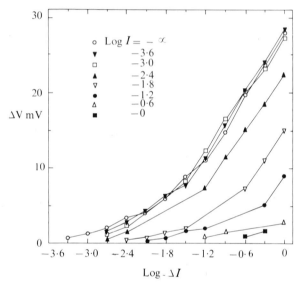

FIG. 4.10. Effect of background light on the curve describing response amplitude as a function of light intensity (ΔV versus log ΔI curve). The amplitude (ΔV) of responses to 25 ms flashes is plotted *versus* the $\log_{10}$ (flash intensity) (ΔI). After determination of a control curve (open circles), the flashes were superposed on background lights of increasing intensity, each represented by a different symbol.

to fall on the extrapolated *V* versus log *I* curve of the dark-adapted cell. The analysis is illustrated in Fig. 4.11. The linear regions of the curves fall on the same line if the curves for the light-adapted cell are shifted upwards on the voltage axis by (1) the amount of depolarization of the retinula cell evoked by the background light, and, in addition, (2) by the potential difference between the initial transient depolarization and the plateau phase to which the background light gives rise. According to these authors, this finding indicates that there would be no decrease in amplitude (relative to the resting potential of the cell) of the response to a flash superimposed on a background

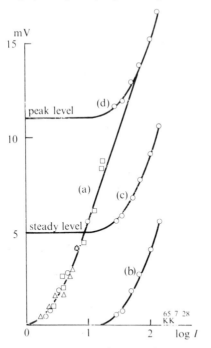

FIG. 4.11. Curve of *V* versus log *I* in dark (squares and triangles) and in the presence of a steady background light (circles). In (a) all points were superimposed. (b) is the position of *V* versus log *I* curve in the presence of a steady background light plotted without any d.c. bias, (c) with a d.c. bias equal to the steady level and (d) with a d.c. bias equal to the peak level of the response by the steady illumination. All continuous curves are the same except for their relative positions. (reproduced from Naka and Kishida 1969).

illumination, i.e. no decrease in sensitivity of the cell, if the membrane potential could be maintained at the level of the initial transient.

An experiment favouring the hypothesis of Naka and Kishida for the mechanism of light adaptation in the drone is illustrated in Fig. 4.12. In this experiment, short test flashes of constant intensity are applied with various delays after the onset of lasting conditioning lights (background illumination) of different intensities. After every test flash the background light is turned off, the preparation allowed to adapt to the dark and the experiment is repeated by applying the same test flash, but with a different delay. All responses to flashes applied on backgrounds of the same intensity are super-imposed photographically. Fig. 4.12a shows that the response to a weak background consists of a maintained depolarization of small amplitude. The responses to the four test flashes are all of the same shape and the same amplitude and they all depolarize the cell, at their peak, to the same potential level. Responses to stronger background lights (*b*, *c*, and *d*) are composed of an initial transient depolarization and a plateau depolarization of lower

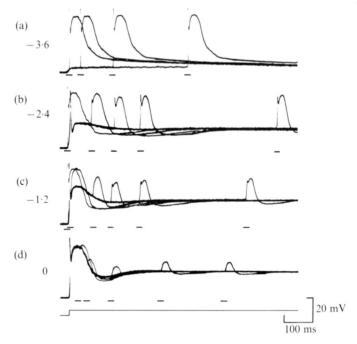

FIG. 4.12. Superimposed recordings of responses to short bright flashes applied during long-lasting background flashes, with various delays the onset of the background light and the test flash. The figures on the left show the intensity of the background light. Only one test flash was applied on each background flash and the preparation was allowed to dark adapt after every background flash. Light flashes are indicated by short lines below each set of recordings; the background flash is represented by the bottom trace.

amplitude. With these backgrounds, the duration of the responses to a test flash shortens and the peak depolarization decreases as the delay increases. The amplitude however (depolarization reached at the peak minus depolarization evoked by the background light) changes relatively little. In *c*, with increasing delays, the amplitudes are 15, 13·5, 17, 15, 16, and 17 mV and in *d* with the strongest background illumination they are respectively 6, 7, 6·5, 7, and 6·5 mV. Hence, for a given background, the amplitude of the responses to test flashes does not decrease. From this experiment therefore (as from experiments of Naka and Kishida) it appears likely that, if the potential change evoked by the background did not drop from the initial transient to the plateau, the peaks of the responses to test flashes would all reach approximately the same level of depolarization and there would be no decrease in sensitivity of the retinula cell.

In other words, the drop of the receptor potential from the initial transient to the plateau is a manifestation of light adaptation. If this interpretation is correct, the question which then arises in trying to understand the mechanism of light adaptation is why the membrane potential in a preparation illuminated with a constant light falls to the plateau level. Naka and Kishida (1969) suggested that this levelling off might be the consequence of a membrane process that is independent of the photochemical part of the visual system.

As shown in the previous section, the difference in amplitude between the transient and the plateau is much reduced when the calcium concentration of the bathing medium is decreased (Fig. 4.4). Hence, if the levelling off of the receptor potential is a determining factor in the process of light adaptation, one might expect to find that responses to flashes of standard brightness light-adapt less in preparations kept in a bathing medium of reduced calcium concentration. Figure 4.13 shows that this is actually the case. In this

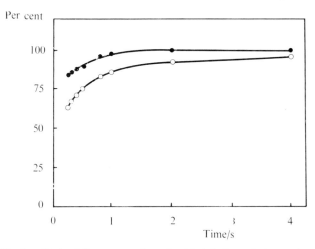

FIG. 4.13. Amplitude of the response to a short test flash applied at various intervals after a short conditioning flash in two different calcium concentrations (closed circles 0·1 mM; open circles 1·8 mM). The intervals between conditioning and test flash are indicated on the abscissa. On the ordinate, the amplitude of the response to the test flash is expressed as a percentage of the amplitude of the conditioning flash.

experiment, light and dark adaptation are studied by applying to a dark-adapted preparation a short conditioning flash (adapting flash) which is followed after a variable interval by a second light flash (test flash). Conditioning flash and test flash are of the same intensity and of the same duration. The amplitude of the test flash which follows the adapting flash after an interval of 200 ms (the first point in Fig. 4.13) is more depressed in normal solution than in the reduced calcium solution; whereas more than four seconds are required for the response to the test flash to attain the amplitude of the conditioning flash in the normal calcium concentration, an interval of only one second is sufficient in low calcium for the response to the test flash to recover completely. Hence, as expected, lowering the calcium concentration in the bathing medium decreases the effect of light adaptation on the amplitude of the receptor potential and decreases the time required for dark adaptation. Conversely, increasing the external calcium concentration increases the time required for dark adaptation.

Recently it has been observed that sodium probably plays a role in visual adaptation in the drone. Changes in the shape and in the amplitude of the receptor potential, similar to those observed by adapting a preparation to light, can be evoked by increasing the intracellular sodium concentration (Baumann 1972). A summary of this finding is illustrated in Fig. 4.14. All

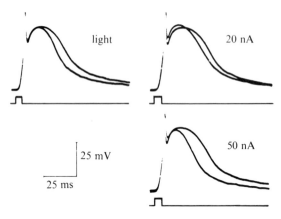

F IG. 4.14. Comparison of the effects of light adaptation and intracellular application of sodium on the receptor potential. On the left, the response of a regularly stimulated cell (control) is superimposed on the response to identical stimulation 10 s after the end of a strong adapting light. On the right, the response recorded 10 s after the injection of sodium is super-imposed on the same control response. Currents of 25 or 50 nA were passed through a micro-pipette containing 2 M NaCl for 30 s. Both light adaptation and injection of sodium reduce the duration of the receptor potential.

recordings in this figure are responses to 5 ms flashes of the same intensity. The preparation is stimulated throughout the experiment at a rate of one flash every 10 s. The superimposed tracings on the left of Fig. 4.14 compare a control response with a response recorded 10 s after the end of a steady adapting light of 10 s duration. As in Fig. 4.8, light adaptation shortens the duration of the receptor potential. The influence of sodium injection on the receptor potential is pictured on the right of Fig. 4.14. Sodium was injected into the retinula cell through an intracellular microelectrode filled with a 2 M NaCl solution during 30 s with currents of 20 and 50 nA. Receptor potentials were recorded 10 s after the end of the intracellular injection. Figure 4.14 shows that sodium injection also reduces the duration of the receptor potential. With an injection current of 20 nA, the shortening of the duration is slightly less than that due to the adapting light. With a 50 nA injection current, it is slightly greater. In addition, there is a slight increase of the amplitude of the receptor potential after the weak sodium current injection.

A similar effect is sometimes observed in weakly light-adapted retinula cells (Baumann 1968, Fig. 6).

A relation may exist between the influence of calcium and sodium ions on visual adaptation, in that the action of an increase in intracellular sodium is mediated by calcium ions. A recent study by Lisman and Brown (1972), working with cells of the *Limulus* ventral eye, has shown that there is a decrement in the size of the receptor potential during injection of sodium or calcium ions into the cell. The decrement was associated with a decrease in the light-induced sodium current and could not be attributed to an alteration of the sodium gradient across the cell membrane. They suggest that light stimulates an inflow of sodium ions which cause an increase in intracellular calcium concentration. This, in turn, is thought to act on the retinula cell membrane, producing the decrease in the light-induced sodium conductance change. They thought that this mechanism could be the cause of the depression of the membrane potential from the initial transient to the plateau phase of the receptor potential. In the drone, as we have seen, this depression is the expression of light adaptation. Consequently, light adaptation might be due to an increase in the intracellular levels of sodium and calcium ions. The recovery of sensitivity of the retinula cell after exposure to an adapting light (dark adaptation) could then be due to recovery of the intracellular sodium and calcium ion concentrations from that induced by light to that found in unstimulated (dark-adapted) cells.

Illumination of a drone retinula cell with an adapting light beam covering only a small fraction of its photoreceptive surface shows that light adaptation is a local phenomenon, i.e. the adapting flash has little or no influence on the sensitivity of the retinula cell beyond a certain distance from the adapting spot. A similar observation has been made by Hagins *et al.* (1962) in the squid and by Hamdorf (1970) in *Calliphora erythrocephala*.

In the experiment illustrated in Fig. 4.15, a single rectangular beam with a width of 20 μm and a length of *c*. 0·5 mm was positioned on a retinula cell perpendicular to its longitudinal axis and, consequently, perpendicular to the direction of incidence of light in the intact eye. This beam could be moved rapidly from one region of the cell to another. The response on the left in Fig. 4.15 was obtained by applying a flash from the small beam on a dark-adapted cell close to the place where the intracellular microelectrode was located. Applying a second flash 6 s later to the same region, gave rise to the response illustrated on the right top of Fig. 4.15. The reduction in duration of this receptor potential is typical for a response of a cell which has been light-adapted. Applying the first flash at increasing distances from the micropipette without moving the second flash decreased the amount of light adaptation. This can be seen in the progressive increase in the duration of the response to the second flash. Hence, the greater the distance between the two regions stimulated on the cell, the less is the degree of light adaptation. In

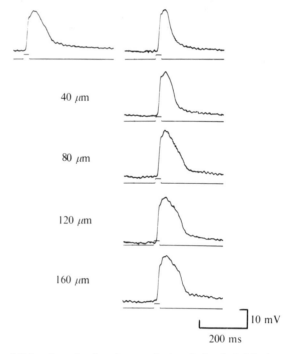

40 μm

80 μm

120 μm

160 μm

10 mV

200 ms

FIG. 4.15. Local light adaptation in a drone retinula cell. On the left is shown the response of a dark-adapted cell to a light flash with a beam of 20 μm width applied at the same point as the intracellular micropipette. On the right, at the top, is the response to a second identical flash applied at the same point after a 6 s interval. The other records on the right are responses to the second flash applied at the electrode after the first (adapting) flash had been displaced by the distances shown.

other words, light adaptation with a flash covering only a small segment of the photosensitive part of the retinula cell does not decrease the sensitivity of the whole cell but acts mainly on the region to which the adapting light is applied. In Fig. 4.15, it can be seen that there is no adaptation if the two flashes are separated by more than 80 μm.

In squid retinula cells, Hagins (1965) has found, by using a narrow beam of light, that the inward membrane current which light induces flows through the cell membrane near where the light is absorbed and does not appear to spread beyond the stimulus beam. This is probably true for drone retinula cells as well (Baumann 1966). Hence, one might expect, if the inward current were responsible for light adaptation, as has been suggested previously, that the region of the cell which adapts to light is restricted to the spot where the adapting flash is applied. As seen in Fig. 4.15, this is clearly not the case. The light spot used, with a width of 20 μm, affected the retinula cell in regions beyond the one to which it was applied. This may be due to a scatter of light laterally from the region where the beam was applied. However, if light adaptation is caused by an increase in the intracellular sodium and calcium concentration, it could be explained by lateral diffusion of these ions inside the retinula cell. Local injection of sodium and calcium ions with an intra-cellular micropipette and measuring the effect on responses evoked by a small beam placed at different distances from the point of injection would be a test of the sodium–calcium hypothesis of light adaptation.

4.5. The pigment cells

Individual retinula cells are almost completely surrounded by accessory pigment cells, for a sleeve of pigment cells surrounds each ommatidium (Fig. 4.1) and thin sheet-like processes from some of the pigment cells penetrate between retinula cells almost as far as the rhabdom. There is, however, a continuous space between retinula cell and pigment cell which is seen in the electron microscope as a regular gap with a width of 15 nm, and which is accessible to extracellular markers (Perrelet and Baumann 1969*b*). Striking features of the pigment cells are their richness in glycogen, few mitochondria, and the presence of microtubules in their cytoplasm. These features, together with their close association with the retinula cells, allow the pigment cells to be considered as the equivalent of glial cells in other nervous structures.

Pigment cells can be impaled with intracellular microelectrodes (Bertrand *et al.* 1972). The identification of the impaled cell was confirmed by ionto-phoretic injection of Niagara Blue and subsequent histological examination. The pigment cells have a resting potential somewhat lower than that of retinula cells, −45 mV on average, and illumination of the retina decreases their membrane potential (Fig. 4.16). Simultaneous recordings in pigment

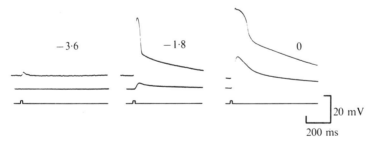

FIG. 4.16. Responses of a retinula cell (upper traces) and a pigment cell (middle traces) to flashes of light of three different intensities. The light intensity is indicated in log units.

and retinula cells (but situated in two different ommatidia), show that more light is needed to produce a depolarization of the pigment cell than to evoke the retinula cell receptor potential. In addition the depolarization of the pigment cell is slower, and reaches its peak later, than the receptor potential. (This is best seen in the response to a flash of medium intensity.)

The most probable cause of the light-evoked depolarization of the pigment cell is an accumulation, within the space between retinula and pigment cell, of potassium ions which leave the retinula cells during the receptor potential. The experimental evidence in favour of this hypothesis (Bertrand *et al.* 1972) is similar to that presented by Orkand *et al.* (1966) to show that depolarization of glial cells in the central nervous system of amphibia, as in the leech ganglion during nerve activity, is due to potassium accumulation. This is that: (1) pigment cells have a membrane potential which depends mainly on the extracellular potassium concentration, and is little affected by changes in the concentration of other ions in the bathing medium; (2) the amplitude of the pigment cell response on illumination decreases when the extracellular potassium concentration is increased; and (3) changing the membrane potential of a pigment cell by applying current through the intracellular micropipette leaves the response to light of the pigment cell unaffected. This last finding indicates also that light does not affect the membrane conductance of the pigment cell (as it does for the retinula cell), and that the retinula and pigment cells are not, or at most very weakly, coupled electrically. The latter eliminates the possibility that the pigment cell response is due to electrical invasion of the receptor potential into the pigment cell. On the other hand, it is noteworthy that the pigment cells of an ommatidium, which are not everywhere separated by extracellular space but, on the contrary, have many 'gap-junctions' between them (Perrelet 1970), are probably electrically coupled. This would explain the observation that all pigment cells respond similarly to illumination of the ommatidium even though only a small number of them interdigitate between retinula cells.

With strong illumination of the retina, the depolarization of the pigment cells can reach an amplitude of more than 20 mV. To produce a similar potential change with potassium, its concentration in the bathing medium has to be increased from the normal value of 3·2 mM to *c.* 30 mM. This makes it likely that, during illumination, the potassium concentration in the space between retinula and pigment cells often reaches values sufficient to produce a depolarization of the retinula cell as well as of the pigment cell. Hence, the depolarization of the retinula cell seen on illumination is probably not determined exclusively by light-evoked sodium current into the cell. The increase in extracellular potassium will have a depolarizing action on the cell towards the end of a response to a strong light when the sodium conductance of the membrane decreases.

Besides illumination, one would expect that experimental conditions which decrease the activity of the Na–K pumping system responsible for the normal ionic gradients across the retinula cell membrane would also lead to a leak of potassium ions from the retinula cell into the extracellular space. The subsequent accumulation of this ion in the extracellular space would then be seen in a depolarization of both retinula cell and pigment cell.

A depolarization of the retinula cell occurs, as mentioned in the section on the membrane potential, during anoxia, at low temperatures, when sodium in the bathing medium is replaced by lithium, or when ouabain is added to the bathing medium. Recording simultaneously from a retinula cell and a pigment cell, it has been found that these experimental conditions also reduce the membrane potential of the pigment cell. The experiment shown in Fig. 4.17 illustrates the effect of anoxia: both types of cell depolarize

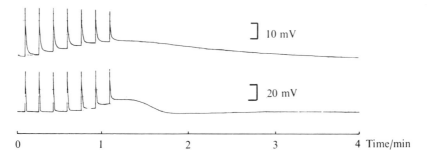

FIG. 4.17. Influence of anoxia on a pigment cell (upper trace) and a retinula cell (lower trace). During the period of anoxia the preparation was stimulated regularly with strong flashes at a rate of one flash every 10 s. Oxygen in the bathing medium was replaced by nitrogen 3 min before the first response to light shown in the figure. Oxygen was readmitted immediately after the last response shown.

during anoxia and both recover their membrane potential on readmission of oxygen. However, the time course of depolarization and repolarization in the two cells is different. The membrane potential of the retinula cell starts to decrease later and recovers more quickly than the membrane potential of the pigment cell. These differences in the behaviour of the two types of cell were also observed for the other experimental conditions which produced depolarization of the retinula cell, and which were expected to inhibit Na—K pumping activity.

At first sight, one would have expected to find a parallel time course for depolarization of the two cells if this were due to an increase in the potassium concentration in the space between them. The differences can be explained if one assumes that the retinula cells have an electrogenic component in their pumping system, but that the pigment cells do not. At the beginning of anoxia, the electrogenic pump of the retinula cell, presumably stimulated by extracellular potassium (*cf.* Ritchie 1971) would compensate for some time the depolarizing action of the increase in extracellular potassium, and thus retard the onset of depolarization. As soon as oxygen is readmitted, the electrogenic pump starts working again, leading to a rapid repolarization of the retinula cell despite the relatively high extracellular potassium concentration which keeps the pigment cell depolarized. Activity of the electrogenic pump also explains the transient overshoot of retinula cell membrane potential which is found after the period of anoxia (postanoxic hyper-polarization; *cf.* Baumann and Mauro 1973). Since the pigment cell is thought to have no, or very little, electrogenic pumping activity, its membrane potential probably reflects changes in the extracellular potassium concentration more precisely than does the membrane potential of the retinula cell.

Finally, one might speculate on the possible functional significance of the changes observed in the drone pigment cell. A strong Na–K pumping action by the pigment cells would assist the retinula cell to restore its ionic balance during and after illumination. One could then think of the pigment cell as a regulator of the extracellular ionic concentration. On the other hand, a high and maintained potassium concentration in the space between retinula and pigment cell could be a means whereby activity in the retinula cell is signalled to the pigment cell, and perhaps acts as a stimulus for a metabolic process within the ommatidium. One wonders, for example, if the depolarization of the pigment cell induces release of glycogen from a store in the pigment cell and transfer to the active or recovering retinula cell, which is very poor in this substance (Figs. 4.1 and 4.2). Such a function is considered possible for glial cells within the leech ganglion (Wolfe and Nicholls 1967).

ACKNOWLEDGEMENTS

The author is very grateful to A. Perrelet who kindly prepared the electron micrographs of this paper and who, together with G. J. Jones, has taken a very active part in preparing the manuscript. For advice and criticism the author also wishes to thank Ch. Bader, A. Mauro, and J. Posternak.

5. Pattern recognition

R. WEHNER

5.1. Introduction

IT is not surprising that during the last decade neurophysiologists, psychologists, and computer scientists have become interested in the question of how sensory systems are able to recognize and differentiate visual patterns. There is strong evidence for features held in common in the processing of visual information at all levels for transduction to pattern abstraction, in different groups of animals. In insects, however, relatively few papers have been published on pattern abstraction, but even the few data available cannot as yet be explained by any broadly applicable concepts of pattern recognition. The same situation holds for the vertebrates, where information processing along the visual pathway has been studied more thoroughly.

Let us begin with a short outline of decisive conditions that an adequate theory of pattern recognition must meet. The central demand to be made on a visual system is that it should generate a structured description of each given input picture. Such a description, in general, must be a transformation of the input into a less redundant form suitable for pattern classification. The central idea of any pattern-recognition task has to be the reduction of information in the pictorial input in order to obtain a simplified representation. Therefore, something more than spatial mapping processes must occur in the neural layers of the visual system. This statement may be illustrated by the number of possible input patterns generated by the eye of the bee. The individual ommatidia do not detect images, but merely represent small pictorial regions of the visual environment (Snyder and Pask 1972; Kirschfeld 1972b; Eheim and Wehner 1972). Let us suppose that each of the 6000 ommatidia (unpublished data) are only capable of discriminating between light and dark, i.e. between an excited and an unexcited state (threshold detectors). Even in that over-simplified case some 10^{1800} possible combinations of light and dark points in an image can be represented by the receptor layer at any instant. If one considers, additionally, that along a broad range of intensities the visual system of the bee is able to discriminate between two light intensities $I_{1,2}$, which only differ by $2(I_1 - I_2)/(I_1 + I_2)$ of c. 15 per cent (Kunze 1961, Labhart 1972), an astronomical number of possible representations of the whole visual field is possible at the outputs of the receptor layer. These considerations make it quite unlikely that in measuring differences between patterns the visual system of the bee uses one commonly applicable strategy—such as, for example, a point-to-point matching device. Pattern discrimination requires particular properties of a pattern to be selected by specific higher-order units and at the same time these must be distinguished from other unspecific parameters. On the other hand, if one-to-one correspondence exists between visual patterns and their neural correlates or between the neural correlates at successive levels of the visual pathway, pattern recognition is not possible in the sense proposed here.

The mechanisms of the classificatory system are the central problem in any work on pattern detection. In a practical approach, one uses a series of known patterns, the so-called training set, which differ in a consistent and readily-definable way from each other. Each feature of a training pattern can be represented quantitatively by a numerical measure. This measure has to be treated as a dimension in a multidimensional stimulus space. When all quantitative aspects of the pattern are plotted in this way lines or surfaces can be drawn around those feature clusters that are discriminated from others. These surfaces define similarity classes or invariance classes of patterns. The basic problem is to find ways of defining the clusters of stimulus parameters in this multidimensional space *that appear to be relevant to the animal*. These are *presumably* the sets of feature detectors that are used by the pattern-discriminating mechanism. In practice one has to find those sets of feature detectors which appear to be used by the visual system and then to seek for quantitative measurements (algorithms) that determine similarity classes. To do this one must explore the effects of progressive quantitative change of each aspect of the pattern. How precise patterns may be recognized by the visual system can be restated as a more exact question: at what regions or points in the appropriate n-dimensional pattern of stimulus parameters do the feature detectors have to sample the input, or more precisely, how many feature detectors are required to perform that task.

Applying this concept to the results so far obtained on pattern recognition in insects, we can say that true pattern detection, so defined, has scarcely been proved so far. We are, as yet, unable to classify patterns according to invariant classes or to specify the number of dimensions along which patterns are spread. Recent work, mostly processing mechanisms, suggests that, in spite of the multiple array of the compound eye, pattern detection in insects is similar to that in other groups, represented so far only by vertebrates. The theory of 'mosaic vision', by which pattern recognition in insects has been described by early compound eye anatomists (Mueller 1826; Kiesel 1894; Exner 1876, 1891), refers only to processing by the dioptric apparatus peculiar to these eyes, and does not imply a mechanism of pattern detection that is characteristic only of compound eyes.

5.2. Pattern modulation

5.2.1. *Image processing by the lens system*

The first step in information processing along the visual pathway is modulation of input patterns by the lens system. Where it has been examined experimentally by optical methods, the lens system of each ommatidium in most species of diurnal insects forms an image within the crystalline cone: in bees, for example 35 μm distal to its pointed tip (Varela and Wiitanen 1970), or just, as in flies, at the tip of the cone. However, this is only relevant in eyes where the rhabdomeres are separate (Snyder 1972). The small diameter of the Fraunhofer diffraction pattern allows each retinula cell to have its own axis and angular sensitivity function in the flies *Musca* and *Calliphora* (Kirschfeld 1965, Seitz 1968a; see also Reichardt (1961) for *Limulus*). Antidromic illumination experiments show that each lens in *Musca* resolves at least the image consisting of the seven rhabdomere tips of a dipteran ommatidium. By means of direct optical methods (Kirschfeld and Franceschini 1968) as well as by geometrical calculations using experimentally-determined refractive indices (Seitz 1968a) the focal planes of the *Musca* and *Calliphora* eye were found to coincide with the plane in which the rhabdomeres terminate in special structures known as caps (Trujillo-Cenoz and Melamed 1966; Langer and Schneider 1970) (Fig. 5.1). The focal length of the dioptric system is very short ($f = 40$–60 μm), and the inverted images formed in the focal plane are of high contrast for all objects within a distance of 6 mm or more from the eye surface.

Measurements of the refractive indices of the optical components have also been made by interference microscopy. In the corneal lenses of the fly *Calliphora* (Seitz 1968a,b), the bee *Apis* (Varela and Wiitanen 1970) and the ant *Cataglyphis* (Mondadori and Wehner 1972) the refractive index varies only in the proximal–distal direction (maximum values $n = 1\cdot473$, $1\cdot490$, and $1\cdot535$ respectively), whereas in some clear-zone eyes, especially skipper butterflies (Horridge, Giddings, and Stange 1972), a specific decrease of refractive indices from the axis to the periphery of the dioptric apparatus has been demonstrated, as assumed by Exner (1891) for his lens-cylinder concept (see Horridge (Chapter 11), Meyer-Rochow (Chapter 12), and Meggitt (Chapter 13)). In the bee and the fly, however, the dioptric effect in each ommatidium depends only on the curvature of the inner and outer lens surfaces and the refractive indices.

In the ommatidia of dipteran eyes the inverted image at the first focal plane of the lens system is divided up by the seven tips of the rhabdomeres, which act as real apertures. Diffraction, which must be taken into account because of the small facet diameters (Barlow 1952, 1965; de Vries 1956; Palka 1965; Kuiper 1965), does not eliminate the influence of the rhabdomere diameters on the size of the receptor apertures. Therefore in dipteran eyes the 'visual unit' is the individual receptor cell, or, to be more exact, the pseudopupil

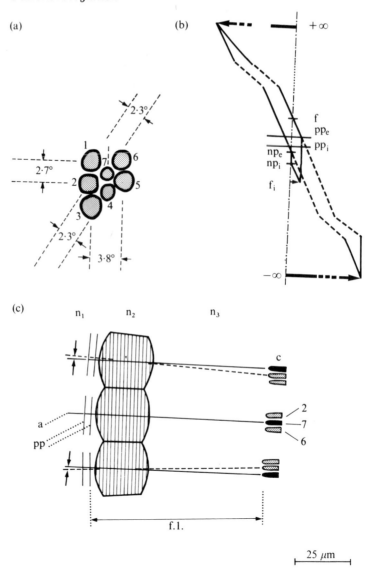

FIG. 5.1. The geometry of the dioptric system in flies (*Musca, Calliphora*). (a) Rhabdomere arrangement within one ommatidium in cross-section; (b) Ommatidial ray traces. With the object in the inner focal plane f; and antidromic illumination (see text), an inverted real image is seen in the plane $+\infty$, and an upright virtual image in the plane $-\infty$. (c) Longitudinal section of three-lens systems cut through the plane of rhabdomeres 2, 7, and 6 (see (a) above). The black rhabdomeres look at the same point in space. (a) optical axis; (c) caps at the distal terminals of the rhabdomeres; f_i internal, and f_e, external focal planes; $f.l.$, focal length; n_1, n_2, n_3, refractive indices; np_i, internal, and np_e, external nodal points; pt_i, internal, and pt_e, external principle planes; $\Delta\varphi$ divergence angle between adjacent rhabdomeres. After Kirschfeld and Franceschini (1968) and Seitz (1971).

group of 6 plus 2 visual cells, which are looking at the same point in space from different facets (Kirschfeld 1967). As the axons of these sets of six peripheral cells converge on identical second order neurons, whereas the two central cells terminate at different lower levels (Trujillo-Cenóz and Melamed 1966; Braitenberg 1967), each point of the environment is represented in the outer visual ganglia in a number of ways.

In the apposition eye of Orthoptera, Hymenoptera, Odonata and others, characterized by fused rhabdoms, the optical axes of the visual cells coincide within each ommatidium (Kuiper 1962; Autrum and Wiedemann 1962; Shaw 1967; Varela and Wiitanen 1970; Kirschfeld 1972*b*). In a fused rhabdom eye the crystalline cone narrows towards the entrance of the rhabdom and as a result of mode propagation along the rhabdom an image is not resolved (Snyder and Pask 1972). So far as it is used for pattern detection, the visual unit of these apposition eyes is the whole ommatidium.

5.2.2 *Resolving power and contrast transfer functions*

The resolving power of a visual system depends on the density of receptor units, which can be described by the divergence angle or interommatidial angle $\Delta\varphi$ between neighbouring units. On the other hand, contrast transfer from the dioptric system is determined by the angular sensitivity function, the half-width of which is $\Delta\rho$. At this point it seems important to define exactly what the terms 'resolving power', 'visual acuity', and 'contrast transfer' shall mean, for in the literature they are used in a variety of ways depending on different formulations of the question: how sharply does an insect see?

The visual resolving power is defined basically as the capacity of a visual system to transfer unambiguously a patterned set of stimuli on the receptor layer. Its lower limit or threshold of resolution is reached when the one-to-one correspondence between the input pattern and the output pattern no more exists. Theoretically, that resolution threshold can be defined as:

$$\lambda_0 = 2\Delta\varphi,$$

where λ is the pattern wavelength and $\Delta\varphi$ the divergence angle between adjacent receptor units (Goetz 1964). All two-dimensional intensity distributions, the spatial frequencies of which do not exceed $1/(2\Delta\varphi)$, are unequivocally transferred and therefore 'resolved' in this terminology. In *Drosophila* and *Musca* and *Tenebrio* beetles, λ_0 and from this an inferred value of $\Delta\varphi$ were determined behaviourally by studying optomotor reactions (Fermi and Reichardt 1963; Goetz 1964, 1965; Goetz and Gambke 1968; McCann and MacGinitie 1965; Thorson 1966*a*; Eckert 1971*b*) and spontaneous pattern preferences (Wehner and Wehner-von Segesser 1973). The values obtained by behavioural tests are in good agreement with anatomical findings (*Drosophila*: Hecht and Wald 1934; von Gavel 1939). As the optical and anatomical axes of an ommatidium do not necessarily coincide, direct optical observations are preferable for the measurement of $\Delta\varphi$. Irrespective of the resolving power or the interreceptor spacing, the contrast transfer depends only on the width of the angular sensitivity curve (acceptance angle curve in Tunstall and Horridge (1967)) defined by means of its half-width $\Delta\rho$ (s in Fermi and Reichardt (1963), φ in Washizu, Burkhardt, and Streck (1964), α in McCann and MacGinitie (1965), θ in Kirschfeld (1965)). This angular sensitivity function was measured at different levels of the visual pathway (Table 5.1) by means of optical and electrophysiological methods and always found to be approximately a Gaussian function in shape. In the contrast transfer function (Fig. 5.2), the contrast attentuation $m_i/m_e < 1$ is related to the acceptance angle $\Delta\rho$ and the pattern wavelength λ by the expression

$$m_i = m_e \times \exp\left[\frac{\pi^2}{4\ln 2}\left(\frac{\Delta\rho^2}{\lambda^2}\right)\right]$$

where m_i is the internal (transferred) and m_e is the external contrast, defined as

$$m = (I_{max} - I_{min})/(I_{max} + I_{min}).$$

The total light flux into a single receptor, characterized by the area under the angular sensitivity curve, is reduced by decreasing $\Delta\rho$, so that $\Delta\rho$ must exceed a certain limit although this sets a limit on m_i/m_e. The same consideration applies to $\Delta\varphi$. Therefore the product $\Delta\varphi.\Delta\rho$ is constant for a given light sensitivity. Because of this relation, contrast transfer and light sensitivity are inversely related and only where light sensitivity is not a limiting factor is the visual acuity able to increase by narrowing of the angular-sensitivity curve.

The angular-sensitivity curve, and therefore also the contrast transfer, is a property of the receptor, irrespective of how the light reaches it. As Shaw (1969*b*) and Horridge, Walcott, and Ioannides (1970) showed by positioning light guides on individual facets, tremendous differences are found between the pathways of light in apposition eyes and some clear-zone eyes (Horridge 1971). Whereas in the bee and locust at least 99 per cent of the light reaching a receptor enters by its own lens system, this value is less than 1 per cent in the dark-adapted *Dytiscus* eye. In skipper butterflies, on the other hand, light is quite sharply focused upon the receptors, which have a fairly *narrow* angular sensitivity curve, although the light enters the eye by a patch fifteen facets wide (Horridge, Giddings, and Stange 1972). Although, in principle, resolving power and contrast transfer can be regarded as independent of each other, Goetz (1964, 1965) emphasized that the most convenient transfer characteristics occur in an optical system where the contrast-transfer

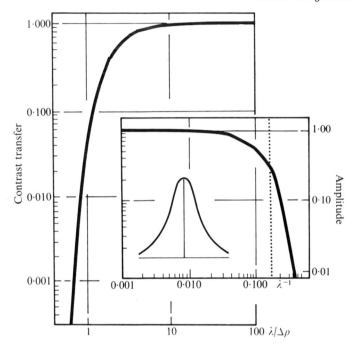

F ɪ ɢ. 5.2. Theoretical contrast transfer function of a receptor assuming only a gaussian angular sensitivity, and different ratios of acceptance angle $\Delta\rho$ to pattern wavelengths.
Inset figure: The relative amplitude at the output level of a dioptric system of acceptance angle $\Delta\rho = 4°$ for different spatial frequencies (λ^{-1}). The resolution threshold is shown by the dotted line for a divergence angle $\Delta\varphi = 3°$ between adjacent lens systems. After Goetz (1964) and Kirschfeld (1965).

function falls off within the range of the resolution threshold. Contrast transfer can then be read off the contrast transfer function (Fig. 5.2) at the value $2\Delta\varphi/\Delta\rho$. This condition would be fulfilled when $\Delta\rho/\Delta\varphi = 0.88$. In most cases, however, the ratio $\Delta\rho/\Delta\varphi$ exceeds unity (Table 5.1) caused by a strong overlap of visual fields of neighbouring ommatidia. In the bee and the ant *Cataglyphis*, there is a better correlation between the acceptance angles and divergence angles in the horizontal plane ($\Delta\rho/\Delta\varphi = 0.9$ and 1.6, respectively) than in the vertical plane ($\Delta\rho/\Delta\varphi = 1.9$ and 2.3, respectively) (Eheim and Wehner 1972). In *Musca*, the central (high acuity) system of receptor cells No. 7 and 8 is inferred to have smaller values of $\Delta\rho/\Delta\varphi$ than the peripheral (high sensitivity) system, because the angular sensitivity function of a visual cell is determined by the interaction (in mathematical terms, convolution) of the diffraction pattern of the lens with the acceptance function of the rhabdomere cap cross-section. As the central rhabdomeres have smaller diameters than the peripheral ones (Eckert 1971, Kirschfeld 1971) the latter show less pronounced contrast-transfer capacities. This difference between

TABLE 5.1

Visual receptor spacing and visual receptor fields in some insect species

The half-widths $\Delta\rho$ of the angular sensitivity functions are presented for different levels of data processing, accompanied by the divergence angles $\Delta\varphi$ between the optical axes of adjacent visual units. The $\Delta\rho$-values are measured at the retina level (by optical methods (1) or intracellular recordings (2), at the lamina level (3) or by behavioural responses (4)). If the $\Delta\varphi$- resp. $\Delta\rho$-values are mathematically calculated by means of anatomical measurements, refractive indices, light-guide properties of rhabdomeres etc., the data are characterized by an asterisk(*). Specifications are given with respect to specific eye regions, types of receptors concerned, directions of scanning basis, adaptation level, spectral composition of stimulating light, etc. In some cases mean values had to be calculated where more detailed data on the spatial distribution of $\Delta\varphi$ are dealt with in the literature, i.e. in Portillo (1936); Washizu, Burkhardt, and Streck (1964); Vowles (1966); Burkhardt, de la Motte and Seitz (1966); Wehner (1972c). Evaluation of $\Delta\rho$ by means of pseudopupil methods (Autrum and Wiedemann 1962; Wiedemann 1965; Burkhardt et al. 1966; Seitz 1968a, Ohly 1968) are not listed in the Table, because they do not correspond to the half-widths of the angular sensitivity distribution as shown by Eheim and Wehner (1972). The data on the beetle Chlorophanus ($\Delta\rho = 12 \cdot 6°$, $\Delta\rho = 1 \cdot 2$; Varju (1959)—left out on purpose—do not fit those obtained from other insect species. As a result isolated in literature, the visual fields of that beetle would be widely separated.

A, histological measurements; B, behavioural responses; c, central eye region; ch, chalky mutant; d.a., dark-adapted; E, electrophysiological recordings; h, horizontal direction; l.a., light-adapted; L 1, 2, monopolar cells in the lamina; M, mathematical calculation; O, optical measurements; opm-2, opto-motor visual mutant; +, wild type; R 1-6, 7, 8, number of receptor cells; se, sepia mutant; v, vertical direction; w, white mutant; w^a, white-apricot mutant.

Animal	Specifications	$\Delta\varphi$		$\Delta\rho$ (1)	(2)	(3)	(4)	Author
Diptera:								
Calliphora	+, ♂, h	2·9°	A					Washizu, Burkhardt, and Streck (1964)
	+, ♂, v	2·5°	A					
	+, ♀, h	3·3°	A					
	+, ♀, v	2·8°	A					
	+	2·5°*	M					
	ch, h	3·0°	O					de Vries (1956)
	ch, v	2·9°	O					Burkhardt, de la Motte, and Seitz (1966); Seitz (1968a)
	+			1·2°*				Seitz (1968a)
	+, h				3·5°			Washizu, Burkhardt, and Streck (1964)
	+, v				3·0°			
	+, c, 360, 495 nm				2·8°			Streck (1972a,b)
	+, c, 625 nm				4·0°			
	ch, w, c, 360, 495, 625 nm				2·5°			
	+, R 1–6, h				4·5°			Zettler and Jaervilehto (1972a,b)
	+, L 1, 2, h						2·0°	

Musca						
h	3·9°	A				Vowles (1966)
v	2·4°	A				Kirschfeld (1965)
h	3·9°	O				Kirschfeld (1971)
v	2·3°	O				
R 1–6			3·0°			Vowles (1966)
R 7,8			1·5°			Scholes (1969)
d.a.				2·5°		Scholes and Reichardt (1969)
			2·7°			McCann and MacGinitie (1965)
			2·5°			
			3·0°			
d.a.					3·5–5·1°	Eckert (1971)
l.a.					2·4–4·0°	
R 1–6					3·5°	
R 7,8					1·7°	
Drosophila						
+ , c	4·2°	A				Hecht and Wald (1934); von Gavel (1939)
+ , w, w^a, se	4·6°	B				Goetz (1964)
+	4·8°	B				Wehner and Wehner-von Segesser (1973)
+ , R 1–6			7·5°			Franceschini (unpublished results)
+ , w, w^a, se					3·5°	Goetz (1964, 1965)
opm-2					14°	Heisenberg (1972)
Hymenoptera:						
Apis						
c, h	2·7°	A				Baumgaertner (1928)
c, v	0·9°	A				
c, h	2·7°	A				del Portillo (1936)
c, v	1·2°	A				
c, h	2·8°	O				Autrum and Wiedemann (1962)
c, v	1·4°	O				
c, h, v			6·8°			Kuiper (1962)
			2·6°			Eheim and Wehner (1972)
			4·6°*			Varela and Wiitanen (1970)
h				2·5°		Laughlin and Horridge (1971)
v				2·7°		
o, c, h	2·0°	E				Shaw (1969a)
o, c, v	0·8°	F				

TABLE 5.1 (*continued*)

Animal	Specifications	Δφ	Δρ (1)	Δρ (2)	Δρ (3)	Δρ (4)	Author
Cataglyphis	c, h	3·4°* A					Wehner, Eheim, and Herrling (1971)
	c	2·9°** A					Wehner (1972c)
	c, l.a., h		6·8°				Eheim and Wehner (1972)
	c, l.a., v		8·8°				
Lepidoptera:							
Toxidia		1·3° E		6–8°			Horridge, Giddings, and Stange (1972)
Epargyrus				2·1°			Døving and Miller (1969)
Hemiptera:							
Lethocerus	d.a.			9·0°			Walcott (1971)
	l.a.			3·5°			
Orthoptera:							
Locusta	c, h	2·4° A, O					Burtt and Catton (1954)
	c, v	1·1° A, O					
	c, h	2·4° O					Autrum and Wiedemann (1962)
	c, v	1·1° O					
	v	1·0° A					Tunstall and Horridge (1967)
	d.a.			6·6°			Tunstall and Horridge (1967)
	l.a.			3·4°			
					4·5°		Shaw (1968)
Schistocerca						2·0–4·5°	Thorson (1966b)
Blattoidea:							
Periplaneta	v, d.a.			6·8°			Butler and Horridge (1973)
	v, l.a.			2·4°			
Odonata:							
Aeschna	c	0·9° A					Hesse (1908)
Libellula		1·2–1·8°					Horridge (1969)

the two receptor systems could be used behaviourally to isolate a *Drosophila* mutant with only the high acuity system functioning (*ebony;* Heisenberg 1972).

In these experiments the flies walk on top of an air-supported ball in the centre of the rotating striped drum. In general, the experimental set-up is similar to the one used by Goetz (1972) (see p. 103). When the flies see pattern wavelengths near the resolution threshold ($\lambda_1 = 18°$, $\lambda_2 = 7.2°$) the high frequency pattern (with λ_2) can be transferred only by the high acuity system of receptor cells No. 7 and 8. On account of their broader angular sensitivity functions the peripheral receptor cells No. 1–6 generate insufficient contrast transfer. By testing movement perception over an intensity range of 4 log units the system, detecting only the 18° pattern, is found to be *c.* 50 times more sensitive to light than the high-acuity system, the output of which is tested by the 7.2° pattern. In the mutant *ebony*, the responses towards both types of moving patterns are equal to the 7.2° responses of the wild-type, so that the high-sensitivity system of *ebony* seems to be blocked for optomotor responses. Until recently, all studies of rhabdom acceptance properties ignored their action as light guides (see Snyder (Chapter 9)).

As shown by the relation between $\Delta\varphi$ and $\Delta\rho$, the visual fields of adjacent receptor units overlap strongly. All angular sensitivity functions known show a smooth Gaussian shape of width greater than the limit given by diffraction theory (Barlow 1952) which cannot, therefore, account for the overlap of the visual fields. The functional significance of this overlap probably lies in the increased mean light flux which accompanies a decrease of the amplitude of the response at higher spatial frequencies (Figs. 5.2 and 5.4). The latter fact, i.e. the strong attenuation of contrast with narrow stripes, may be responsible for the suppression of counter-directed turning tendencies (Hassenstein 1953; Goetz 1964; Kirschfeld 1965; Thorson 1966a), which are predicted in optomotor experiments in the region of the resolution threshold.

In any array of lens systems with overlapping visual fields the amount of contrast modulation can be calculated by means of the light-flux ratio $(\phi_{A,C}-\phi_B)/\phi_{A,C}$ (*Limulus:* Reichardt (1961); *Apis, Cataglyphis:* Eheim and Wehner (1972)). In order to obtain these data, in three adjacent ommatidia A, B, and C, the light-flux values ϕ, induced by two point light sources (separation angle α) are read off the corresponding angular sensitivity curves. Besides contrast attenuation, a contrast reversal occurs when the light sources are separated by $\alpha < \alpha_0$ (at α_0 the light flux ratio becomes zero because of $\phi_{A,C} = \phi_B$; Fig. 5.3). As there are no zeros in the amplitude frequency characteristic (Fourier components) of the angular sensitivity function (Kirschfeld and Reichardt 1964), the overlap of the visual fields does not result in any loss of visual information, even within the range $\alpha < \alpha_0$. Therefore the overlap of visual fields does not influence the resolution power,

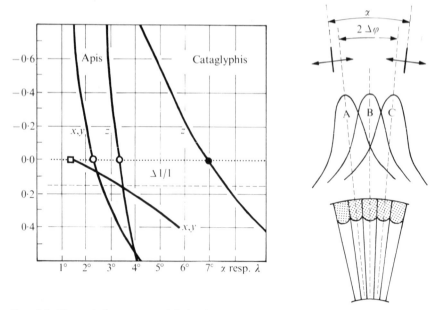

FIG. 5.3. Theoretical contrast modulation by an array of three-lens systems in the worker bee, and the ant, *Cataglyphis*, calculated from the angular sensitivity curve. *Abscissa;* angular separation α between two point sources (curves O, ●) or alternatively the pattern wavelength λ of equidistant stripes (curve □).
Ordinate: Relative light flux difference between ommatidia A and B, calculated as

$$(\phi_{A,C} - \phi_B)/\phi_{A,C}.$$

In the calculations for two point light sources (O, ●) the sums of the light flux values are read off the angular sensitivity-curves (ASC) graphed for the ommatidia A, B, and C. For the calculation of the curve □ (contrast transfer of the two-dimensional equidistant stripe pattern) the sum of the solid segments cut off by the pattern in the ASC rotated around its central axis was calculated. $\Delta\varphi$, divergence angle between the optical axes of adjacent ommatidia, x, y, and z, axes according to the hexagonal array of lens systems. The threshold value $\Delta I/I$ is taken from Labhart (1972). After Eheim and Wehner (1972).

which only depends on receptor spacing. For all spatial frequencies which are resolved by the receptor array the one-to-one correspondence between input pattern and optically-transferred pattern is ensured.

5.3. Pattern detection

5.3.1 *Preprocessing*

5.3.1.1. *Contrast transfer.*

    (1) *Temporal frequency modulation.* Whether optical contrast modulation allows the detection of resolved patterns depends on the overall transfer characteristics of the subsequent stages within the nervous system. In this context, the term 'visual acuity' may be introduced to

refer to the neural transfer of optically-modulated patterns. At the level of the retina, the temporal transfer characteristics are one of the most decisive factors in determining the neural contrast transfer. The high temporal resolution of retinula cells as recorded intracellularly is of functional significance (Fig. 5.5) in that shortening the time constants of the visual scanning process results in contrast enhancement (*Calliphora*: Zettler 1969; Gemperlein and Smola 1972; *Acheta*: Pinter 1972). It must be emphasized, however, that the spatial resolution

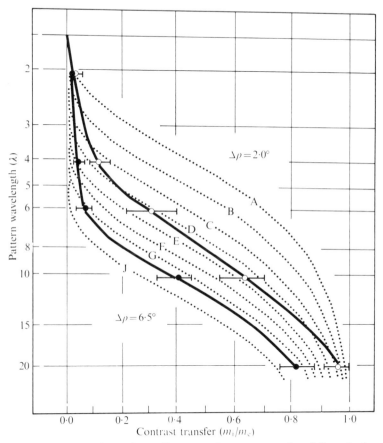

FIG. 5.4. Calculated and experimentally measured contrast transfer of the retinula cells of *Locusta*. The responses of retinula cells to moving stripes of various pattern wavelengths were recorded for the dark-adapted (●) and light-adapted (○) locust retinula cells penetrated by microelectrode. The dotted lines show theoretical curves for gaussian angular sensitivity curves of various acceptance angles $\Delta\rho$ = 2·0°(A), 2·5°(B), 3·0°(C), 3·5°(D), 4·0°(E), 4·5°(F), 5·0°(G), 6·5°(J), calculated by means of the contrast transfer function (Fig. 5.2). From the lack of correspondence between the experimental and theoretical curves it was inferred that the angular sensitivity curves are sharper at the peak than a gaussian function. After Tunstall and Horridge (1967).

cannot be shifted towards smaller pattern wavelengths by increasing
the temporal resolving power. It is an old, but incorrect opinion, that
in insects a high temporal resolution would compensate for a poor
spatial one.

Unfortunately, only little is known on the temporal transfer characteristics of the whole system. In open-loop experiments, where tethered

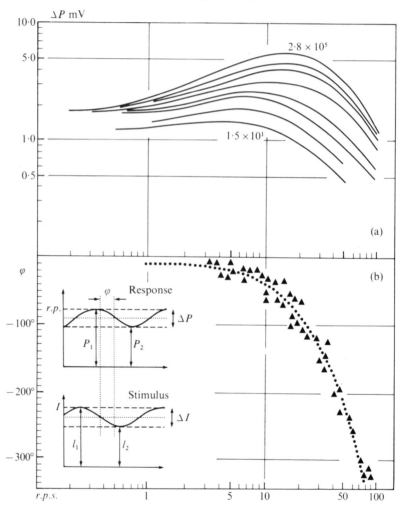

FIG. 5.5. Experimentally determined temporal transfer characteristics in visual cells of *Calliphora*. The stimulus, the frequency of which is shown in the abscissa, was a sinusoidally oscillating intensity. (a) Amplitudes of receptor potentials (ΔP; mV) at various mean light intensities
($15...2 \cdot 8 \times 10^5$ f.c.) (b) Phase-shift (φ) as a function of stimulus frequency. The insets define the
use of symbols for the stimulus (lower) and response (upper). *I.*, light intensity; *r.p.*, receptor
potential; *r.p.s.*, frequency. After Zettler (1969).

Musca flies can fix upon a black stripe on a white screen, the flies show a marked difference in turning tendencies induced by progressive (forward to backward) and regressive (backward to forward) pattern movement at $f < 30$ Hz (Reichardt 1973, (Fig. 5.6)) (for similar results in optomotor experiments of *Drosophila* see Goetz (1972); for *Apis* see Kunze (1961)). The most relevant work on the frequency dependence of a behavioural response has been done on the neck-torque response of locusts to small sinusoidal pattern motion (Thorson 1966b). Systematic gain and phase measurements are described over 3·5 decades of frequency (Fig. 5.7). A comparison of the intensity-dependent properties of the reflex with related electrophysiological studies at the retina level (Tunstall and Horridge 1967) implies that the frequency dependence of the optomotor response is dominated by receptor dynamics. A more elaborate mechanism of flicker frequency processing is required to explain the reactions of the water strider *Velia* to flickering prey dummies (Meyer 1972b). Here intensity modulation of frequencies within the range of 1·6–8·0 Hz increases the stimulus value of a dummy, whereas flicker at 0·4–3·0 Hz inhibits it. Low-frequency movements are also most effective in releasing the predatory strike in dragonfly

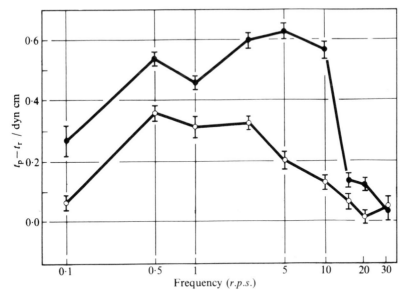

FIG. 5.6. Temporal frequency domain of induced flight torque responses of tethered flies *Musca*. The fixation of single black stripes within a white cylindrical drum is measured by means of the induced torque response $(t_p - t_r)$. The angle between the stripe position and the longitudinal body axis of the fly varies from 15° (●, upper curve) to 90° (○, lower curve). t_p, and t_r torque response to progressive (from front to back) and regressive (from back to front) motion respectively, with respect to the fly's eye. After Reichardt (1973).

larvae (Vogt 1964; Etienne 1969). Frequency discrimination training by sinusoidally modulated lights has been performed only in bees (Vogt 1966, 1969). A constant discrimination level according to Weber's law is reached within the range 40–140 Hz. Most remarkable, at different flicker frequencies, the discrimination is independent of the amplitude of modulation.

(2) *Intensity characteristics.* Thresholds of intensity discrimination, determined behaviourally, provide information on how the receptor characteristics are interpreted by the succeeding stages of data processing. Although many intensity characteristics are known for single

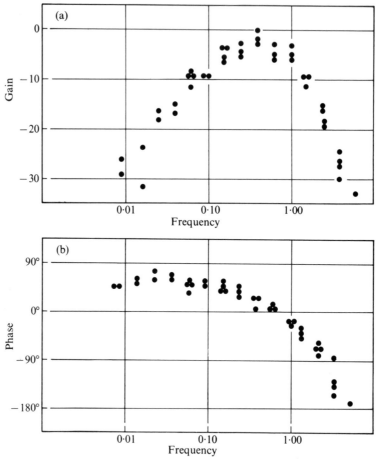

FIG. 5.7. Temporal transfer characteristics of optomotor responses of the head in *Schistocerca*. Amplitude (a) and phase (b) *versus* frequency of sinusoidal neck-torque about the longitudinal axis, for a constant-amplitude sinusoidal oscillation of the striped drum. Gain (*db*): 20 log (peak-to-peak torque/constant reference level). After Thorson (1966*b*).

retinula cells in insect eyes (for example *Apis:* Autrum and von Zwehl 1964; Baumann 1968; *Calliphora:* Zettler and Jaervilehto 1970; Jaervilehto and Zettler 1971; Gemperlein and Smola 1972; *Locusta:* Bennett, Tunstall, and Horridge 1967; *Lethocerus:* Walcott 1971; Fig. 5.8a), relatively few precise and complete measurements of the intensity dependence of behavioural responses are available (*Apis:* Heintz 1959; Labhart 1972; Labhart and Wehner 1972; von Helversen 1972; *Cataglyphis:* Wehner and Toggweiler 1972; *Musca:* Chmurzynski 1967; *Calliphora:* Chmurzynski 1969; *Drosophila:* Wehner and Schuemperli 1969; Schuemperli 1972; Fig. 5.8b). For the optomotor torque response, the intensity discrimination functions correlating pattern contrast or light intensity differences with amplitude have been investigated in flies (Fermi and Reichardt 1963; Goetz 1964, 1965; Hengstenberg and Goetz 1967; Eckert in preparation), bees (Kunze 1961), and locusts (Thorson 1966a). For an evaluation of the intensity transfer capacities of the whole system from these two sets of data, the relative intensity discrimination threshold can be inferred from the final output level, i.e. behaviourally. In bees the intensity discrimination threshold obtained by optomotor responses ($\Delta I/I = 0.18$, Kunze 1961)) or by discrimination between two illuminated screens ($\Delta I/I = 0.13$, Labhart (1972)) is much larger than in flies ($\Delta I/I = 0.005$, Fermi and Reichardt (1963)), so that further experiments are required to show whether this difference is due to a real difference between these groups or to the experimental conditions used. In the neck-torque response of *Schistocerca* an even smaller value $\Delta I/I = 0.003$ was found (Thorson 1966a). Alternatively, by bringing together the optic contrast-transfer function and the intensity-discrimination threshold (Fig. 5.3), a prediction can be made about the visual acuity for regular stripes or other patterns. From the intensity discrimination threshold of the bee (Wolf 1933; Daumer 1961; Labhart 1972) and the intensity characteristics of single visual cells, as recorded intracellularly in the bee's retina (Laughlin and Horridge 1971), the threshold of the behavioural response is equivalent to a change of *ca.* 1 mV in the receptor potential (calculated on the on-peaks), which implies a change of 2 per cent of the saturation level of intensity. From more detailed calculations of this kind it should be possible to infer which slopes of receptor and neural characteristics are relevant. For the interpretation of the whole behaviour pattern, however, intensity discrimination thresholds for the same responses must first be known (Stevens 1972).

(3) *Lateral inhibition.* Visual systems can be regarded as sets of spatial and temporal frequency filters, and it has been shown in general mathematical terms that the spatial and temporal distributions of inhibitory

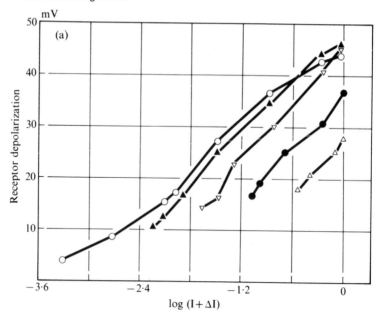

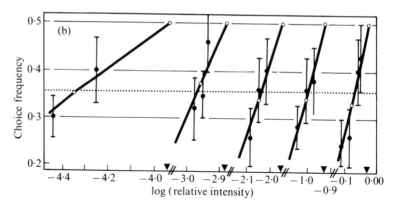

FIG. 5.8. Intensity-characteristics of visual cells and behavioural responses, in the honey bee. (a) Amplitudes of intracellularly recorded receptor potentials showing the actions of background lights (I) on the initial transient responses to long light pulses (ΔI). $\log I = -\infty(\bigcirc)$, $-2 \cdot 4$ (▲), $-1 \cdot 8$ (▽), $-1 \cdot 2$ (●), and $-0 \cdot 6$ (△). (b) Spontaneous discriminations between two illuminated screens of differing intensities. The constant intensity I_c of one screen is marked by the black arrow (▼) at the abscissa.
Ordinate: choice frequencies to the screen which receives a stepwise decreased variable light intensity I_v. For $I_c = I_v$ the choice frequency is 0·5. The lower confidence limit for the choice frequency of 0·5 is indicated by the stippled line ($n = 50, p = 0 \cdot 05$). (a) is from Baumann (1968), (b) from Labhart (1972).

influences can form powerful mechanisms of contrast enhancement. As shown in Table 5.1, in most insect eyes so far studied the half-width $\Delta\rho$ of the angular sensitivity function of receptor cells exceeds the interommatidial angle $\Delta\varphi$, so that there is a marked overlap in the visual fields of adjacent receptor units. According to Fig. 5.3 an external distribution of light intensity, for example, of two single intensity peaks, is changed drastically by optical-contrast modulation. However, if lateral inhibitory connections are present between the input channels—as described classically for the *Limulus* eye (Hartline, Wagner, and MacNichol 1952; Kirschfeld and Reichardt 1964)—the original 'picture' can be restored. If the matrix of the overlap values coincides with the matrix of the inhibition values, the outside intensity distribution will be restored unambiguously at the level following the inhibitory network. Thus lateral inhibition can markedly influence visual acuity. The resolution power, of course, cannot be enhanced by lateral inhibition.

Lateral inhibitory interactions of some form have been demonstrated by electrophysiological methods (Horridge, Scholes, Shaw, and Tunstall 1965; Arnett 1971; Zettler and Jaervilehto 1972; Mimura 1972).

Lateral inhibition has also been inferred by an interesting behavioural method (Meyer 1971*a*, 1972). The spontaneous preferences of the water bug *Velia* are tested with circular black or white discs (subtending 3·4° or 4·0°), to which the animal orients as if to prey. The immediate stimulus for the prey-catching response is a pair of needles which produce similar wave patterns as they vibrate together (Wiese 1972). The animal makes its choice between a test stimulus placed over one needle and a constant disc over the other. For tests of lateral inhibitory influences, the test stimulus consists of two identical discs beside each other, one being over the first needle and the other further laterally. The frequency with which this pair of discs is selected, as against the single disc on the other needle, depends on the distance between the two discs of the test stimulus. Large inhibitory effects as in Fig. 5.9 are found at regular repeats of about 10°, which coincides with the inter-ommatidial divergence angle of *Velia*, suggesting that one visual unit is inhibited by the excitations of neighbouring units. Referring to the maxima of the inhibitory influences, the inhibitory effect is an exponential function of the interstimulus distance. By electrophysiological studies and mathematical considerations Reichardt (1961) and Hartline and Ratliff (1972) propose the same relationship for *Limulus* eyes.

An exponential spatial decline of inhibitory influences is also shown for retinal ganglion cells of cats (Gruesser, Vierkant, and Wuttke 1968; Creutzfeldt, Sakmann, Schaich, and Korn 1970).

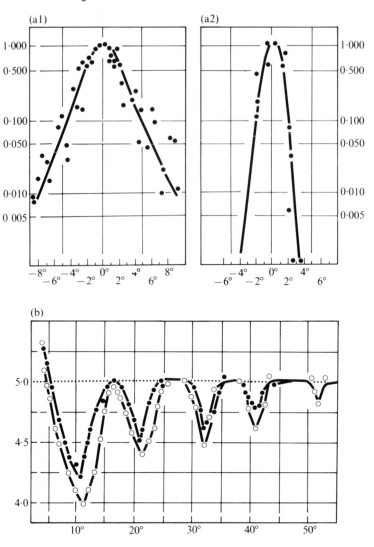

FIG. 5.9. Evidence for lateral inhibitory processes in the visual system of the fly *Calliphora* electrophysiological and of the water strider *Velia* behavioural. (a) Intracellularly recorded angular sensitivity distributions of retinula cells (*a1*) and of monopolar cells of the lamina (*a2*) of *Calliphora*. The optical axes of the cells recorded from are indicated by 0°. (b) Relative stimulus effectiveness of black (●) and white (○) circular discs, when a second disc is horizontally located at the side of the standard figure. Correlation between the stimulus effectiveness is plotted on a probit scale on the ordinate, and the horizontal angular distance between the centres of the two discs on the abscissa. The diameters of the black and white discs were 4·0° (●) and 3·4° (○). With increasing horizontal interstimulus distance the stimulus effectiveness changes periodically. The inhibitory influence of the second laterally shifted disc is correlated with multiples of the interommatidial divergence angle $\Delta\psi = 10\cdot3°$. (a) is from Zettler and Jaervilehto (1972a), (b) from Meyer (1972).

Lateral inhibitory systems are supposed generally to be of much wider distribution in the visual systems of insects. Off-types of discharges, centre-surround organization of receptive fields, spatial and temporal frequency characteristics, and directional selectivity, all of which depend on other arrangements of inhibitory connections, are found in insects and vertebrates. Motion-sensitive units with preferred directions have been described for deep optic-lobe and protocerebral regions of flies (McCann and Foster 1971), bees (Kaiser and Bishop 1970), and moths (Swihart 1968; Collett 1971). Models of movement perception (Goetz 1968, 1972; Kirschfeld 1972a) all rely on inhibition of some form. As motion-detecting systems seem to be involved in pattern recognition (Reichardt 1973) it is likely that spatial frequency filtering by means of inhibitory interactions is an important mechanism for pattern perception, but the actual mechanisms still elude us.

5.3.1.2. *Mapping.* Pattern recognition can be defined as the detection of geometrical correlations within spatially-patterned stimuli. As the visual pathways of arthropods and vertebrates consist of many systems in series, the transfer of information inwards may preserve the topology of the input to a certain degree. In a detailed study of the projection pattern in several groups of insects including bee, *Calliphora*, dragonfly, locust and several butterflies (which is not yet published fully), Meinertzhagen and Horridge (1971) showed that the outside world is mapped in perfect reverse order upon the cartridges of the optic medulla. Reflecting on pattern recognition problems the question has to be, how far in the inward information channels does an orderly point-to-point representation persist beyond that, and by what mechanism can the position-invariance necessary for pattern detection be extracted from the topologically-mapped data.

In the visual system of the bee the existence of some form of topological mapping is proved functionally by a special training procedure. Pattern recognition tests using a simple pattern (Figs 5.10 and 5.11) reveal a well-established discrimination of stripes or black-and-white discs only varying by their orientation in space, but coinciding in all other parameters (Fig. 5.10). In other words, according to this test there is no rotation invariance in the action of the visual system of the bee (Wehner and Lindauer 1966; Wehner 1967a, 1968). The test pattern is more easily discriminated from the training pattern the more the distribution of black and white areas within the visual field differs from that during the training (Fig. 5.11). On account of the varying extent of overlapping and non-overlapping areas between test pattern and training pattern, the two-dimensional cross-correlation demonstrated by these experiments cannot be described by a linear model. Discrimination clearly depends on the spatial positions of the contrasts within the visual field, and upon how the effectiveness of the training pattern is

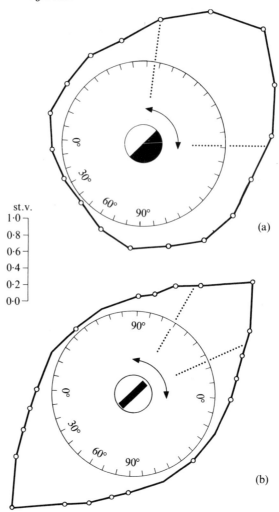

st.v.
1·0
0·8
0·6
0·4
0·2
0·0

(a)

(b)

FIG. 5.10. Evidence for spatial representation in the visual system of bees (see also Figs. 5.10–5.12). Sensitivity towards an antisymmetrical (a) and a symmetrical (b) black-and-white pattern presented upon a vertical screen. Some bees were trained to the 45° inclination of the half-black half-white disc, others to the black stripe. In successive discrimination tests the stimulus values (*st.v.*) of varying inclinations α were determined as a fraction of the stimulus value of the training inclination (see scale inset). The half-widths of these direction–sensitivity curves are $(\Delta\alpha)_{0.5} = 82°$ (a) and 36° (b). After Wehner (1972a).

decreased by inserting a contrasting black or white area into the original pattern. Therefore different parts of the visual field are of unequal significance in this pattern recognition. As shown in Fig. 5.12 the effectiveness of a black-and-white disc rotated by $\Delta\alpha$ from its training position can be approximated by the stimulatory effects of the corresponding insertions of black and white areas tested separately (Wehner 1972a). Two main conclusions can be drawn from these results: (1) an outside intensity distribution is mapped topographically to some extent in the visual system of the bee, and (2) there is some form of specific weighting of each spatial position (Fig. 5.13).

From the few electrophysiological data available on the central nervous representation of the visual field, we know that spatial weighting of mapped visual information occurs because there are units with fields of intermediate sizes, as described by Mimura (Chapter 19). On the behavioural side, Reichardt (1972, 1973) has shown that in *Musca*, vertical black stripes are fixed only when they are projected on the lower visual field (Fig. 5.14). The reaction frequencies of *Drosophila* flies to black areas of a given size and

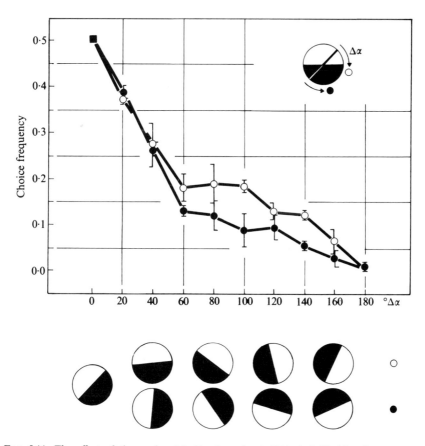

FIG. 5.11. The effect of the angle of inclination of a half-black half-white disc presented vertically to the bee. The frequencies of choice towards the reference pattern (*abscissa*) presented simultaneously with the training pattern (see *inset figure*) are shown for clockwise (O) and counter-clockwise (●) rotations of the reference pattern. Δα angle between the training pattern and the reference pattern. After Wehner (1972a).

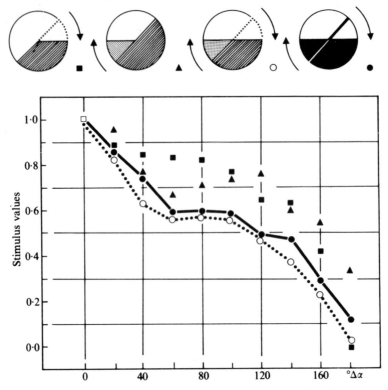

FIG. 5.12. Comparison between the experimentally determined effect of pattern angle (●; from the O-curve in Fig. 5.11) and the theoretical curve obtained by two-dimensional cross-correlation (O). All stimulus values (ordinate) are calculated as a fraction of the stimulus of the training pattern ($\Delta\alpha = 0°$). In the calculation, the positive (black to white, ■) and the negative (white to black, ▲) contrast changes that are simultaneously caused by the rotation of the black-and-white disc away from its training position ($\alpha_{tr} = 45°$) were separately determined in pattern discrimination experiments. The stimulus values of the rotated black-and-white discs are then calculated by multiplying by each other the relative stimulus values (■, ▲) of the separately tested insertions of the black and white areas in the training disc. The resulting O-curve coincides fairly well with the experimental effect of pattern angle (●-curve). For further detail see Wehner (1972*a*).

configuration also depend on the position in the visual field (Wehner 1972*d*), but apparently these weighting functions are not part of the motion detection system in *Drosophila* (Goetz 1964). In the water strider *Velia*, detection of prey dummies is done only by two horizontal rows of ommatidia in the lower middle part of the eye near the equator, and the capacity to discriminate prey decreases progressively from the front to the lateral parts of the eye (Meyer 1971*a*).

The aspects of pattern recognition that have been most intensively studied by measuring behavioural responses of bees and flies are all found not to be isotropic, meaning that data processing between two visual input channels depends not only on their relative separation, but also on their absolute positions in space. In bees and flies, the central lower parts of the visual fields are especially significant for pattern recognition, i.e. those parts which during flight are directed normally towards landmarks near the horizon. It is to be hoped that studies on the fields of single units, such as those outlined by Mimura (Chapter 19), Kien (Chapter 18), and Northrop (Chapter 17) will explain ultimately the behavioural data.

Having established that mapping occurs in some form one has to ask whether the effect of position is derived from the positions of the receptors on the head or whether it is related via the head position to a system of absolute co-ordinates in space. Only when a receptor array is kept spatially constant during pattern recognition do both systems of co-ordinates coincide. Otherwise information about the position in space requires additional information about the position of the head on the body, which can be done by using either afferent inputs (Wiersma and Yamaguchi 1967; Wiersma and

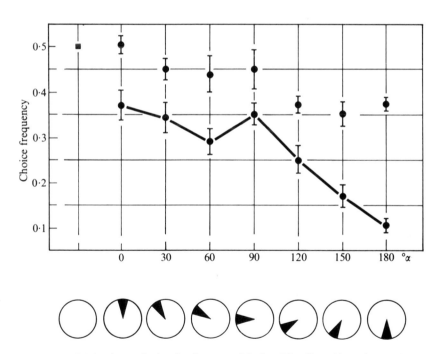

FIG. 5.13. Weighting factors in the visual system of the bee. The effect of inserting a 30° sector at various angles, α, in a white training disc. The frequencies of choice of the black sector discs (*abscissa*) simultaneously presented with the white training disc are shown as a function of the position of the black sector in the visual field. The strongest effects are caused with the black sector in the lower part of the visual field. The values that are plotted separately (●) show the effect of white sectors within a black training disc. After Wehner (1972a).

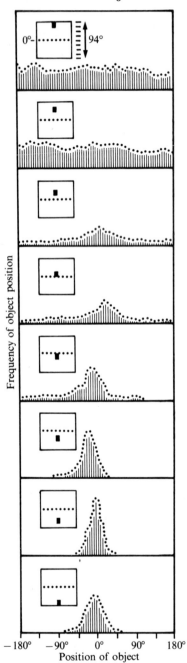

FIG. 5.14. Fixation of black stripes that are placed in various parts of the visual field of the fly *Musca*. By a closed-loop device the fixation of a black stripe could be studied in tethered flies with head fixed to the thorax. During fixation the fly experimentally coupled to its visual surround is able to move the object (a vertical black stripe) towards its flight direction (0°). The fixation tendency can be measured as the frequency (*ordinate*) of object positions (*abscissa*).

Inset figures: vertical positions of the stripe varying between ±47° about the equator (0°). Pronounced fixation occurs only in the lower part of the visual field. After Reichardt (1973).

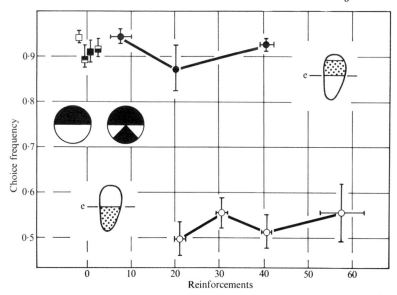

F IG. 5.15. Pattern discrimination in bees with eyes partially covered by paint. The two discs shown in the inset were presented in a vertical plane. The difference between them is in the lower part of the visual field. Before eye covering each bee had fulfilled sixty reinforcements on the training pattern (the left-handed pattern in inset figure). Bees perform as well as the controls when the upper part of the eye is covered with paint (●). The extreme region of the eye was left free to allow sun-compass orientation. Controls (□, ■, ▣, ▤) involving pattern discrimination by bees with uncovered eyes. When the lower parts of both eyes are covered with paint (O) the discrimination ability is totally lost. *Abscissa:* number of training flights by bees with partially covered eyes. *Ordinate:* frequency of choice of the training pattern. After Wehner (1972b).

Yanagisawa 1971) or internal signals such as a measure of the motor outflow (for vertebrates see Wurtz and Goldberg (1971)). When the upper or the lower half of the eye of the freely-moving bee is covered with paint it detects patterns in that region of the visual field that corresponds with the unpainted part of the eye (Fig. 5.15). Black-and-white areas presented in the lower part of the visual field are only processed by the lower ommatidia of the bee's eye and *vice versa* (Wehner 1972b) i.e. it does not tip its head and compensate for the tilt. These results are likely to be interpreted by the first mechanism: dependence on the geometry of the receptor system. According to high-speed photography, bees flying before the test apparatus during the pattern discrimination keep their head (and therefore eyes) at a very constant inclination to the horizontal (Wehner 1972b). Data are not yet available for rotations around the longitudinal and dorsoventral axes of the flying bee, but the patterns which are projected upon the frontal eye surface seem to fulfil one main condition for pattern detection, namely a controlled position in the visual field. The bee in flight is stabilized with respect to rotation, and the translatory flight components apparently do not influence the space-constant mapping of pattern positions. Therefore, there seem to be no significant differences in these respects between a freely-flying bee and one in tethered flight.

A constant relation between the co-ordinates of the receptors and those of real space cannot be maintained when the patterns are presented on a horizontal plane, as is usually done (Schnetter 1968, 1972; Mazokhin-Porshnyakov 1969b; von Weizsaecker 1971; Cruse 1972a,b; Anderson 1972) (Fig. 5.18b). When patterns are shown on a horizontal surface, gravity cues are not available to show their location in space, and the exact shape of a pattern must depend on the angle from which it is seen. Theoretically, that information could be drawn from celestial cues or by use of landmarks on a distant horizon. The latter have often been found to be relevant (Wehner 1967b; Lauer and Lindauer 1971), and we have unpublished evidence from our own work with bees that celestial information is not used for adjusting the shapes of patterns which differ in spatial position. If terrestial cues are lacking it is possible that figures may be compared by estimation of overlapping and non-overlapping areas. This would require a method of calculating the maximum common area for both patterns.

5.3.1.3. *Fixation.*   An assumption about visual mapping according to a space-constant coordinate system concerns fixation. One feature of the method used by the bee in discriminating the vertically presented black-and-white screens of Fig. 5.12 is that fixation processes are indispensible. As shown by covering parts with paint, certain areas of the eye must maintain constant spatial relations to the pattern for it to be recognized. Although visual fixation is already described in insects, mainly in prey detection or the estimation of catching distance in mantids or dragonfly larvae (Zaenkert 1939; Mittelstaedt 1950, 1954; Maldonado and Barros-Pita 1970; Barros-Pita and Maldonado 1970; Levin and Maldonado 1970; Meyer 1971a; Kirmse and Laessig 1971a,b), more detailed information is available only in *Musca* (Reichardt and Wenking 1969; Reichardt 1970, 1971, 1973). By means of a servosystem the fly can control the angular velocity of its surround by its own torque response and so operate under closed-loop conditions. The fixation of a single vertical black stripe which is mounted at the inner surface of a white cylinder surrounding the fly results in the fly moving the stripe into its flight direction, where a stable fixation position is reached. Stripes that are stationary and light flux oscillations which cause no effect of movement are unable to cause stripe fixation. This suggests for one thing that an external eye muscle which causes the rhabdomeres to move perpendicular to the ommatidial axes (Hengstenberg 1971, 1972), is not important in scanning the eye or causing relative movement of the visual axes. Fixation, however, occurs when the stripe oscillates symmetrically about a mean position. Therefore the fixation process is dependent upon (1) relative movements of the eye and pattern, and (2) an induced torque response that is not the same for progressive (from front to back) and regressive (from back to front) motion. The torque signal responsible for fixation strongly depends on the original position of the stripe in the visual field, i.e. on the mismatch between flight direction and stripe position (Fig. 5.6), and on the stimulus frequency (Goetz 1972).

The latter result is obtained by open-loop experiments with walking *Drosophila* flies. The servo-system apparatus, where the fly walks for several hours on top of a ball but is maintained in stationary positions to the moving pattern stimuli, is described in Götz and Gambke (1968). In these experiments the flies tend to minimize the horizontal components of the movement stimulus, as they do for progressive and regressive motion studied alone. Maximum responses to rotation (rev./min) are found for stimulus frequencies of 5 s^{-1}. Although the movement stimulus is characterized by the angular velocity ω and the pattern wavelength λ, the fly's reaction does not depend on either parameter alone, but on the frequency of contrast changes ω/λ $[\text{s}^{-1}]$. To study fixation processes, however, Reichardt (1973) uses *Musca* flies coupled to their visual surround under closed-loop conditions, and has found the same progressive and regressive movement detectors.

Besides the description of the fixation process these closed-loop data of the tethered fly provide evidence for the involvement of motion detection in pattern recognition. Movement detectors are also used for control of direction in freely walking *Drosophila* flies (Goetz 1972; Wehner and Wehner-von Segesser 1973).

5.3.2 *Classification*

As outlined at the beginning of this paper, pattern discrimination implies a classification with reference to features which are extracted by special processing mechanisms. Feature extraction, however, does not necessarily involve neurons that are specific for particular objects. On the contrary, from the study of vertebrate visual systems it is unlikely that specific detector neurons exist there (Gruesser and Gruesser-Cornehls 1970; Eysel and Gruesser 1971; Creutzfeldt, Poeppl and Singer 1971). An object in the visual field may be represented by the excitation of a combination of millions of nerve cells, each of which may also code other symbols, so that the visual information is in the form of a spatial and temporal distribution of activity between all these neurons. Some basic parameters extracted from the input must not be mistaken for classified objects. On the other hand, classification has to result in an n to 1 relationship between the visual input and the 'spatial and temporal distribution of activity' within a nervous subsystem. The n input patterns belonging to the one activity level in the subsystem form an invariance class of patterns. For example, considering the patterns represented graphically in Fig. 5.16 the question is: which patterns are confused by the bees and may thus form an invariance class, i.e. a cluster of values in n-dimensional feature space?

From this point of view, pattern recognition can be studied exclusively by

FIG. 5.16. Set of patterns composed of white squares and triangles on a black background.
After Anderson (1972).

discrimination tests. Spontaneous pattern preferences are only valid as a rough estimate of the types of patterns which may be differentiated by the insect in question (Wehner 1973). For example, by recording frequencies of spontaneous choices of bees towards the sets of patterns shown in Fig. 5.16, it can be shown that some patterns are discriminated and others are not (Fig. 5.17), but no further conclusions can be drawn from spontaneous preferences. If one succeeds in deriving a function which describes the frequencies of choice with reference to one pattern parameter, for example, the relative length of contours as in Fig. 5.17, one cannot deduce that the visual system really uses this particular parameter, or even that only one parameter is used for this set of patterns. Therefore, first of all it is absolutely necessary to evaluate the number of parameters varying independently in a given set of patterns. A strong coupling of pattern parameters, for example, can be deduced from the experiments of Schnetter (1968, 1972), who succeeded in training bees to circular discs and *n*-pointed stars (Fig. 5.18b). As one parameter is inevitably correlated with all the other parameters varied in the tests (i.e. pattern area, diameter, contour length, etc.), the choice frequencies can be described as functions of any of these parameters (Fig. 5.19).

In spite of this ambiguity one has to seek for a method, which enables one to test systematically whether the frequency of choice depends upon one or more independently varying parameters. The latter can be eliminated by demonstrating that the patterns in question lie on a straight line in the feature space. This implies that the frequencies of choice *p* tested for all three members A, B, and C of a given pattern set fulfil the equation:

$$p(AC) = p(AB) + p(BC).$$

In general, additivity of frequencies of choice is demonstrated always when the function relating choice frequency and stimulus parameter of a given set of patterns does not depend on the varying reference patterns used in at least two different test series. In Figs. 5.20 and 5.21 the frequencies of choice between multiple point and multiple stripe configurations are plotted against the stimulus parameter. In each case all patterns are tested with respect to

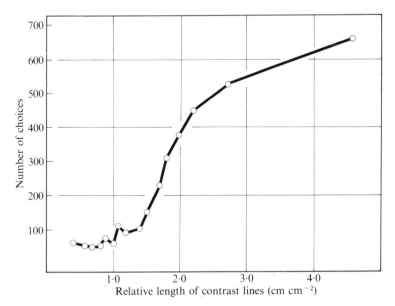

FIG. 5.17. Spontaneous pattern preferences of bees to the examples illustrated in Fig. 5.16. The patterns, which were presented simultaneously to the bees, are graphed according to their relative lengths of contour. After Anderson (1972).

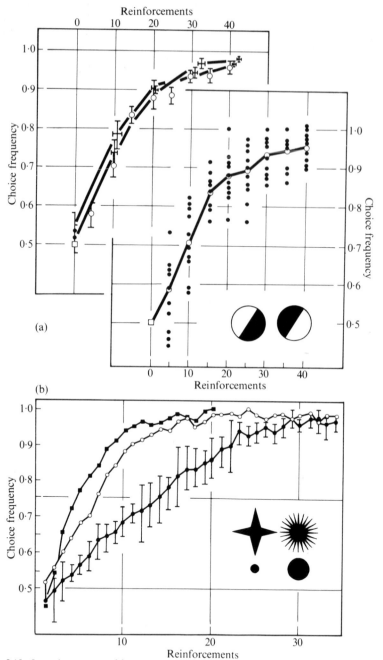

FIG. 5.18. Learning curves of bees to (a) vertically and (b) horizontally presented patterns, which are shown in the inset figures. (a) Learning curves of ten individually-trained bees (*lower graph*) and groups of ten simultaneously trained and tested bees (*upper graph*); the mean values (O) of the lower graph are shown for comparison. (b) Discrimination training between four- and twenty-four-pointed stars as well as between circular discs of different sizes (■). Reciprocal training: training patterns are either the four-pointed star (●) or the twenty-four-pointed star (O). After Wehner (1972a) and Schnetter (1972).

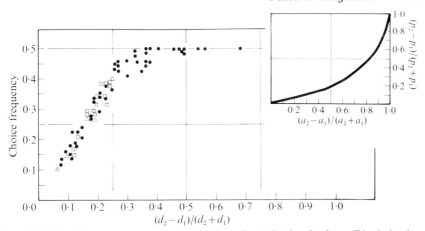

FIG. 5.19. Coupled pattern parameters in pattern discrimination by bees. Discrimination training between geometrically-similar patterns: circular discs (●), four-pointed stars (△), and eight-pointed stars (□). The choice frequencies (*ordinates*) are plotted against the relative differences of diameters d_1 and d_2 of the pairs of patterns in question (*abscissa*).
Inset figure: correlation between the relative differences of areas of the two patterns (a_1 and a_2) and three other relative parameter differences (p_1 and p_2, p = contour length, size, and diameter, respectively). The relations between parameter differences and choice frequencies are the same whatever parameter of the pattern is selected for calculation. After Schnetter (1972).

two different reference patterns. If the ordinate is shifted, both curves coincide within the limits of error. An advantage of this method is that it can easily be applied to spontaneous preference tests (Fig. 5.20) as well as discrimination training tests (Fig. 5.21). However, the method reveals nothing about the quality of the parameter involved when two or more parameters are not independently varied within the test series.

The concept of a given set of feature detectors involved in pattern recognition in bees only holds when identical discrimination frequencies are obtained for reciprocal training experiments. This is always the case when the patterns A and B are equally discriminated irrespective of whether pattern A or pattern B is chosen as training pattern by the experimentor. When this condition is fulfilled each pair of patterns is characterized by a distinct distance in feature hyperspace. Although Schnetter (1972) succeeded in finding invariance of the discrimination frequencies against reciprocal training, other workers did not (Wehner 1968, 1969; Cruse 1972*a*), which implies that the bees choose parameters in pattern classification according to the training situation.

A general finding in all discrimination experiments using a large variety of pattern types consists in the influence of the 'type' of pattern. If geometrically-similar shapes belonging to such a type of pattern (i.e. circles, four-pointed stars, equidistant stripe patterns, checkerboards, concentric angular rings) are varied along one parameter scale, e.g. in relative contour length, a difference function can be found which describes the choice frequencies of the bees. Schnetter (1972) shows that in experiments with two out of four parameters kept constant, the results can be described by one or the other of the remaining variable parameters. The four parameters changing in his star-like patterns are area, diameter, number and tip angle of the points of the stars. If, however, only one parameter is kept constant, more than one of the three remaining parameters must be used for a complete description of the results. In general, those descriptions are often no longer possible, when patterns of different types are compared by means of the parameter previously selected in one-type-experiments. Whatever distance function (p. 110) may be used, it reveals only a very poor description of the choice frequencies between patterns which are either very similar or very different with respect to qualitative terms (Cruse 1972a). Especially those patterns, which are qualified as *good* patterns in human psychology (Garner 1970), e.g. squares, crosses, and stars, seem to be characterized also for bees by a

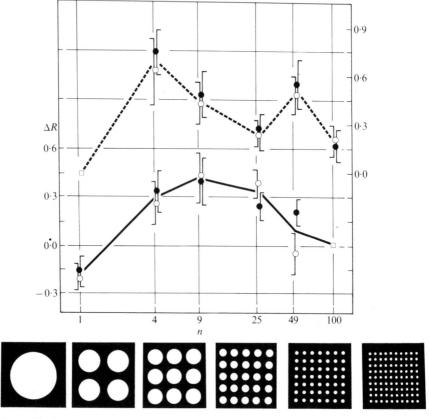

FIG. 5.20. Spontaneous pattern selection in *Drosophila*. In successive tests all patterns graphed at the abscissa are simultaneously presented together with either the one-point-pattern (*upper curve, right ordinate*) or the 100-point-pattern (*lower curve, left ordinate*). ΔR, differences in frequencies of choice; n, number of white points within the black areas. After Goetz (1971).

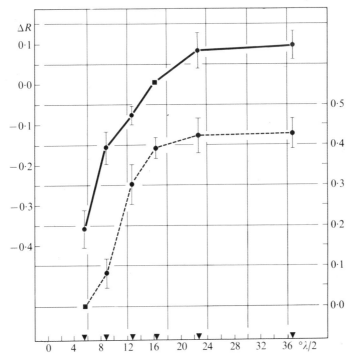

FIG. 5.21. Trained pattern discrimination in bees. The points show the frequencies of choice to the striped patterns shown below the abscissa when the bees are trained to pattern wavelengths of $\lambda_1 = 11\cdot0°$ (*dashed line, right ordinate*) and $\lambda_2 = 32\cdot6°$ (*solid line, left ordinate*). ΔR, differences in frequencies of choice; λ, pattern wavelengths. As in Fig. 5.20 the two curves coincide when shifted along the ordinate scale, so that the patterns are lying along a straight line in parameter space.

common parameter. Until now, however, these common parameters can be only described qualitatively.

Nevertheless, one has first to look at the significance of parameter variations within sets of one-type patterns, however these sets may be characterized. Even in the few cases where evidence is available that only a single stimulus parameter is involved in the pattern recognition task, no example of such a parameter has been deduced unambiguously. From the early findings of Hertz (1934, 1935) and Wolf (1935) (for summary of literature see Wehner (1973)) the relative contour length, i.e. the contour length per unit area, was supposed to be the decisive parameter for pattern detection in bees (see also Fig. 5.17). Although Schnetter (1968) described the frequencies of choice within pattern sets of four- or six-pointed stars by means of the distance function

$$d = \log(\Delta C / C),$$

where ΔC = difference between the contour lengths of the two patterns, C = contour length of the pattern characterized with shortest contours, this simple calculation does not hold when four-pointed stars are tested against six-pointed stars. Furthermore, as von Weizsaecker (1970) has pointed out in detail, no four-pointed star is completely mistaken for any six-pointed star *vice versa*. That important discovery shows unambiguously that each type of star-like patterns is processed by at least two independent pattern parameters.

All attempts to define the criteria of classification used by bees—or any other insect species—having failed, Cruse (1972*a,b*) returned to the method of calculating overlapping and non-overlapping areas between two figures by using a two-dimensional cross-correlation. As previously shown by Wehner (1968), the method can be applied to the frequencies of choices to black stripes which vary in angular position when represented on a vertical screen (Fig. 5.12). The cross-correlation term is defined as

$$\Delta A = f(A, R^+, R^-) = \frac{R^+ + R^-}{A}(A + R^+)$$

where A is the common area, and R^+ and R^- are the remaining areas of the training and the reference patterns, respectively. The quantity ΔC is defined as the log of the ratio of length of contours C^+ and C^- in the two patterns

$$\Delta C = \log(C^+ / C^-).$$

Thus the combined difference function

$$d = a\Delta A + b\Delta C,$$

where a and b are weighting factors, is based on the hypothesis that pattern recognition in bees depends on two processes, one of which (ΔA) can be reduced to a two-dimensional cross-correlation of the two shapes, whereas the other (ΔC) might be the result of an effect of contour.

The frequencies of choice of bees for a set of various figures, e.g. squares, triangles, stars, checkerboards and concentric angular rings, can be adequately described by this difference function (Fig. 5.22), if the weighting factors a and b can vary. On the other hand, the formula fails to explain some experiments of Cruse (1972), and some described by other authors (Mazokhin-Porshnyakov 1969*b*; Wehner 1971), in which the parameters used by the bees for pattern recognition change with the nature of the problem to be solved.

It is unlikely from another point of view that pattern detection can be done

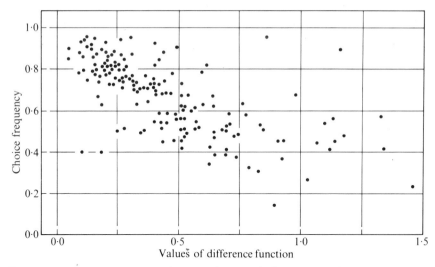

FIG. 5.22. Pattern discrimination in bees using many different types of patterns e.g. stars, checkerboards, concentric circular rings. The frequencies of choice (*ordinate*) are plotted against the values of the difference function *d* (*abscissa*) as calculated for each pair of patterns tested. Further details about the difference function *d* will be found in the text. From Cruse (1972*a*).

merely by cross-correlation methods. As outlined above (p. 95), the spatial mapping of the visual input which has been demonstrated by use of vertical pattern screens (Wehner 1972*a*) is an indispensable preprocessing mechanism. The method of comparing two patterns by means of overlapping and non-overlapping areas, i.e. by use of the term ΔA above, would describe the patterns theoretically by n features in an n-dimensional feature space, where n is the number of sampling stations. Each of these parameters (features) would then vary along the light intensity scale. But as that procedure does not lead to any decrease in pattern information, an immense storage capacity of the visual system would be required, in contrast to more reasonable methods of classification.

Solving the problem of pattern recognition consists in finding a classification scheme, by which patterns are clustered by $n_p < n$ pattern parameters. Therefore, many-to-one correspondences must be postulated if pattern detection is to be done by the visual system. In flies, Reichardt (1973) has shown that the correlation between a given stripe pattern and the frequencies of choice induced by that pattern is preserved even under reversed contrast conditions (for bees see Wehner (1971)). It is difficult to fit this fact into a theory of cross-correlation. Some evidence that parameters are chosen according to the problem to be solved comes from special training experiments in which bees previously trained to a set of patterns were tested with new transfer patterns (Mazokhin-Porshnyakov 1968; 1969*a,b*). Complex parameters such as 'multicoloured', 'checkered', or 'position of the test object with respect to other objects', are inferred to be used by the bees. By spontaneous preferences of several species Jander (1964, 1971), Jander and Volk-Heinrichs (1970), Jander and Schweder (1971) also tried to evaluate the features involved in pattern detection, for example 'contrasting area in a typical direction', 'texture', or 'edge with a typical slant'. Whether these 'features', however, are really used as parameters in an *n*-dimensional classification scheme, can only be decided by appropriate training tests. First, one has to find out by tests what patterns are mistaken for others, then a strategy can be developed for testing a proper classificatory model.

5.4. Conclusions

As for vertebrates, no scheme for classification of patterns has been found experimentally in insects. In recent years, however, it has become acceptable that the processes behind compound eyes do not necessarily differ from those in vertebrates. When one tries to define the functional differences between a lens eye and a compound eye system, one may only refer to the difference in light gathering devices, i.e. to the multiplication of the dioptric apparatus in insects. As the radius of curvature of the outer eye surface can be decreased more easily, the field of view in insects generally exceeds that of a vertebrate eye, but we have no *a priori* reasons for concluding that the processing of visual information should be similar or different in the two systems. With respect to picture transfer by the array of lens systems we stress that the strong overlap of visual fields found in most insect species studied (see Table 5.1) does not influence the resolution power of the eye. In accordance with the classical mosaic theory of insect vision (which, however, entirely disregarded visual field overlap), the resolution power depends only on the spatial density of sampling stations. In most species examined so far, the contrast transfer functions (Fig. 5.2) do not decline at lower frequencies than the resolution threshold frequencies, and the latter can be determined by measuring behavioural responses (Goetz 1964; Wehner and Wehner-von Segesser 1973). On the other hand, the angular sensitivity functions that are decisive for contrast transfer have also been deduced from behavioural work (see Table 5.1). In bees as well as flies some data are available showing the angular sensitivity functions at successive levels of information processing, at the lens systems, visual cells, and lamina neurons (Table 5.1). Lateral inhibition, so well known for the *Limulus* compound eye, has been proposed only in some recent electrophysiological work on the fly (Zettler and Jaervilehto 1972*a*; Mimura 1972) and behavioural work on the water strider (Meyer 1971*a*, 1972).

Excluding the peculiarities of the dioptric system in insects (pattern modulation) successful pattern discrimination has to solve the same classificatory problems in insects as in vertebrates. In general, the problem is to selectively throw away most of the information. Let us consider *n* parameters which can vary over v separately identifiable levels (intensity values) along each parameter scale, then $N = v^n$ different patterns can be presented at the inputs, say to a bee's visual system. It depends on the clustering of these N points in the n-dimensional feature space, which and how many of them are discriminated by the bee. The actual mechanisms used to perform particular pattern recognition tasks may be a function of the training procedure. Therefore invariance classes of patterns are only in part attributable to fixed clustering mechanisms. By appropriate 'transfer training' experiments (p. 111) either the strategy of feature clustering may be changed, or the pattern-recognition processes will become invariant for certain features of all the patterns. In the latter case the number of independently varying parameters will decrease, thus reducing the total amount of classified patterns by multiples of powers of v. At the present time, however, one can only state that in certain recognition tasks only one, in others at least two independently varying parameters are used by bees. Therefore, first we have to specify qualitatively the parameters involved in pattern detection. In the immediate future, the study of processing or filtering mechanisms such as transfer of spatial frequencies and contrast transfer, is necessary before one tries to solve the classificatory problem.

ACKNOWLEDGEMENT

The own experiments dealt with in that paper were financially supported by the Fonds National Suisse de la Recherche Scientifique and the Deutsche Forschungsgemeinschaft. The author wants to thank Dr. T. Labhart and Dr. R. Schinz for help and assistance.

6. Investigations on the vision of ants

G. A. MAZOKHIN-PORSHNYAKOV

THE unusual way of life and complex behaviour of ants have attracted attention for many years, and nowadays the interest in ants is not only a purely scientific one. These insects are considered to play an important part in effective biological control of forest pests in many European countries.

So far, effort has been concentrated upon the visual behaviour of ants and less attention has been paid to visual mechanisms. Therefore, in the present review I shall dwell upon the structure and function of the ant's compound eyes rather than the behaviour of ants, which is better known to a wide range of readers. The data to be presented on the eye's electrical response, with reference to spectral and contrast sensitivity are wholly original.

6.1. The structure of the compound eye

The number of ommatidia in different species of ants ranges from 9 to 1300. Moreover the number varies even between individuals (workers) of the same nest. In *Cataglyphis bicolor* the number of ommatidia is correlated with the body size. It is minimal (*ca.* 600) for the smallest workers, which are confined to the nest, and reaches 1300 for the largest workers which serve as hunters (Menzel and Wehner 1970). In workers of *Formica integroides* the larger the surface of the head, the larger is the number of ommatidia in the eye, the longer is the eye and the larger is the diameter of each individual facet (Bernstein and Finn 1971). Hence larger individuals may have better vision. These authors consider this one of the reasons for another interesting phenomenon, that the larger foragers learn faster and orient better to visual stimuli.

There is a correlation between number of ommatidia and way of life. Species which live in the open and presumably use visual orientation, e.g. the desert species *C. bicolor* (Wehner 1969), have more facets than species which tend to hide e.g., *Lasius niger* which has *ca.* 120 facets (Mazokhin-Porshnyakov and Trenn 1972). The forest ant *Formica rufa* has 460 facets (Mazokhin-Porshnyakov and Trenn 1972) whereas *F. polyctena* has 750 facets (Menzel 1972).

The fine structure of the eye has been studied recently in *F. polyctena*

(Menzel 1972) and in *C. bicolor* (Wehner and Eheim 1971). The dioptric apparatus in *F. polyetena* is formed by a rather flat corneal lens and a crystalline cone. The latter is enclosed by two principal pigment cells, while eight additional cells surround the ommatidium over its whole length from cornea to basement membrane. Each ommatidium contains eight retinula cells in its distal parts, of which six are large and two small, which lie opposite to each other. A ninth basal retinula cell lies in the proximal part of the ommatidium. Microvilli of all nine retinula cells contribute to the fused rhabdom. In the distal part of the rhabdom the microvilli are arranged in three directions subtending 120° to each other. At the proximal end of the rhabdom lie four basal pigment cells.

In *C. bicolor* the interommatidial angles in the horizontal direction are 4·3°, in the longitudinal (vertical) direction −3·3°. Visual fields of central ommatidia, measured optically in eye slices, are as large as 12·5° in the horizontal and 15·5° in the vertical direction. The half-widths of the angular admittance curve behind the crystalline cone in the same directions, are 6·8° and 8·8°, respectively. In agreement with these values the narrowest stripes that provoke a response have a repeat period of 6·6–7·0° (Eheim und Wehner 1972).

6.2. Electrical responses

As far as I know, the ERG has been recorded only in *Messor structor* (Campan *et al.* 1965), *L. niger* and *F. cunicularia* (Mazokhin-Porshnyakov

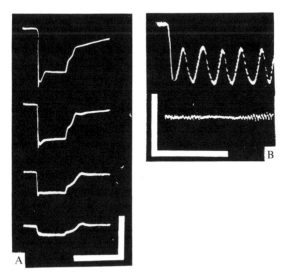

Fig. 6.1. ERG of worker of *Formica cunicularia*. (A) at four intensities of light with 100, 49, 16 and 0·5 per cent transmission of the light filters; (B) retinal responses to flashes at 5 Hz (upper) and at 40 Hz (lower). Vertical calibration: 10 mV, time 1 s.

and Trenn 1972). The latter, which are known in most detail, have a mono-phasic corneal-negative electroretinogram (ERG) (Fig. 6.1a). With high intensities of light the plateau of the ERG is preceded by a characteristic peak of 12–15 mV, and an afterpotential follows the plateau when the light is switched off. A logarithmic dependence of ERG amplitude against stimulus intensity can be demonstrated over a tenfold range in light intensity. With a flickering light the ERG has a saw-tooth form, with teeth smaller and closer as the flashes increase in frequency (Fig. 6.1b). Flicker fusion occurs at 73 Hz in *F. cunicularia* to 79 Hz in *L. niger*. The critical flicker frequency for the larger ant *M. structor* is essentially lower at 40 Hz (Campan *et al.* 1965). This difference in speed of response makes sense in relation to the body size. For a small body with an eye in a low position the angular velocity of move-ment relative to substrate is much higher for *Lasius* or *Formica* than for the larger and leggy *Messor*. It seems practical that small species should have more rapidly acting eyes to perceive visually the substrate details during their movements.

6.3. Adaptation

The velocity and extent of the dark-adaptation process have been deter-mined in *F. cunicularia* (Mazokhin-Porshnyakov and Trenn 1972). Using white light flashes, the ERG threshold was measured on an eye that could be held in the dark (Fig. 6.2). During the first 5 min of dark adaptation the sensitivity sharply increases by approximately tenfold, then increases more slowly. The maximum sensitivity is reached in 25 min, with a total gain in sensitivity of *ca.* 50–70.

Changes in the eye structure in *F. polyctena* during the course of adaptation have been described (Menzel and Langer 1971). According to these authors,

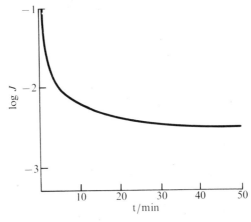

FIG. 6.2. The time course of dark adaptation in *Formica cunicularia* worker. Ordinate; log of threshold light intensity; abscissa, time in the dark in minutes.

the vacuoles in the cytoplasm of the visual cells concentrate in the dark-adapted retina around the rhabdom and form a layer of palisade around it, while the screening pigment grains lie at the periphery of the cells. When illuminated the pigment grains drift distally and gather in the immediate vicinity of the rhabdom, replacing the vacuoles. The position of other pigment grains in the principal and accessory pigment cells remains almost unchanged. The phenomena described necessarily control the sensitivity level of the receptors by changing the waveguide properties of the rhabdom (Kuiper 1962; Horridge and Barnard 1965; and Snyder (Chapter 9).

6.4. Contrast sensitivity

Contrast sensitivity is defined here as the ratio $\Delta I/I \times 100$ per cent where ΔI is the minimal quantity of light, addition of which to I produces the threshold ERG. In the best results with *L. niger* and *F. cunicularia* the contrast sensitivity was *c.* 5 per cent (Mazokhin-Porshnyakov and Trenn 1972).

6.5. Spectral sensitivity

I used two electrophysiological methods to measure the spectral sensitivity in *F. cunicularia* and *L. niger*. The first was the method of equal responses, in which intensities of different monochromatic rays eliciting a constant threshold electroretinogram in the dark-adapted eye were measured. The second method was that of flickering colorimetry or colorimetric pull-method. The curves of equal spectral responses in the above ant species differ conspicuously in position of the maximum and in ultraviolet sensitivity. The curve for *L. niger* is maximal at $\lambda \sim 500$ nm but in the u.v. region ($\lambda > 365$ nm) it declines practically to zero. The curve for *F. cunicularia* ($\lambda_{max} \sim 510$ nm) declines less abruptly and the sensitivity at $\lambda = 365$ nm is still 20 per cent of the maximal value. Both species are insensitive to rays with $\lambda > 650$ nm at the red end of the spectrum. In both species, the position of the maximum can be shifted by *c.* 20 nm by adaptation to coloured light, and another small peak appears (Fig. 6.3). The shift of the sensitivity maximum to longer wavelengths after adaptation to blue light and its displacement in the opposite direction by adaptation to yellow indicate that at least two visual pigments are present.

A similar result was obtained by the flickering colorimetry method. An estimate of the number of receptor types was made from the number of basic colours necessary to reproduce the responses to all spectral colours (Mazokhin-Porshnyakov 1969). Absence of an ERG response when a monochromatic colour was flickered with a proper mixture of basic colours was used as the criterion that they are equivalent. The ratio (converted to a percentage of its maximum value) of each basic colour energy in the mixture to the

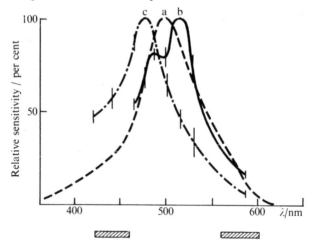

FIG. 6.3. Effect of colour adaptation in *Lasius niger* worker. (a) the spectral sensitivity curve measured by constant response. (b) the same after adaptation to blue; (c) the same after adaptation to yellow. Ordinate: relative (per cent of maximum) sensitivity; abscissa; wavelength of light, in nm. Vertical lines are standard deviations. The spectral composition of the adapting light is shown by the shaded stripe. *Editor's note:* I believe that measurements of intensities were in energy units, not photons.

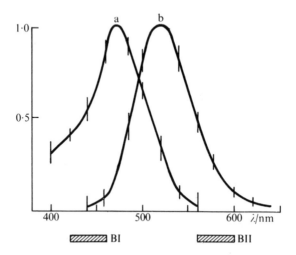

FIG. 6.4. Colour composition curves for the eye of *Formica cunicularia* worker. (a) curve of colour composition of basic light of short wavelength (BI); (b) the same for basic light of long wavelength (BII). Ordinate: ratio values of each basic colour intensity in the mixture to the intensity of visual non-distinguished monochromatic colour; maximal ratio value is taken as 1·0. Abscissa: wavelength of light in nm. Vertical lines are standard deviation. The spectral composition of each basic colour is shown by the length of the shaded stripe. *Editor's note:* I believe that measurements were in energy units, not photons.

energy of the indistinguishable monochromatic colour is plotted as the so-called colour composition curve. Such curves (within various limitations of the method) represent the spectral sensitivity of each spectral type of receptors.

Measurements over the range $\lambda = 400-650$ nm in *F. cunicularia* (Fig. 6.4), show at least two spectral types of receptors: one with $\lambda_{max} = ca.$ 470 nm, the other with $\lambda_{max} = ca.$ 520 nm. This result in good agreement with the threshold method, favours the assumption of colour vision in ants.

6.6. Colour vision

The results of behavioural experiments on colour discrimination by ants are contradictory and not always properly understood. Following Lubbock (1885) the opinion has been affirmed that ants are very sensitive to u.v., but in his experiments the ants could have avoided u.v. because of its harmful effect on pupae and not because the eye is most sensitive to it. Similarly, that u.v. is fifty-eight times more attractive for *C. bicolor* does not imply a fifty-eight times greater sensitivity to u.v. (see Wehner und Toggweiler 1972). The attraction of monochromatic rays is *not proportional* to the sensitivity of the eye to them (Mazokhin-Porshnyakov 1969).

Tsumeki (1950, 1953) supposed that the Japanese species *Camponotus* and *Leptothorax* can distinguish colours, but he failed to train ants in colour discrimination. Since Moller-Racke (1952) found no optomotor response of *Formica* in the rotating two-coloured cylinder he concluded this species to be colour-blind. But Kiepenheuer (1968) showed that *F. polyctena* can learn their way home using colour clues and he arrived at the opposite conclusion. Strictly speaking, neither method permitted the conclusions these authors reached. Kaiser (1968) has proved that the absence of the optomotor reaction to a colour contrast cannot be the criterion for colour blindness in either test. Finally, discrimination between two colour rays compared at *any brightness* relation was not shown in the experiment of Kiepenheuer (1968). It would be the necessary and sufficient explanation of all results to admit that ants do see colours.

Nevertheless, taking into account the electrophysiological evidence, one may be almost confident that ants have colour vision, but the final conclusion must await a decisive behavioural experiment.

6.7. Visual orientation and learning capacity

Many well-known works deal with orientation and individual habit acquisition in ants, so I shall deal with these only briefly.

Landmarks and the sun are both used as reference points by *Formica*, *Lasius*, *Cataglyphis* and other ants (Jander 1957; Voss 1965; Wehner 1969; Wehner and Menzel 1969). When the sun is obscured by cloud the directional reference can be supplied by the polarization of the blue sky (Vowles 1950,

1954; Carthy 1951; Jander 1957). The moon can serve as a reference point at night (Jander 1957). As a rule ants can make corrections for the change in azimuth of celestial bodies used as reference points. They tend to use celestial orientation when they find themselves in unknown territory (Wehner 1968).

Ants are quick to learn their way to the nest or through a maze using artificial reference points (Chauvin 1964; Vowles 1965; Dlusski 1967). The corpora pedunculata, which are highly developed in ants (especially in *Formica and Lasius*), are considered to be the site of memory (Pandazis 1930; Vowles 1955). *F. incerta* remember a labyrinth path for up to two months (Schneirla 1960), and in *F. rufa* and other species of this genus the memory trace of the path referred to visual reference points is preserved through the whole winter (Rosengren 1971). It seems possible to teach *F. rufa* (Voss 1967) and *F. truncorum* (Schulze-Schencking 1970) to distinguish crosses, triangles, and circles of equal area, thus implying an ability to recognize form.

6.8. Conclusion

Ants that wander in the open (*Cataglyphis, Formica* and others) have well-developed compound eyes with up to 1300 ommatidia, and possess high visual capabilities, at least as compared to workers of the termite *Anacanthotermes* which has a similar way of life (Mazokhin-Porshnyakov *et al.* 1967). Larger workers of the same nest have more ommatidia and seem to have better vision than their smaller fellows. The eye in *C. bicolor* can resolve 6·6–7·0°. The minimum effective intensity increment (*Formica, Lasius*) is *c.* 5 per cent. The ERG of ants is monophasic, with retina negative. Flicker fusion in small species (*Formica, Lasius*) occurs at a frequency of *c.* 75 Hz, in large species (*Messor*)—at 40 Hz. Dark adaptation (*Formica*) is completed in 25–30 min. During this time, but mainly in the first few minutes, the threshold sensitivity increases fifty–seventy times. The adaptation mechanism is related to changes in the rhabdom waveguide properties. The maximum of the spectral sensitivity curve of *F. cunicularia*, measured by constant threshold response, is located at 510 nm, and that of *L. niger* at 500 nm. The relative sensitivity to u.v. with $\lambda = 365$ nm in *F. cunicularia* was as high as 20 per cent of that at peak λ, and in *L. niger* it is less than 5 per cent. Both species are blind to red rays with $\lambda > 650$ nm. Colour adaptation shifts the maximum of the constant response spectral curve, suggesting at least two visual pigments. By means of the flickering colourimetry method, two receptor types were found in *F. cunicularia*: one with $\lambda_{max} \sim 470$ nm and the other with $\lambda_{max} \sim 520$ nm. The electrophysiological and behavioural experiments taken as a whole suggest the probable existence of real colour vision in ants. There is some evidence that ants recognize the form of objects. For orientation they use either landmarks or the celestial compass reaction. Ants can learn readily and remember visual references for several months.

7. Colour receptors in insects

R. MENZEL

SINCE the first measurement of spectral sensitivity of a single visual cell of the fly twelve years ago (Autrum and Burkhardt 1960), the action spectra of the retinula cells of a number of insect species have been examined. Photometry of extracted visual pigments and microspectrophotometry of rhabdomeres and rhabdoms have shown that a Retinal$_1$–protein complex is the only photopigment present, but the spectral sensitivity of single retinula cells and the absorption properties of these rhodopsins rarely coincide. The discrepancies between the absorption spectra of photopigments and the spectral sensitivity of the functioning cell can be attributed to a number of factors. These factors will be discussed first and followed by a comparison between the spectral sensitivity of photoreceptors and the extinction spectrum of photopigments for a number of insects.

Recently, the identification of colour receptors by light and electron microscopy has become possible. These techniques provide information on the location of colour receptors with known action spectra which together with the behavioural analysis of colour discrimination is the first step towards the analysis of the interactions and neuronal connections of colour receptors.

This review tries to sample the data necessary for a comparative analysis of the process of colour coding in insects and is not primarily concerned with the biochemical and biophysical aspects of action spectra. As only recent work is covered this review supplements some other excellent reviews on colour vision in insects (e.g. Autrum 1958; Burkhardt 1964; Carlson 1972; Goldsmith 1961, 1964, 1972; Hamdorf *et al.* 1972; Kirschfeld 1969; Kuiper 1966; Langer 1972; Moody 1964).

7.1. Factors influencing spectral sensitivity

7.1.1 *Photopigments*

The spectral sensitivity of a photoreceptor depends mainly on the absorption properties of its photopigment or photopigments and also to a variable extent upon other factors which are discussed later. As there are a number of recent reviews on visual pigments in arthropods (Carlson 1972; Goldsmith

1972; Hamdorf, Paulsen, Schwemer, and Taeuber 1972; Langer 1972) only a short summary of the important points will be given here. The chromophore molecule is always *retinal* (retinene₁), never dehydroretinal (retinene₂), and in the unbleached state this is bound to a protein as 11-*cis*-retinal. During bleaching a thermostable metarhodopsin containing bound all-trans retinal is produced. It is not known if free *all-trans-retinal* is formed *in vivo*, and if so in what quantity, or if most of the metarhodopsin is photoreisomerized.

Rhodopsin has been isolated from the heads of bees with a maximum of the difference spectrum at 440 nm (Goldsmith 1958), from the heads of cockroaches with a maximum at 500 nm (Wolken and Scheer 1963) and from the heads of house-flies with a maximum at 510 nm (Marak, Gallik, and Cornesky 1970).The method of extraction of the bee photopigment probably altered the protein structure (Carlson 1972). The extracted visual pigments of cockroaches and flies show an increase in absorbance at 375 and 380 nm respectively during bleaching and the difference spectrum in the longer wavelengths is in good agreement with the spectral sensitivities of the green receptors of the two species. Langer and Thorell (1966*b*) measured the spectral absorptance of the rhabdomeres in the retina of the fly by microspectrophotometry (msp) (Fig. 7.1). The six outer rhabdomeres have a maximum at 510 nm and a secondary maximum at 360 nm. The central rhabdomere has a maximum at 476 nm and also a smaller maximum in the ultra-violet region.

The green sensitive photopigment of the fly *Calliphora* (Langer 1972) and the yellow-green sensitive pigment of the butterfly *Deilephila elpenor* (Hamdorf, Hüglund, and Langer 1972) both produce with green light a thermostable photoreisomerized rhodopsin which changes back to the unbleached pigment on illumination with blue light. Such a thermostable photore somerizable metarhodopsin was also found for the u.v. photopigment of *Ascalaphus* (Neuroptera) (Gogala, Hamdorf, and Schwemer 1970) and *Deilephila* (Hamdorf, Hüglund, and Langer 1972). The rhodopsin itself absorbs maximally at 345 nm (*Ascalaphus*) and 350 nm (*Deilephila*), while the photoproduct after u.v. illumination has an absorbance maximum at 483 nm (*Ascalaphus*) and 453 nm (*Deilephila*) (Fig. 7.1). These results indicate that the u.v. receptors often found in insects have a retinal containing photopigment and that the different sensitivities of photoreceptors depend on the different absorption properties of retinal photopigments (see also Goldsmith 1972).

A theoretical analysis by Snyder and Miller (1972) shows that in flies the shift of the α-maximum of rhabdomeres 7 and 8 is due to its small diameter (i.e. due to waveguide effects, see below). Snyder (1973) and Snyder and Pask (1973) have also shown that waveguide effects greatly enhance the u.v. spectral sensitivity peak in comparison to the visible maximum. However, the sensitivity in the u.v. of fly retinula cells is so high (Fig. 7.17), that Snyder and Pask (1973) and Snyder (1973) need an absorption of 50 per cent in the

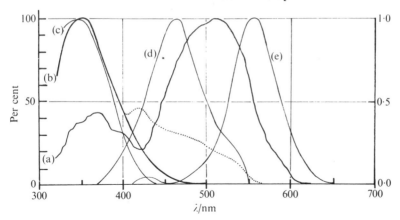

FIG. 7.1. Spectral properties of some insect photopigments. Curves (a) and (b): left ordinate, absorption in per cent of maximal absorption. (a) gives the spectral absorptance of one of the six outer rhabdomeres of the fly *Calliphora erythrocephala* measured by microspectrophotometry (msp) (Langer and Thorell 1966). (b) is the extinction spectrum of the u.v.-photopigment of *Ascalaphus macaronius* in aqueous solution. The dotted part of the curve above 400 nm gives the extinction actually measured, the solid line in this region the expected extinction of a pure solution (Gogala, Hamdorf, and Schwemer 1970). Curves (c), (d) and (e) are difference spectra (right ordinate) which give the change in extinction after illumination at different wavelengths in relation to the maximal change $\Delta E_{var}/\Delta E_{max}$. (c) and (d): difference spectra of both components of the u.v.–visual pigment system of *Deilephila elpenor* (Lepidoptera; Hamdorf, Höglund, and Langer 1972). (c) saturating illumination with 453 nm, then illumination with different wavelengths in the u.v. (abscissa) and measurement of the absorptance changes. This difference spectrum gives mainly the spectral absorption of the photoreisomerized, u.v. photopigment. (d) saturating illumination with 371 nm and measurement of the absorptance changes at different wavelengths in the blue and green regions of the spectrum. This curve shows mainly the spectral absorption of the thermostable metarhodopsin resulting from isomerization of the u.v.-photopigment. (The original curve of Hamdorf, Höglund, and Langer (1972) shows a small negative part between 550 and 600 nm which is omitted here.) (e) difference spectrum of the yellow–green photopigment after saturating illumination with 453 nm. Measurement of the absorptance changes at wavelengths between 400 and 650 nm.

u.v. for their calculation. This is much more than the absorption of a rhodopsin with a green maximum (see later). The relatively high absorption at 360 nm therefore in all probability indicates the presence of a special u.v. photopigment together with a green pigment in the same cell. This would account for the small change in absorptance on bleaching because the absorption decrease during bleaching of the u.v. photopigment would be compensated for by the production of free retinal with an absorption at 360 nm. A rhodopsin with a maximal difference spectrum at 515 nm was also found by msp in the ocelli of larvae of *Aedes aegypti* (Brown and White 1972). This rhodopsin has a high absorptance in the u.v. (350 nm) which shows no change during bleaching. In this species also a u.v. photopigment may exist, together with a green photopigment, particularly since a u.v. peak in the spectral sensitivity was found electrophysiologically (Seldin, White, and Brown 1972).

7.1.2. *Spectral transmission of the dioptric apparatus*

7.1.2.1. *Cornea reflection.*   Light strikes the corneal surface before reaching the photosensory structures of the retina. The reflection at this surface is diminished in many species, especially in those which orientate behaviourally at low light intensities, e.g. nocturnal Lepidoptera, by a regular array of protrusions (Bernhard 1967; Bernhard and Miller 1968; Bernhard, Miller, and Møller 1963, 1965; Bernhard, Gemne, and Møller 1968; Bernhard, Gemne, and Sallström 1970). These protrusions or cornea nipples project 50–250 nm, i.e. less than a wavelength of light and function as a broad-banded anti-reflection surface maximally effective for wavelengths corresponding to about twice the nipple height. In contrast smooth corneas are found in the diurnal insects *Aeschna*, *Apis*, *Bombus* and *Coccinella* (Bernhard, Gemne, and Sallström 1970) except in the case of the butterfly *Vanessa* and the mosquito which possess a nipply array while other flies have flat protrusions (Miller, Møller, and Bernhard 1966).

7.1.2.2. *Cornea transmission.*   The nipple structure of the cornea increases the light reaching the dioptric apparatus without significantly changing its spectral content (Bernhard, Miller, and Møller 1963, 1965). However, obvious chromatic selective effects are seen from the bright orange, red, yellow, green, and blue reflection patterns on the compound eyes of many Tabanidae and twenty-three other dipteran species (Bernhard and Miller 1968; Steyskal 1957). These are thought to be produced by the alternate dense and light bands seen in the cornea under the electron microscope. They have a periodicity of approximately a quarter of a wavelength of the reflected light and are thought to act as interference filters. The half band width of the reflected light has been measured as 60–100 nm (Fig. 7.2) (Bernard 1971; Bernard and Miller 1968) and the position of the maxima varies between species (for example the range of the blue green facets is 55 nm and that of the red facets is 80 nm). Although the positions of the reflection maxima are variable in closely related species the maximum for one of the facet colours always has a wavelength at which the other has its first reflection minimum. This suggests that the interference filters may be used for colour vision by the consecutive comparison of contrast between adjacent facets. Trujillo-Cenóz and Bernard (1972) have found evidence for this hypothesis by analysing the connections of the six short retinula axons in the lamina. The normal pattern of connections (Braitenberg 1967; Kirschfeld 1968; Trujillo-Cenóz 1966) is changed so that only those retinula cells beneath corneal lenses with the same colour converge upon the same cartridge.

Eyes with metallic sheens of different colours have been found in some Neuroptera (Chrysomelidae and Myrmeleontidae: Bernard 1971) and Coleoptera: Friederichs 1931; see also Meyer-Rochow, Chapter 12). Neither the corneal spectral transmission nor the spectral sensitivities of single

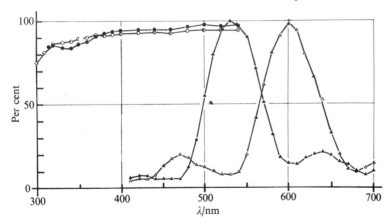

FIG. 7.2. (▲, △) Spectral transmission of two neighbouring facets in the ventral region of the eye of *Condylostylus* (Dolichopodidae, Diptera) calculated from normalized reflectance characteristics (Bernard 1971; Bernard and Miller 1968).
Spectral transmission of the cornea (○) and the crystalline cone (●) of fresh, unfixed preparations of *Aeschna cyanea* (Kolb, Autrum, and Eguchi 1969).

retinula cells have so far been studied in these species with corneal interference filters.

All those insects in which colour discrimination has been studied also have a layered cornea but these have no specific colour and the layers have a wider separation (e.g. 800–1200 nm for drone bee, Perrelet 1970; and the wood ant *Formica polyctena*, Menzel 1972a). The colour of light reflected from the eye in these insects depends only on the colour of the screening pigments, for instance the different eye colours of *Drosophila* mutants (see later) (Strother and Superdock 1972). Measurements of corneal absorption in dragonfly (Kolb, Autrum, and Eguchi 1969), show that the absorption decreases slowly from long to short wavelengths and increases rapidly below 330 nm (Fig. 7.2). The same result is found in the moths *Smerinthus ocellata* and *Endromis versicolora* (Bernhard, Miller, and Møller 1965), *Manduca sexta* (Carlson and Philipson 1972) and the hymenopterans *Apis mellifera* and *Xylocopa violacea* (Carricaburu and Chardenot 1967), the housefly *Musca domestica* (Goldsmith and Fernandez 1968), the dungbeetle *Euoniticellus* (Meyer-Rochow, Chapter 12) and the fruitfly *Drosophila melanogaster* (Strother and Superdock 1972). In both dipterans the transmission was measured down to 240 nm, and showed a minimum at 280 nm. Therefore with the exception of the dipterans (many Tabanidae), Neuropterans and Coleopterans which have corneal lenses with interference filters, for all other insects it can be presumed that light reaches the rhabdom through the dioptric apparatus without any chromatic change. The transmission in the far u.v. (280–320 nm) is more strongly attenuated and may influence the spectral sensitivity of u.v. receptors. It is important that msp transmission measurements are carried out with preparations of fresh cornea because the transmission changes considerably during dehydration by alcohol or drying (Kolb, Autrum, and Eguchi 1969).

The spectral transmission of the crystalline cone is known only in the dragonfly *Aeschna cyanea* (Kolb, Autrum, and Eguchi 1969) and the lepidopteran *Manduca sexta* (Carlson and Philipson 1972). There is no large difference in absorption compared with the cornea (Fig. 7.2).

7.1.3. *Selective reflection of the tapetum*

Some butterflies with apposition eyes and long fused rhabdom possess a tracheal tapetum between the proximal end of the rhabdom and the basement membrane which gives rise to the glow of dark adapted eyes. Miller and Bernard (1968) found that this glow has different colours in different parts of the eye in some butterflies (e.g. *Euptychia, Anartia, Danaus*). Typically the dorsal facets have a blue glow, anterodorsal a green–yellow glow and anteroventral an orange–red glow. This is because there are four tracheoles at the proximal end of each rhabdom with a large number of plates arranged very regularly, perpendicular to the long axis. In orange facets the plates are 106 nm thick and are separated by 148 nm of air, whereas in blue–green facets they are 123 nm thick and with a separation of 88 nm. As the refractive index of the cytoplasm within the plates is *c*. 1·4, these parallel plates function as interference filters which reflect maximally in the first example at 590 nm (bandwidth 125 nm) and in the second example at 490 nm (bandwidth 104 nm). The non-reflected light is absorbed by the underlying screening pigment. As the reflected light passes through the rhabdom twice the sensitivity of the receptor cells is selectively increased in this part of the spectrum.

Like the interference filters of corneal lenses and the coloured oil droplets in bird retinae this system may have a contrast filtering function in colour vision. In contrast, however, to the distribution of oil droplets in the pigeon retina the sensitivity is increased for blue in the dorsal part of the eye and for yellow and red in the ventral part. Therefore the background illumination of the blue sky is not selectively reduced to allow better observation of small objects as was suggested for the pigeon retina by Walls (1963). The tapetal system in butterflies works by selectively increasing the sensitivity for light predominantly represented in different parts of the visual field.

The first indication of the selective increase in sensitivity by tapetal reflection was found from ERG responses by Bernhard, Boethius, Gemne, and Struwe (1970) and Struwe (1972) as the small secondary maximum measured cannot depend on the transmission of the screening pigments because these have a high absorption up to 700 nm. Direct measurement of the effect of the tapetal reflection has not yet been made.

7.1.4. *Spectral transmission of the screening pigment*

Pigment cells and retinula cells contain granules with dark brown, red, or yellow pigment which function as a shield between individual ommatidia and which regulate the light flux in the rhabdomere (see Walcott, Chapter 2). In Lepidoptera, Hymenoptera, and most other insects the pigment granules predominantly contain ommines (*Celerio euphorbiae*, Lepidoptera, and *Vespa* sp. Hymenoptera, Höglund, Struwe, and Thorell 1970; *Heliconius*, Lepidoptera, Langer and Struwe 1972; *Apis mellifera*, Hymenoptera, Langer and Schneider 1972). In Diptera usually the granules contain predominantly xanthommatines (Langer 1967; Langer and Hoffmann 1966; Strother and Casella 1972; Strother and Superdock 1972).

Different absorption curves of single pigment granules in a single eye have been found, both because several different pigments may be found in the same eye and even in the same pigment granule. Langer and Struwe (1972) distinguished between three groups (Fig. 7.3). Group B_2 shows a maximal absorption at 560 nm (50 per cent absorption at 680 nm) and was found in the moth *Celerio euphorbiae*, the wasp *Vespa* sp. (Höglund, Struwe, and Thorell 1970), honey bee *Apis mellifera* (Langer and Schneider 1972), and the butterfly *Heliconius* (Langer and Struwe 1972). These pigment granules probably contain only ommines. Another group (B_1) has an absorption maximum at 450 nm (50 per cent absorption at 620 nm) and contains oxidized

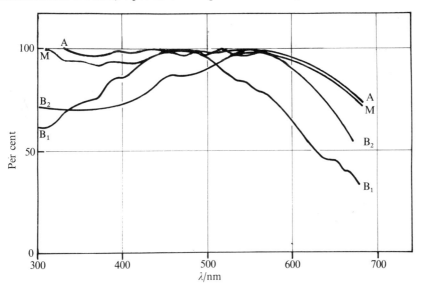

FIG. 7.3. Normalized mean spectral absorptance of screening pigment granules in slices from fresh eyes of *Heliconius* (Heliconidae, Lepidoptera). M: granule mass in the compound eye of *H. erato*. A: single pigment granule of group A in the eye of *H. erato*. B_1: single pigment granule of group B_1 in the eye of *H. sara*. B_2: single pigment granule of group B_2 in the eye of *H. erato*. (After Lange and Struwe 1972.)

xanthommatines. It has been found in *Heliconius* and *Calliphora* (Langer 1967; Langer and Struwe 1972). Most pigment granules in the eye of *Heliconius* have a uniform absorption between 300–700 nm (group A). These granules consist of a mixture of oxidized and reduced xanthommatines and ommines. The absorption of all these pigments cuts off sharply at *c*. 600 nm so that the absorption between 600–700 nm results from an unknown pigment.

The pigments of group A and B_2 should cause little or no effect on the spectral sensitivity of the eye in the light-adapted compared with the dark-adapted state. Höglund and Struwe (1970) found only small differences in the spectral sensitivity of the light- and dark-adapted eye in *Manduca sexta*, which may depend upon the spectral absorption of the screening pigments.

The screening pigment of the eye of flies contains xanthommatines which in the case of the red pigment granules have an absorption cut-off at wavelengths up to 605 nm (50 per cent absorption) of *Calliphora* and *Drosophila*. Long wavelength cut-off in other screening pigments occur, for example, at 490 nm (yellow pigment granule in *Calliphora*) or 560 nm (brown pigment granule in *Drosophila*) (Langer 1967; Strother 1966; Strother and Casella 1972; Strother and Superdock 1972). Therefore the screening pigment is transparent in the red-wavelength region in which the spectral sensitivity of the photopigments is considerable. For this reason ERG measurements demonstrate an increase in sensitivity at 610–620 nm as a result of the increase of the number of excited visual cells and the broadening of their acceptance angles due to passage of light between ommatidia (screening pigment effect) (Autrum 1955; Burkhardt 1962, 1964; Burkhardt and de la Motte 1972; Goldsmith 1965; Streck 1972). This peak in the orange has been interpreted incorrectly as a red receptor (Mazokhin-Porshniakov 1960). At high light intensities this effect has also been observed with intracellular recordings particularly if the stimulating light is not on-axis (Burkhardt 1962, Fig. 8).

Streck (1972) measured angular sensitivity curves in *Calliphora* of $\Delta\rho = 2\cdot8°$ at 360 and 495 nm and $\Delta\rho = 4\cdot0°$ at 615 nm. Mutants lacking the red pigment of the wild-type are more light sensitive and have the same acceptance angle at different wavelengths.

A reversed screening pigment effect was found in the mutant *white*. This mutant contains mainly pterins which have a low absorption above 460 nm and the absorption minimum at 350 nm characteristic of xanthommatine pigments is absent. Off-axis stimulation is therefore less effective in the u.v. than in the visible range (Streck 1972) which results in a relatively low u.v. maximum (Langer and Hoffman 1966). The pterins are also found in the eye of the wild-type and their u.v. absorption characteristics, which are complementary to those of xanthommatines, mask the effect on spectral sensitivity of the transmission of xanthommatines at 350 nm. The absorption of the pterins in the brown eye of the wild-type *Drosophila* on the other hand increases below 320 nm, an absorption minimum is therefore found at 350 nm (Strother and Superdock 1972) which gives rise to a higher u.v. sensitivity of the whole eye. In fact Wehner and Schuemperli (1969) found in phototaxis experiments in *Drosophila* an increase of the u.v. sensitivity at higher intensities.

Light is reflected from pigment granules and may subsequently reach photopigment containing structures. The reflected light is greenish in flies because of selective reflection below 500 nm from the red pigment granules (Kirschfeld and Franceschini 1969). In apposition eyes the light flux in the rhabdomeres is controlled by radial movement of pigment granules within the retinula cells. When the pigment granules lie very close to the rhabdomeres in the light-adapted state there is a higher probability that reflected light is absorbed by the photopigment. This may cause a relatively stronger excitation of the green receptors which would then show a broader spectral sensitivity. This effect has not been observed because no spectral sensitivity measurements have been made in the light-adapted state. In the wasp eye there is only a very small change in the relative spectral sensitivity during light adaption (Menzel 1971), but the absorption of the screening pigment in the eye of this insect varies less between 300–600 nm than in that of the fly.

7.1.5. *The rhabdom as a dielectric wave guide*

In several theoretical studies Snyder and his colleagues (see his article in this volume) has shown that the physical properties of the rhabdom influence the spectral sensitivity of retinula cells. In the list of factors influencing the spectral sensitivity these effects will be described here only briefly.

(1) Snyder and Miller (1972) and Snyder and Pask (1973) have shown that the effect of confining a photopigment to within a narrow rhabdomere (e.g. in flies) is to (a) shift the visible absorption peak to lower wavelengths and (b) to increase the u.v. peak absorption relative to the visible. The smaller the diameter the larger the effect. For a refractive index of 1·349 of the rhabdomere and 1·336 for the surrounding cytoplasm (Seitz 1968) only a very small change occurs in the action spectrum of rhabdomeres 1 μm or larger in diameter (retinula cells 1–6 in the fly). In thinner (0·5 μm) rhabdomeres (retinula cells 7 and 8 in the fly), however, the light travelling inside the rhabdomere decreases at 600 nm to a quarter of that at 300 nm. This filtering effect causes a shift of the sensitivity maximum from 515 to 478 nm and an enhancement of the u.v. maximum. Furthermore an increase of the refractive index of the rhabdomeres from 1·349 to *c*. 1·388 would cause the same effect for a thin (0·5 μm) rhabdomere as an increase in diameter and this effect may possibly be found in the eye of *Condylostylus*.

(2) The refractive index of photosensitive structures depends on the wavelength. Near the absorption maximum the refractive index decreases to shorter wavelengths and increases to longer wavelengths. As very small changes of the refractive index influence the total internal reflective properties of the rhabdomere, a decrease of refractive index causes a decrease of the proportion of modal light intensity within the rhabdomere (Snyder and Richmond 1972, 1973).

Figure 7.4 shows the depression of modal light intensity at 470 nm calculated for a 1·25 per cent change of refractive index and an absorption maximum at 500 nm. Anomalous dispersion effects such as these may be seen for example in Langer and Thorell's (1966b) spectrophotometric curves (a small inflection was measured in the extinction curve at wavelengths just less than that of maximum extinction).

(3) Snyder and Pask (1972) showed by analysing the measurements of Eguchi (1971) on dragonflies that the action spectrum of modal light intensity within the rhabdom depends on the angle of incidence (Fig. 7.4). For an angle of incidence to the rhabdom axis of 4° or 6° the light intensity in the rhabdom decreases strongly with increase in wavelength from 400 to 500 nm, and causes a shift of the sensitivity maximum to shorter wavelengths. As the angle of incidence to the rhabdom

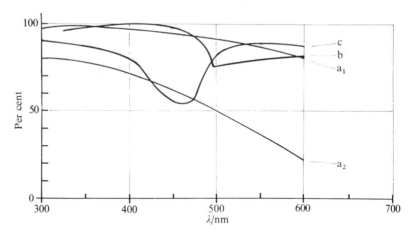

FIG. 7.4. (a) Fraction of modal light within the rhabdomeres of the fly *Calliphora* as a function of the wavelength, a_1 for the large rhabdomeres of cells 1–6, a_2 for the small rhabdomeres of cells 7 and 8. (From Snyder and Miller 1972.) (b) Fraction of modal light entering the rhabdom for off-axis light equal to $\theta_R = 4°$ (θ_R is the angle of incidence to the rhabdome). For worker bee θ_R is half the size of the corresponding angle of incidence to the corneal lens. (From Snyder and Pask 1972.) (c) Fraction of modal light within the rhabdomere 2 μm in diameter and with a 2 per cent difference of refractive index between the rod and its surrounding medium. The curve shows the effect on the wavelength for a 25 per cent change of refractive index near the absorption maximum of the photopigment (see text). Calculations are made for a rhabdomere containing a rhodopsin photopigment with a peak absorption at 500 nm. (From Snyder and Pask 1972.)

is always greater than that to the corneal lens (Snyder 1972; Varela and Wiitanen 1970; Walcott, Chapter 2) small changes in the angle of incident light are sufficient to cause a shift of the sensitivity maximum. Snyder and Pask (1972) were able to explain with these calculations the observations of Eguchi (1971) that the peak sensitivity for the green receptor of the dragonfly *Aeschna* lies at a shorter wavelength with off-axis stimulation.

(4) Parallel light is focused by a perfect dioptric apparatus as an Airy disc on the distal end of the rhabdomere (Kuiper 1966). The diameter of the Airy disc depends strongly upon the wavelength because of the diffraction properties of the lens. In flies the diameter is 1·5 μm at 300 nm and 3·0 μm at 600 nm. As the rhabdomeres 1–6 have a diameter of 2 μm different proportions of light reach the rhabdomere at different wavelengths. Therefore a reduction of the illumination of the rhabdomeres can be expected at wavelengths longer than 450 nm. This effect is counteracted by the diffraction properties of the rhabdomere which decrease with decreasing wavelength. The effect of decreasing illumination density in the Airy disc with increasing wavelength therefore has no great consequence for the light travelling in the rhabdomere (Snyder, Chapter 9).

7.1.6. *Self-absorption of photopigments*

Comparing the spectral sensitivity curves of single cells with the absorption spectrum of rhodopsin one has to consider that Dartnall's (1953) curves are calculated for infinitely thin layers. The rhabdom on the other hand is sometimes very long, e.g. up to 800 μm in dragonflies. In such rhabdoms the self-absorption of photopigment in superimposed layers must be significant. Supposing that $c. 0·8-1$ per cent of the incident light is absorbed per micron in passing through the rhabdom column, more and more light is selectively absorbed at deeper levels in the rhabdom. The spectral distribution of the light reaching these deeper levels is changed because of the absorption by the layers above. Thus a family of curves can be constructed using the following equation (Fig. 7.5).

$$S_\lambda = I_\lambda(1 - e^{-\alpha_\lambda x})$$

Where S_λ = spectral absorption of the whole rhabdom; I_λ = spectral distribution of the incident light; α_λ = spectral absorption of the photopigment in an infinitely thin layer; $x = c \times l$ where c = concentration of photopigment l = length of rhabdom.

The spectral absorption curve flattens more and more as the light absorbed in the rhabdom increases. The position of the maximum shifts a little towards longer wavelengths because of the asymmetric absorption function of rhodopsin.

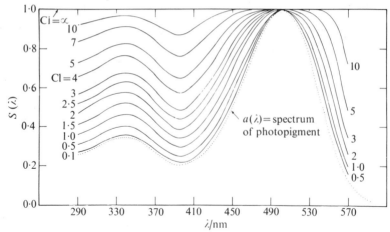

F IG. 7.5. Influence of self-absorption of photopigment on the spectral absorption properties of rhodopsin. The lowest curve (.........) gives the absorption of a rhodopsin measured in a thin layer (Dartnall curve: Dartnall 1953). Spectral absorption in thick layers (S_λ) depends not only on the spectral distribution of the incident light (I_λ) and the absorption properties of photopigment in a thin layer (α_λ) but also on the concentration (C) of the photopigment and the length of the layer (l), where $x = cl$. The total spectral absorption S_λ is therefore given by the formula:

$$S_\lambda \approx I_\lambda[1 - \exp(-\alpha_\lambda x)]$$

These curves give the change of S_λ for different $x = cl$ (by courtesy of A. Snyder).

During bleaching of rhodopsin the photoproducts formed have absorption maxima lying either towards shorter wavelengths or towards longer wavelengths (Hamdorf, Paulsen, Schwemer, and Taeuber 1972). As a result, in the deeper layers of the rhabdom the maximum of the resultant spectral sensitivity of, for example, a green photopigment is shifted to longer wavelengths compared to the absorption maximum of the photopigment. The influence of the self-absorption effect should be larger with stronger illumination because light reaches deeper parts of the rhabdom and more photoproducts are accumulated.

The self-absorption effect could be most significant in tiered retinae where the rhabdomeres of deep retinula cells are connected to those of overlying cells, as for example in cell 8 in the fly ommatidium, cell 9 in the bee ommatidium or the four proximal cells in the dragonfly ommatidium. Snyder and Pask (1973) (see also Snyder, Chapter 9) show that the lack of a u.v. maximum and the shift of the green maximum to longer wavelengths in one of the spectral sensitivity curves in the fly (Burkhardt 1962) can be explained by supposing that this curve was recorded in cell 8. Cell 7 selectively filters both u.v. and blue light because of its high absorption in these wavelengths. Therefore relatively more long wavelength light penetrates down to the rhabdomere of cell 8. Horridge (1969) inferred that the flat spectral sensitivity curves recorded in dragonflies originated from the proximal retinula cells and are caused to some extent by the self-absorption in the overlying rhabdom and in the length of the rhabdomeres themselves.

7.1.7. *Electrical coupling between retinula cells*

If neighbouring retinula cells contain different photopigments and these retinula cells are electrically coupled, the spectral sensitivity measured in one cell may reflect the sensitivities of both photopigments depending on the strength of the coupling. A strong coupling has been found between retinula cells of the same ommatidium of the drone bee with a coupling ratio of 0·53 measured between neighbouring cells and 0·21 measured between cells furthest apart (Shaw 1967, 1969b). From their low sensitivity to the plane of polarized light Horridge (1969) inferred a strong coupling between the retinula cells in dragonfly but it is more likely that this results from self absorption in long rhabdoms (Snyder, Chapter 9). The broad sensitivity curves in *Locusta* are probably not due to coupling between cells with different photopigments (Shaw 1967). It is supposed in other insects from which single cell recordings have been obtained, *Musca, Calliphora, Phaenicia, Aeschna, Periplaneta, Lethocerus, Notonecta, Apis* worker, that the retinula cells are electrically isolated but in none of these species has this been proved with double electrode recordings. The spectral sensitivity curves with two peaks, e.g. in *Apis* worker (Autrum and von Zwehl 1964) and *Aeschna* (Autrum and Kolb 1968), may arise by coupling and not necessarily by the artifactual location of the electrode tip (see later).

The fact that different colour receptors are sampled in the same ommatidium (bee: Gribakin 1969, 1972; Gribakin, Chapter 8; wood ant: Menzel 1972b; cockroach: Butler 1971) and that the same insect has colour vision shows that coupling between all the retinula cells of an ommatidium cannot be very strong. Shaw's observations are the only ones available, so it is not known if retinula cells with the same photopigment are coupled. In connection with this there is no ultrastructural evidence for exclusive coupling between cells of the same spectral sensitivity. The u.v. retinula cells in the ant *Formica* (Menzel 1972b) lie opposite each other in the ommatidium and have no common area of membrane contact, except the tips of their microvilli in the proximal part of the rhabdom.

7.1.8. *State of light adaptation*

The adaptation of the receptors to large changes in light intensity occurs at different levels, by pigment movement (see Walcott, Chapter 2), isomerization of photopigments and change in membrane resistance. There are only a small number of observations which demonstrate an influence of the state of light-adaptation on the spectral sensitivity of either single retinula cells or the whole eye.

7.1.8.1. *Pigment movement.*   In the superposition eye of moths the threshold after light adaptation is 10^5-10^6 times as high as the value in complete dark-adaptation (Bernhard, Höglund, and Ottoson 1963) but no significant change in the spectral sensitivity was found because the screening pigments act as neutral density filters (Höglund and Struwe 1970). Similarly in the apposition eye of the wasp *Paravespula germanica* no change in the spectral sensitivity was found after light adaptation (Menzel 1971).

7.1.8.2. *Photopigment bleaching.*   The equilibrium between the photopigment and its products may change during isomerization and during photoreisomerization of the products, and this may cause changes in the spectral sensitivity. Hamdorf, Gogala, and Schwemer (1971) found in the u.v. receptor of *Ascalaphus macaronius* (Neuroptera) that the sensitivity increases by 2 log units after illumination by blue light. Similarly Autrum and Kolb (1972) found an increase in the rate of dark adaptation in long wavelength receptors ($\lambda_{max} = 494$, 519 nm and $\lambda_{max} = 458$, 495 nm) in the eye of *Aeschna cyanea* with higher light intensity test-flashes than with lower.

These changes in spectral sensitivity may be of great importance under natural light conditions. Hamdorf, Gogala, and Schwemer (1971) calculated roughly for the u.v. receptor of *Ascalaphus* an increase in sensitivity when the sky was overcast and at sunset, which depends solely on the relatively more intense illumination with blue light. This may be an important general phenomenon for the regulation of sensitivity in photoreceptors in insect eyes which primarily affects the u.v. receptor because of its stronger or perhaps exclusive photoreisomerization. For example, colour training experiments with wasps (Menzel, unpublished observations) showed that two coloured papers differing only slightly in the u.v. are not distinguished by the wasps under bright illumination but are well distinguished after sunset.

7.1.9. *The experimenter and his methods*

Comparison of experimental data from different workers is often very difficult because of differences in methods, calibration of stimulating light sources, and the definition of essential terms. My studies of the experimental data of the last ten years, suggest that if spectral sensitivity measurements of single cells are to be of value, special attention must be paid to the following points.

7.1.9.1. *Stimulation.*   The stimulation should be made with a point light source (usually with an angle below $1°$) which should be centred accurately on the optical axis of the cell under investigation. The state of adaptation of the animal is very important and needs to be carefully controlled. Complications due to isomerization must be guarded against.

7.1.9.2. *Calibration.*   As photochemical processes are quantal effects the radiation flux of monochromatic light needs to be expressed in quanta $m^{-2} s^{-1}$. Relative measurements do not allow a comparison of absolute sensitivity from different experiments. The use of energy units e.g. $W\ cm^{-2}$ has the disadvantage that a conversion to the number of quanta is necessary. The spectral transmission properties of interference filters or monochromators have to be controlled very carefully, because sidebands may have a strong effect on special receptors. For example, a 0·01 per cent transmission in a u.v. sideband of a green filter can stimulate the u.v. receptors (Wehner and Toggweiler 1972) and sidebands at long wavelengths lead to inaccuracies during calibration. When the stimulus is not a monochromatic light the illumination has to be measured in lux ($= $ candela $m^{-2} sr^{-1}$). This measurement is only meaningful for insects if the light is similar to daylight because the definition of lux depends upon the spectral sensitivity of the human eye and does not take into account the u.v. sensitivity of the insect eye. Tungsten lamps give inaccurate results because they lack u.v.

7.1.9.3. *Sensitivity measurement.*   First of all each recorded cell must be calibrated by measuring the response as a function of the intensity of monochromatic light. As the absolute response heights are different from cell to cell whereas the relative response heights are similar it is generally most useful to give the relative response as a ratio of the maximal saturated response. Spectral sensitivity should be calculated from the response intensity functions measured at all test wavelengths as the number of quanta necessary to elicit a constant response in the middle part of these functions. If there is evidence for response/intensity functions being parallel at all wavelengths single responses at different wavelengths can be measured providing that their amplitudes fall in the linear region of the response–log intensity curve. The reciprocal of the number of quanta plotted against the wavelength gives the absolute sensitivity for the chosen constant response, for example the absolute sensitivity for 50 per cent response. As the exact form of the curve near its sensitivity maximum is important, spectral sensitivity must be expressed on a linear scale.

7.1.9.4. *Spectral efficiency.*   Spectral efficiency is the amplitude of the response measured with equal quanta test stimuli of different wavelengths. Spectral efficiency is only meaningful if the responses to all test wavelengths are in the same part of the intensity function. It is however almost impossible to provide an efficiency function by using test flashes which stimulate the cell within its dynamic range at all wavelengths.

7.2. The spectral sensitivity of insect colour receptors
7.2.1. *u.v. receptors*
The sensitivity to u.v. is very high in insects; often much higher than to blue or green light (Autrum 1955); Daumer 1956; Menzel 1967; Helversen 1972; Wehner and Schuemperli 1969; Wehner and Toggweiler 1972). It has

been found that the part of the eye looking upward contains more u.v. sensitive receptors (Walther and Dodt 1959, *Periplaneta;* Autrum and von Zwehl 1964, drone-bee; Bennett 1967, *Dineutes;* Gogala 1967, *Ascalaphus;* Burkhardt and de la Motte 1972, *Bibio*). In some cases the eye is divided and one part may contain only receptors which are sensitive between 300–400 nm, for example the frontal eye of *Ascalaphus macaronius* (Neuroptera) (Gogala 1967) and the principal eye of *Bibio marchi* (Diptera), (Burkhardt and de la Motte, 1972). This convenient situation was utilized by Hamdorf, Paulsen, Schwemer, and Taeuber (1972), Hamdorf, Gogala and Schwemer (1971), and Schwemer, Gogala, and Hamdorf (1971) to isolate a pure u.v. photopigment from the frontal eye of *Ascalaphus.* The u.v. photopigment is not a hydroxy-indole derivate–protein complex as supposed by Kay (1969), but a thermo-stable retinal$_1$ pigment system with an extinction maximum of the unbleached pigment at 345 nm, a half bandwidth of 85 nm and no other maxima at longer wavelengths (Fig. 7.1). The shoulders at 435 nm and 480 nm were attributed to contamination of the extract. This extinction spectrum fits very well with msp measurements of the rhabdom and the electrophysiologically measured sensitivity of the eye of *Ascalaphus.* During bleaching two products are formed (B and C) but only B can be found by msp *in vivo*, it absorbs maximally at 480 nm. On illumination with blue light B is photoreisomerized. The same u.v. photopigment system was meanwhile found by msp measure-ments of difference spectra after u.v. illumination in the rhabdom of *Deilephila elpenor* (Hamdorf, Höglund, and Langer 1972).

   The photoreisomerization of the bleached pigment can be measured directly in electrophysiological experiments. The dark adaptation rate following illumination with u.v. is increased by blue light (Hamdorf, Gogala, and Schwemer 1971, *Ascalaphus;* Nolte and Brown 1969, repolarizing effect of blue light on u.v. receptors in the median ocellus of *Limulus*) and the depolarization in response to u.v. illumination is reduced by illumination with superimposed long-wavelength light (Nolte and Brown 1972b, *Limulus;* Hamdorf, Paulsen, Schwemer, and Teuber 1972, on *Calliphora*). The analysis of the continuing depolarization after u.v. illumination and the repolarization by long-wavelength light in the u.v. receptors of *Limulus* suggests the following mechanism (Fig. 7.6). The existing u.v. pigment (VP 360) is connected to either or both efflux of Na$^+$ or to a membrane impermeable to Na$^+$ ('membrane closed'). The change to or the existence of the thermostable metarhodopsin (M 480) opens the membrane for Na$^+$ influx (i.e. depolarization, 'membrane open'). In the dark M 480 may convert back slowly to VP 360 under the influence of the cell's metabolism. Blue illumination causes a more rapid reconversion by photoreisomerization. During low-intensity u.v. illumination the concentration of M 480 is low, the cell potential returns to the resting level and blue-light illumination has no effect. At high intensities of u.v. illumination the capacity of the metabolic

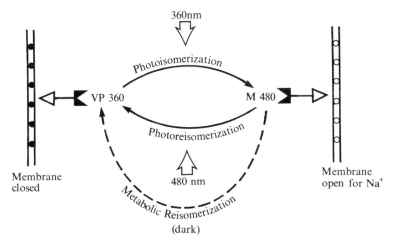

FIG. 7.6. Hypothetical relationships between the different components of a u.v. photopigment system and the generator potential in the receptor. The u.v.-photopigment VP 360 is coupled in an unknown manner to the efflux of Na$^+$ or its low intracellular concentration. At u.v. illumination (360 nm) VP 360 changes to the thermostable metarhodopsin M 480 (photo-isomerization). M 480 is coupled to the permeability increase of the membrane to Na$^+$. The reisomerization goes through retinol in the dark or by photoreisomerization on illumination with blue light (480 nm) directly back to 11 *cis*retinal. This model supposes that the generator potential of the cell depends on the equilibrium between VP 360 and M 480 and describes the phenomena of sustained depolarization after u.v. illumination, repolarization with long-wavelength illumination, and the decrease in depolarization to u.v. illumination when illumination by long wavelengths is superimposed.

conversion system is exceeded and the concentration of M 480 increases. Illumination with 480 nm light during or after u.v. illumination removes M 480 and hyperpolarizes or repolarizes the cell. In *Calliphora* the situation is more complex because the same cell may contain a long-wavelength absorbing photopigment which is probably connected in the same way as the u.v. photopigment to the mechanism of conductance change in the membrane.

u.v.-receptors with a single maximum: Spectral sensitivity curves of single retinula cells which depend on this kind of photopigment system have been measured in a number of insects. With the exception of the u.v. receptor of *Notonecta glauca* (Bruckmoser 1968) these are given in Fig. 7.7. Further examples in other arthropods are the u.v. receptor in the median ocellus of *Limulus*(Nolte and Brown 1969, 1972*a,b*) and the anteromedian eye of the wolf spider (De Voe 1972). In the drone bee besides u.v. receptors with λ_{max} at 341 nm a similar number of receptors were found with an increasing sensitivity to the shortest u.v. wavelengths measured down to 318 nm. The maximal sensitivity of the other u.v. receptors lies at 332 nm in *Ascalaphus*, 336 nm in *Aeschna* larva, 341 nm in drone bee, 350 nm in *Notonecta*, 356 nm in *Aeschna* adult, 365 nm in *Periplaneta*. By spectral adaptation in the u.v.

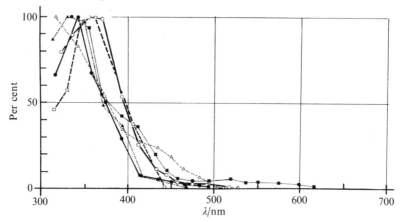

FIG. 7.7. Relative spectral sensitivity of those u.v.-receptors with sensitivity only in the u.v. region. All curves of single cell recordings. —●— drone bee, λ_{max} = 341 (Autrum and von Zwehl 1964); --△-- drone bee, λ_{max} < 318 nm (Autrum and von Zwehl 1964); ---▲--- *Ascalaphus macaronius*, lateral eye, λ_{max} = 332 nm (Gogala 1967); ——■—— *Aeschna cyanea*, larva, λ_{max} ≤ 336 nm, measurement of only one cell (Autrum and Kolb 1968); —○— *Aeschna cyanea*, adult dorsal eye part, λ_{max} = 356 nm (Eguchi 1971); --□-- *Periplaneta americana*, λ_{max} = 365 nm (Mote and Goldsmith 1970).

and measurement of the ERG, u.v. receptors have been described in *Peripla-neta* (Walther 1958a; Walther and Dodt 1959), *Dineutes ciliatus* (Bennett 1967), *Apis mellifera* (Goldsmith 1960, 1961), *Limulus* (Wald and Krainin 1963), *Colias eurytheme* (Post and Goldsmith 1969), *Paravespula germanica* (Menzel 1971) and *Formica polyctena* (Roth and Menzel 1972). Figure 7.8 shows the difference curves between spectral sensitivity in the dark and after u.v. adaptation in the wasp and ant. The difference is maximal at 374 nm in the wasp and at 361 nm in the ant and the shapes of the curves agree well with the single cell recordings and the extinction spectrum of the u.v. photo-pigment.

7.2.2. *Additional factors*

(1) *u.v. receptors with a secondary maximum.* Very often, however, spectral sensitivity curves have been measured which have in addition to a maximum in the u.v. a secondary maximum at longer wavelengths. Some examples are given in Figs. 7.8, 7.10, and 7.12. Unlike the drone no u.v. receptor was found in the worker bee which was exclusively u.v. sensitive (Autrum and von Zwehl 1964) but a maximum of equal or lower sensitivity was always found at 435, 463, or 540 nm. In *Notonecta*, cells with a u.v. maximum have secondary maxima at 414 or 414 and 567 nm, in *Calliphora* with a secondary maximum at 470 or 490 nm (the last one with an equal sensitivity to that of the u.v. maxi-mum) and in *Heliconius* with a secondary maximum at 500 nm. More

frequently, sensitivity curves were measured with a primary maximum at long wavelengths and a secondary maximum in the u.v. (see Fig. 7.10).

Autrum and von Zwehl (1964) supposed that in sensitivity curves with two or more maxima the electrode records simultaneously from two cells. It is known however that between retinula cells there is an extracellular space (Perrelet and Baumann 1969a) so that at least one cell should deteriorate after a short time because of the ion influx through the hole caused by the electrode. Cells with double peaks are however no less unstable than those with a single peak (Menzel, unpublished observations). Bruckmoser (1968) shows one set of records, the sensitivity curve of which changes from a double peak to a single peak in the u.v. This change is shown by several sensitivity curves measured at different times which decline slowly in their sensitivity to long wavelengths. This observation may be explained as a double recording in the sense of Autrum and von Zwehl but it cannot be excluded that the recording was made from two electrically-coupled cells one of which was slowly dying.

The low u.v. maximum in sensitivity curves with primary peaks at longer wavelengths has often been attributed to the β-absorption of rhodopsin (for example, Burkhardt 1962). Recent studies with very pure extracts of rhodopsin from vertebrate retinae demonstrate however that unbleached rhodopsin has no β-absorption (Heller 1968a,

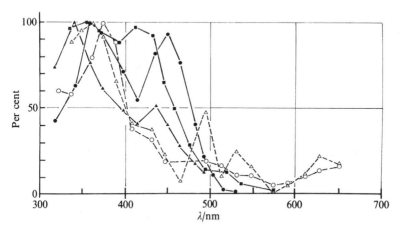

FIG. 7.8. Three curves (▲, ●, ■) show the relative spectral sensitivity of single u.v.-receptors with additional sensitivity maxima outside the u.v.-region. —■— and —●— drone bee primary maxima at 341 and 358 nm, secondary maxima at 435 or 450 nm, single cell recording by Autrum and von Zwehl (1964); —■— *Aeschna cyanea*, larva, high sensitivity between 336 and 445 nm (Autrum and Kolb 1968). The two remaining curves (△, ○) give the difference spectral sensitivity of dark and u.v. adapted eyes of the ant *Formica polyctena* (--△--) and the wasp *Paravespula germanica* (--○--) measured with the ERG (Menzel 1971; Roth and Menzel 1972).

1968*b*) and that the absorption between 310–400 nm can be explained as the absorption of a photoproduct (retinol). As in this wavelength range rhodopsin has circular dichroic absorption which disappears during bleaching (Shaw 1972) it must have some absorption in the u.v. which is probably lower than 20 per cent and without a secondary maximum (Dartnall 1972). In most cases the u.v. sensitivity is higher, exceptions being some green receptors of *Apis* and *Periplaneta* and the blue receptor of *Apis* and *Notonecta*. It is not yet possible however to decide if sensitivity curves with a double peak having a secondary maximum in the u.v. arise from two photopigments in one cell or electric coupling between a u.v. and a long-wavelength cell.

(2) *Two pigments in one cell*. There is preliminary evidence for a u.v. photopigment in addition to a green photopigment in single receptors of *Calliphora*. Langer and Thorell (1966) observed that after long-wavelength adaptation the sensitivity to u.v. increases, while after short-wavelength adaptation the sensitivity to long wavelengths increases. Furthermore, the u.v. sensitivity is higher than would be expected if only a rhodopsin with $\lambda_{max} = 500$ nm is present (Snyder and Pask 1973) even if the waveguide effects are considered. With chromatic adaptation experiments (Burkhardt 1962) no selective effect upon a u.v. and green photopigment could be distinguished, probably because the mechanism generating the potential is affected to a greater extent than the differential bleaching of the photopigments. Also Autrum and Kolb (1968) were unable selectively to adapt the u.v. secondary maximum of the green receptor in *Aeschna*. De Voe (1972) however was able to adapt selectively either the u.v. or the green peak of one of the sensitivity curves with a double peak recorded intracellularly from the retinula cells of the anteromedian eye of the wolf spider. He concluded that this was evidence for the presence of two photopigments in one cell although electric coupling between a green and a u.v. cell cannot be excluded.

(3) *Coupling*. Electric coupling between retinula cells in the same ommatidium was found by Shaw (1969*a*,*b*) in the eye of the drone bee. In the dorsal and median parts of the eye which do not contain green receptors no blue and u.v. sensitive cell has a secondary maximum in the green, while the blue cells of the whole eye and the rarely-found green receptors of the ventral part very often have, however, a high u.v. sensitivity (up to 50 per cent at 360 nm). Therefore, it may be inferred that double peaked sensitivity curves in the drone arise from coupling between receptors of different spectral sensitivity. Electric coupling, but only between u.v. cells, is described in the median ocellus of *Limulus* by Nolte and Brown (1972*a*). The scarce u.v. and green sensitive receptor

cell type (u.v.–vis cells) in the ocellus of *Limulus* are probably u.v. sensitive cells which are coupled to a green sensitive cell.

From these few results one has to anticipate that both mechanisms may coexist and provide sensitivity maxima in u.v. and the long wavelengths: the simultaneous existence of two photopigments in one cell which are both connected to the mechanism of the generator potential and the electric coupling between retinula cells with different spectral sensitivities. The advantage in both cases would be to broaden the spectral sensitivity curve and therefore to increase the absolute sensitivity.

7.2.3. *Green receptors*

Only a few receptors with a sensitivity maximum in the green to orange range (490–560 nm) have been found which have a spectral sensitivity comparable to the absorption properties of rhodopsin absorbing maximally in the green. Examples of receptors which agree well with Dartnall curves are the green receptor of *Periplaneta* (Mote and Goldsmith 1970) with $\lambda_{max} = $ 507 nm, some green receptors in the worker and drone bee (Autrum and von Zwehl 1964) with $\lambda_{max} = 527$ nm, and the green receptor of adults and larvae of the dragonflies *Aeschna* and *Anax* (Autrum and Kolb 1972; Eguchi 1971; Horridge 1969) with $\lambda_{max} = 519$ nm. In most cases, however, the maximum is broader than the absorption of retinal photopigment, moreover the position of the maximum deviates in a single cell or between different cells and also a secondary maximum in the u.v. is found which is higher than the u.v. absorption of rhodopsin (see above). Causes for the differences between spectral sensitivity and absorption properties of relevant photopigment may be as follows:

1. The deviation of the α-maximum of the same rhodopsin. For example Langer and Thorell (1966) found by msp measurements a range of deviations between 470–550 nm for the most common receptor in *Calliphora*; Autrum and Kolb (1968) measured with single cell recordings in the eye of *Aeschna* a range of 475–550 nm (mainly 530–560 nm) for the green receptor. In *Notonecta* Bruckmoser (1968) found a range of 505–567 nm (mainly 537–567 nm) for the green receptor. If such curves with a large deviation are averaged the resultant curve is broader and has a varying position of the maximum. The reason for this distribution is unknown.

2. If two photopigments with relatively close absorption maxima occur in the same receptor cell the superposition of both causes a broader sensitivity curve. The position of the peaks would be changed by a variation of the concentration ratio of the two pigments. Bennett, Tunstall, and Horridge (1967) use this explanation for the broad spectral sensitivity curves of *Locusta* retinula cells which have peaks varying between 430–515 nm (Figs. 7.10 and 7.12). The two invoked photopigments may have their λ_{max} at 430 and 515 nm

and occur mainly in the concentration ratio of 1:3 (more green sensitive), 1:1 or 2:1 (more blue sensitive). Also the broad sensitivity curve of the green receptor in *Aeschna* with λ_{max} at 536, 550, or 458 nm may be caused by two photopigments with varying concentration ratios (Figs. 7.9 and 7.10). This interpretation is supported by the double-peaked sensitivity curves (λ_{max} = 458, 519 nm) found by Eguchi (1971) in the dorsal part of the eye of *Aeschna* (Figs. 7.10 and 7.12).

Struwe (1972) measured very broad sensitivity curves with varying secondary maxima in the butterfly *Heliconius numata* (Fig. 7.10). It is not known which of the possibilities mentioned underlies the broader sensitivity, the varying relative heights of the lower maxima, however, point to the existence of more than one photopigment in the same cell.

3. The effect of self-absorption of the photopigment and its products on the sensitivity of long rhabdoms has been discussed above (see Fig. 7.5). As the concentrations of photopigment and its products change with the state of adaptation of the cell the variability of the sensitivity of green receptors may be attributed to different states of adaptation. Although all workers supposedly used dark-adapted eyes, the state of dark-adaptation probably varied in different experiments. No control experiments are yet available.

4. Another possibility could be the stray light through the screening pigment. Bruckmoser (1968) attributes the variability of the green receptor of *Notonecta* to this effect.

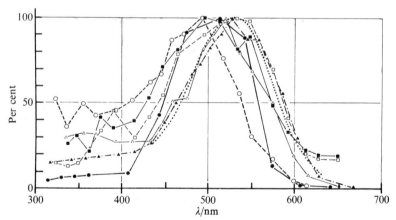

FIG. 7.9. ▲, ●, △, ○: Relative spectral sensitivity of single receptors sensitive in the green region. —▲— worker bee, λ_{max} = 530 nm (Autrum and von Zwehl 1964); —●— *Periplaneta americana*, λ_{max} = 510 nm (Mote and Goldsmith 1970); —○— *Aeschna cyanea*, adult, λ_{max} = 494 nm (Eguchi 1971); —△— *Aeschna cyanea*, larva, λ_{max} = 519 nm (Autrum and Kolb 1968). ■, □: Difference spectrum between spectral sensitivity of dark- and orange-adapted eyes of the ant *Formica polyctena* (—■—) and the wasp *Paravespula germanica* (--□--) measured with the ERG (Menzel 1971; Roth and Menzel 1972). ... Dartnall curve for λ_{max} = 530 nm.

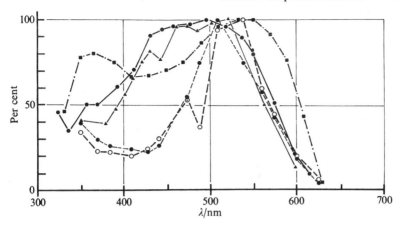

F IG. 7.10. Relative spectral sensitivity of single receptors sensitive in the green region. --O--
Libellula needhami, ventral part of the eye λ_{max} = 538 nm (Horridge 1969); --●-- *Anax junius*,
anterior ventral part of the eye λ_{max} = 500 nm (Horridge 1969); —▲— *Locusta migratoria*
average of the curves with maxima in the long wavelength, λ_{max} = 525 nm (Bennett, Tunstall,
and Horridge 1967); —●— *Aeschna cyanea*, imago, dorsal part of the eye, λ_{max} = 494 nm
(Eguchi 1971); —■— *Heliconius numata*, λ_{max} = 535 and 365 nm (Struwe 1972).

5. Off-axis stimulation causes a shift in spectral sensitivity to shorter wavelengths (Snyder and Pask 1972; Eguchi 1971) and averaging different series
of measurements made with inconsistent angle of stimulation relative to the
optic axis may also broaden the spectral sensitivity function.

Green receptors with λ_{max} at *c.* 520 nm are the most frequent and the most
widely encountered receptors in insects. It is astonishing and so far unclear
why these receptors tend to broaden their spectral sensitivity, possibly by
combination with another or several other photopigments. It may be that in
addition to their contribution to colour vision green receptors are used as a
sensitive system for general light sensitivity throughout the spectrum.

Unlike the u.v. photopigment the reisomerization of the green pigment
may depend mainly on metabolism, but photoreisomerization with blue light
(Hamdorf, Höglund, and Langer 1972) and increase in the rate of dark
adaptation by light (Autrum and Kolb 1972) have been described recently.

7.2.4. *Other colour receptors*
Besides green and u.v. receptors, blue receptors are occasionally found.
Drone bee is unique in having a predominance of blue receptors with λ_{max} at
450 nm and a shoulder at 358 nm. In the worker bee two types of blue
receptor have been described (λ_{max} at 430 and 462 nm) of which the most
frequently encountered type has λ_{max} at 430 nm. Possibly both these are of
the same receptor type and the shift in the maximum is caused by measurements with off-axis stimulation as described by Snyder and Pask (1972) and
discussed above. If so, most of the receptors would have been stimulated

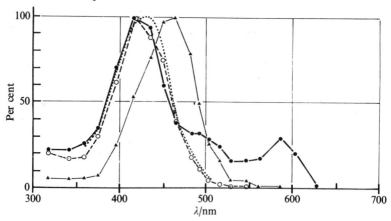

FIG. 7.11. Relative spectral sensitivity of blue receptors of the bee worker. —▲— λ_{max} = 463 nm, --○-- λ_{max} = 414 nm, —●— λ_{max} = 414 nm and secondary maximum at 586 nm; ... Dartnall curve for λ_{max} = 430 nm.

off-axis (eight cells measured) and only two cells on-axis, because this light guide effect shifts the maximum to shorter wavelengths (Fig. 7.4); this explanation therefore seems unlikely. Bruckmoser (1968) also describes two blue receptor types (λ_{max} 420, 464 nm), which fit a Dartnall curve well if the shoulder in the u.v. and the occasional secondary maximum in the green are not considered. All other blue receptors have broad sensitivity curves which cannot be explained by the absorption function of a single photopigment, for

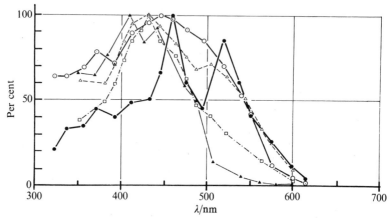

FIG. 7.12. Relative spectral sensitivity of blue receptors with broad or double peaked sensitivities. —●— and —○— *Aeschna cyanea*, dorsal part of the eye, λ_{max} = 458 nm with a secondary maximum at 519 nm (●), λ_{max} = 445 nm (○), (Eguchi 1971); —▲— *Libellula needhami*, dorsal part of the eye, λ_{max} = 410 nm (Horridge 1969); --△-- and --□-- *Locusta migratoria* average of curves with maxima in blue (λ_{max} = 430 nm) and high (△) or low (□) sensitivity in the green region (Bennett, Tunstall, and Horridge 1967).

example *Aeschna* larva λ_{max} 412–32 nm, *Aeschna* adult λ_{max} 445–75 nm (Autrum and Kolb 1968; Eguchi 1971); *Libellula* larva λ_{max} 420 nm, *Libellula* adult λ_{max} 415–30 nm (Horridge 1969); *Heliconius* λ_{max} 450–70 (Struwe 1972) and *Locusta* λ_{max} 430 nm (Bennett, Tunstall, and Horridge 1967) (Figs. 7.11 and 7.12).

In the last few years there has been some discussion of the existence of red receptors in insects. In Diptera it is inferred that no red receptor exists (Burkhardt 1962; Goldsmith 1965) but that the sensitivity peak at 620 nm measured with the ERG is caused by the transmission of screening pigments (see earlier). On the basis of ERG measurements and recordings from neurons from the photocerebrum of the butterflies *Heliconius* and *Papilio*, Swihart (1970, 1971) infer a red receptor with λ_{max} at 600 nm. Post and Goldsmith (1969) measured the spectral sensitivity of the butterfly *Colias eurytheme* up to 700 nm, but as the red sensitivity could be increased by the tapetal interference system of the tracheae behind the receptors (see earlier) it is possible that no special red photopigment is present. The only red receptor recorded intracellularly was found by Autrum and Kolb (1968) in a male larva of *Aeschna cyanea*. More recordings are necessary to decide if other insects, mainly nocturnal species, have a red sensitive photopigment.

7.3. Distribution of colour receptors within and between ommatidia

The general distribution of different colour receptors over the compound eye is known in some species but no detailed analysis is available. U.V. receptors are more frequent in the dorsal part of the eye and in some species constitute the only colour type present (*Ascalaphus macaronius*, frontal eye: Gogala 1967; *Bibio marci*, male principle eye: Burkhardt and de la Motte 1972). In adult dragonflies (*Aeschna* and *Anax*) u.v. receptors are confined exclusively to the dorsal part of the eye (Autrum and Kolb 1968; Eguchi 1971; Horridge 1969), while blue receptors are only found in the ventral region.

Green receptors are more frequent in the ventral and frontal eye, in the drone bee for example they occur only in the most ventral part. In *Notonecta glauca* it was supposed that only the posterodorsal region of the eye has colour vision but the anteroventral does not (Rokohl 1942). However, ERG measurements following selective adaptation (Bennett and Ruck 1970; Ruck 1965) showed that both eye regions contain all three colour types as described by intracellular recordings from the posterodorsal part (Bruckmoser 1968). Also in *Dineutus ciliatus* (Bennett 1967) and *Paravespula germanica* (Menzel 1971) no differences between eye regions were found. Walther (1958a) on the basis of ERG measurements supposed that u.v. receptors were only situated in the dorsal part of the eye of *Periplaneta* but this was not confirmed by the single cell recordings of Mote and Goldsmith (1971), nor by the analysis of selective pigment movement in retinula cells (Butler

1971), which show that the ratio of three u.v. to five green receptors per ommatidium is constant throughout the whole eye.

With a constant pattern across the eye, ommatidia usually contain all colour receptors in different numbers. The lack of data on intraommatidial distribution of colour types does not allow further general conclusions to be drawn at present, but all available data are presented below. The different colour receptors are not distributed in separate ommatidia but are concentrated in varying numbers in each ommatidium. Mote and Goldsmith (1971) marked u.v. and green receptors in the same ommatidium in the eye of *Periplaneta*. The retinula cell groups ('ommatidia') in the median ocellus of *Limulus* similarly contain both u.v. and green receptors (Nolte and Brown 1972a). From msp measurement on the fused rhabdom of *Deilephila* (Hamdorf, Höglund, and Langer 1972) u.v., blue, and green photopigments were demonstrated in one rhabdom. As the difference curve after long wavelength adaptation varies in different rhabdoms at c. 460 nm these authors conclude that there are a varying number of blue receptors present in one ommatidium.

Gribakin (1969) describes distortions of the microvillar structure in the rhabdomeres of ventral worker bee ommatidia after strong chromatic adaptation and osmic acid fixation. After u.v. illumination the microvilli of retinula cell type I (two cells per ommatidium) are swollen and shortened, after long wavelength illumination ($\geqslant$ 480 nm) the remaining six retinula cells (type II and III) are similarly affected. He therefore supposed that type I are the u.v. and type II and III the green receptors in the ventral bee ommatidia. In a recent paper Gribakin (1972) showed that the ventral part of the eye contains six (type II and type III) cells and the dorsal part four (type II) cells, the rhabdomeres of which are more electron dense if postfixed with osmic acid during long wavelength illumination ($\geqslant$ 480 nm). On the assumption that osmium reacts with retinal aldehyde liberated during bleaching of the green-sensitive pigment, he concludes that the method provides selective identification of green receptors. There is no evidence that the remaining cells are u.v.-sensitive (two cells in the ventral part) or u.v.- and blue-sensitive (two cells each in the dorsal part) (see Gribakin, Chapter 8).

The light flux is regulated in fused rhabdom apposition eyes by a radial movement of pigment granules in the retinula cells and by a change of the refractive index of the cytoplasm surrounding the rhabdom (Snyder and Horridge 1972 and Walcott, Chapter 2). With selective chromatic adaptation the pigment movement in particular cells reveals the predominant spectral range absorbed by that cell's rhabdomere. With this method Butler (1971) demonstrated that each ommatidium in the eye of *Periplaneta* contains three u.v. and five green sensitive cells. Two of the u.v. cells have a smaller rhabdomere and one a large rhabdomere; of the green sensitive cells three contain large rhabdomeres and two small rhabdomeres.

Independently similar experiments with the wood ant *Formica polyctena* were carried out by Menzel (1972*b*) and the pigment distribution studied with the electron microscope. Unlike the other authors (Gribakin, Butler) the radiation intensity of the spectral adaptation lights (337, 447, 591 nm) was measured and varied over 3 log units to determine a response/intensity function (Fig. 7.13). As can be seen in Fig. 7.14 each ommatidium contains eight retinula cells sectioned distally, two small cells lying opposite each other and six large cells. The small retinula cell which lies in the dorsal half of the ommatidium is numbered 1, the other cells are numbered in a clock-wise sequence (Menzel 1972*a*, 1972*b*). In the dark-adapted state, the rhabdom is surrounded by large intracellular vacuoles (palisade zone) and the pigment granules in the retinula cells lie in the outer part of the cell. After light-adaptation the intracellular vacuoles disappear and the pigment moves close to the rhabdom. As Horridge and Barnard (1965) suggested, and Snyder has since proved, the change of the refractive index of the cytoplasm surrounding the rhabdom for a distance of 0·5 μm is important for the regulation of the light flux within the rhabdom. Therefore each cell was defined as 'adapted' if it showed in any micrograph one or more pigment granules within 0·5 μm of the rhabdom. The rate of adaptation is given as the percentage of adapted cells within the sample. All measurements were carried out in the distal region 2–5 μm proximal to the end of the crystalline cone.

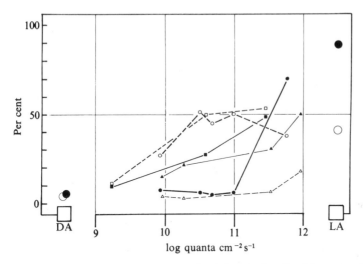

FIG. 7.13. Dependence of pigment movement in the eight retinula cells of *Formica polyctena* (Hymenoptera) on the intensity of spectral light. Ordinate: percentage of adapted cells; abscissa: light intensity in log numbers of quanta cm^{-2} s^{-1}. DA: percentage of adapted cells after 12 h dark adaptation; LA: percentage of adapted cells after exposure to white light of 40·000 L$_x$ (xenon lamp). Adaptation with u.v.-light 337 nm (●, ○), blue light 447 nm (■, □), yellow light 591 nm (▲, △). The averaged data for cells Nos. 1 and 5 are given by open symbols (○, □, △) and are connected by dashed lines, that for cells Nos. 2, 3, 4, 6, 7, 8 by closed symbols (●, ■, ▲) and solid lines (Menzel 1972 and unpublished observations).

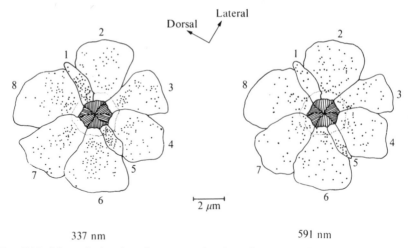

337 nm   591 nm

FIG. 7.14. Schematic drawing of a cross-section through a ommatidium of the ant *Formica polyctena* 2–5 μm proximal the end of the crystalline cone. The distribution of pigment granules in the eight retinula cells is shown after u.v.- (337 nm) and yellow- (591 nm) adaptation in the median part of the intensity function (10^{11} quanta cm^{-2} s^{-1}). Each drawing represents five cross-sections superimposed and the middle point of each pigment granule is given by a black dot. The dotted line surrounding the rhabdom indicates the distance of 0·5 μm. Each cell containing 1 or more pigment granules 0·5 μm or nearer to the rhabdom is called an adapted cell.

The percentage of adapted cells increases with the intensity of the adaptating light and reaches a maximal value of *c.* 50 per cent in the small cells and *c.* 90 per cent in the large cells (Fig. 7.13). This difference depends upon the different width of the two retinula cell types. After u.v. adaptation the intensity curve for the small cells (Nos. 1 and 5) increases at lower intensities while after yellow adaptation the intensity curve for the six large cells (Nos. 2, 3, 4, 6, 7, 8) increases at lower intensities. Blue-light illumination causes an equal increase of the intensity function for both cell types. If the average percentage of adapted cells is calculated in the middle part of the response–intensity function (10^{10}–10^{11} quanta cm^{-2} s^{-1}) and is given in a histogram (Fig. 7.15) it is clear that the small cells are the u.v. receptors and the large cells the green receptors. Blue light adapts all cells but the small cells react relatively more strongly. A statistical analysis (t-test) yields highly significant differences between small and large cells, but no differences between the large cells, after u.v. and yellow adaptation and no differences between all eight cells after blue adaptation.

In addition ERG recordings after selective adaptation showed that the eye of *Formica polyctena* contains only u.v. and green receptors and no other colour receptor type (Roth and Menzel 1972). Behavioural experiments demonstrated that these ants distinguish between u.v. and green as different colours but not between blue and green (Kiepenheuer 1968; Menzel, unpublished results). Furthermore, these experiments show that the retinula cells

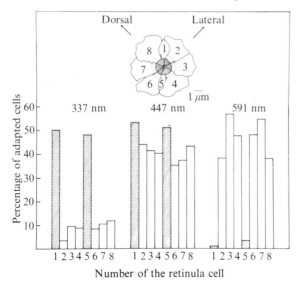

FIG. 7.15. Percentages of adapted cells after exposure to u.v. (337 nm), blue (447 nm) and yellow (591 nm) light with about the same radiation density (10^{10}–10^{11} quanta cm^{-2} s^{-1}) given separately for each of the eight retinula cells. The inset shows the numbers of retinula cells. Cell Nos. 1 and 5 are marked by hatched bars.

are independent of each other, at least the large and small cells, and that the pigment movement is regulated by the photopigment probably through the generator potential (Kirschfeld and Franceschini 1969; Menzel 1972b).

Recent electrophysiological and structural studies show clearly that the number of different colour receptors differs across the eye and that these receptors are contained within single ommatidia in different combinations. The significance of the distribution of the colour receptors is not yet substantiated in adequate detail for the visual capacity of any insect. It is uncertain if the ratio of the colour receptors is for example independent of the season, the age or the physiological state of the animal. In *Procambarus* (Crustacea) Nosaki (1969) found only green receptors in the winter but blue and orange receptors in the summer.

7.4. Behavioural experiments in colour vision

Whether or not insects use the different spectral sensitivity of their receptors for colour vision can only be decided by studying the behaviour of the intact animals. In these experiments the animal must be demonstrated to be able to distinguish colours seen either successively or simultaneously solely on the basis of the wavelength differences of the test colours and not by the different brightnesses of the colours. In the following account colour vision is defined as the ability of an animal to distinguish between wavelengths independent of their intensity. On this definition colour specific responses, e.g. the different steepness of response–intensity functions at different wave-length (Schuemperli 1972) or the different action spectra to various patterns in an optomotor experiment (Eckert 1971) do not necessarily demonstrate colour vision.

The honey bee *Apis mellifera* is still the insect for which colour vision is best documented (Daumer 1956; von Frisch 1914; von Helversen 1972; Kühn 1927; Menzel 1967; Thomas and Autrum 1965). In other species colour vision has been shown more qualitatively (for older literature: see Autrum 1961; *Formica polyctena:* Kiepenheuer 1968, Menzel, unpublished results; *Paravespula germanica:* Beier and Menzel 1972, *Papilio troilus,* Lepidoptera: Swihart 1970).

The results from colour-mixing experiments with the honey bee are most economically explained on the basis of a trichromatic colour vision system (Daumer 1956). In a very careful analysis von Helversen (1972) measured by training experiments the spectral sensitivity (the spectral threshold function) and the spectral discrimination function ($\Delta\lambda$ function) in the same experimental situation for both. The spectral threshold function shows three maxima at 345, 440, and 550 nm in which the maxima have the proportions of 16:2·7:1 (Fig. 7.16). Comparison with earlier experiments on spectral phototaxis in bees, training experiments (Bertholf 1931; Daumer 1956;

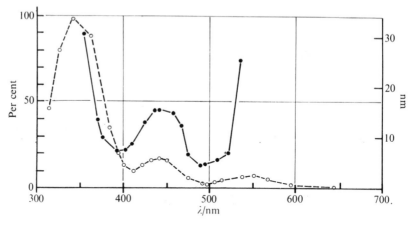

FIG. 7.16. ---O--- Relative spectral sensitivity of the honey bee measured as the spectral threshold function in a training experiment (v. Helversen 1972). The bee was trained to an unilluminated ground glass disc and during the test the illumination with the test wavelength of one of two alternative ground glass discs was decreased up to the point where the test animal failed to distinguish between the illuminated and unilluminated ground glass discs. The reciprocal of the number of quanta at threshold gives the spectral sensitivity (left ordinate normalized to the maximum). —●— Spectral discrimination function $\Delta\lambda$ for a reliability of correct response of 70 per cent (right ordinate). The test animal was trained to a wavelength given on the abscissa. In the test situation the wavelength of the alternative colour was approached to the learned wavelength until the correct discrimination dropped to 70 per cent. The discrimination sensitivity is maximum in violet ($\Delta\lambda = 8$ nm) and bluishgreen ($\Delta\lambda = 5$ nm).

Menzel 1967; Sander 1933) and with electroretinogram measurements (Goldsmith 1960) shows that the relative heights of the maxima depend greatly on the experimental methods used. This indicates that during information transfer within the visual system a selective amplification of distinct parts of the spectrum can occur. The wavelengths of these maxima in von Helversen's experiment are in good agreement with the sensitivities of the different colour receptors (Autrum and von Zwehl 1964) (Figs. 7.7, 7.8, 7.9, and 7.11), except that electrophysiologically two blue receptor types have been found and that single cells often have two maxima. The two blue receptor types were not found in the same eye and may also lie in different parts of the eye so that in behavioural experiments only one receptor type in the anteroventral part of the eye is used.

As may be expected from the position of the maxima of the three colour types, the discrimination sensitivity is highest in the violet, around 400 nm, and in the bluish green, around 490 nm (Fig. 7.16). For discrimination with a reliability of 70 per cent correct responses for violet the test wavelengths need to be separated by 8 nm and for bluish-green by 5 nm. Colour discrimination starts at 350 nm and ends at 540 nm.

These behavioural studies prove that bees have a trichromatic colour system as supposed by Daumer in which the receptors are maximally sensitive in the u.v. blue and green. There is close agreement with the single cell recordings of Autrum and von Zwehl only if the sensitivity curves found frequently with double peaks are artefactual (see earlier). If the retinula cells often contain more than one photopigment, or if receptors of different colour type are coupled, then the possibility of inhibitory connections between the receptors or their interneurons has to be entertained to explain the behavioural experiments. The three primary colours in the trichromatic system would then correspond more to neuronal integration steps and less to receptors. As a result of these studies the understanding of the bee colour vision is as good as that of the best understood vertebrate systems (e.g. De Valois 1965) and appears to be more promising for further electrophysiological analysis.

Action spectrum and discrimination behaviour has also been measured in the desert ant *Cataglyphis bicolor* but only for a small number of test wavelengths (Wehner and Toggweiler 1972). The three receptor types may have their maxima at 350, 500–520, and 600 nm. U.V. light is, for the dark-adapted ants, *c.* 80 times more effective than bluish-green or orange light. The wavelength discrimination is best at *c.* 380 nm ($\Delta\lambda = 22$ nm) and at 550 ($\Delta\lambda = 17$ nm).

Whether or not flies have colour vision is as yet unknown. The action spectrum is described from behavioural studies by phototaxis (*Drosophila*: Bertholf 1932; Wehner and Schuemperli 1969) and optomotor studies (Eckert 1971; Kaiser 1968; McCann and Arnett 1972). In optomotor analysis flies respond only to intensity contrast and not to spectral contrast (Kaiser 1968). This result seems to exclude colour vision but these experiments were carried out at relatively low illumination intensities. Eckert (1971) showed, however, that *Musca* has a receptor system with high sensitivity (D-system) and one with low sensitivity (L-system) which can be distinguished by their spectral sensitivity and acuity. The D-system has a high u.v. (360 nm) and low green (486 nm) sensitivity (Fig. 7.17), while the L-system has only one maximum in the blue (464 nm). As the D-system can be stimulated by broad stripes (10·6° stripe width) and the L-system by narrow stripes (3° stripe width) to elicit an optomotor response it is supposed that the D-system is identical with the receptors 1–6 and the L-system with the receptors 7 and 8. This interpretation is not consistent with the single cell recording of Burkhardt (1962), the recording and marking experiments of McCann and Arnett (1972) and the theoretical analysis of Snyder and Pask (1973). The receptor type most frequently recorded by Burkhardt was without doubt retinula cells 1–6 and has an equal sensitivity in u.v. and green (*cf.* McCann and Arnett 1972). A receptor was rarely found with a u.v. maximum 2·5 times higher than the maximum at 450 nm (probably retinula cell 7). It may therefore be assumed that the systems D and L measured by Eckert do not represent directly the two receptor systems but the neuronal interactions between them. Basically it cannot be concluded from different action spectra in the D and L systems that flies have colour vision because

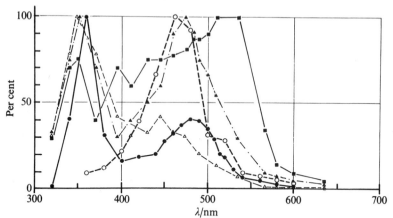

FIG. 7.17. Electrophysiological and behavioural measurements of the relative spectral sensitivity of receptors in the eye of the fly. ▲, △, ■: relative spectral sensitivity of three groups of receptors measured by single cell recordings (Burkhardt 1962). Most frequently found was the receptor type with equal sensitivity in u.v. and green (—▲—), rarely the receptors with a high u.v.-sensitivity (--△--) or a higher green sensitivity (—■—). ●, ○: relative spectral sensitivity measured with the optomotor response to moving stripe pattern of 10·6° stripe width and low illumination intensity (—●—) or to 3° stripe width and high illumination intensity (--○--). (After Eckert 1971.) The curve with the high peak in the u.v. is thought to represent receptors Nos. 1–6, the one without u.v. sensitivity receptors Nos. 7 and 8 (but see text).

there is no indication that the two systems are compared with each other to make possible the simultaneous analysis of colour contrast.

ACKNOWLEDGEMENTS

I wish to thank my wife for drawing the figures, Dr. I. Meinertzhagen for his help with the translation, and Dr. A. Snyder and Dr. S. Laughlin for many discussions and the reading of the manuscript, and Prof. Horridge for technical and laboratory facilities as a visitor in Canberra.

8. Functional morphology of the compound eye of the bee

F. G. GRIBAKIN

8.1. Introduction

I N the following account, I have three main objectives in view. First, some results of morphological methods, which have been worked out in my own laboratory with reference to the honey bee compound eye, can be extended to other insects. In particular, some details of technique are discussed. Secondly, a number of electron microscopic studies of bee compound eye have been published recently (Varela and Porter 1969; Perrelet 1970; Skrzipek and Skrzipek 1971; Kolb and Autrum 1972; Gribakin 1972). This new data from different laboratories with a similar objective make it possible to refine general ideas on the structure and function of the eye not only in bees but also in other insects. Thirdly, my intention is to attract attention to some problems which have usually remained in the background but which probably can be solved within this decade.

The honey bee was the first insect in which vision was studied physiologically in detail. von Frisch (1914) found that bees are capable of colour discrimination, and in 1949, he demonstrated the orientation of honey bees to the plane of polarization of light. Colour vision in bees proved to be by a trichromatic system, as demonstrated by behavioural observation (Daumer 1956; Maznkhin-Porshnyakov 1956, 1959). Colorimetric experiments showed the same to be true for the bumble-bee *Bombus* (Mazokhin-Porshnyakov 1962). At the same time, study of the honey bee electroretinogram, with selective adaptation to different coloured lights, suggested a role for retinol (vitamin A) in the visual process (Goldsmith 1958*a,b*, 1960; Goldsmith and Warner 1962, 1964). Finally, in 1964 the spectral sensitivity curves of single cells of the honey bee eye with maxima at 340, 420, and 530 nm were obtained by intracellular recording (Autrum and von Zwehl 1964). Thus, in the early sixties something was known of the physiology and chemistry of honey bee vision, whereas only one light-microscopic study (Phillips 1905) provided a relatively poor picture of the structure of the eye. The electron microscopic study, which had become an urgent matter, was carried out by Goldsmith

(1962). Nevertheless, the morphological basis of specific sensitivity of the honey bee to different colours as well as to linearly polarized light remained to be cleared up. With this as the aim, an examination of the honey bee compound eye by electron microscopy was initiated in 1965 at the Laboratory of Evolutionary Morphology of the Sechenov Institute in Leningrad under the guidance of Professor Ya. A. Vinnikov.

8.2. Formulation of the problem

Irrespective of the type of compound eye, it seems most reasonable to take the dark-adapted state as the starting point of an electron microscopic study. Almost no attention was paid to this matter in early descriptions of the compound eye. At about the time that we started work it was found that light produces alterations in the fine structure of visual cells and that these changes can be used as evidence of receptor cell activity, e.g. movement of 'palisade' in locust (Horridge and Bardnard 1965), swelling of the microvilli of the rhabdomeres in *Daphnia* (Röhlich and Török 1965). Recently an increase in number of some organelles in crab (Eguchi and Waterman 1967, 1968), movement of retinula cell screening pigment in cockroach and bee (Butler 1971; Kolb and Autrum 1972) and in the fly (Boschek 1971) has been reported. Thus, it became apparent that an electron microscopic study of photoreceptor cells has to begin with a description of the cell in the dark-adapted state. Then, by illuminating with light of controlled intensity, spectral composition, polarization plane, etc., one could hope to detect morphological or electroncytochemical changes, which can be referred back to the process of light reception. In doing this, the results may be dependent on the experimental methods, which therefore must be thoroughly formulated beforehand on the basis of repeated physiological tests on the eye under study.

8.3. Cytology of retinula cells in the dark-adapted state

A conventional description of the ultrastructure in the dark (see Gribakin 1969) is the first step towards a better understanding of probable cellular mechanisms of vision; from this, one can select those aspects which are worthy of further study. Besides, we found by comparing the visual cells of one ommatidium, that they may be divided into groups with hypothetically different physiological properties. Details of fine structure of interest are given below, with special reference to the honey bee eye, and an attempt towards a functional interpretation of the fine structure is made. A comprehensive description of the bee ommatidium can now be found in the literature (Phillips 1905; Goldsmith 1962; Gribakin 1967a,b and 1969a,b; Varela and Porter 1969; Perrelet 1970; Skrzipek and Skrzipek 1971; Kolb and Autrum 1972; Gribakin 1972).

8.3.1. *The distribution of mitochondria in the retinula cells*

When studying the distribution of mitochondria in a visual cell, some principles of their localization in other tissues should be borne in mind (Lehninger 1964):

(1) In epithelial cells mitochondria are often arranged in relation to the active transport or secretion processes, which are characteristic of these cells.

(2) Mitochondria are often arranged near the sources of the substances used in respiration.

(3) Mitochondria can usually be found in regions where a demand may arise for ATP, which they produce.

8.3.1.1. *Axial distribution of mitochondria.*   The distal part of the cell can be compared with the ellipsoid of vertebrate photoreceptors in that mitochondria are aggregated at the end of the photoreceptor structure (rhabdom or outer segment), which faces the source of light. The distribution is presumably related to the absorption law. Assuming that the absorption by visual pigment in the honey bee rhabdom is 1·3 per cent per μm of rhabdom length for non-polarized light, as in the crab *Libinia* (Hays and Goldsmith 1969), the distal part of the rhabdom, from the cone tip to the level of the nuclei *ca* 100 μm away, apparently absorbs 70 per cent of the incident light. From these observations we can infer that the distal part of the cell is the most active metabolically, which is in line with data on the ellipsoid (Tsirulis 1966). Along the retinula cell from distal to proximal the number of mitochondria progressively decreases and near the basal membrane only occasional mitochondria occur within the axons. The gradient in the density of mitochondria along the length of the retinula cells, characteristic of many insects (Hymenoptera, Orthoptera, Odonata, Diptera, etc.) possibly indicates an exponential 'gradient of action current' along the length of the cell according to absorption of light along the length of the rhabdom. The smaller number of mitochondria in the axon may serve as indirect evidence for a higher resistance of the axon membrane compared with that of the cell soma (*cf.* Shaw 1972), since the greater the membrane resistance the less ionic transport is needed in an axon.

8.3.1.2. *Cross-sectional distribution of mitochondria.*   There are two types of distribution of mitochondria in cross-sections of ommatidia having fused rhabdomeres. In the retinula cells of Odonata, Blattoidea and Orthoptera, the mitochondria are disposed near the centre of the cross-section around the 'palisade' zone (Fig. 8.1a,b). They are capable of moving towards the rhabdom under the influence of light, as in the locust (Horridge and Barnard 1965). It is possible that this movement serves other functions beside an optical one, for mitochondria are presumably involved in metabolic processes such

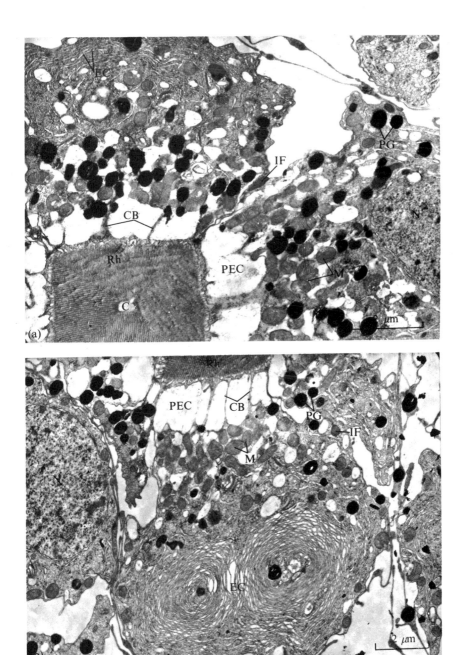

FIG. 8.1(a) Transverse section through the distal part of an ommatidium of the cricket *Gryllulus domesticus* (dark-adapted). The cone apex (C) lies in the middle of the square rhabdom (Rh), which is surrounded by palisade, or principal endoplasmic cysternae (PEC). Other organelles are: mitochondria (M), endoplasmic channels (EC), pigment granules (PG), nuclei (N), extensions of cone cells (IF), and cytoplasmic bridges (CB) penetrating the PEC. Only four visual cells form the distal layer of the cricket ommatidium and the other four contribute to the triangular proximal rhabdom. (b) Detail of retinula cell cytoplasm showing whorled ER channels.

as the restoration of an ionic gradient, the re-isomerization of retinal and resynthesis of visual pigment in the rhabdom. Within the axons in these insects mitochondria are tightly pressed against the axon membrane (see, for example, Fig. 9 and 10 in Ninomiya *et al.* (1971)), as found also by us in several insects.

A second type of transverse distribution of mitochondria has been found in honey bee worker and drone. Here the mitochondria are in close contact with the outer membrane of the retinula cell, as in some other receptors (Vinnikov 1969) and as in many axons. In the honey bee retinula cell the mitochondria do not migrate in the light. The pattern of mitochondria against the outer membrane suggests active transport and transfer of metabolites between the sheet-like processes of the secondary pigment cells and the cytoplasm of the retinula cell.

The full physiological significance of the two types of distribution of mitochondria cannot be derived from morphological observations alone. The transverse arrangement of mitochondria in the adaptation of other compound eyes, including those with open rhabdomeres, also deserves to be examined.

8.3.2. *Endoplasmic reticulum in the retinula cell*

Channels of the endoplasmic reticulum spread through all the retinula cell cytoplasm, but their most interesting formation is the 'palisade' (Fig. 8.1), as first described by Horridge and Barnard (1965) in the locust. In fact the endoplasmic cysternae known as 'palisade' in all the arthropods studied is simply a giant endoplasmic cysterna, which extends along the full length of the rhabdomere and may occupy all the perirhabdomeric space. For this reason this formation is better referred to as the principal endoplasmic cystern (PEC) (Gribakin 1967a).

In honey bee retinula cells the PEC is annular in shape. The inner surface outlines the rhabdom with a thin layer of cytoplasm, which is connected to the rest of the cytoplasm by thin cytoplasmic bridges 0·1–0·2 μm in diameter and 1–2 μm in length penetrating the PEC. These bridges have a remarkable regularity along the axis of the retinula cell. With a spacing of 0·5–0·6 μm each bridge caters for ten layers of rhabdomeric microvilli (Gribakin 1967a). An extremely thin layer of cytoplasm (a total of 20 nm), the cell membrane and the endoplasmic membrane of the PEC, separate the PEC from the extracellular space (Fig. 8.2a,b). Sometimes continuity can be traced between the PEC and other endoplasmic channels (Gribakin 1967a; Varela and Porter 1969) and the endoplasmic membrane may be involved closely with mitochondria.

The composition of the substance filling the PEC and the channels is as yet unknown. From the refractive index of 1·339 compared to 1·347 for the rhabdom, and 1·343 for the cytoplasm (Varela and Wiitanen 1969) one can infer that the PEC contents are dilute. The double membrane system of the PEC may have a function related to extracellular ions.

Light-induced permeability changes of the rhabdomere may allow ions to enter microvilli. When diffusing along the cytoplasmic bridges these ions may penetrate the PEC, since the side surface area of the bridges is 15–30 times greater than their cross-section. At the same time, the depolarizing ionic current must decline exponentially along the length of the cell as a consequence of the light absorption law. Therefore an intracellular potential gradient arises and current flows in the PEC parallel to the rhabdom axis. If the resistance of the PEC is smaller than that of the cytoplasm, the voltage drop would be brought closer to the axon. Such a consideration is in line with the data on electrotonic spread along the retinula axons of arthropods (Shaw 1972).

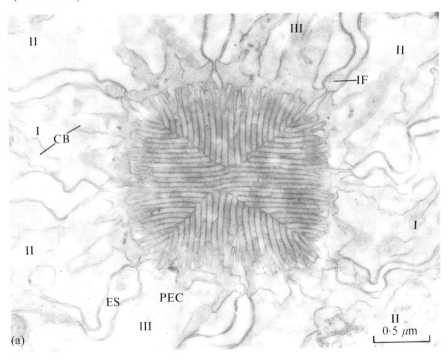

FIG. 8.2. Cross-sections of the bee rhabdom in the ventral part of the eye in (a) distal and (b) proximal regions (dark-adapted). The rhabdomeres of the type I cells have a common border at the centre of the rhabdom (cf. Fig. 8.5). There are eight visual cells in (a) but only seven in (b) since two type III cells do not contribute to the rhabdom at this level. Instead, the ninth cell replaces one type III cell. Pigment granules occur in the expansions of the cone cell extensions. At the bottom of (b) is the sheet-like process of a secondary pigment cell (SLP). A double membrane separates the PEC from extracellular space (ES). The directions of diffusion or other transfer phenomena between SLP and cytoplasm of the visual cell, and between ES and PEC, are indicated by arrows. Types of retinula cell are marked by Roman numerals. Other organelles labelled as in Fig. 8.1.

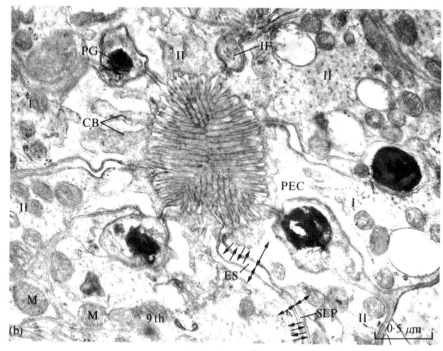

F IG. 8.2. (*Cont.*)

In some insects the cross-sectional area of the PEC decreases in the light (Horridge and Barnard 1965). This phenomenon is one intracellular mechanism of light adaptation. Transmission of light along the rhabdom progressively deteriorates as the mitochondria approach and modify the refractive index of the surrounding media (Horridge and Barnard 1965). The decrease in the PEC may cause an increase in the longitudinal resistance of the cell which, in turn, could reduce the depolarizing current in the axon. Also the PEC is replaced by mitochondria moving towards the rhabdom, so that ions entering the cell in the light are possibly removed more easily.

Eyes with mitochondria grouped around the PEC (locust, dragonfly, cricket) have numerous regularly-arranged endoplasmic channels (Fig. 8.1b, see also Horridge and Barnard (1965); Ninomiya *et al.* (1969)). Channels may be parallel as in damsel-fly; spiral as in cricket; or form concentric rings called 'onion bodies' in locust. There are no mitochondria within these endoplasmic systems. So far no continuity has been found between endoplasmic channels and extracellular space.

8.3.3. *Screening pigment granules; interretinular fibres*

Granules of screening pigment occur in primary and secondary pigment cells, basal pigment cells (if any) and retinula cells, of which the last two groups are of the most interest. In the honey bee the retinula cell granules are distinguished from the rest by their range in size, from 0·03 to 1·0 μm in diameter (Gribakin 1967a). This range in size is unlike that found in primary (2.33 ± 1·11 μm) and secondary pigment cells (0·42 ± 0·14 μm) (Varela and Wiitanen 1969) and may indicate a continuous genesis of granules. Unfortunately, the nature and fate of these granules is not known. As Phillips (1905) has found, in the honey bee the pigment granules first occur embryologically in the retinula cells, then in primary pigment cells, and finally in secondary pigment cells, in distal and proximal parts simultaneously. No basal pigment cells or their nuclei have been found in the honey bee compound eye. As shown by Gribakin (1967a), the pigmented regions of the proximal part of the honey bee ommatidium in fact consist of the pigment-filled expansions of the four processes of the cone (Fig. 8.2a,b). These processes, first described by Goldsmith (1962), were traced to cone cells in Diptera (Waddington and Perry 1963; Brammer 1970) and bee (see Fig. 5b in Skrzipek and Skrzipek 1971). Similar extensions of the cone occur in many insects (Horridge 1966; Perrelet 1970; Ninomiya *et al.* 1971; Eguchi 1971).

The pigmented basal expansions of the cone cell processes form an absorbing layer at the end of the rhabdom (Fig. 8.5) possibly to protect the nerve centres from ultraviolet radiation. The nature and role of the thin fibrils extending along the cone cell extensions is unknown, but where the extensions widen out one can see pigment granules of irregular form and lower density (Fig. 8.2b, also Fig. 11a in Skrzipek and Skrzipek 1971). Possibly the cone cell extensions are capable of transporting compounds to synthese pigment granules in the basal expansions, or possibly pigment granules are metabolized when 'worn out'. Therefore we must admit that cone cells can produce pigment and the basal pigment cells in the honey bee eye can be fully explained in this way.

The form of the cone cell extensions suggests a role for them in the optics of the ommatidium. Light not absorbed by the rhabdom and consequently reaching the basal pigment grains, could possibly excite the cone cell extensions in some way and thereby influence the optics of the cone by small changes in refractive index (which is 1·348 according to Varela and Wiitanen (1969)). The refractive index of primary pigment cells gradually changes from 1·311 distally to 1·351 near the cone apex (Varela and Wiitanen 1969). Light-induced transport of material along the cone cell extensions might cause small changes in refractive index of the cone tip, and in turn cause a sharp change in the intensity of light entering the rhabdom. Most of the primary pigment is concentrated just around the cone apex. In other words, the cone cells and their extensions, together with the primary pigment cells could act as a pupil. It seems worthwhile to keep a sharp look-out for a function of these structures in adaptation to light but so far no changes in fine structure have been observed.

8.3.4. *Relations with secondary pigment cells*

The secondary pigment cells of the bee form thin sheet-like processes separating the retinula cells (Fig. 8.2b). This separation of the retinula cells may be interpreted as an electrical isolation, or the enlarged area of contact could facilitate metabolic exchange (Gribakin 1967a). Whether such assumptions are valid and extend to other insects remains to be seen.

8.3.5. *The rhabdomeres and the rhabdom*

Visual pigments of insects have retinal as prosthetic group (Goldsmith 1958; Briggs 1961; Goldsmith and Warner 1962, 1964; Goldsmith 1970). Sometimes the microvilli have a dark axial fibre seen in transverse sections as a central dark spot, but possibly this fibre is an artefact of fixation and not related to visual pigment distribution. Microvilli are 50–60 nm in diameter, and all arthropods studied have an identical three-layered structure of their photoreceptor membrane (Fig. 8.3). The membrane of the microvilli is that of the photoreceptor cell, but is not uniform along its length. At the base of the microvilli the membrane looks the same as the cell plasma membrane with the inner osmiophylic layer thicker than the external (Fig. 8.3d). However, where the microvilli are in contact the dark membrane layers have the same thickness. We can give no interpretation of this fact since little is known of the mechanisms of osmic fixation and uranyl acetate staining. Sometimes osmiophylic compounds, possibly mucopolysaccharides, can be seen between microvilli (Fig. 8.3b,c). Thus, the rhabdom is not so simple as is usually considered. It may be interesting to search for distinguishing features of cells that are sensitive to ultraviolet light, which are known to occur in the same ommatidium as other cells in the bee (Gribakin 1969, 1972), cockroach (Mote and Goldsmith 1971; Butler 1971) and presumably other insects. The problem is to find a suitable histological treatment.

8.3.6. *Summary*

This brief and incomplete survey reveals many unsolved problems in cytology of insect retinula cells. It should be remembered that the electron micrograph is only a snap of a living cell and that complicated processes are going on in the structures studied. Extension of our knowledge in this field strongly depends on the development and use of new techniques in electron microscopy.

8.4. Types of visual cells

As is well known, the retinula cells of one ommatidium can differ sharply from each other in position (two or three tiered retinulas), length, shape and cross-sectional area of rhabdomeres (see for example Goldsmith 1964; Mazokhin-Porshnyakov 1965; Horridge *et al.* 1969, for *Dytiscus*; Schneider and Langer 1969 for *Gerris*; Perrelet 1970 for drone bee; Walcott 1971 for

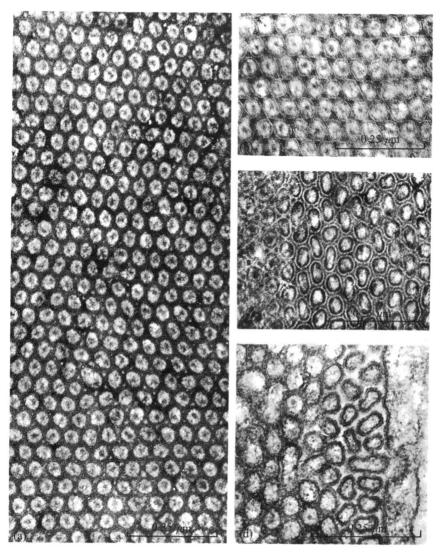

FIG. 8.3. Sections of microvilli, (a) cockroach *Periplaneta americana* (glutaraldehyde and osmium tetroxide with uranyl acetate stain) (b) as in (a) but a scattered dark substance lies between the microvilli. (c) honey bee *Apis mellifera* (mercuric chloride fixation). The pattern is different on the left owing to more dense packing of microvilli towards the centre of the rhabdom. (d) as in (c) but with conventional fixation in glutaraldehyde, osmic post-fixation and uranyl acetate staining. Note that the inner osmiophylic layer of the trilaminar membrane is the same as that of the cell membrane at the base of the microvilli, but different in the microvilli themselves.

Lethocerus; Eguchi 1971 for *Aeschna*). Such structural differences must presumably correlate with as yet unknown physiological features and *vice versa*. In the honey bee, three kinds of spectral sensitivity curves have been obtained by intracellular recording (Autrum and von Zwehl 1964) but previous to that, no morphological differences between the cells had been described (Phillips 1905, Goldsmith 1962). Subsequently differences have been found (Gribakin 1967*b*) but they are not very distinct and so were unfortunately overlooked in later studies (Varela and Porter 1969; Skrzipek and Skrzipek 1971; Kolb and Autrum 1972). The specific features of the three types are given below (Gribakin 1967*b*) as well as the history and characteristics of the ninth cell (Gribakin 1972).

Type I cells (two per ommatidium, Figs. 8.2, 8.4, and 8.5)

(1) In the ventral part of the eye the type I cells have rhabdomeres with a common border at the centre of the rhabdom first described by Goldsmith (1962) as hour-glass rhabdomeres. They occupy up to 40

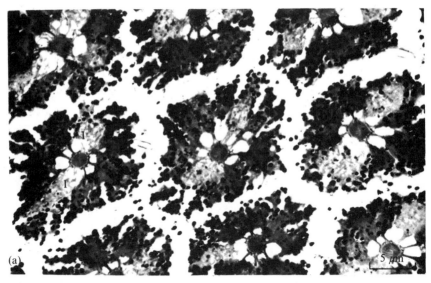

FIG. 8.4. (a) Thick transverse section through a distal level of ommatidia of the ventral region of the bee eye to demonstrate the difference between the type I cells and the others. The dark-adapted eye was fixed with glutaraldehyde in the dark, then rinsed in phosphate buffer and post-fixed in OsO_4 while illuminated ($\lambda > 480$ nm). The section was not stained with lead citrate or uranyl acetate. The absence of selective osmic staining in the rhabdomeres of type I cells is seen, but not so well as in (b). (b) a normal thin section of the rhabdom (uranyl acetate, lead citrate). (c) a thin section of the ventral part of an eye of a living bee adapted to strong light ($\lambda > 480$ nm) for 15 min. Then the animal was cooled gradually to 4 °C under continuous illumination (15 min more). This was followed by dissection and osmic fixation at 4 °C in the dark and embedding in methacrylate. The procedure resulted in swelling of microvilli of the type II and III cells, while type I cells remained intact. Similar results with u.v. radiation (Gribakin 1969*a,b*) showed type I cells to be receptors-340.

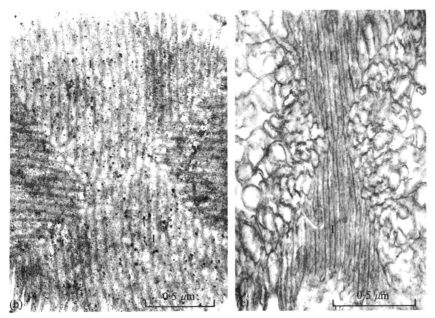

FIG. 8.4. (*Cont.*)

per cent of the cross-sectional area of the rhabdom (Fig. 8.6). In the dorsal part of the eye they have no common border and are smaller in cross-section.

(2) The distal region of a type I cell contains fewer mitochondria and pigment granules than the distal parts of the other cells (Fig. 8.4a; also Fig. 1 of Varela and Porter 1969). By this feature the type I cells are distinguishable by conventional light microscopy (Fig. 8a in Skrzipek and Skrzipek 1971).

(3) The nuclei of the type I cells lie at one level, forming a distal nuclear layer in the ommatidium, as first noticed by Phillips (1905).

(4) Axons of the type I cells have the largest areas in cross-section (Fig. 8.11a in Skrzipek and Skrzipek 1971).

Type II cells (four cells per ommatidium alternate with type I and III cells). The nuclei lie almost at one level, forming the middle nuclear layer in the ommatidium, and the rhabdomeres are of the same length as those of the type I cells.

Type III cells (two cells per ommatidium)

(1) The rhabdomeres of the Type III cells extend only as far as the distal part of the ninth cell and are 150–200 μm long, compared with those of type I and II cells which are 250–300 μm long (the length of the ommatidium).

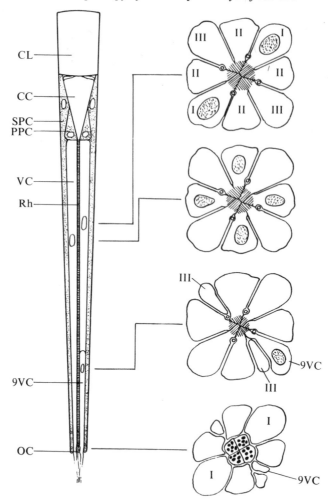

FIG. 8.5. The worker bee ommatidium slightly modified on the left from Phillips (1905), compared to present knowledge shown on the right. CL corneal lens; CC crystalline cone; SPC secondary pigment cell; PPC primary pigment cell; VC retinula cell; Rh rhabdom; 9VC the ninth visual cell; OC the optical stopper formed by pigmented extensions of the four cone cells. The cell types (Gribakin 1967b, 1972) are given by Roman numerals. Note the different cross-sectional areas at the basement membrane level.

(2) Nuclei of the type III cells often lie proximal to those of the type II cells, but a distinction between proximal and middle nuclear layers is less than that between middle and distal layers.

(3) The type III cells lie on the diameter at right angles to that on which the type I cells lie.

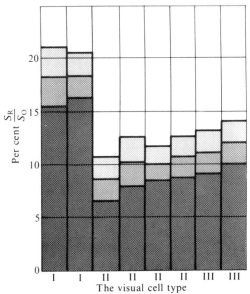

FIG. 8.6. Cross-sectional areas of rhabdomeres (S_R) relative to the cross-section area of the whole rhabdom (S_0) for the ventral part of the worker bee eye (Gribakin 1967*b*), based on measurements of eighteen ventral ommatidia of different eyes. Rhabdomeres are largest in type I cells.

The ninth (basal, proximal, or eccentric) cell

All authors dealing with the honey bee eye at one time agreed that the ommatidium of the worker contains eight retinula cells (Phillips 1905; Goldsmith 1962; Gribakin 1967*a,b*), except that nine were occasionally found (Phillips 1905). Although Varela and Porter (1969) indicated one of the retinula cells to be eccentric (mentioned also by Skrzipek and Skrzipek 1971) they did not distinguish this cell as additional to the other eight. Meanwhile, Phillips had emphasized that one retinula cell nucleus is always situated midway between the other nuclei and the basal membrane. Phillips proposed this cell to be the ninth, as subsequently confirmed (Gribakin 1972). Its specific features are as follows:

(1) The ninth cell is located in the proximal part of the ommatidium with its rhabdomere and cell body in the location of a type III cell.
(2) The ninth cell is the shortest of the ommatidium (*ca.* 100 μm or less) with the shortest rhabdomere.
(3) There are no pigment granules in this cell, so it is the only non-pigmented cell of the ommatidium.

Thus the previous observation (Gribakin 1967*a*; Varela and Porter 1969) of nine axons of the ommatidium is readily explained without dichotomy of axons (Gribakin 1967*a*; Varela and Porter 1969) which was with good reason criticized by Perrelet (1970). The drone ommatidium also consists of nine cells (Perrelet and Baumann 1969; Perrelet 1970), of which the 'three small cells' are similar to the type III and ninth cells in the worker bee.

The above classification of the honey bee retinula cells is illustrated by Fig. 11a from the study of Skrzipek and Skrzipek (1971). Axons are numbered by us in this figure clockwise from the cell marked RZ, and their cross-sectional areas are written as a line in Table 8.1. The excellent agreement between the alternation of the cell types and the measured areas demands no comment (Table 8.1).

TABLE 8.1

The alternation of retinula cell types in the honey bee ommatidium (Gribakin 1967b, 1972) compared with cross-sectional areas of the axons measured from Skrzipek and Skrzipek (1971)

Cell number	1	2	3	4	5	6	7	8	9
Cell type	I	II	III	II	I	II	III	9th	II
Cross-sectional area of the axon (μm^2)	8·5	3·3	1·5	3·3	8·1	3·3	1·5	1·5	3·4

8.5. Light-induced structural changes in the rhabdomere

Findings concerning the structural changes induced by light in the rhabdomere are to some extent contradictory. It is not clear whether the changes observed in some studies (Röhlich and Török 1962; Röhlich and Töro 1965; Gribakin 1969a,b) are artefacts of the fixation procedure since similar changes were not found by others (Kolb and Autrum 1972). The question arises as to how photoreceptors should be treated chemically to reveal the action of light upon them.

After selective adaptation of the honey bee to bright light with a wavelength of more than 480 nm the microvilli of type II and III cells in the ventral part of the eye show drastic swelling, whereas those of type I cells remain intact (Fig. 8.4c). Upon illumination by ultraviolet light with maximum at 365 nm the microvilli of type I cells become shorter and disintegrate while rhabdomeres of type II and III cells remain intact (Gribakin 1969a,b). These experiments were carried out under conditions which must be considered in more detail.

Light intensity for each colour was raised to the intensity of direct sunlight in the same spectral region. To do this, a comparison of intensities was made by a light-meter supplied with the coloured filters. The small aperture of the lamp relative to the field of view of the whole eye meant that only a small part of the eye (ventral in this study) was available for experiment.

It should be emphasized that light intensity measurements in lux cannot be correct when using a narrow bandwidth. Similarly the measurements of Kolb and Autrum (1972) cannot be taken as correct since they measured in lux light which contained a strong u.v. component. The possibility of errors arising from the divergence between relative luminous efficiency curves for bee and man has been indicated by van Praagh and Velthuis (1971).

Gradual cooling during adaptation to the light is essential because the time needed to complete dark-adaptation is comparable to that for the fixation procedure. A gradual cooling during light adaptation to a temperature of 2–4 °C reduces the rate of possible restoration upon subsequent fixation in the dark. This technique, first used in our study (Gribakin 1969a,b) has been employed successfully in later experiments (Kolb and Autrum 1972). Another possibility is to use boiling water (Butler 1971) which preserves adequately for conventional light microscopy but is not suitable for electron microscopy.

Fixing solution is the main item in experiments of this kind. As a rule one strives for good tissue preservation but this may be at variance with an unconventional requirement, e.g., the revealing of Na^+ and K^+ ions in vertebrate retina by the freeze-substitution method (Govardovskii 1971). At present optimum preservation is obtained using glutaraldehyde, alone pure or in combination with formaldehyde, as pre-fixative. But as early as 1965 Röhlich and Töro indicated aldehyde pre-fixation to be unsatisfactory for the eye of *Daphnia* after adaptation to bright light, whereas use of osmium tetroxide solution (OsO_4) revealed a swelling of the microvilli. F. Varela (personal communication) and Kolb and Autrum (1972) saw no changes in honey bee rhabdomeres after adaptation to bright light (despite the gradual cooling used by Kolb and Autrum). This failure is, in our opinion, due to aldehyde pre-fixation, for apparently only OsO_4 is a suitable 'developer' of photo-induced changes. A reason for this specific effect is difficult to imagine, but we recall that the photoreceptor membrane of the honey bee releases retinal in the light (Goldsmith and Warner 1964; Goldsmith 1970). Since retinal is an aldehyde, it may be that glutaraldehyde replaces retinaldehyde in the light-adapted photoreceptor membrane and in some way prevents the electron microscope visualization of light-induced changes. At any rate the result resembles that found by Kabuta, Tominaga, and Kuwabara (1968) who kept several arthropods in total darkness for a prolonged time. Subsequently aldehyde–osmic fixation revealed no change in the rhabdomeres whereas direct osmic treatment revealed swelling of microvilli and a great number of vesicles, as we found in our experiments with bright light. Sensitivity of the photoreceptor membrane of planaria to OsO_4 is increased after a prolonged period in total darkness (Röhlich 1967) and a 'strange' action of OsO_4 on photoreceptor membranes of the freshwater lamprey has been reported (Ohman 1971, see his Fig. 5).

So, despite poor fixation, we found a test which would reveal differences between retinula cells. In the ventral part of the honey bee eye, type I cells respond to light of wavelength 340 nm, and the type II and III cells to longer wavelengths, 530 nm. We do not say that the observed changes are real, but we simply use as a test the visible change in the sensitivity of the photoreceptor membrane to OsO_4. Certainly rhabdomeres of receptors-530 and receptors-340 disintegrate in different ways, which must have some real basis.

Illumination of the ventral part of the honey bee eye with blue light (420 nm) induced no changes in fine structure, and suggests that receptors-420 are absent from this region of the eye.

The most important conclusion is that cells with different spectral sensitivities occur within one ommatidium in the bee (Gribakin 1969a,b), as confirmed for the cockroach (Mote and Goldsmith 1971; Butler 1971). Therefore there must be different visual pigments in different rhabdomeres, whereas in the open rhabdomere eye a difference in spectral sensitivity may depend on rhabdomere diameter with only one visual pigment involved (Snyder and Miller 1972).

8.6. Further use of osmic acid to reveal receptor types

The technique described above has disadvantages, as follows. A source of bright light and a cooling installation is necessary; the lamp illuminates only a part of the eye (as in the above experiments) and only one eye can be treated at once. To avoid these limitations another technique has been developed (Gribakin 1972). In previous experiments the sensitivity of light-adapted rhabdomeres to OsO_4 played an important role; later an attempt was made to detect the products of bleaching of visual pigment with the same reagent, OsO_4, in an isolated and pre-fixed eye.

Osmium tetroxide is a well known reagent for double-bonds in organic chemistry (Criegee 1936; Cook and Schoental 1948; Badger 1949, 1951; Milas *et al.* 1959). On the other hand, visual pigments usually incorporate retinaldehyde, which has five or six double bonds. Eakin and Brandenburger (1970) were the first to demonstrate that upon fixing photoreceptor structures in OsO_4 solution at a high temperature (40 °C) for a prolonged period (three days), the organelles containing retinal showed a blackening in the electron microscope (see also, Ohman 1971). In their case the blackening seems to be of retinal that is released by termal decay from visual pigments with simultaneous denaturation of membranes by prolonged action of OsO_4 (*cf.* Porter and Kallman 1953). Anyway, this 'thermal osmic fixation' apparently reveals retinal and retinol, either incorporated in visual pigments or stored in the eye as esters of unsaturated fatty acids (Krinsky 1958).

Visual pigments of vertebrates and some arthropods are not bleached when the dark-adapted eye is fixed in aldehyde solution (Collins 1953; Waterman *et al.* 1969; Hays and Goldsmith 1969). Moreover, the absorption maxima and extinction values are not greatly modified. Receptors fixed with aldehyde in the dark can easily be bleached upon illumination. Thus it was possible to fix a number of dark-adapted eyes in conventional aldehyde fixative in the dark and subsequently choose the illumination and osmic post-fixation regimes to provide suitable conditions for interaction between light-released retinal and osmium tetroxide. The 'osmium black' deposit proved to be easily visualized under the electron microscope. There are two points of importance in these experiments. First, the level of illumination is adjusted on the assumption that at low intensities the fixation process, limited by OsO_4 diffusion, can be completed before the receptors are totally bleached (almost no osmium black). Thereby the products of bleaching are fixed before they can move away from their release points, which may not be possible at high-light levels. Secondly, the spectral composition of the bleaching light must be selected to give a minimal effect upon receptors of other spectral types. In these experiments (Gribakin 1972), additional staining by uranyl acetate or lead citrate is needed in ultrathin sections, although the blackening is clearly seen alone when thicker sections are used (Fig. 8.4a). At first the long wave receptors were stimulated with wavelengths greater than 480 nm. Receptors-340 are still outside the scope of this study because they possibly have a different type of visual pigment, although first efforts were made recently (Schwemer *et al.* 1971). The results (Fig. 8.4a,b) support our previous data on rhabdomere structual changes in living animals. In the dorsal ommatidia there are only four receptors-530 whereas there were six of them in the ventral region. As the ninth cell showed no staining we infer it to have maximum sensitivity in the blue or ultraviolet, or to be relatively insensitive to all wavelengths.

Type III cells in the dorsal part of the eye appear to be receptors-420, but further experiments are necessary to demonstrate them directly, and also to show that blue-sensitive cells have not been affected by long wavelengths or osmic staining under the above experimental conditions.

8.7. The basal cell and analysis of polarized light

Tiered rhabdoms are found in many insects (Grenacher 1879; Jörschke 1914; Horridge 1969; Melamed and Trujillo-Cenóz 1968; Schneider and Langer 1969; Ninomiya *et al.* 1969; Eguchi 1971; Walcott 1971), and the small basal rhabdomeres of the drone bee (Perrelet 1970) and worker (Gribakin 1972) can be included here. No true basal cells have been recorded in a functional analysis, but it is possible that the proximal rhabdomeres have a special function in the analysis of polarized light, as follows.

The molecules of vertebrate visual pigment are oriented randomly in the plane of the membrane (Schmidt 1938; Denton 1959; Wald, Brown, and Gibbons 1962). If this membrane is rolled up, the tubule so produced will have a dichroic ratio of two (Moody and Parriss 1961), and photometric measurements yield this value for some arthropods (Waterman, Fernandez, and Goldsmith 1969; Hays and Goldsmith 1969). Excitation of the receptor depends on the number of quanta absorbed, so that a dichroic rhabdomere responds to rotation of the polarization plane by a change in depolarization. Thus, different directions of dichroic absorption in different rhabdomeres is the basis of polarized light perception. However, the longer is an isolated rhabdomere, the less its dichroic ratio (Shaw 1969) and if the rhabdomere were long enough, it would show 100 per cent absorption irrespective of the plane of polarization (Fig. 8.7).

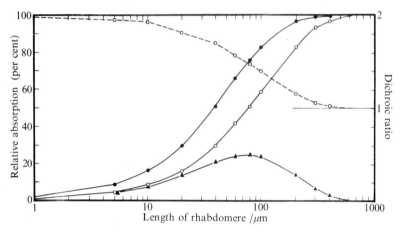

FIG. 8.7. The dependence of absorption of linearly polarized light on the length of the rhabdomere. —●—, with e-vector parallel to the microvilli; —○—, with e-vector perpendicular to the microvilli; —▲—, the difference between the above two curves, showing an optimal length at 60–100 μm. --○--, the dichroic ratio taken as the ratio of the first curve (parallel absorption) to the second one (perpendicular absorption). The dichroic ratio tends towards one with increasing length of rhabdomere, with no dichroism and therefore no PL-perception in the limit.

The optimal length of a rhabdomere which is made of microvilli with known dichroic ratio (two in this case) can be defined as the length at which the difference in absorption with electric vector parallel to the microvilli and perpendicular to them is a maximum. Let the concentrations of visual pigment molecules within a microvillus causing absorptions $A_{\parallel}$ and $A_{\perp}$ be $c_{\parallel}$ and $c_{\perp}$. Then $d = c_{\parallel}/c_{\perp}$ is the dichroic ratio of a single microvillus. The difference in absorption (or transmission) for the two extreme positions of the electric vector is:

$$\Delta I = I_0(e^{-2\cdot3\varepsilon c_{\parallel} z} - e^{-2\cdot3\varepsilon c_{\perp} z}) \qquad (8.1)$$

where: I_0 is intensity of incident light, ΔI is the difference in absorption (transmission), ε is the molar extinction, $c_{\parallel}$ and $c_{\perp}$ are the concentrations of oriented molecules of visual pigment, and z is the length of the rhabdomere. for ΔI to be a maximum

$$2\cdot3\varepsilon c_{\parallel} e^{-2\cdot3\varepsilon c_{\parallel} z} + 2\cdot3\varepsilon c_{\perp} e^{-2\cdot3\varepsilon c_{\perp} z} = 0 \qquad (8.2)$$

then, if
$$\ln \frac{c_\parallel}{c_\perp} = 2.3\varepsilon z(c_\parallel - c_\perp),$$

and
$$z_{opt} = \frac{\ln c_\parallel/c_\perp}{2.3\varepsilon c_\perp(c_\parallel/c_\perp - 1)} = \frac{\ln d}{2.3\varepsilon c_\perp(d-1)}, \tag{8.3}$$

or, using $c_\parallel$:
$$z_{opt} = \frac{d \ln d}{2.3\varepsilon c_\parallel(d-1)} = \frac{d \log d}{\varepsilon c_\parallel(d-1)}. \tag{8.4}$$

The specific absorption of a rhabdomere when the electric vector is parallel to the microvilli is c. 1·7–1·8 per cent μm^{-1} (i.e. an optical density or absorbance of $a = 0.0078 \ \mu m^{-1}$ (Hays and Goldsmith 1969). Since specific absorbance (i.e. per unit of length) is equal to $a = \varepsilon c_\parallel z$ with $z = 1$, the value of z_{opt} will be:

$$z_{opt} = \frac{2 \log 2}{0.0078(2-1)} = 77 \ \mu m \tag{8.5}$$

The optimal length of the rhabdomere lies between 40 and 100 μm for corresponding dichroic ratios of 1·4–1·6 (Fig. 8.7). On the assumption that threshold sensitivity is to a change of intensity of 5 per cent, a rhabdomere with dichroic ratio of only 1·05 could detect polarized light, and its length could reach 200–300 μm.

Another way to analyse linearly polarized light is to have two rhabdomeres with identical spectral characteristics disposed so that one is the continuation of the other (Fig. 8.8). Let the relative absorption of the distal rhabdomere R_d be A_{xd} and A_{yd} in directions x and y, with $A_{xd} > A_{yd}$, and the dichroic ratio $d_d = A_{xd}/A_{yd} > 1$. The relative transmissions will be, respectively, $T_{xd} = 1 - A_{xd}$ and $T_{yd} = 1 - A_{yd}$, with $T_{xd} < T_{yd}$. If the proximal rhabdomere R_p has its own dichroic ratio $A_{xp}/A_{yp} = d_p$, the effective dichroic ratio is:

$$d_\parallel = \frac{T_{yd}}{T_{xd}} \times \frac{A_{yp}}{A_{xp}} = \frac{T_{yd}}{T_{xd}} \times \frac{1}{d_p} \tag{8.6}$$

for microvilli arranged in parallel, and:

$$d = \frac{T_{yd}}{T_{xd}} \times \frac{A_{xp}}{A_{yp}} = \frac{T_{yd}}{T_{xd}} \times d_p \dagger \tag{8.7}$$

if the microvilli are perpendicular. For example, if the distal rhabdomere is 200 μm long the following values are obtained: $A_{xd} = 0.97$; $A_{yd} = 0.83$; $T_{xd} = 0.03$; $T_{yd} = 0.17$ and $d_d = A_{xd}/A_{yd} = 1.16$; $T_{yd}/T_{xd} = 5.7$.

† This equation agrees with that derived quite independently by Snyder (J. comp. Physiol. **83**, 331–360, 1973). The constant 2·3 is derived from changing from extinction coefficient $k = \varepsilon c_x$ to absorption coefficient $k = 2.3ec_x$, as these terms are used here.

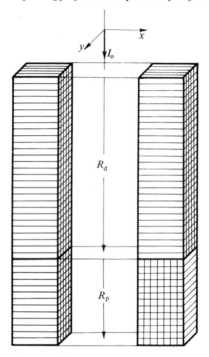

F IG. 8.8. Models of a two-tiered rhabdom, with R_d the distal rhabdomere and R_p the proximal rhabdomere, oriented parallel (left) or perpendicular (right) (see text for details). I_0 represents the direction of incident linearly polarized light.

For a proximal rhabdomere of 60 μm long $d_p = A_{xp}/A_{yp} = 1\cdot57$ so that

$$d_{\parallel} = \frac{T_{yd}}{T_{xd}} \times \frac{1}{d_p} = 5\cdot7 \times \frac{1}{1\cdot57} = 3\cdot6$$

and

$$d_{\perp} = \frac{T_{yd}}{T_{xd}} \times d_p = 5\cdot7 \times 1\cdot57 = 9\cdot0$$

Thus, starting with relatively small dichroic ratios, one can construct an analysing system which is much more effective than the component units alone by using the first rhabdomere in the light path as a selective filter. It is of interest that mutually perpendicular microvilli have been found in fly central retinula cells (Melamed and Trujillo-Cenóz 1968). Possibly this analysis applies not only to a single rhabdomere but also to laterally fused rhabdomeres as in the bee, when the whole fused rhabdom is taken as the filtering system and the basal cell as the detector. Also, a short proximal rhabdomere could not provide a high sensitivity on account of its small absorption but an efficient filter would confer a sensitivity to the plane of polarization even if the basal cell has no polarization sensitivity.

Unfortunately, basal cells are small and inaccessible to electrophysiological recording. Nevertheless, if the model suggested is true to fact, one can expect that approximately tenfold effective dichroism will be found in the eighth visual cell in Diptera (with much more absorption *across* the seventh cell microvilli) and possibly an increase for the ninth cell of the honey bee.

Preservation of the orientation of the electric vector of the light reaching the basal rhabdomere is essential for the above model and this orientation is preserved even in a cylindrical waveguide (see Snyder, Chapter 9). The above waveguide considerations refer only to a narrow spectral band, and the situation must be more complicated for white light.

At present, only indirect data supports this model. First, the morphology combined with the photometric measurements lead to the above calculations. Secondly, the long visual fibres of Diptera extend to the medulla (Melamed and Trujillo-Cenóz 1968), which may be evidence of special signal processing. Thirdly, the spectral sensitivities of the ninth cell and the type III cell in honey bee could be the same (Gribakin 1972). Fourthly, as von Frisch showed (1965), orientation by polarized light is effective in the bee only at the ultraviolet end of the spectrum, in line with our finding that the ninth cell is probably a shortwave receptor. Fifth, in the bee the basal cell must function at intensities much less than optimal for the distal cells, so that signals from basal cells must perhaps be summed. In fact it has been found that 25–30 ommatidia are necessary for threshold orientation to polarized light and practically faultless orientation is reached only when *ca.* 150 ommatidia are involved (Zolotov and Frantzevich 1973) in agreement with von Frisch (1965).

Perhaps the most reasonable attitude towards the possibility of the rhabdom acting as a filter for improving selection by the basal cell, however, is to attempt the experimental demonstration of something which seems inevitable from the morphology and physical properties of the components. (Menzel and Snyder 1974).

8.8. Summary

(1) The description of the ommatidium and especially the state of dark adaptation is the first stage of the analysis of functional morphology of the compound eye. A functional interpretation is attempted for the distribution of mitochondria, the principal endoplasmic cysternae and their relations with endoplasmic channels and rhabdom, the extensions of the cone and the pigment layer near the basement membrane, in the bee.

(2) Three types of retinula cells in the bee are described. The main features are the positions of nuclei, the cross-sectional area of the rhabdomere, the number of pigment granules and mitochondria and the axon diameter.

(3) Structural changes induced by light in rhabdomeres are discussed. The

action of osmium tetroxide in developing light-induced changes is
compared with the stabilizing effect produced by aldehyde fixation on
photoreceptor membrane. As revealed by osmic fixation, type I cells
appear to be the receptors-340 and the type II and III ones in the
ventral part of the eye are the receptors-530. Compare the situation in
ants (article by Menzel in this volume).

(4) Cells with different spectral sensitivities are found in a single ommatidium, and the fused rhabdom is therefore heterogeneous.

(5) A cytochemical method, apparently based on a reaction between
osmium tetroxide and retinal, confirms the results in (3). The ninth or
basal cell in the bee appears to be a receptor for short wavelengths.

(6) Calculations show that an efficient analyser of polarized light can be
constructed from relatively poor components if the distal rhabdomeres
act as selective filters for polarized light passing through to the proximal
or basal cells.

Part III
Optics

9. Optical properties of invertebrate photoreceptors

A. W. SNYDER

9.1. Introduction

THERE is a considerable literature on the optics of compound eyes. Hitherto the main effort has been directed at understanding the dioptric apparatus of the ommatidium, e.g. see the recent reviews of Miller *et al.* (1968) and Kirschfeld (1969). Much less quantitative information is available on the optical properties of photoreceptors, i.e. the rhabdomeres. Quite apart from the dioptric system, the optical properties of rhabdomeres (determined by their size, shape, and index of refraction) have a marked influence on the spectral, polarization, angular, and absolute sensitivity of an ommatidium. It is the intent of this paper to elucidate the role of photoreceptor optics, to describe recent developments in the field and to indicate the direction of future research.

Photosensitive cells contain the photopigment which upon light absorption initiates the visual response. Most photoreceptors of those vertebrates and invertebrates with a high degree of acuity and sensitivity can be described roughly as long narrow (not necessarily circular) cylinders with a diameter of the order of the wavelength of light in the visible spectrum. Examples of these are the rods and cones of the vertebrates and the rhabdomeres of the arthropods. Our general concern here is with the consequences of confining the photopigment within long narrow cylinders.

The refractive index of rhabdomeres is larger than that of the surrounding medium, e.g. Seitz (1968) and Varela and Wiitanen (1970). A cylinder with this property is capable of transmitting or guiding electromagnetic energy; however, to attribute the phenomena to light rays undergoing a succession of total internal reflections is an oversimplification (Snyder, Pask and Mitchell 1972) and Snyder and Pask (1973a). Due to the small cross-sectional area, interference effects prevent the light intensity from being distributed uniformly in cross-section. Instead the light is transmitted in patterns known as modes or more precisely as dielectric waveguide or optical fibre modes (Snitzer 1961 and Snyder and Pask 1972a).

Modes like those of Fig. 9.1 have been observed on vertebrate outer segments of the rat, rabbit, rhesus monkey, the squirrel monkey, frog, and man (Enoch 1963) as well as on the rhabdom of the worker bee (Varela and Wiitanen 1970). In these examples the photoreceptors were bleached fully from exposure to bright light, i.e. the photoreceptors were dead. However, modes have also been observed *in situ*, *in vivo* on the outer segments of the rabbit (Enoch 1967) and on the rhabdomeres of the fly *Drosophila* (Franceschini and Kirschfeld 1971). These observations furnish direct proof that

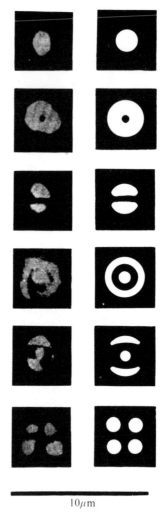

10μm

FIG. 9.1. Examples of modal patterns photographed at, or near, the terminations of vertebrate retinal receptors (Enoch 1963). On the right side of the figure the theoretical mode patterns for a circular fibre are shown.

photoreceptors are dielectric waveguides. *Thus dielectric waveguide theory must be used to describe light absorption and transmission within rhabdomeres.*

In order to interpret the retinula cell output, we must fully understand its input. The input is related to the light absorbed by the visual photopigment rather than the light impinging on the ommatidium. We are primarily concerned with the contribution of dielectric waveguide effects to the sensitivity *S* of the retinula cells. However, we emphasize that *S* also depends on the effects of the dioptric apparatus, i.e. the lens and any intervening light-channelling structures.

In summary, light propagates along rhabdomeres as patterns (in cross-section) known as modes. We, therefore, have the fundamental result, that all information about the environment must be extracted from modes. This fact raises some interesting questions. For example, does the strong dependence of modes on the size and refractive index of the rhabdomeres manifest itself in the visible response? Have certain photoreceptors evolved special properties (e.g. arrangement of microvilli in the bee) to extract information contained within the modes? We answer such questions in this paper.

For example, we show that the spectral sensitivity of a retinula cell can depend on the size and index of refraction of its rhabdomere. The smaller the diameter of the rhabdom the more its visible peak response is shifted to lower wavelengths and the greater its u.v. peak response becomes relative to that of the visible peak. We provide the physical principles necessary to understand the mechanism for the polarization sensitivity of any retinula cell. In particular we derive a simple criterion to determine when a rhabdom is an effective polarization analyser. We show that the acceptance property of an ommatidium depends on the waveguide and diffraction characteristics of the rhabdom.

A photoreceptor is essentially an absorbing optical fibre. We, therefore, begin by discussing the light-transmission properties of optical fibres. These concepts are then applied to various insect photoreceptors including the ant, bee, cockroach, dragonfly, and fly (Diptera) considering their spectral, polarization, and angular sensitivity. The reader is advised to see Snyder (1974), and the appendicies of Snyder (1973), Snyder and Pask (1973c), and Snyder, Menzel, and Laughlin (1973) for more recent results.

9.2. Properties of dielectric waveguides

As noted in the introduction dielectric waveguide theory must be used to calculate the light absorbed or transmitted within structures like rhabdomeres. This requires solving Maxwell's electromagnetic field equations within a cylindrical/absorbing anisotropic medium. Snyder (1969a, 1972b) and Snyder and Pask (1972a) have presented simple mathematical solutions specially adapted to the study of photoreceptors like rhabdomeres. We make use of their results in this section to describe several fundamental properties of dielectric waveguides necessary for our discussion.

9.2.1. *Light transmission* (*the concept of modes*)

Light is transmitted along optical fibres of arbitrary cross-section in the form of patterns known as dielectric waveguide modes (Fig. 9.1). A mode by definition preserves its cross-sectional pattern as it is transmitted along the fibre. The number of modes that can propagate depends on the wavelength. Decreasing the wavelength increases the number of modes.

A fundamental property of a mode is that only a fraction (which depends on the wavelength) of its total light energy is transmitted *within* the optical fibre. The remaining portion is transmitted along, but *outside* it. We define n_i to be the fraction of a mode's power within the optical fibre, i.e.

$$\eta_i = \frac{\text{Power of mode } i \text{ within the fibre}}{\text{total power of mode } i}, \tag{9.1}$$

where power is defined as the total integrated light intensity passing through a particular cross-section, i.e. that measured by a photocell.

Since many photoreceptors are approximately circular in cross-section, we apply the above concepts to the special case of the circular fibre.

The number of modes that can propagate and their η_i characteristics are specified when the dimensionless parameter V is known. V is defined as

$$V = \frac{\pi d}{\lambda}\{n_1^2 - n_2^2\}^{\frac{1}{2}} \tag{9.2}$$

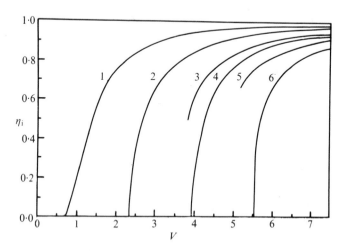

FIG. 9.2. The fraction (η_i) of light power within the fibre for each of the six mode types that exist for $V < 6\cdot38$ where V is defined by eqn (9.2).

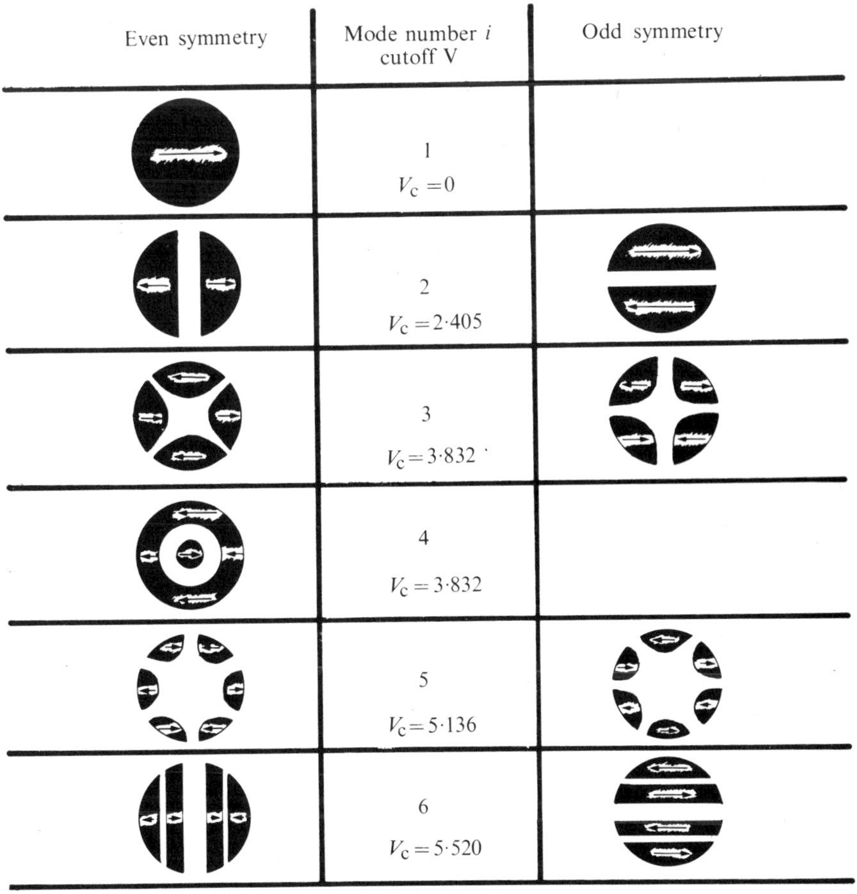

Even symmetry	Mode number i cutoff V	Odd symmetry
	1 $V_c = 0$	
	2 $V_c = 2 \cdot 405$	
	3 $V_c = 3 \cdot 832$	
	4 $V_c = 3 \cdot 832$	
	5 $V_c = 5 \cdot 136$	
	6 $V_c = 5 \cdot 520$	

FIG. 9.3. Schematic of the intensity patterns for the various mode types. The arrows are for x-polarized light. Rotating the arrows by 90° provides the y-polarized set.

where λ is the wavelength of light in vacuum, d is the diameter of the photoreceptor and n_1, n_2 are the index of refraction of the fibre and its surrounding medium, respectively.

For photoreceptors the parameter V is acutely sensitive to the values of the index of refraction because $n_1 \approx n_2$. This means that three significant figure accuracy for the values of the index of refraction is required. Even when such measurements are possible on excised retina, there is evidence that they may not be representative of *in situ* conditions (Enoch 1967). Thus there is a certain degree of freedom in choosing the third significant figure for the values of the index of refraction to be substituted into eqn (9.2).

Each mode has a particular cut-off V called V_c below which it no longer propagates. In Fig. 9.2 we show η_i curves for all the mode types for $V < 6\cdot38$. The field-intensity patterns associated with these modes is given in Fig. 9.3.

Note that as λ increases, the fractional power of a mode within the fibre (η_i) decreases. The modes are labelled in order of their increasing cut-off values (V_c), i.e. $i = 1$ is the mode for which $V_c = 0$, $i = 2$ is for the mode for which $V_c = 2\cdot405$ and so on. This is not the usual notation but it is the simplest for application to photoreceptors.

The most convenient set of modes for our discussion is that given in Fig. 9.3. From this set any other observed mode pattern can be constructed (see Fig. 9.4 as an example).

FIG. 9.4. A linear combination of the even and odd symmetric type 2 modes produces the circularly symmetric pattern (known as the TE_{01} or TM_{01} mode) frequently observed on visual photoreceptors.

9.2.2. *Excitation of modes (modal power* P_i)

Until now we have been discussing the properties of modes without regard to their formation. Figure 9.2 gives the *fraction* of a mode's power within a fibre. The absolute power is determined by the properties of light impinging on the fibre, i.e. on the dioptric apparatus.

The light illuminating the photoreceptor depends on the properties of the light impinging on the dioptric apparatus in addition to the diffraction properties of the dioptric apparatus and intervening light-channelling structures. Thus it is difficult to specify accurately the light illuminating the photoreceptor. There are, however, some general statements that can be made about light in the focal plane of a diffraction-limited system. Light is not focused to a point but rather to a circular non-uniform disc (known as the Airy disc) of diameter d_a where

$$d_a \approx 2\cdot44\left(\frac{\lambda}{n}\right)\left(\frac{f}{d_p}\right) \tag{9.3}$$

d_p is the diameter of the entrance pupil (cornea lens), n the refractive index of the medium between the dioptric system and the focus, and f is the focal length, i.e. the distance from the focus to the nearest portion of the exit pupil. The factor f is in general found only from experiment. Changes in d_A with λ

must be taken into consideration when investigating the causes of the spectral sensitivity of a receptor, and lens aberration must be considered.

The light (of a narrow bandwidth) in the region of the Airy disc is approximately coherent, i.e. consists of parallel in phase rays (Born and Wolf 1965). For on-axis illumination the focus is centred on the optical axis; however, for illumination at angle θ to the axis the focus moves off the optical axis. Thus, if the focus is centred at the entrance to the photoreceptor as in the Diptera compound eye (Kirschfeld 1969) illumination is described by a disc of light displaced from the optical axis.

Many invertebrate dioptric systems focus light above (closer to the lens) the entrance to the photoreceptor; for example, the worker bee (Varela and Wiitanen 1970). This produces a spot size on the photoreceptor larger than the diameter d_A of the Airy disc. Furthermore, the light rays are no longer parallel, although the rays that enter the fibre most effectively are nearly parallel. Light incident at angle θ to the lens axis then converges approximately at angle θ_p on the photoreceptor. The relationship between θ_p and θ follows from a ray-tracing analysis of the dioptric system.

9.2.2.1. *Acceptance property of an optical fibre.* The acceptance property of an optical fibre is determined by its response to a uniform beam of light obliquely incident to the axis of the fibre.

Consider a uniform light beam incident as illustrated in Fig. 9.5. The mathematical details are presented by Snyder (1969*b*) and Snyder and Pask

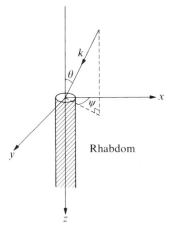

FIG. 9.5. The angles used to define light incident obliquely on the rhabdom from direction $\hat{k}$.

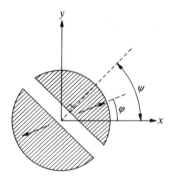

FIG. 9.6. The orientation of mode 2 due to light illustrated in Fig. 9.5.

(1972a). We find that the orientation of a mode is related to the direction, ψ, of off-axis incident light as illustrated in Fig. 9.6 for the $i = 2$ mode. The null (or low energy line) is perpendicular to ψ for the $i = 2$ mode.

The power, P_i, of the modes due to the incident light depends on the properties of the fibre, i.e. on V defined by eqn (9.2), on the diameter of the light beam and on the angle (θ) of light incidence.

For $\theta = 0$ (on axis illumination) only the $i = 1$ and $i = 4$ modes are excited as shown in Fig. 9.7. For $\theta > 0$ all the modes can be launched if V and θ are large enough. In Fig. 9.8 we show the total light power transmitted within the fibre, i.e. the sum of the power of each mode. We can use Fig. 9.8 to determine the light-acceptance property of any fibre. The acceptance properties of a dielectric waveguide depend acutely on its diameter and refractive index in addition to the wavelength of incident light. The ray optical result is only valid for fibres with a large V. Further details are available from Snyder, Pask, and Mitchell (1973).

When light is polarized at angle ϕ to the x axis, the power can be split into x- and y-polarized modes. We use the notation $P_{i,x,e} P_{i,x,o} P_{i,y,e}$, and $P_{i,y,o}$ to indicate the x- or y-polarized even or odd modes. Then from the analysis of Snyder and Pask (1972a)

$$P_{i,x,e} = \cos^2 \phi P_i(v, \theta) \tag{9.4a}$$

$$P_{i,y,e} = \sin^2 \phi P_i(v, \theta), \tag{9.4b}$$

for the $i = 1$ and 4 modes and for the other modes

$$P_{i,x,e} = \cos^2 \phi \cos^2 p\psi P_i(v, \theta) \tag{9.5a}$$

$$P_{i,x,o} = \cos^2 \phi \sin^2 p\psi P_i(v, \theta) \tag{9.5b}$$

$$P_{i,y,e} = \sin^2 \phi \cos^2 p\psi P_i(v, \theta) \tag{9.5c}$$

$$P_{i,y,o} = \sin^2 \phi \sin^2 p\psi P_i(v, \theta) \tag{9.5d}$$

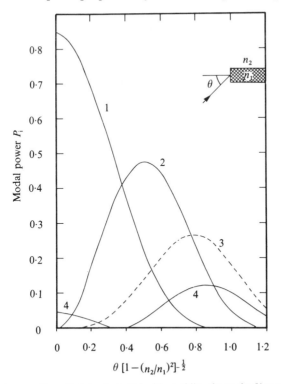

FIG. 9.7. Modal power P_i excited due to light incident obliquely on the fibre at angle θ. $V = 5$ where V is defined by eqn (9.2). The numbers on the curve refer to the mode type illustrated in Fig. 9.3. Since $V < 5\cdot136$ the fifth and sixth modes are not excited.

where $P_i(v, \theta)$ is the power of mode i determined from Snyder (1969b), $p = 1$ for $i = 1$ and 6 and $p = 2$ and 3 for $i = 3$ and 4. When illuminated by unpolarized light we replace $\cos^2 \phi$ and $\sin^2 \phi$ by $\frac{1}{2}$.

9.2.3. *Light-absorbing optical fibre*

A rhabdomere or fused rhabdom differs from the optical fibres so far considered here in that it is absorbing. Furthermore, this absorption is anisotropic or directionally sensitive to the electric vector of the illuminating light. Maximum absorption occurs when the electric vector is parallel to the longitudinal axis of the microvilli shown as the parallel dark lines in the representative photoreceptors of Fig. 9.9. Rhabdomeres 7 and 8 of the Diptera eye are an exception (Kirschfeld 1969). The molecular basis for this directional sensitivity is the dichroism of the photopigment molecules on, or within, the microvilli.

We begin by deriving an expression for absorption in a photoreceptor neglecting its dichroic property, e.g. the vertebrate rods and cones. Then we generalize the expression to account for the alignment of the microvilli.

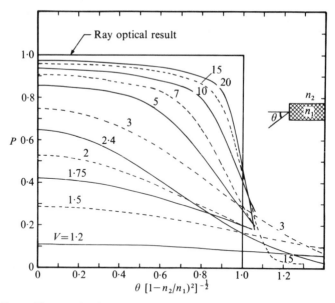

FIG. 9.8. Power (P) transmitted within the fibre due to light obliquely incident at angle θ. The results are found by summing the power of all modes that can propagate at each value of V, which is defined by eqn (9.2).

9.2.3.1. *Isotropic fibre (vertebrate photoreceptor)*.

For an absorbing cylinder of arbitrary cross-section the power of a mode decreases as it propagates. The fraction dP_i^A/P_i^A of light power absorbed from mode i in a differential length dl is proportional to the fraction of power, η_i, of mode i within the cylinder in addition to the extinction coefficient or spectral absorption property $\{\alpha(\lambda)\}$ of the photosensitive material and its concentration (c). Thus for each mode i (Snyder 1972b)

$$\frac{dP_i^A}{P_i^A} = \gamma_i(\lambda)\,dl, \qquad\qquad (9.6)$$

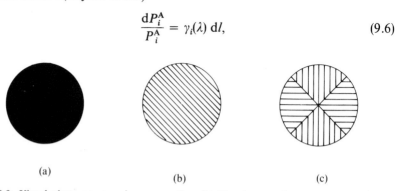

(a)   (b)   (c)

FIG. 9.9. Visual photoreceptors in cross-section. (a) Vertebrate rod or cone outersegment. (b) a rhabdomere in Diptera. (c) one type of fused rhabdom. The parallel dark lines represent the microvilli in which the pigment molecules are ordered.

where the modal absorption coefficient $\gamma_i(\lambda) = c\alpha(\lambda)\eta_i(\lambda)$. The power P_i^A absorbed from the ith mode for a cylinder of length l is found by integrating eqn. (9.6) leading to

$$P_i^A(\lambda) = P_i(\lambda)\{1 - e^{-\gamma_i(\lambda)l}\} \tag{9.7}$$

where $P_i(\lambda)$ is the power of the ith mode discussed in the last section.

If the fibre is tapered, η_i is a function of the position z. For this case $l\gamma_i(\lambda)$ is replaced by an average η given by Snyder and Pask (1973).

If the medium outside the photoreceptor is also absorbing, e.g. retinular cell screening pigment, the parameter γ_i in eqns. (9.6) and (9.7) is replaced by $\gamma_i + c_0\alpha_0(\lambda)\{1 - \eta_i(\lambda)\}$ where the subscript 0 refers to the medium surrounding the photoreceptor and $1 - \eta_i$ is the fraction of modal power outside the photoreceptor.

Equation (9.7) is the power absorbed by the fibre due to the ith mode. In general many modes propagate so that we must sum an expression equivalent to eqn. (9.7) for each mode. This summed result leads to a theoretical expression for the spectral sensitivity of the photoreceptor $S(\lambda)$ defined as

$$S(\lambda) = \sum_i P_i(\lambda)\{1 - \exp - \gamma_i(\lambda)l\} \tag{9.8}$$

where the summation is taken over each propagating mode. We have assumed that S is proportional to P_i^A. Equation (9.8) forms the basis for a study of vertebrate photoreceptors. We mention that the vertebrate photoreceptor is anisotropic in a longitudinal section but for light from the physiological direction this need not be considered.

9.2.3.2. *Anisotropic fibre (rhabdomeres of the diptera eye)*. The cross-section of the fly rhabdomeres schematically illustrated in Fig. 9.9 is uniformly anisotropic. More light is absorbed when the electric vector is parallel to microvilli (x-axis) than when it is perpendicular to the microvilli (y-axis). Thus more x-polarized modal power $P_{i,x}(\lambda)$ than y-polarized modal power $P_{i,y}(\lambda)$ is absorbed. Following the steps leading to eqn. (9.8) we have (Snyder and Pask 1973c; Snyder and Sammut 1973).

$$S(\lambda) = \sum_i [P_{i,x}(\lambda)\{1 - \exp - \gamma_i(\lambda)l\} + P_{i,y}\{1 - \exp - \gamma_i l/\Delta\} \tag{9.9a}$$

$$= \sum_i P_i(\lambda)[\{1 - \exp - \gamma_i(\lambda)l\}\cos^2\phi + \{1 - \exp - \gamma_i(\lambda)l/\Delta\}\sin^2\phi] \tag{9.9b}$$

where Δ ($\Delta \geq 1$) expresses the dichroic sensitivity of the medium within the rhabdomere. For the unpolarized case, we replace $\cos^2\phi$ and $\sin^2\phi$ by $\frac{1}{2}$.

9.2.3.3. *Fused rhabdoms*. The fused rhabdom illustrated in Fig. 9.9 is an example of a non-uniformly-anisotropic absorbing cylinder. The light energy of every mode of Fig. 9.3 is absorbed differently by the rhabdom (Snyder and Pask 1972a), so that the modal power is found by summing the even, odd, x- and y-polarized modes. However, the result of this summation does not in general correspond to the measured spectral sensitivity of a retinular cell. This is found by determining the absorbed power in each of the rhabdomeres. We do this for a general rhabdom in the next section.

9.2.4. *Power absorbed by individual rhabdomeres of fused and laminated rhabdoms*

In this section we derive expressions for the light power absorbed by an individual rhabdomere of a fused rhabdom of arbitrary shape, size, and consistency. The results of this section form the mathematical basis of our discussion of polarization sensitivity introduced in a later section. The analysis is valid for structures for which waveguide effects are not important or when only one mode propagates as is the case for on-axis illumination.

In general, the light power P absorbed by *all* the rhabdomeres in a length z along the rhabdom is given by Snyder (1972b) and Snyder and Pask (1972a) as

$$P(z) = 1 - \exp\{(1/A) \int_V \gamma(x,y,z)\, \mathrm{d}V\}$$

where A is the cross-sectional area of the rhabdom, V is the volume of the rhabdom enclosed in a length z and $\gamma(x,y,z)$ is the absorption coefficient of the medium. For simplicity we use P rather than P^A for absorbed power. In general, γ includes the properties of the photopigment and waveguide effects as discussed above.

The light power P_i absorbed in the ^{i}th rhabdomere of volume V_i is given as

$$P_i = \frac{1}{A} \int_{V_i} \gamma(x,y,z)\{1 - P(z)\}\, \mathrm{d}V \tag{9.11}$$

where $P(z)$ is defined by eqn (9.10).

(a) *Rhabdoms uniform in length.*   When the rhabdom is uniform in length the integral in eqn (9.10) becomes

$$\frac{1}{A} \int_V \gamma(x, y, z)\, \mathrm{d}V = \frac{z}{A} \sum_{j=1}^{8} \gamma_j A_j \tag{9.12}$$

where γ_i is the absorption coefficient and A_i the area of the ith rhabdomere (Fig. 9.10a). Then from eqns. (9.10)–(9.12) we find the power absorbed in the ith retinula cell of length l leading to (Snyder 1973).

$$P_i = \gamma_i \left(\frac{A_i}{A}\right) \int_0^l \{1 - P(z)\}\, \mathrm{d}z \tag{9.13a}$$

$$P_i = \left(\frac{\gamma_i A_i}{\sum_{j=1}^{8} \gamma_j A_j}\right)\left\{1 - \exp\left(-\frac{l}{A} \sum_{j=1}^{8} \gamma_j A_j\right)\right\} \tag{9.13b}$$

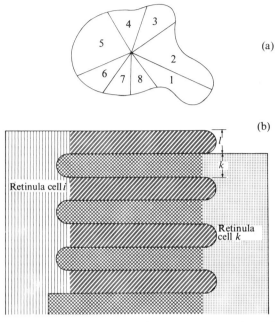

F IG. 9.10. (a) illustrates a cross-section of a fused rhabdom. The numbers refer to the rhabdo-meres each of area A_i with absorption coefficient γ_i. (b) represents a longitudinal section of the layered or laminated rhabdom typical of Crustacea. The cross-section of (b) is that of (a).

(b) *Rhabdom with a laminated structure.*   We now assume that the fused rhabdom discussed earlier is laminated or layered along its length as depicted in Fig. 9.10b in part by the rhabdomeres of the ith and kth retinula cells. The object is to derive an expression for the power absorbed by retinula cell i, i.e. for the light power absorbed by all of the dark layers of Fig. 9.10b. Following Snyder (1973) we have from eqn. (9.11)

$$P_i = \gamma_i \left(\frac{A_i}{A}\right) \sum_{q=1}^{n} \int_{l_q} \{1 - P(z)\} \, \mathrm{d}z \tag{9.14}$$

where the integration is carried out over the length (l_q) of the qth layer of the ith retinula cell and n is the number of these layers. A more physical deriva-tion is to sum the power absorbed in each of the n layers. The power absorbed in the ith rhabdomere of the qth layer (assumed made up in part by the ith rhabdomere) is

$$P_i^{(q)} = P_t^{(q)} \left(\frac{\gamma_i A_i}{\xi_i A}\right) \{1 - \exp - l_i/\xi_i\} \tag{9.15}$$

where $P_t^{(q)}$ is the power transmitted into the qth layer and ξ_i the absorption coefficient for layers formed in part by the ith rhabdomere

$$\xi_i = \frac{1}{A} \sum_{j=1}^{N_i} \gamma_j^{(i)} A_j^{(i)} \tag{9.16}$$

N_i is the number of rhabdomeres in the layers formed in part by retinula cell i. A similar expression for retinula cell k is found by replacing i by k.

Either procedure leads to

$$P_i = \left(\frac{\gamma_i A_i}{\xi_i A}\right)\{1 - \exp - l_i\xi_i\}\{1 + \exp - (\xi_i l_i + \xi_k l_k) + \exp - 2(\xi_i l_i + \xi_k l_k)$$
$$+ \ldots \exp - n(\xi_i l_i + \xi_k l_k)\} \tag{9.17}$$

using the identity

$$(1 - \exp x)(1 + \exp x + \ldots \exp nx) = 1 - \exp nx$$

leads to an expression for the power absorbed by the ith retinula cell of Fig. 9.10.

$$P_i = \left(\frac{\gamma_i A_i}{\xi_i A}\right)\frac{\{1 - \exp - l_i\xi_i\}}{\{1 - \exp - (l_i\xi_i + l_k\xi_k)\}}\{1 - \exp - n(\xi_i l_i + \xi_k l_k)\} \tag{9.18}$$

This expression is simplified by noting that the absorption in any given layer is small, i.e. $l_i\xi_i \ll 1$ and $l_k\xi_k \ll 1$ so that

$$P_i \approx \left(\frac{A_i}{A}\right)\frac{\gamma_i l_i}{(l_i\xi_i + l_k\xi_k)}\{1 - \exp - n(\xi_i l_i + \xi_k l_k)\} \tag{9.19}$$

When the path length of the i and k layers are equal, i.e. when $nl_k = l_i n$ we have $nl_i = l/2$ where l is the length of the rhabdom. Thus

$$P_i = \left(\frac{A_i}{A}\right)\left(\frac{\gamma_i}{\xi_i + \xi_k}\right)\left\{1 - \exp -\frac{l}{2}(\xi_i + \xi_k)\right\} \tag{9.20}$$

9.2.5. *Information contained within modes*

Modes have the fundamental properties of light, i.e. wavelength, intensity, and polarization. Furthermore, certain illumination conditions can sometimes be transmitted by modes, e.g. simple image information and the direction of off-axis light. Image information is possible when more than one mode propagates. For example, a linear combination of the $i = 1$ and 2 modes approximate to a semi-circular image. We have already shown in Section 9.2.2 that the $i = 2$ mode contains information about the direction of off-axis light.

We conclude that modes contain wavelength, intensity, and polarization information and in some cases image and direction of off-axis incident light information.

9.2.6. *Optical coupling between fibres*

Light energy is transferred between parallel fibres. This phenomenon is discussed in detail by Snyder (1972c) and McIntyre and Snyder (1973). As an example consider two identical optical fibres with their centres separated by a distance d_s. Then 100 per cent of the light power of one fibre is transferred

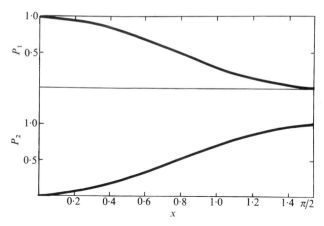

F IG. 9.11. The transfer of light power between two identical parallel fibres in which only mode 1 is propagated. It is assumed that only fibre 1 is illuminated (with unity power). The mode power for fibre 1 (P_1) and fibre 2 (P_2) are illustrated. In a distance $x = \pi/2$ all the light power in fibre 1 is transferred to fibre 2. $x = 2zK[1-(n_2/n_1)^2]^{1/2}/d$ where z is the length along the fibre and K, n_1, n_2 and d are defined in Fig. 9.12.

to the other fibre in a distance l along the fibre where

$$\frac{l}{d} = \frac{\pi}{4}\left\{1-\left(\frac{n_2}{n_1}\right)^2\right\}^{-\frac{1}{2}}\left(\frac{1}{K}\right)$$

as shown in Fig. 9.11 is the fibre diameter. The parameter K is found from Fig. 9.12 when V (eqn 9.2) and d_s/d are specified. n_1, n_2 are the refractive index of the fibre and surround, respectively. The results are modified if the fibres are different or if there are more than two fibres (Snyder 1972c; McIntyre and Snyder 1973).

9.3. Application of dielectric waveguide theory to insect photoreceptors

Armed with the basic concepts of dielectric waveguide theory we are in a position to study the optical properties of a photoreceptor. We begin by considering the sensitivity of the photoreceptor. The reader is encouraged to see Snyder, Menzel and Laughlin (1973) for more recent findings.

9.3.1. *Sensitivity of a photoreceptor—S*

Only the light absorbed by the photopigment leads to the visual response. We make the assumption that the sensitivity, S, of a retinula cell is proportional to the light absorbed by its rhabdomere. To obtain S experimentally the receptor potential measurements are converted into receptor sensitivity, S, curves by using the voltage–intensity characteristics of the retinula cell. The parameters of the sensitivity depend on the physical system as follows:

9.3.1.1. *Spectral sensitivity—$S(\lambda)$.* For simplicity of presentation we consider the mathematical expression for the rhabdomeres of the Diptera

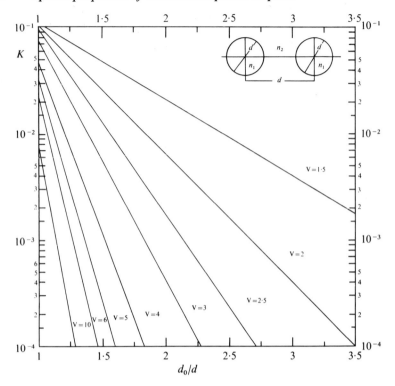

FIG. 9.12. Parameter K *versus* the separation between two parallel rods as shown in the figure. The results are only for mode 1 and are modified from Snyder (1972c). V is defined by eqn (9.2).

compound eye. As shown in Section 9.2.2, for on-axis illumination only one mode is excited. Thus from eqn (9.9b) the spectral sensitivity is

$$S(\lambda) = P(\lambda)[\{1 - \exp - \gamma(\lambda)l\}\cos^2\phi + \{1 - \exp - \gamma(\lambda)l/\Delta\}\sin^2\phi] \quad (9.21)$$

where $P(\lambda)$ is the power of mode 1 found from eqn (9.34a) of Snyder (1969b) when the spot size defined by eqn (9.3) and rhabdom parameters are specified. The absorption coefficient is defined as

$$\gamma(\lambda) = C\alpha(\lambda)\eta_1(\lambda) \quad (9.22)$$

where C is the concentration and $\alpha(\lambda)$ the spectral absorption (extinction coefficient) of the photosensitive materials within the rhabdomere. The fraction of the power of mode 1 within the rhabdomere is η_i (Fig. 9.2). $\Delta(\Delta \geqslant 1)$ is the dichroic sensitivity of the pigment medium and is discussed in detail in section 9.3.1.3 below. l is the length of the rhabdomere and ϕ is the angle of the electric vector with the long axis of the microvilli. For unpolarized light $\cos^2\phi$ and $\sin^2\phi$ are replaced by $\frac{1}{2}$.

It is clear from eqn (9.21) that *the spectral sensitivity* $S(\lambda)$ *of a rhabdomere* is in no simple way related to the spectral sensitivity $\alpha(\lambda)$ of the photosensitive pigments within it. The physical properties of the rhabdom (diameter and index of refraction) determine $\eta_1(\lambda)$. The diffraction properties of the lens and rhabdomere are reflected in $P(\lambda)$. If we ignore waveguide and diffraction effects $P(\lambda) = \eta_1(\lambda) = 1$ in eqn (9.21).

The spectral sensitivity depends in a rather complicated manner on Cl, Δ, ϕ, the diameter, and refractive index of the rhabdomere expressed through $\eta_1(\lambda)$. It is therefore useful to study separately the dependence of $S(\lambda)$ on each of these variables. We do this next.

Dependence of $S(\lambda)$ on the dichroic properties of the rhabdom. Equation (9.21) shows that, in general $S(\lambda)$ depends on the dichroic nature of the rhabdomere expressed through Δ. $\Delta = \infty$, 1 corresponds to S_{max}/S_{min} equal to ∞, 1, respectively. It is interesting to see if the shape of the $S(\lambda)$ curve is affected by changes in the value of Δ. Figure 9.13 shows that for $\Delta = \infty$ or 1 there is no change in $S(\lambda)$ and that for other values of Δ, for example, $\Delta = 2$, there is little change. Our results are for unpolarized light with $Cl = 1.5$.

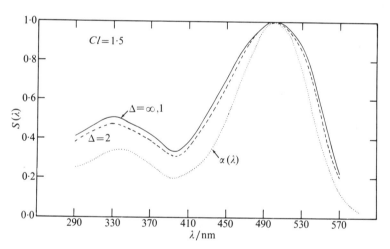

FIG. 9.13. The spectral sensitivity $S(\lambda)$ *versus* λ for several values of the dichroic sensitivity Δ of the photopigment. The curve is for unpolarized light. The dotted line is the extinction property of the photopigment. $S(\lambda)$ is normalized to a maximum of unity.

Decreasing Cl decreases the variation of $S(\lambda)$ with Δ. We have ignored diffraction and waveguide effects ($\eta = P = 1$) for these results.

We conclude that $S(\lambda)$ depends weakly on the dichroic property (Δ) of the rhabdom. Thus $S(\lambda)$ given by eqn (9.21) can be simplified by setting Δ equal to any convenient value, e.g. $\Delta = 1$ which leads to

$$S(\lambda) \approx P(\lambda)\{1 - \exp(-Cl\alpha(\lambda)\eta_1(\lambda))\} \qquad (9.23)$$

Dependence of S *(λ) on length and concentration of the photopigment*. In Fig. 9.14 we show the effect of increasing the product of the pigment concentration and the length of the rhabdomere. *There is a broadening or flattening effect in* S(λ) *as Cl increases.* We have ignored again diffraction and waveguide effects, i.e. $P(\lambda) = \eta(\lambda) = 1$.

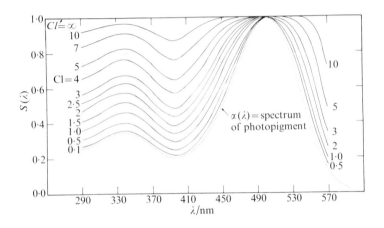

FIG. 9.14. The spectral sensitivity $S(\lambda)$ *versus* λ for values of pigment concentration C multiplied by the length of the photoreceptor l. The dotted line $\alpha(\lambda)$ is the extinction property of the photopigment. $S(\lambda)$ normalized to a maximum of unity. Diffraction and waveguide effects have been neglected, i.e. $\eta = 1$ (Fig. 9.2) and $P(\lambda) = 1$ in eqn (9.23). For results on dragonfly attributed to this effect, see Horridge (1969).

As a direct consequence of this effect, $S(\lambda)$ due to light polarized along the microvilli axis will have a broader spectrum than light polarized perpendicular to the microvilli axis.

Dependence of S(λ) *on the diameter and index of refraction of the rhabdom.* Here we investigate the effect of confining the photopigment to within a narrow cylinder, i.e. the waveguide effects. The diffraction effects of the dioptric apparatus and rhabdom are intentionally neglected, i.e. $P(\lambda) = 1$. Figure 9.15 shows the result for rhabdomeres typical of the Diptera compound eye.

We conclude that the effect of confining the photopigment to within a narrow cylinder is (a) to shift the visible absorption peak to shorter wavelengths and (b) to increase the u.v. peak absorption relative to the visible. The smaller the diameter of the rhabdom the larger the effect. This result is due to the dielectric waveguide properties of the rhabdom as determined from $\eta_1(\lambda)$ of Fig. 9.2. In Fig. 9.16 we show the results for a family of different photoreceptors characterized by their V values (eqn (9.2)). The curve can be used to determine the importance of waveguide effects on $S(\lambda)$ for many photoreceptors.

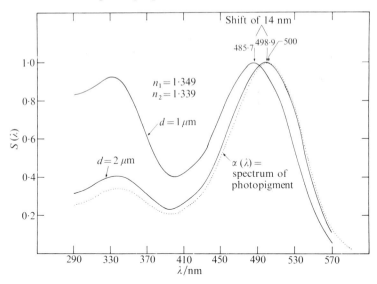

FIG. 9.15. Spectral sensitivity $S(\lambda)$ *versus* λ when the photopigment (with extinction $\alpha(\lambda)$) is confined within a rhabdomere of diameter $d = 1$ or 2 μm. n_1 and n_2 are the index of refraction for the rhabdomere and its surround respectively. $C/\eta \leqslant 0.5$. Curves are normalized to unity in the visible. Diffraction effects have been ignored i.e. $P(\lambda) = 1$ in eqn (9.23).

In the next section we show that the enormous u.v. sensitivity of rhabdomeres of the Diptera compound eye is consistent with the enhanced u.v. sensitivity caused by waveguide effects. We also show that the different spectral sensitivities of rhabdomeres 1–6 and 7, 8 are due to their diameter differences rather than different photopigments.

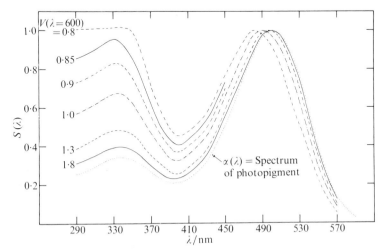

FIG. 9.16. Spectral sensitivity $S(\lambda)$ *versus* λ curves for an arbitrary rhabdomere characterized by its V value defined by eqn (9.2). $C/\eta \leqslant 0.5$. Curves are normalized to unity in the visible. Diffraction effects have been ignored, i.e. $P(\lambda) = 1$ in eqn (9.23).

Retinula cells of the dipteran eye. This section has several objectives:

(1) To further elucidate the influence on spectral sensitivity (particularly in the u.v. region) of confining a photopigment within a rhabdomere.

(2) Trujillo-Cenóz and Melamed (1968) have shown that rhabdomere 8 lies under rhabdomere 7 as schematically illustrated in Fig. 9.17. This arrangement leads one to speculate that rhabdomere 7 acts like a colour filter for rhabdomere 8.

(3) We wish to calculate the spectral characteristics of the photosensitive material within the rhabdomeres.

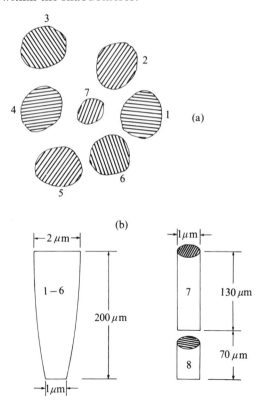

FIG. 9.17. Rhabdomeres of the Diptera compound eye. (a) is a schematic of a cross-section in the distal end. The parallel dark lines represent the microvilli. (b) represents a schematic of the rhabdomeres in a longitudinal section. Rhabdomeres 1–6 are tapered with the largest diameter at the distal end. The drawings are from the electron micrographs of Melamed and Trujillo-Cenoz (1968) and Boschek (1972).

Spectral sensitivity curves for individual retinula cells of the fly were first obtained by Autrum and Burkhardt (1961) and Burkhardt (1962) using intracellular electrodes. *Three* receptor types, green (G), blue (B) and yellow–green (Y–G), were found in the ventral half of the fly *Calliphora*. The most numerous type (G) was recorded five times more frequently than the other types. The peaks were distributed statistically with G-486 nm, B-470 nm and Y–G-521 nm. G and B have very large u.v. peaks as shown in Fig. 9.18. The results were found to be independent of the absorption property of screening pigments (Burkhardt 1962). In these experiments it was not known which of the retinula cells was recorded.

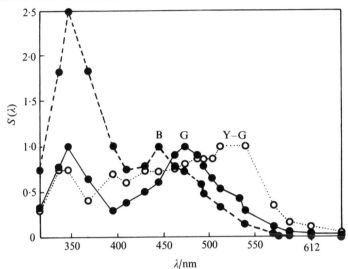

FIG. 9.18. Typical spectral sensitivity curves of individual retinula cells. The maxima in the visible is normalized to unity. The circles represent test points. B = blue, G = green and Y–G = yellow–green. This curve does not show the statistical averaged results discussed in the text. (Modified from Burkhardt (1962)).

McCann and Arnett (1972) marked the retinula cells from which they recorded by using intracellular staining techniques. They found that the spectral sensitivity of *Calliphora* retinula cells 1–6 was the G type. They also found similar results when recording from on–off units and sustaining units; however, they were unable to record directly from cells 7 or 8.

This evidence provides direct proof that retinula cells 1–6 are the G type discussed above.

Microspectrophotometric measurements by Langer and Thorell (1966a,b) have established that rhabdomeres 1–6 have a different absorption spectra than rhabdomeres 7 and 8 *measured together*. They find a green absorption in the large rhabdomeres and a blue absorption in the central rhabdomeres.

Spectral sensitivity measurements have been made by Eckert (1971) based on the optomotor reaction. Dim light at threshhold intensities and a long spatial wavelength of 10·6° gives an $S(\lambda)$ curve with a visible peak at $\lambda =$ 486 nm and an enormous maximum (2·5 times that at $\lambda = 486$ nm) in the u.v. at $\lambda = 360$ nm. Using higher light intensities and a spatial wavelength of 3° gives an $S(\lambda)$ curve with only one peak (at $\lambda = 464$ nm) and very poor sensitivity in the u.v. Eckert's results are presented on a linear plot by Menzel (1973). Eckert (1971) alleged that the curve with $\lambda_{max} = 486$ nm was due to rhabdomeres 1–6. This is not possible as it clearly violates the unequivocable evidence of McCann and Arnett (1972) that rhabdomeres 1–6 are identically the G-type measured by Burkhardt (1962). Eckert's (1971) result has a u.v. peak 2·5 times as large as the correct G-type receptor. (His original results plotted on a log scale disguise this fact.) The u.v. peak looks remarkably similar to Burkhardt's (1962) B-type u.v. curve and suggests that optomotor responses are a product of the outputs of rhabdomeres 1–6 and either or both of 7 and 8. Eckert further alleged that the single peaked curve with $\lambda_{max} = 464$ nm is that of rhabdomere 7 and 8.

In the light of the incorrect result for rhabdomeres 1–6 we dismiss the validity of spectral sensitivity curves based on optomotor responses. Furthermore, due to the difficulties of end-on microspectrophotometry (Liebman 1972; Snyder and Miller 1972) we are cautioned in the detailed interpretation of Langer and Thorell's (1966a,b) results. Thus only Burkhardt's (1962) measurements are used for comparison with the theory that follows.

We have already discussed in the last section the effects on the spectral sensitivity $S(\lambda)$ of confining photopigment to within a long narrow receptor like fly rhabdomere. Using a similar theory Snyder and Miller (1972) interpreted the results of microspectrophotometry. They showed that the smaller more centrally located rhabdomeres 7 and 8 *measured together* have a different spectral absorption property than the larger rhabdomeres 1–6 because of the size difference rather than because of differences in photopigments within them. Their analysis was for the difference spectrum and is different from $S(\lambda)$ determined by single cell electrophysiology. Here we present the theoretical results for $S(\lambda)$, i.e. the spectral sensitivity of a cell as measured electrophysiologically.

Rhabdomeres 1–6 have the form of a frustrum of a cone (Fig. 9.17) with a diameter of c. 2 μm at the distal ends and 1 μm at the proximal ends while rhabdomeres 7 and 8 are of a constant diameter 1 μm (Boschek 1971). Rhabdomere 7 is twice as long as rhabdomere 8 (Melamed and Trujillo-Cenóz 1968). The length of the rhabdomeres 1–6 is c. 200 μm, which is equal to the sum of the lengths 7 and 8.

The index of refraction of the rhabdomeres is $n_1 = 1·349$, of the region between the rhabdomeres $n_0 = 1·336$ and of the retinula cells $n_R = 1·341$. To determine the dimensionless parameter V defined by eqn (9.2) we take n_2 to be the average of n_0 and n_R, i.e. $n_2 \approx 1·339$ leading to

$$V = 0·52(d/\lambda), \tag{9.24}$$

where d is the diameter of the rhabdomere and λ the wavelength of the light in vacuum.

The absorption coefficient $\gamma = C\alpha(\lambda)\eta(\lambda)$ has been estimated for visible light by Kirschfeld (1969) to be $\gamma \approx 7\cdot5 \times 10^{-3}~\mu m^{-1}$ with a dichroic sensitivity $\Delta \approx 2\cdot3$ (see also Shaw 1969*b*). Because of the findings of Snyder and Miller (1972) we make the assumption that all rhabdomeres have the identical photosensitive materials within them; however, we do not as yet know their spectral sensitivity or extinction $\alpha(\lambda)$ characteristics.

Because of the complexity of the dioptric apparatus it is difficult to have a precise knowledge of the light illuminating the rhabdom. The work of Kuiper (1966), Seitz (1968), and Kirschfeld (1969) indicates that the focus is at or near the distal ends of the rhabdomeres. A representative ommatidium of the Diptera eye has a cornea lens of 25 μm and a focal length of *c.* 50 μm (Seitz 1968). Thus the diameter d_a of the non-uniform patch known as the Airy disc is from eqn (9.3)

$$d_a \approx 3\cdot64\lambda \qquad (9.25)$$

where λ is in μm. The result depends on the region of the eye. Fortunately our conclusions are not very sensitive to the accuracy of this approximation.

Before we present our theoretical results it is useful to discuss first the philosophy of what we expect to happen. From the last section we learned that by confining a photopigment to within a narrow cylinder the peak absorption in the visible was shifted to lower wavelengths and the u.v. sensitivity peak was increased. The effect was larger, the smaller the rhabdom diameter (Fig. 9.16). We now add to this effect the properties of diffraction of the rhabdom and lens $\{P$ in eqn (9.23)$\}$ and the effect of length (l) and concentration (C) of photopigment.

Our results for rhabdomeres 1–6 are shown in Fig. 9.19. The dotted line represents the extinction of the photopigment (or pigments) used to obtain the result. On the same curve we show the theoretical results for rhabdomere 7. The enormous u.v. peak is in complete agreement with Autrum and Burkhardt's (1961) and Burkhardt's (1962) single cell electrophysiology for the B receptor. Thus we assume that their B receptor is rhabdomere 7, an assumption which is reasonable since it was only measured $\frac{1}{5}$ as often as the G type.

Rhabdomere 8 is below a rhabdomere (NR7) which absorbs an enormous amount of the incident u.v. light. Thus it is expected that much less u.v. light enters rhabdomere 8 than 7 so that rhabdomere 7 u.v. sensitivity is considerably reduced. Burkhardt's (1962) Y–G type curve shows a u.v. response lower than its visible peak and very much lower than the u.v. peak of the B receptor which we have shown to be consistent with rhabdomere 7. The photosensitive material (or materials) within the rhabdom that is consistent with all these results is shown as the dotted line in Fig. 9.19.

In conclusion we have shown theoretically that there are three spectral receptor types in the Diptera eye consistent with the single cell measurements of Autrum and Burkhardt (1961) and Burkhardt (1962). The spectral types arise

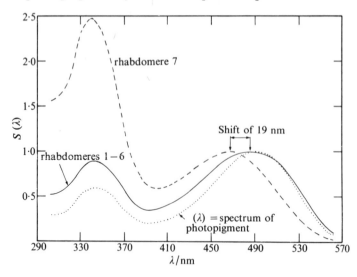

FIG. 9.19. Theoretical spectral sensitivity $S(\lambda)$ of rhabdomeres of the fly *Calliphora*. The dotted curve is the extinction spectrum of the photosensitive material within the rhabdomeres used to obtain the results shown in the figure. The photosensitive material in each rhabdomere is assumed identical. Agreement with Burkhardt's (1962) statistical results is excellent both for the relative positions of the visible absorption peaks as well as the magnitude of the u.v. absorption peaks. Thus we assume rhabdomeres 1–6 are the green (G) type and rhabdomere 7 is the blue (B) type of Fig. 9.18.

due to physical differences in the rhabdomeres not due to different photopigments. The effect of confining the photopigment to within a fly rhabdomere is to (a) shift the visible peak to lower wavelengths and (b) increase the u.v. peak relative to the visible. The smaller the rhabdom the greater the effect. Thus the enormous u.v. sensitivity of the fly is due in a large part to the physical properties of the rhabdom rather than the sensitivity of the photopigments. Rhabdom 8 is below rhabdom 7, which has an enormous u.v. absorption. Thus rhabdomere 7 acts as a u.v. colour filter reducing the u.v. sensitivity of rhabdomere 8 and altering its $S(\lambda)$ curve. Spectral sensitivity curves based on optomotor response characteristics are incorrect at least in the u.v. region of the spectrum. A more thorough analysis of this problem is given by Snyder and Pask (1973c).

Changes in $S(\lambda)$ due to off-axis light in dragonfly. In a study of the dragonfly *Aeschna*, Eguchi (1971) showed electrophysiologically that the spectral sensitivity of retinula cells differ for on- and off-axis light, as shown in Fig. 9.20. Snyder and Pask (1972c) demonstrated that Eguchi's measurements were consistent with the diffraction properties of the dragonfly rhabdom. We present a simplified form of their analysis here.

The mathematical expression $S(\lambda)$ for a retinula cell of a fused rhabdom like the dragonfly is complicated. We can avoid the mathematical details by

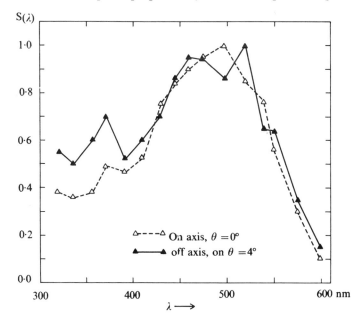

FIG. 9.20. Spectral sensitivity $S(\lambda)$ of the dragonfly *Aeschna* due to on- and off-axis light. Modified from Eguchi (1971).

noting that $S(\lambda)$ only depends on the angle of incidence θ_i through the modal power $P_i(\lambda)$ and not through the absorption coefficient γ_i. Thus if we ignore all other effects of the dragonfly rhabdom shape except for the one which depends upon θ_i we simplify our analysis.

The diameter of the dragonfly rhabdom is 1·5 μm (Eguchi 1971). We take 1·35 and 1·33 as the approximate index of refraction for the rhabdom and its surround respectively.

With these parameters the light transmitted into the rhabdom $T(\theta)$ is found using the analysis of Snyder (1969b). By light transmitted we mean the sum of all modal power. This is equivalent to determing the light acceptance property of the dragonfly rhabdom as discussed in Section 9.2.2. Our results are shown in Fig. 9.21. For light incident normally on the rhabdom, $T(\theta)$ is relatively uniform. However, for light off-axis ($\theta_R > 0$) there is a dip in the curve at c. $\lambda = 490$ nm.

We have been discussing the angle θ_R of light incident on the rhabdom rather than the angle θ_i of light incident on the ommatidium. The relation between θ_i and θ_R depends on the dragonfly's dioptric apparatus and is determined by a ray-tracing analysis like that on the worker bee (Varela and Wiitanen 1970). At present there is no detailed analysis of the dragonfly's dioptric apparatus: however, we should keep in mind that $\theta_R = 2\theta_i$ for the

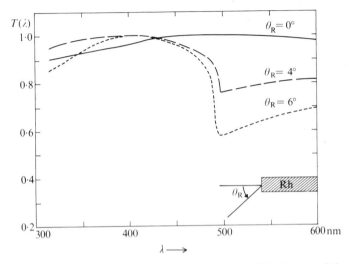

FIG. 9.21. The light power $T(\theta)$ transmitted within the rhabdom (Rh) determined theoretically for light on axis ($\theta_R = 0$) and off axis at 4° and 6°.

worker bee (Snyder 1972a) and that it is likely that $\theta_R > \theta_i$ for the dragonfly as it is for most lens systems (Born and Wolf 1965).

With these concepts it is now possible to interpret Eguchi's (1971) measurements. We know that the dragonfly retinula cell spectral sensitivity is proportional to the light $T(\lambda)$ transmitted along the rhabdom multiplied by the absorption property of the rhabdom.

For normal incidence $T(\lambda)$ is relatively uniform and $S(\lambda)$ depends only on the absorption property of the rhabdom leading to Eguchi's (1971) $\theta_i = 0$ curve (Fig. 9.20). For light off axis the product of $T(\lambda)$ multiplied by Eguchi's (1971) $\theta_i = 0$ curve leads to an $S(\lambda)$ with a dip at approximately 490 nm.

We conclude that changes in spectral sensitivity due to off-axis light in dragonfly are consistent with the diffraction properties of the rhabdom. Further details are given by Snyder and Pask (1972c).

Anomalous dispersion (changes of n *with* λ). The refractive index of a light absorbing material fluctuates in the wavelength region of its peak absorption. The relationship between the refractive index, $n(\lambda)$, and the absorption property of the pigment characterized by its extinction coefficient, $\alpha(\lambda)$, is shown in Fig. 9.22. $n(\lambda)$ increases for values of λ greater than λ_0 and decreases for values of λ less than λ_0. This behaviour is known as anomalous dispersion and is discussed by Lipson and Lipson (1969).

The change in the refractive index is very slight so that it has little effect on the absorption property of the photopigment in solution. However as shown by Snyder and Richmond (1972 and 1973) the effect becomes evident when the pigment is confined within a photoreceptor. This is because the

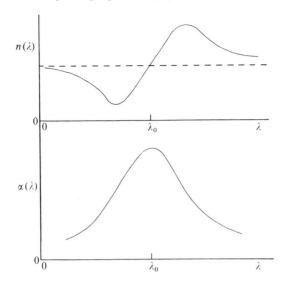

FIG. 9.22. The relationship between the refractive index n and the extinction coefficient α for a light-absorbing material. The upper curve represents the refractive index of the material. The dashed curve is the refractive index of the solvent or background substance in which the light absorbing material is dissolved. The lower curve represents the corresponding extinction coefficient or absorption property of the medium.

difference in index of refraction between the photoreceptor and its surrounding medium is small so that a slight decrease in refractive index of the photopigment can alter the total internal reflection property of the photoreceptor and hence its absorption spectrum. Since the decrease in refractive index occurs at values of $\lambda < \lambda_0$, we anticipate a perturbation of the absorption spectrum for $\lambda < \lambda_0$.

Since we wish to compare our theoretical calculations with those found by microspectrophotometric measurements, we use the expression for the difference spectrum (ΔE) derived by Snyder and Miller (1972)

$$\Delta E(\lambda) \approx C\eta_1^2(\lambda)\alpha(\lambda) \tag{9.26}$$

where C is a constant.

For $\alpha(\lambda)$ we take Dartnall's (1962) normalized optical density spectrum with $\lambda_0 = 500$ nm. For the photoreceptor we take a diameter of 2 μm and a 2 per cent index of refraction difference between it and the surrounding medium. The rhabdomeres 1–6 of the fly *Calliphora* closely resemble this model.

The fraction of modal light, $n_1(\lambda)$, within a photoreceptor depends on the dielectric difference between the photoreceptor and its surrounding medium, i.e. on V defined by eqn (9.2). If the refractive index changes with wavelength, then the amount of modal light within the photoreceptor must also change.

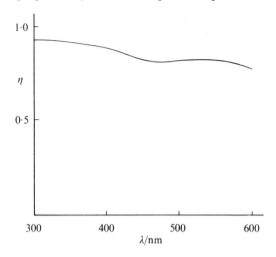

FIG. 9.23. The fraction of modal light intensity within a photoreceptor 2 μm in diameter with a 2 per cent index of refraction difference between the rod and its surrounding medium. Curve A is for a 0·25 per cent maximum change in index of refraction discussed in connection with Fig 9.22. The fibre is assumed to contain a photopigment, like the normalized optical density spectrum given by Dartnall (1962), with a peak absorption at $\lambda_0 = 500$ nm.

Our calculations show that $\eta(\lambda)$ changes with wavelength as shown in Fig. 9.23. The curve is based on a 0·25 per cent maximum change in refractive index discussed in connection with Fig. 9.22. The depression in the curve at $\lambda \approx 470$ nm is due to the decrease in refractive index shown in Fig. 9.22.

Having determined $\eta_1(\lambda)$ it is a simple matter to find the photoreceptor absorption difference spectrum, $\Delta E(\lambda)$, from eqn (9.26). The solid curve of

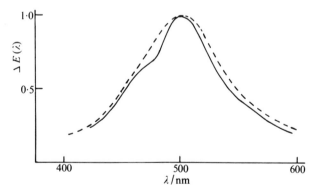

FIG. 9.24. Spectral sensitivity expressed as a difference extinction spectrum $\Delta E(\lambda)$ for the photoreceptor described in Fig. 9.23. The dashed curve is the absorption spectrum of the visual pigment. The solid curve is the result of the theoretical analysis.

Fig. 9.24 illustrates the results normalized to unity. The dashed curve represents the assumed absorption property, $\alpha(\lambda)$, of the visual photopigment in solution. The changes in refractive index are so small that they would be very difficult to measure. Yet, as we see from Fig. 9.24, they are sufficient to produce a noticeable perturbation in the absorption spectrum of the photo-receptor. A general feature of the result is that the perturbation always manifests itself as a depression in the absorption curve at wavelengths lower than the peak absorption. Because the expected effect is small, it is difficult in searching the spectrophotometric literature to cite undoubted changes in extinction resulting from anomalous dispersion. Examples must be chosen from end-on microspectrophotometric measurements of single photoreceptor organelles because the side-on measurements do not emphasize waveguide effects. The reader is referred to Fig. 9.4 of Langer and Thorell (1966), the mean curve for thirty-nine single end-on extinction curves of fly rhabdomeres 1–6 which shows a slightly less prominent inflection but similar to that of Fig. 9.24 and in the same location. Because of the similarity between pre-viously reported measurements and our predictions, we believe that future end-on microspectrophotometric measurements should be examined care-fully in the critical region of wavelengths shorter than the peak absorption in order to evaluate the role of anomalous dispersion in photoreception.

Rhabdomeres with small V values have the largest bumps in the absorption curve. This is because the η *versus* V curve (Fig. 9.2) is steepest for small V, where small changes in V produce the greatest changes in η. Although our analysis and discussion was directed at microspectrophotometric measure-ments only, anomalous dispersion should also appear in single cell electro-physiology.

Effects of lens diffraction. The ability of an optical system to concentrate light is limited by the wave nature of the light, the smaller the aperture, the larger the size of the focused spot. The size of this spot (eqn (9.3)) is directly proportional to wavelength, i.e. it is twice as large at 600 nm as it is at 300 nm. Unless the spot size at $\lambda = 600$ nm is smaller than the receptor cross-section it would appear that light at 600 nm is less effective than at $\lambda = 300$ nm because the area of the spot is larger than the geometrical cross-section of the photoreceptor.

This indeed would be the case if it were not for the fact that the photo-receptor has an effective light capture area which is larger than its geometric cross-section (Snyder and Hamer 1972). Furthermore, the longer the wave-length the greater the effective capture area. This means that as the spot size due to diffraction becomes larger the capture area of the photoreceptor also increases. The effects work in concert to compensate each other (Snyder and Pask 1973b). Stated in another way, *the diffraction properties of the photo-receptor compensates for diffraction of the dioptric apparatus.*

9.3.1.2. *Polarization sensitivity* S_{pol}. The polarization sensitivity of a retinula cell, S_{pol}, depends in a complicated manner on the physical parameters of the rhabdomeres and the concentration, excitation, and dichroic properties of the photosensitive material within the rhabdomere. Thus we first study the effects on the S_{pol} of each of these variables separately to develop physical insight into the mechanism of polarization sensitivity. This allows us to make a comparative study of polarization sensitivity in various insect photoreceptors.

We begin by discussing rhabdomeres of the Diptera. Then we consider other rhabdoms including those of Crustacea.

The polarization sensitivity of Diptera rhabdomeres is given by eqn (9.21). Maximum absorption occurs at $\phi = 0°$ while minimum absorption at $90°$. We define the polarization sensitivity S_{pol} *of a retinula cell* as

$$S_{pol} = \frac{S(0°)}{S(90°)} = \frac{1 - \exp(-Cl\alpha(\lambda)\eta_1(\lambda))}{1 - \exp(-Cl\alpha(\lambda)\eta_1(\lambda)/\Delta)} \qquad (9.27)$$

where l is the length of the rhabdomere, C the concentration and $\alpha(\lambda)$ the extinction coefficient of the photosensitive material and $\eta_1(\lambda)$ is the fraction of modal light power within the rhabdom (Fig. 9.2).

$\Delta(\Delta \geqslant 1)$ is the polarization (dichroic) sensitivity of the material within the rhabdomere. The dichroism of this material is due to the dichroism of the individual pigment molecules which are aligned along the microtubules. The greater the alignment the greater Δ. The S_{pol} of Diptera type rhabdomeres eqn (9.27) approaches Δ when $Cl\alpha\eta \rightarrow 0$, i.e. when the absolute sensitivity of the rhabdomere becomes small. The S_{pol} of a retinula cell must be less than or at most equal to Δ.

Dependence of the S_{pol} on Cl and Δ. The dependence on γl and Δ of the S_{pol} is given by Fig. 9.25 for the situation in which $\alpha(\lambda) = 1$ and waveguide effects are ignored ($\eta = 1$). Increasing either the length or the concentration of photopigment decreases the S_{pol}. A particular S_{pol} can be achieved by an infinite variety of Δ, γl combinations. We show later that the largest dichroic sensitivity Δ yet measured in insects is $\Delta \approx 10$. From Fig. 9.14 we see that for $Cl \geqslant 10$ the spectral sensitivity $S(\lambda)$ is very flat, flatter than most $S(\lambda)$ measurements. Thus Fig. 9.25 probably applies to most photoreceptors.

Knowing $S(\lambda)$ and the S_{pol} at $\alpha(\lambda) = 1$ we can estimate Cl and Δ. Cl is estimated from Fig. 9.14 and Δ from Fig. 9.25.

We conclude that decreasing the product (Cl) of photopigment concentration (C) and receptor length (l) increases the polarization sensitivity S_{pol} of the photoreceptor. The relationship between S_{pol}, Cl and the dichroic property of the photopigment (Δ) is shown in Fig. 9.25.

Dependence of S_{pol} on $\alpha(\lambda)$. The absorption coefficient γ is proportional to $\alpha(\lambda)$ thus we obtain the result that the maximum S_{pol} occurs at the minimum

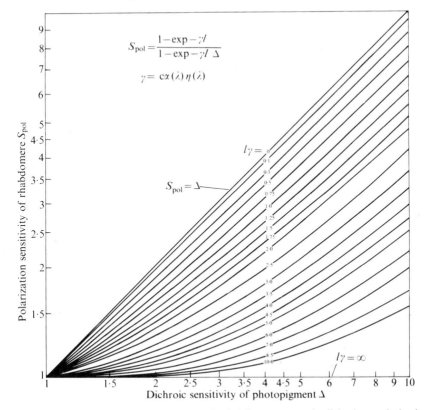

FIG. 9.25. The polarization sensitivity (S_{pol}) of a rhabdomere *versus* the dichroic or polarization sensitivity Δ of the photopigment for various values of the absorption coefficient times the photoreceptor length γl. $\gamma = C$ when waveguide effects $\eta = 1$ (Fig. 9.2) and when the extinction coefficient $\alpha(\lambda) = 1$. Note that the S_{pol} equals Δ only when $l\gamma = 0$, i.e. when the cell has negligible absolute sensitivity.

sensitivity of the receptor. This is illustrated for a particular example in Fig. 9.26. At $\lambda = 500$ nm, $S_{pol} \approx 1\cdot5$ while at $\lambda = 400$ nm, $S_{pol} \approx 2$. The curve looks very similar to the unpublished results of Langer in Kirschfeld (1969) for the S_{pol} of *Calliphora* determined by microspectrophotometric methods. Thus at λ values corresponding to a minimum receptor sensitivity the S_{pol} is more closely that of the dichroic sensitivity of the photopigment (Δ) than at λ values corresponding to maximum sensitivity. In our theoretical work we have ignored any dependence of the dichroic nature of the photopigment on λ; however, the agreement of Fig. 9.26 with the results of Langer in Kirschfeld (1969) appears to show that the dependence of Δ on λ is small.

We conclude that the maximum S_{pol} occurs at λ values corresponding to the minimum absolute sensitivity of the rhabdomere.

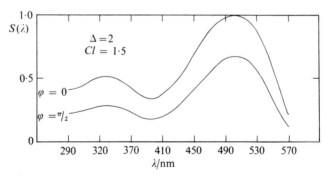

FIG. 9.26. Spectral sensitivity $S(\lambda)$ *versus* λ for light polarized along the microvilli ($\phi = 0$) and perpendicular to the microvilli ($\phi = \pi/2$). The S_{pol} is defined as $(S$ at $\phi = 0°)/(S$ at $\phi = 90°)$ and is larger at $\lambda = 400$ nm than at $\lambda = 500$ nm. We have ignored waveguide effects, i.e. η_i from Fig. 9.2 is equal to unity.

Dependence of the S_{pol} on the diameter and index of refraction of the rhabdomere. The effect of confining the dichroic photopigment within a narrow rhabdomere is found by examining the absorption coefficient $C l \alpha(\lambda) \eta_1(\lambda)$. Knowing the physical parameters (diameter and index of refraction), $\eta_1(\lambda)$ can be obtained from Fig. 9.2.

Since $\eta_i(\lambda) < 1$ its presence decreases the absorption coefficient and thus increases the S_{pol} of the rhabdomere. For example, let us confine a photosensitive material with $\gamma l = 1$ and $\Delta = 2.5$ to within a rhabdomere of diameter $d = 1\ \mu$m, $n_1 = 1.339$, $n_2 = 1.349$ and $\lambda \approx 500$ nm ($V = 1.04$) for which we take $\alpha(\lambda) \approx 1$. From Fig. 9.2, $\eta \approx 0.17$. Without waveguide effects ($\eta = 1$) the S_{pol} is found from Fig. 9.25 to be 1.9. With waveguide effects $\eta C l = 0.17$ and the $S_{pol} = 2.4$.

We conclude that the effect of confining a dichroic photopigment within a narrow rhabdomere is to increase its S_{pol} and lower its absolute sensitivity.

Changes in the S_{pol} due to light off-axis. We showed in Section 9.2.2 that only one mode is significantly excited for on-axis illumination of rhabdomeres. Off-axis light excites higher order modes (Fig. 9.7). The $\eta_i(\lambda)$ for higher order modes (Fig. 9.2) is less than that of mode 1 so that the S_{pol} due to higher order modes is greater than for mode 1. Based on this fact *we predict that the S_{pol} due to off-axis light in Diptera is larger than the S_{pol} due to light on-axis.* The magnitude of this difference cannot be easily determined without a thorough analysis of the dioptric apparatus.

S_{pol} for a rhabdomere below another rhabdomere. It is interesting to investigate the consequences on the S_{pol} of a rhabdomere when it is below another rhabdomere. Such a situation exists for rhabdomeres 7 and 8 of the Diptera and the tiered rhabdom of the dragonfly (Horridge 1969; Eguchi 1971).

We consider the rhabomeres of Fig. 9.27 with their microvilli arranged perpendicular to each other.

The light power transmitted from the top (t) into the bottom (b) rhabdomere is given as

$$P^x = \exp - \gamma_t l_t / \Delta_t; \ x\text{-polarized light} \tag{9.28a}$$

$$P^y = \exp - \gamma_t l_t; \ y\text{-polarized light} \tag{9.28b}$$

The light power absorbed by rhabdomere 8 is

$$P^x_A = \exp - \gamma_t l_t / \Delta_t \{1 - \exp - \gamma_b l_b\} \tag{9.29a}$$

$$P^y_A = \exp - \gamma_t l_t \{1 - \exp - \gamma_b l_b / \Delta_b\} \tag{9.29b}$$

S_{pol} is then given as

$$S_{pol} = \frac{P^x_A}{P^y_A} \tag{9.30a}$$

$$= \exp \gamma_t l_t \left(1 - \frac{1}{\Delta_t}\right)$$

$$\times (\text{the } S_{pol} \text{ for the bottom rhabdomere alone}) \tag{9.30b}$$

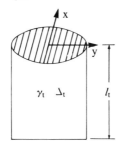

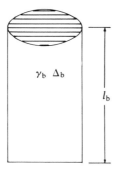

FIG. 9.27. Two rhabdomeres one on top of the other. The upper rhabdomere is characterized by an absorption coefficient γ_t, dichroic photopigment sensitivity Δ_t and length l_t. The bottom rhabdomere has microvilli that are perpendicular to the top rhabdomere and is characterized by an absorption coefficient γ_b, length l_b, and dichroic sensitivity Δ_b.

Since $\exp \gamma_t l_t (1 - 1/\Delta_t)$ is greater than unity, the S_{pol} of the bottom rhabdomere is greater than if it were alone. Furthermore, the longer the top rhabdomere the more the S_{pol} of the bottom rhabdomere is enhanced.

As an example take $\gamma_t l_t = 1\cdot2$ and $\Delta_t = 2\cdot5$. Then the S_{pol} of rhabdomere 8 is twice as large as it would be alone.

We conclude that the effect of placing one rhabdomere on top of another is to increase the S_{pol} *of the lower rhabdomere. The longer the top rhabdomere the greater the effect. Furthermore, as discussed in section 9.3.1(a), the upper rhabdomere acts as a colour filter altering the spectral sensitivity of the lower.*

S_{pol} *for retinula cells of Diptera.* Knowing the parameters of the Diptera rhabdomeres (e.g. *Calliphora*) we can use Fig. 9.25 to determine a theoretical S_{pol}. Conversely we can start with the measured S_{pol} and from Fig. 9.25 provide combinations of the absorption coefficient $Cl\alpha\eta$ and dichroic ratio Δ consistent with the S_{pol}.

We start with the measured $S_{pol} = 2$ from the single cell electrophysiology of Autrum and von Zwehl (1962) and Scholes (1969). This value is assumed to apply to rhabdomeres 1–6 because they are more numerous than 7 and 8.

There are two inferred estimates of γ_1 and Δ based on these S_{pol} measurements. Kirschfeld (1969) gives $\gamma_1 \approx 7\cdot5 \times 10^{-3}$, $\Delta \approx 3$ and Shaw (1969*a,b*) gives $\gamma_1 \approx 1\cdot8 \times 10^{-2}$, $\Delta \approx 5\cdot3$. With the assumption that all rhabdomeres have photopigments with the same dichroic (Δ) property, we estimate the S_{pol} of rhabdomeres 7 and 8 from these numbers.

Rhabdomeres 7 and 8 are both shorter and of smaller diameter than 1–6. We have already discussed in Sections (1) and (3) above how this increases their S_{pol}. For rhabdomere 7

$$l_7\gamma_7 = \frac{l_7\,\eta_7}{l_1\,\eta_1}\,\gamma_1 l_1. \tag{9.31}$$

We know from the anatomy that $(l_7/l_1) \approx \frac{2}{3}$ (Melamed and Trujillo-Cenóz 1968). From eqn (9.24) $V = 2\,d$ at $\lambda = 500$ nm.

The diameter of rhabdomeres 7 and 8 is $d \approx 1\,\mu$m. Rhabdomeres 1–6 are tapered $d \approx 2\,\mu$m at the distal end at $d \approx 1\,\mu$m at the proximal end (Boschek 1972). As we explained in Section 9.2.3, the value of η must be integrated throughout the tapered length. The result of this at $\lambda \approx 500$ nm is as if we used an effective $d \approx 1\cdot6\,\mu$m.

Thus for rhabdomeres 1–6, $V \approx 3\cdot2$ and for 7 and 8, $V \approx 2$. From Fig. 9.2 we determine the appropriate η's for substitution into eqn (9.32). These values lead to

$$l_7\gamma_7 \approx 0\cdot52\gamma_1 l_1$$

Using Fig. 9.25 with $\gamma_1 = 7\cdot5 \times 10^{-3}$ ($\Delta = 3$) we obtain $S_{pol}, 7 = 2\cdot4$ and $\gamma_1 = 1\cdot8 \times 10^{-2}$ ($\Delta = 5$) leads to $S_{pol}, 7 = 2\cdot65$, i.e. a substantial increase from the original $S_{pol} = 2$.

Rhabdomere 8 is one-third the length of rhabdomeres 1–6. Furthermore it is below rhabdomere 7 a property which enhances its S_{pol} as discussed above. Thus

$$\gamma_8 \gamma_8 = \frac{l_8}{l_1} \frac{\eta_7}{\eta_1} \gamma_1 l_1 \tag{9.32a}$$

$$\approx 0.26 \gamma_1 l_1 \tag{9.32b}$$

Using Fig. 9.25 this leads to a S_{pol}, 8 = 2.65 when $\gamma_1 = 75 \times 10^{-3}$ and S_{pol}, 8 = 3.5 when $\gamma_1 = 1.8 \times 10^{-2}$ neglecting the fact that it is below rhabdom 7. From section (5) above we found that

$$S_{pol}, 8 = \exp \gamma_7 l_7 \left(1 - \frac{1}{\Delta}\right) S_{pol}, 8 \text{ alone} \tag{9.33}$$

We find that the correct S_{pol}, 8 = 3.5 when $\gamma_1 = 7.5 \times 10^{-3}$ and S_{pol}, 8 = 5.6 when $\gamma_1 = 1.8 \times 10^{-2}$.

In conclusion it appears that there are three classes of S_{pol} receptors in the fly. The least sensitive is the 1–6 system while the most sensitive is rhabdomere 8. This is consistent with Autrum and von Zwehl (1962) measurements in that they infrequently measured the S_{pol} of *Calliphora* to be as high as 5.

We mention here that our theoretical findings should only be compared with measurements based on single cell electrophysiology and not on measurements of light transmitted through the rhabdom. These latter results can be misleading particularly for rhabdomeres 7 and 8.

S_{pol} *of ninth cell of ant and bee.* In the proximal portion of the fused rhabdom of the ant (Menzel 1972*a*) and the bee (Perrelet 1970; Gribakin 1972) a ninth retinula cell with its associated rhabdomere is formed. At this level the cross-section of the rhabdom shows microvilli arranged at 120° rather than 90° to each other. The ninth cell bears the same relationship in its length to the total rhabdom length as rhabdomere 8 in the fly. We showed in Section (5) above that the S_{pol} for a rhabdomere below another rhabdomere was enhanced. For this reason rhabdomere 8 in the fly may have a greater S_{pol} than 7. Thus it would appear that the ninth retinula cell may also have greater S_{pol} than the other retinula cells. Gribakin (1972) states that the ninth retinula cell of the worker bee is either blue and or u.v. sensitive which is consistent with the suggestion that it is suitable for S_{pol}. Furthermore, Kirschfeld (1972) has shown that three reference directions rather than two are necessary for navigation by polarized light. In the region of the ninth cell the microvilli are in three directions 120° apart rather in two directions 90° apart in the dorsal part. See also Snyder, Menzel and Laughlin (1973), and Menzel and Snyder (1974).

Polarization sensitivity for retinula cells of crustacea.   A salient characteristic of the polarization sensitivity of fly retinula cells is its dependence on the absolute sensitivity (via γl), i.e. the lower the absolute sensitivity of the cell the greater its S_{pol}. Furthermore, the S_{pol} of fly retinula cells is necessarily less than the dichroic sensitivity (Δ) of the photopigment. As the absolute sensitivity becomes negligible, polarization sensitivity approaches Δ.

We now study the crustacean rhabdom. Our purpose is to provide the physical principles necessary to explain the significance of the rhabdoms layered or laminated-structure on the S_{pol}. In particular we show that the crustacean retinula cells have a large S_{pol} identically equal to the dichroic sensitivity (Δ) of the photopigment, without any sacrifice of absolute sensitivity.

Rhabdoms typical of crustacea show alternating layers of microvilli along their length. The microvilli of each layer are perpendicular to the microvilli of their neighbours as illustrated in Fig. 9.28.

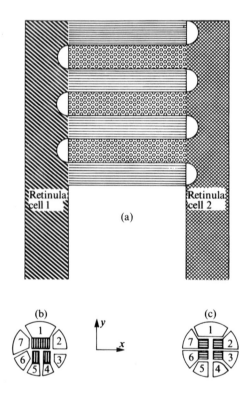

FIG. 9.28. Schematic of a typical crustacean rhabdom. (a) shows a longitudinal section through the rhabdom illustrating how the microvilli of each layer are perpendicular to the microvilli of the neighbouring layers. (b) and (c) show cross-sections of retinula cell 1 and 2 respectively, modified from Eguchi and Waterman (1966).

The mathematics necessary for analysis of this structure is presented in complete detail in Section 9.2.4. If the layers are all of approximately equal path length, then the light power absorbed, i.e. the absolute sensitivity of retinula cell 1 is found from eqn (9.20) of Section 9.2.4 to be

$$S_1 = \frac{A_1}{A}\left(\frac{\gamma_1}{\xi_1 + \xi_2}\right)\left\{1 - \exp -\frac{l}{2}(\xi_1 + \xi_2)\right\} \tag{9.34}$$

where A_1 is the area in cross-section of rhabdomere 1, A is the area in cross-section of the rhabdom, ξ_1 and ξ_2 are the rhabdom absorption coefficients for the layers made up in part of rhabdomeres 1 and 2 respectively, γ_i is the absorption coefficient for rhabdomere 1 and l is the length of the rhabdom.

From the symmetry of the rhabdom of Fig. 9.28 we see that the absorption coefficient of a layer for x-polarized light must equal the absorption coefficient of the neighbouring layers for y-polarized light, i.e. $\xi_1^x = \xi_2^y$ and $\xi_1^y = \xi_2^x$ so that for x-polarized light

$$S_1^x = \frac{A_1}{A}\left(\frac{\gamma_1^x}{\xi_1^x + \xi_1^y}\right)\left\{1 - \exp -\frac{l}{2}(\xi_1^x + \xi_1^y)\right\} \tag{9.35a}$$

and for y-polarized light

$$S_1^y = \frac{A_1}{A}\left(\frac{\gamma_1^y}{\xi_1^x + \xi_1^y}\right)\left\{1 - \exp -\frac{l}{2}(\xi_1^x + \xi_1^y)\right\} \tag{9.35b}$$

Thus the polarization sensitivity (S_{pol}) for retinula cell 1 is

$$S_{\mathrm{pol}} = \frac{S_1^x}{S_1^y} = \frac{\gamma_1^x}{\gamma_1^y} \equiv \Delta_1 \tag{9.36}$$

where Δ_1 is the dichroic sensitivity of the microvilli medium of rhabdomere 1. The result is independent of the arrangement of spectral cell types that form the rhabdom.

In other words the theoretical S_{pol} of a retinula cell of a crustacean-type rhabdom is identical to the dichroic sensitivity of the photopigment within the rhabdomere. Furthermore, as seen from eqn (9.35a), this enormous S_{pol} is achieved without any sacrifice of absolute sensitivity. The result is in general different when the layers are of unequal length.

A rather profound consequence follows from the fact that $S_{\mathrm{pol}} = \Delta$. This is that the dichroic sensitivity of the photopigment within Crustacean rhabdomeres can be measured in the live animal by single cell electrophysiology. Measurements of S_{pol} of fly rhabdomeres only lead to consistent pairs of γ and Δ using Fig. 9.25. Unless the microvilli have different properties in rhabdomeres of the various arthropods, the measurement of Δ in Crustacea may apply to a large class of photoreceptors. Shaw (1969) measured Δ as high as 11 in Crustacea.

In summary, retinula cells of the crustacean type of rhabdom have a S_{pol} equal to the dichroic sensitivity (Δ) of the photopigment without sacrificing their absolute sensitivity.

S_{pol} *for retinula cells of fused rhabdoms uniform in length*. Here we consider fused rhabdoms like the bee (Varela and Porter 1969) and the ant (Menzel 1972a) which unlike the crustacean rhabdoms are uniform along their length. We attempt to elucidate the function of the fused rhabdom structure as a polarization analyser. In particular we find that the S_{pol} of fused rhabdoms with a special symmetry, to be described below, have a S_{pol} equal to the dichroic sensitivity Δ of the photopigment, as in the crustacean retinula cells. We derive a simple criterion for determining which fused rhabdoms have this property.

Let us begin by considering a rhabdom uniform in length with an arbitrary cross-section as illustrated in Fig. 9.29. The mathematical derivation is presented in detail in Section 9.2.4. The absolute sensitivity of retinula cell 1 S_1 is related to the power absorbed by rhabdomere 1. From eqn (9.13b) of Section 9.2.4 we find that

$$S_1 = \left(\frac{\gamma_1 A_1}{\sum\limits_{j=1}^{8} \gamma_j A_j}\right)\left\{1 - \exp\left(-\frac{l}{A}\sum_{j=1}^{8}\gamma_j A_j\right)\right\} \qquad (9.37)$$

where A_1 is the area of rhabdomere and A the area of the rhabdom in cross-section. The γ_j values are the absorption coefficients of each rhabdomere and l is the length of the rhabdom.

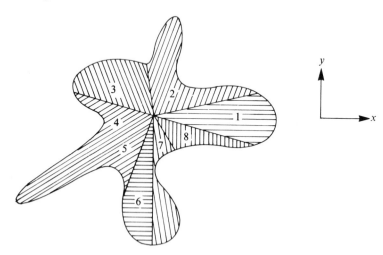

FIG. 9.29. The eight rhabdomeres of a fused rhabdom assumed uniform in length. Each rhabdomere is characterized by an absorption coefficient γ_i and cross-sectional area A_i. The area of the rhabdom in cross-sections is A.

In order to develop an expression for the S_{pol} of retinula cell 1 we want to know its sensitivity to light polarized parallel (x) and perpendicular (y) to its microvilli. We use the superscript x or y on γ_j to denote the absorption coefficient for x- or y-polarized light. Thus

$$S_{pol} = \frac{S_1^x}{S_1^y} = \Delta_1 \left(\frac{\displaystyle\sum_{j=1}^{8} \gamma_j^y A_j}{\displaystyle\sum_{j=1}^{8} \gamma_j^x A_j}\right) \frac{\left\{1 - \exp -\dfrac{l}{A}\displaystyle\sum_{j=1}^{8} \gamma_j^x A_j\right\}}{\left\{1 - \exp -\dfrac{l}{A}\displaystyle\sum_{j=1}^{8} \gamma_j^y A_j\right\}} \tag{9.38}$$

where $\Delta_1 = \gamma_1^x/\gamma_1^y$ and is the dichroic sensitivity of the photopigment within rhabdom 1. For the general rhabdom illustrated in Fig. 9.29 no further simplifications are possible. Unless the microvilli of a rhabdomere are in the x- or y- direction, γ_i^x and γ_i^y are complicated logarithmic functions.

In general, the S_{pol} of a fused rhabdom is an extremely complicated expression which depends on all rhabdom parameters.

We next discuss rhabdoms with a special type of symmetry. This symmetry has a profound consequence on the S_{pol} of fused rhabdoms.

Criterion for determining when a fused rhabdom is a maximum polarization analyser. A study of the S_{pol} for fused rhabdomeres given by eqn (9.38) reveals that when

$$\sum_{j=1}^{8} \gamma_j^y A_j = \sum_{j=1}^{8} \gamma_j^x A_j \tag{9.39}$$

the S_{pol}, 1 for retinula cell 1 of the fused rhabdom equals that of the dichroic sensitivity (Δ_1) of rhabdomere 1, i.e.

$$S_{pol}, 1 \equiv \Delta_1 \tag{9.40}$$

This result (Snyder 1973) has a profound consequence on the S_{pol} of rhabdomeres. We therefore consider the physical conditions for its validity in detail; see also Snyder, Menzel and Laughlin (1973).

The physical meaning of eqn (9.39) is that the light power absorbed in the *entire* rhabdom due to light polarized parallel to the microvilli of the measured cell must equal that due to light perpendicular to the microvilli (Snyder 1973). In Fig. 9.29 this means that the power absorbed by the rhabdom (all retinula cells) is the same for x- or y-polarized light.

The criterion eqn (9.39) is most simply satisfied when the rhabdomeres have a certain basic symmetry to be discussed next.

We start with the rhabdom of Fig. 9.30 like the worker bee (Goldsmith 1962). The shape of the rhabdom in cross-section is not critical, only the symmetry of microvilli is necessary. We assume simple examples in which the areas of all retinula cells are equal. At wavelengths where all retinula cells have approximately the same sensitivity, the criterion for maximum absorption eqn (9.39) is satisfied and the $S_{pol} = \Delta_1$. It is not necessary for all

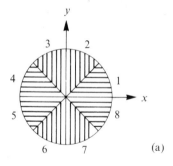

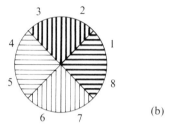

F IG. 9.30. A cross-section of the rhabdom of the worker bee.

cells to have the same spectral sensitivity if they are arranged as in Fig. 9.31 where the different shades indicate cells of different spectral sensitivities. The ant and worker bee rhabdomeres have absorption properties which are different from the symmetry of Figs. 9.30 and 9.31 so that the criterion for maximum polarization sensitivity fails and we would expect a poor polarization sensitivity (Gribakin 1972) There is no published data on the polarization sensitivity of the worker bee retinula cells. However, Autrum and von Zwehl (1964) infer it to be large. Shaw (1969) measured the polarization sensitivity of the drone bee to be less than 1·4. A partial explanation for this low polarization sensitivity is intracellular coupling (Shaw 1969). Other possibilities are the factors mentioned above.

We conclude that fused rhabdoms with a certain basic symmetry in cross-section have retinula cells with a polarization sensitivity equal to the dichroic sensitivity Δ of the photopigment. These rhabdoms, as far as their S_{pol} property, behave like rhabdoms of the Crustacea. The criterion for the basic symmetry is that the light power absorbed in the entire rhabdom due to light polarized parallel to the microvilli of the measured cell must equal that due to light perpendicular to the microvilli. For a complete discussion, with examples, see Snyder (1973).

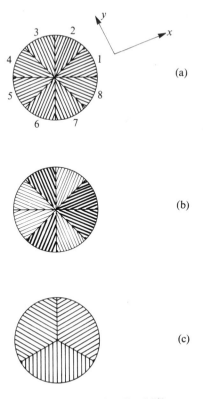

FIG. 9.31. The shading represents retinula cells of different spectral sensitivity.

9.3.1.3 *Angular sensitivity of the worker bee ommatidium.* We showed in Section 9.2.2 that a dielectric waveguide has acceptance characteristics which depend acutely on its diameter and index of refraction in addition to the wavelength of incident light. As an example we presented in Fig. 9.8 the acceptance characteristics for an optical fibre when the illumination is uniform over its entrance. Thus the field of view of an ommatidium can be limited by the field of view of the rhabdom. We next show that the field of view of the worker bee ommatidium is equal to a product of the dioptric apparatus and rhabdom angular sensitivities.

The functional unit of vision in the apposition compound eye is the ommatidium (Fig. 9.32). A fundamental property of an ommatidium is its visual field or angular sensitivity.

Two independent studies of the visual field of a single ommatidium of the worker bee have been reported recently. Varela and Wiitanen (1970) analysed an ommatidium's optical properties as far as the distal end of the rhabdom by means of a ray-tracing procedure. Their angular sensitivity results are

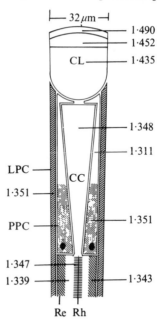

FIG. 9.32. Schmatic representation of the light-receptive portion of a bee ommatidium redrawn from Varela and Wiitanen (1970). Abbreviations are CL cuticular lens; CC crystalline cone; PPC principal pigment cell; LPC long pigment cells, Re retinula cells and Rh rhabdom. The numbers in the figure refer to the index of refraction measurements of Varela and Wiitanen (1970).

illustrated as curve $A_D(\theta)$ in Fig. 9.33. Laughlin and Horridge (1971) determined angular sensitivity by electrophysiological measurements which include the properties of the rhabdom. Their results are also illustrated as curve $A(\theta)$ in Fig. 9.33. We see from Fig. 9.33 that the visual field determined by electrophysiological measurements is approximately half that found from the ray-tracing procedure.

The intent of this section is to resolve the discrepancy between the two different estimates of angular sensitivity, as well as to provide a more complete description of what determines the angular sensitivity. This is accomplished by an analysis of the light-reception properties of the bee ommatidium combining dielectric waveguide theory and ray-tracing techniques. We follow the arguments of Snyder (1972a).

The ray-tracing procedure of Varela and Wiitanen (1970) determines the amount of light incident on the rhabdom for any given angle of light incidence on the ommatidium. They assume that all this light propagates along the rhabdom. This is equivalent to letting $V = \infty$ in Fig. 9.8 and completely neglects the waveguide properties of the rhabdom.

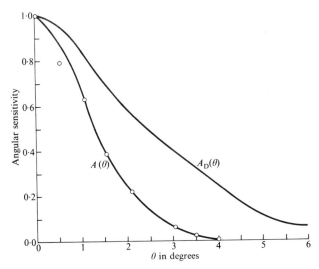

F I G. 9.33. Angular sensitivity of an ommatidium normalized to unity response at $\theta = 0$. θ represents degrees from the optical axis of light incident on the ommatidium. Curve $A_D(\theta)$ is from Varela and Wiitanen's (1970) ray-tracing procedure. Curve $A(\theta)$ is drawn from the data points, shown as circles, of the electrophysiological measurements of Laughlin and Horridge (1971).

Ray-tracing procedures can be used to calculate the amount of light incident on the rhabdom; however, to determine the amount of light that propagates through the rhabdom requires a more detailed knowledge of the light at the end of the crystalline cone than is found from ray tracing. We can approximate the electromagnetic fields at the rhabdom by knowing how the focal point changes with varying angles of light incident on the ommatidium. Varela and Wiitanen (1970) demonstrate that the thick lens formulation of Born and Wolf (1965) gives the approximate focal point but provides little information about ray paths. We therefore use the thick lens formulation only to determine the change in the height (h) of the focus from the optical axis. This is given approximately as

$$h \approx F\theta/n \tag{9.41}$$

where θ (assumed small) is the angle of light incident on the ommatidium, F is the distance of the focal plane from the principal plane as shown in Fig. 9.34 and n is the refractive index of the crystalline cone (CC) medium.

Using the formulation eqn (9.3) for the diffraction pattern at the focal plane, we find that more than 85 per cent of the light energy is within a radius of 0·85 μm so that a relatively sharp focus is formed. Therefore, we can project rays through the focal point as shown in Fig. 9.34 to find that a small,

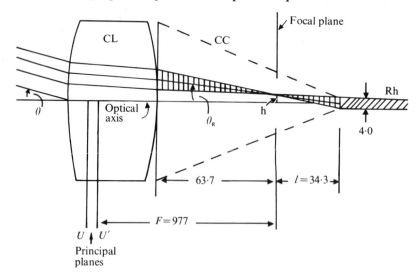

FIG. 9.34. Construction of light incident on the rhabdom due to parallel light rays at angle θ striking the ommatidium. Only the rays reaching the rhabdom are illustrated. The drawing is not to scale. h is the distance of departure from the optical axis of the focal point due to changing θ. θ_R is the angle from the optical axis to the ray that crosses the axis of the rhabdom. The construction of the principal and focal planes follows from Born and Wolf (1965). CC, Rh, and CL are explained in the legend of Fig. 9.32. All numbers refer to lengths in μm.

quasi plane-wave, bundle of light is incident on the rhabdom at approximately angle θ_R. We, of course, have no knowledge from our analysis of the amount of light rays that behave as depicted in Fig. 9.34, since the bee dioptric apparatus is very complicated, but this information is already known from Varela and Wiitanen's (1970) ray-tracing results. The relationship between θ_R and θ, both assumed small, follows from the geometry of Fig. 9.34 and eqn (9.42) to be

$$\theta_R \approx \frac{h}{l} = 2\cdot11\theta. \tag{9.43}$$

Knowing the wavelength of light and the parameters of the rhabdom given in Figs. 9.32 and 9.34 we can calculate, from eqn (9.22) of the electromagnetic theory analysis of Snyder (1969b), the amount of modal light propagating along the rhabdom due to a small bundle of plane-waves incident at angle θ_R with an amplitude independent of θ. We call this quanity $A_R(\theta)$, the angular sensitivity of the rhabdom. To obtain the angular sensitivity of the entire ommatidium, we must determine the fraction of light incident on the ommatidium that eventually propagates along the rhabdom. This is easily found multiplying $A_R(\theta)$ by Varela and Wiitanen's (1970) $A_D(\theta)$ curve from Fig. 9.33 for the amount of light incident on the rhabdom. We call $A_D(\theta)$ the angular sensitivity of the dioptric apparatus which includes everything but the rhabdom. Therefore the visual field or angular sensitivity of the ommatidium, $A(\theta)$, is equal to a product of the dioptric apparatus' and rhabdom's angular sensitivities, i.e.

$$A(\theta) = A_D(\theta)A_R(\theta). \tag{9.44}$$

Since Laughlin and Horridge (1971) showed that no u.v. sensitive cells were included in their electrophysiological measurements, we use a value for λ of 500 nm as an average between the 460 and 530 nm peak receptor responses measured by Autrum and von Zwehl (1964) for the worker bee receptors. The dimensionless parameter V, defined by eqn (9.2), is $V \approx 3.7$. Then from a curve like Fig. 9.8 we obtain $A_R(\theta)$. Figure 9.8 is slightly in error for the worker bee because it is based on incident light confined to the aperture of the rhabdom. Instead calculations for a beam of light 1·5 that of the rhabdom diameter are necessary. The results of the calculations for $A_R(\theta)$ are compared with Varela and Wiitanen's (1970) data for $A_D(\theta)$ in Fig. 9.35. We see that the rhabdom has a narrower visual field, $A_R(\theta)$, than the dioptric apparatus, $A_D(\theta)$. The angular sensitivity of the ommatidium, $A(\theta)$, is found as the product $A_R(\theta)A_D(\theta)$ and is shown in Fig. 9.35 to be in good agreement with the experimental data points. Changing λ by ± 50 nm about $\lambda = 500$ nm, produces approximately a ± 10 per cent change in half-width sensitivity

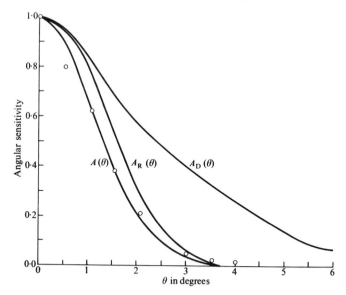

FIG. 9.35. A comparison of angular sensitivity curves for the worker bee. $A_D(\theta)$ is the angular sensitivity of the dioptric apparatus as given by the ray-tracing procedure of Varela and Wiitanen (1970). $A_R(\theta)$ is the angular sensitivity of the rhabdom as given by the electromagnetic analysis of Snyder (1969*b*). $A(\theta)$ is the angular sensitivity of the ommatidium found by multiplying $A_R(\theta)$ by $A_D(\theta)$. The circles represent data points from the electrophysiological experiments of Laughlin and Horridge (1971) as given in Fig. 9.33. Due to reflection symmetry about the $\theta = 0$ axis of the theoretical results, negative angles of θ are not presented.

which is well within the range of values measured by Laughlin and Horridge (1971).

Thus we have found that the angular sensitivity of an ommatidium is limited by its rhabdom as well as its dioptric apparatus and that the discrepancy between electrophysiology measurements and ray tracing is because the field of view of the rhabdom is smaller than that of the optical system in front of it.

When the focus of the dioptric system is at the distal end of the rhabdom as in the Diptera compound eye (Kuiper 1966), our analysis no longer directly applies. For this situation the acceptance properties of the rhabdomeres must be found by evaluating the modal excitation due to a beam of light parallel to the rhabdom but not centred on the optical axis. The same principles are involved because the asymmetric illumination excites the higher order modes.

9.3.1.4. *Effects of changes in the medium surrounding the rhabdom on absolute and angular sensitivity.*   Recent electrophysiological studies on the compound eye of the cockroach *Periplaneta americana* have shown that there are consistent differences in the angular sensitivity and absolute sensitivity of retinula cells in the dark-adapted and light-adapted states (Butler 1972). These changes were shown to be accompanied by movement of retinula cell screening pigment towards the rhabdom in the light, and the formation of a clear palisade around the rhabdom in the dark. Snyder and Horridge (1972) analysed these effects using waveguide theory. Their analysis follows.

The ommatidium of *Periplaneta* has a long fused rhabdom which is contributed by all of the eight retinula cells. The general form is as described by Nowikoff (1932) with details added by Butler (1971).

In the dark, cisternae of the endoplasmic reticulum develop in a few minutes and congregate close to the rhabdom, forming a palisade. This is the 'Schaltzone' of the older literature. The retinula cell screening pigment particles are separated by a distance of up to 2 μm from the rhabdom, as shown schematically in Fig. 9.36. In the light-adapted state the pigment particles lie close to the rhabdom, as schematically illustrated in Fig. 9.36.

The average radius of the circular rhabdom is measured as 2·2 μm. The grains of the screening pigment are assumed to be highly absorbing, with a coefficient to be calculated. The wavelength is taken as 500 nm, since the light source of Butler's (1972) experiments generated insufficient ultraviolet light for measurements on u.v. cells, and the larger cells in the cockroach are maximumly sensitive to light at 507 nm (Mote and Goldsmith 1970). The refractive index of the rhabdom is taken as $n_1 = 1·347$ and the surrounding medium as $n_2 = 1·336$.

To calculate the theoretical angular sensitivity of the rhabdom, i.e. the

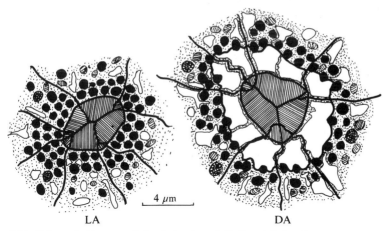

FIG. 9.36. Schematic diagram of the dark-adapted (DA) and light-adapted (LA) cockroach rhabdom. The black circles represent retinula cell screening pigment, Rh represents the rhabdom and P is the clear region known as the palisade ('Schaltzone' in old literature).

angular sensitivity of the ommatidium without the dioptric apparatus, we first calculate the light power $P(\theta_R)$ that is accepted by the rhabdom at various angles of the incident light on the rhabdom θ_R. The electromagnetic analysis of Snyder (1969b) with the dark-adapted rhabdom parameters discussed above, leads to the DA curve of Fig. 9.37. For the cockroach

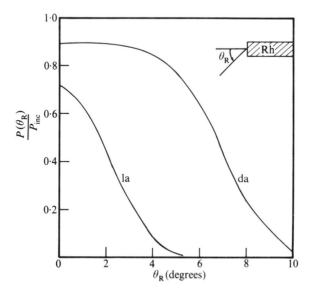

FIG. 9.37. The fraction $P(\theta_R)/P_{inc}$ of incident light power captured by the rhabdom at various angles θ_R of incident light. The results are obtained by the analysis of Snyder (1969b). For the dark-adapted (da) curve $n_2 = 1{\cdot}334$; for the light-adapted curve $n_2 = 1{\cdot}345$ and $n_1 = 1{\cdot}347$.

rhabdom, calculation shows that 10 per cent of the incident light fails to be captured when $\theta = 0°$. Furthermore, we note that the theoretical acceptance angle of 14° *for the rhabdom* is approximately twice that measured $(6·7 \pm 1·5°)$ by Butler (1972) in the vertical plane *for the entire ommatidium* when dark adapted.

The angular sensitivity S_A of the whole ommatidium is determined by the angular acceptance function of the dioptric apparatus and the angular acceptance function of the rhabdom, as previously shown for the worker bee in section 9.3.1.(c).

$$S_A \text{ (Ommatidium)} = S_A \text{ (Dioptric Apparatus)} \times S_A \text{ (Rhabdom)} \quad (9.44)$$

where S_A indicates angular sensitivity. The angular acceptance function of the bee rhabdom turned out to be narrower than that of the dioptric apparatus in front of it, i.e. the rhabdom limited the field of view of the ommatidium.

In order to use eqn (9.44), we need the relationship between the angle θ_R of light incident upon the rhabdom and the angle θ of light incident on the ommatidium. For the worker bee $\theta_R = 2\theta$ (Varela and Wiitanen 1970). The general form of the cornea and cone, and the values of facet radius of curvature and refractive indices in the cockroach are different from those of the bee, but we assume for simplicity that $\theta_R = 2\theta$ in the cockroach. Later we show that this assumption does not affect the description of the optical function of the pigment migration. Our assumption that $\theta_R = 2\theta$ means that the scale of Fig. 9.37 is halved, resulting in a theoretical acceptance angle of the rhabdom of 7°.

As a consequence of eqn (9.44), we can now make a statement about the angular sensitivity of the dioptric apparatus of the cockroach without going through a ray-tracing analysis. We have already found (letting $\theta_R = 2\theta$) that the theoretical angular acceptance function of the rhabdom is the same as that measured electrophysiologically by Butler (1972) for the whole ommatidium. From eqn (9.44) this means that the angular sensitivity of the dioptric apparatus is considerably larger than that of the rhabdom, and can be ignored, so the field of view for the cockroach ommatidium is determined principally by the rhabdom. In the worker bee, the contributions from the two components were more nearly equal.

When the eye is light-adapted, the retinula cell screening pigment moves close to the rhabdom, as shown in Fig. 9.30. We are unable to calculate the theoretical acceptance function of the dark-adapted rhabdom because the average index of refraction of the area including the pigment grains is not known. However, we know from Butler (1972) that the angular sensitivity of the light-adapted ommatidium is $c.$ 2·4°. Therefore, using the analysis of Snyder (1969b) we can determine the theoretical index of refraction n_2 of the medium surrounding the rhabdom that gives rise to the measured value of $2·4 \pm 0·75°$ for the acceptance angle of the light-adapted eye. We find that a hypothetical value of $n_2 = 1\cdot345$ gives rise to the measured angular acceptance function for the rhabdom. The measured values of the dark-adapted and light-adapted angular sensitivity of the whole ommatidium are plotted in Fig. 9.38. On the same graph are plotted the calculated DA curve derived from the refractive index measurements (which shows that the rhabdom properties dominate the angular sensitivity) and the light-adapted curve if we assume n_2 to be $1\cdot345$, all normalized to a maximum of 100 per cent as is usual for physiological data based on relative sensitivities.

The theoretical curves of Fig. 9.37 show that 90 per cent of the axial light enters the rhabdom in the dark-adapted ommatidium, but only $c.$ 70 per cent in the light-adapted state. But the electrophysiological measurements show that absolute sensitivity to axial light is $c.$ ten times greater in the dark-adapted state. From this discrepancy we can calculate the attenuating effect of the pigment around the rhabdom. Six modes theoretically exist on the dark-adapted cockroach rhabdom and one mode on the light-adapted rhabdom; the difference is due to the difference in refractive index outside the rhabdom in the two cases. The modal light power $P(z)$ at position z along the rhabdom is given from Section 9.2.3 as

$$P(z) = P(0)\exp -\gamma z \tag{9.45}$$

where γ, the absorption coefficient, is equal to the loss of the material in bulk multiplied by the fraction of the total modal light passing through the material (Snyder 1972b). This means for the dark-adapted (da) eye, when screening pigment is far from the rhabdom, that γ_{da} is due to the losses α_{vp} of the visual pigment (vp) multiplied by the fraction η of light within the rhabdom. Thus

$$\gamma_{da} = \alpha_{vp}\eta \tag{9.46}$$

However, for the light-adapted (la) eye there is an additional loss due to attenuation by the screening pigment (sp) grains. Thus from Section 9.2.3.

$$\gamma_{la} = \alpha_{vp} + \alpha_{sp}(1-\eta) \tag{9.47}$$

where $(1-\eta)$ is the fraction of light outside the rhabdom.

Electrophysiological measurements are proportional to the light power P_R absorbed by the rhabdom. Mathematically, P_R is found from eqn (1) of Snyder and Pask (1972a). We find that the ratio P_R^{la}/P_R^{da} of power absorbed by the light-adapted to dark-adapted rhabdom is

$$\frac{P_R^{la}}{P_R^{da}} \approx \left(\frac{\gamma_{da}}{\gamma_{la}}\right)\frac{P_{la}(0)}{P_{da}(0)} \approx \left[\frac{\alpha_{vp}\eta}{\alpha_{sp}(1-\eta)}\right]\frac{P_{la}(0)}{P_{da}(0)}, \tag{9.48}$$

where $P_{la}(0)$ and $P_{da}(\theta)$ are the values of the la and da curves of Fig. 9.37 at $\theta = 0$. We have assumed that most of the light is absorbed and that the visual pigment absorption coefficient $\alpha_{vp} \ll \alpha_{sp}$.

From Snyder (1969b), we find for the la rhabdom parameters that

$\eta = 0.73$. From Fig. 9.37, $P_{\text{la}}(0) = 0.7$ and $P_{\text{da}}(0) = 0.9$. Therefore eqn (9.48) is

$$\frac{P_R^{\text{la}}}{P_R^{\text{da}}} \approx 7 \frac{\alpha_{\text{vp}}}{\alpha_{\text{sp}}}. \tag{9.49}$$

The dark-adapted retinula cell is approximately ten times more sensitive to incoming light than when the light-adapted (Butler 1972). This means that $P_R^{\text{la}}/P_R^{\text{da}}$ of eqn (9.49) equals 0.1, so that from eqn (9.49) the screening pigment material must be seventy times more absorbing than the visual pigment, i.e. $(\alpha_{\text{sp}}/\alpha_{\text{vp}}) \approx 70$.

Because of eqn (9.44) the ratio of light to dark-adapted angular sensitivities is independent of the angular sensitivity of the dioptric apparatus so that by changing the relationship $\theta_R = 2\theta$ only the value of n_2 and $\alpha_{\text{sp}}/\alpha_{\text{vp}}$ are altered but not the basic mechanism that describes the optical function of pigment migration.

Using an electromagnetic analysis, we have shown quantitatively how the morphological changes around the rhabdom in the cockroach *Periplaneta americana* control both the angular sensitivity and absolute sensitivity of the retinula cells.

The analysis of the dark-adapted (da) rhabdom is shown in Fig. 9.38 compared to the electrophysiological results. In the light-adapted state (la), retinula cell screening pigment moves close to the rhabdom. For this situation we predict that, on account of the pigment grains, the medium external to the rhabdom is both highly absorbing (seventy times more absorbing than the visual pigment) and of higher index of refraction ($n_2 = 1.345$) than the medium surrounding the dark-adapted rhabdom. The changes in the refractive index around the rhabdom (n_2) cause the change in field size; whereas the attenuation by the pigment is responsible for most but not all of the

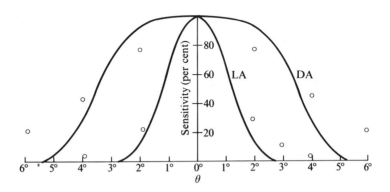

FIG. 9.38. Per cent angular sensitivity as a function of angle θ of light incident upon the facet. The results are normalized to unity at $\theta = 0$. Solid curves are the theoretical electromagnetic results. The circles represent data points from Butler's electrophysiological measurements.

difference in sensitivity. Approximately 25 per cent of the change in sensitivity is caused by the reduction in the number of modes which can propagate along the rhabdom as the value of $(n_1 - n_2)$ is reduced. Furthermore, the angular sensitivity of the ommatidium is determined principally by the aperture of the rhabdom.

9.3.2. *Information that a fused rhabdom can extract from modes*

9.3.2.1. *Bee rhabdom—a mode detector.*   In this section we consider only the worker bee rhabdom schematically illustrated by Fig. 9.30.

By superimposing the intensity patterns for the second mode type given in Fig. 9.39 we see qualitatively that the fine structure of the rhabdom enhances the absorption of the even x- and the odd y-polarized modes while it discriminates against the odd x- and even y-polarized modes. For example Snyder and Pask (1972a) showed that rhabdomere 1 absorbs three times more light energy from the even x-polarized mode than from the odd x-polarized mode. *We conclude that the worker bee rhabdom is theoretically a*

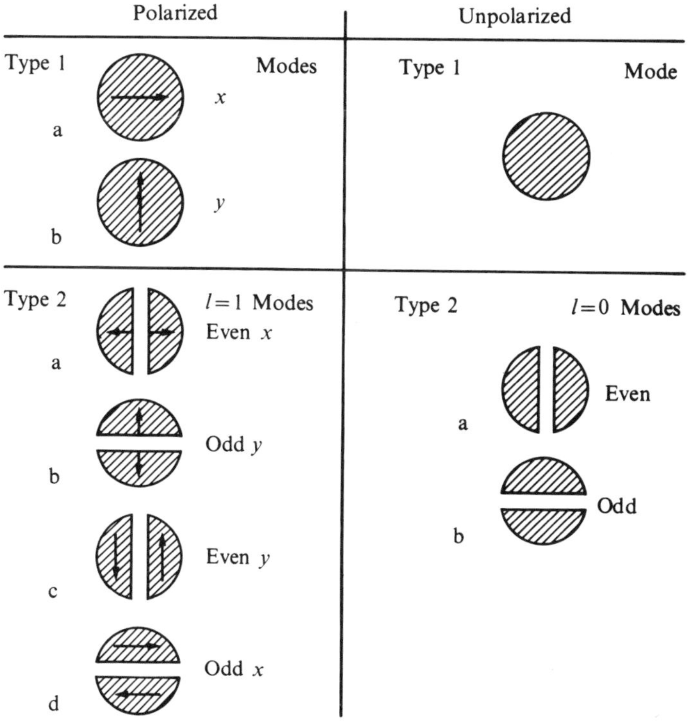

FIG. 9.39. The most commonly observed modes on the worker bee rhabdom (Varela and Wiitanen 1970). Arrows represent electric field vector (polarization), shaded areas represent regions of high energy density.

mode detector (*in the absence of electrical coupling*). In the following sections several consequences of this mode detection are discussed.

9.3.2.2. *Rhabdomeres do not have the same field of view.* We stated in Section 9.2.2 that mode 2 possessed information about the direction ($\hat{k}$) of incoming light in that its null energy band is initially perpendicular to the projection of $\hat{k}$ on the rhabdom cross-section (Fig. 9.6). By rotating the source of incoming light in a circle at an angle θ with the receptor axis, certain combinations of the first and second mode types are excited. In Section 9.2.2 we showed (Fig. 9.7) that for light nearly on axis, $\theta \ll 1°$ and only the first mode is strongly excited. However, for larger θ values, depending on the wavelength (λ) there is an angle θ_0 for which only the second mode is strongly excited. For example, when $\lambda \approx 300$ nm, $\theta_0 \approx 1.5°$, while for $\lambda \approx 600$ nm, $\theta_0 \approx 0.75°$. Therefore θ_0 is well within the range of angular sensitivity measured by Laughlin and Horridge (1971). Figure 9.40 illustrates the second mode type energy absorbed by rhabdomeres 1 and 2, assuming a polarization sensitivity of 2·5 to 1, when θ is fixed and the source is rotated around the axis, i.e. ψ in Fig. 9.6 is varied. Both polarized and unpolarized light results are shown. *The results clearly show that the rhabdomeres absorb different amounts of light as ψ is changed so that the directional information contained in the modes is detected by the fine structure of the rhabdom.* For further details see Snyder and Pask (1972b).

9.3.2.3. *Image detection by the fused rhabdom.* An interesting question is whether or not an individual ommatidium of the worker bee can detect an

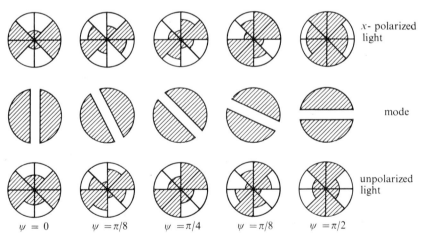

$\psi = 0 \qquad \psi = \pi/8 \qquad \psi = \pi/4 \qquad \psi = \pi/8 \qquad \psi = \pi/2$

FIG. 9.40. The middle line of pictures indicates the mode excited on the rhabdom as the angle of incidence ψ is varied (see Figs. 5 and 6). The corresponding diagrams below (above) the mode pictures represent the light absorbed by the rhabdom when the light is unpolarized (polarized with electric vector in x-direction). The sectors are shaded in proportion to the amount of light absorbed by the corresponding rhabdomeres, the sectors corresponding to the rhabdomeres absorbing most being fully shaded in each case. PL sensitivity is 2·5 to 1.

image. It has in fact been suggested by Miller, Bernard, and Allen (1968) that the fused rhabdom of the tobacco hornworm moth may detect image information from the modes that propagate along it. Here we show that even if the worker bee's dioptric apparatus were to provide a crisp image at the entrance to the rhabdom, the detection of this image is theoretically impossible due to a phenomenon of mode propagation analogous to acoustic beating or interference. We follow the physical discussion of Snyder and Pask (1972*d*). A more complete mathematical analysis is given by Pask and Snyder (1973).

Suppose the image of Fig. 9.41a is projected on the entrance of the worker bee rhabdom. We know that the modes convey light information to the rhabdom. Due to the asymmetry of their pattern, these modes obviously contain image information. In fact, as shown in Fig. 9.41b, superimposing the odd mode 2 plus mode 1 provides a reasonable approximation to the image given in Fig. 9.41a. The next question is whether or not the rhabdom can detect this image information conveyed by the two modes.

The two modes propagate along the rhabdom each with a distinct phase

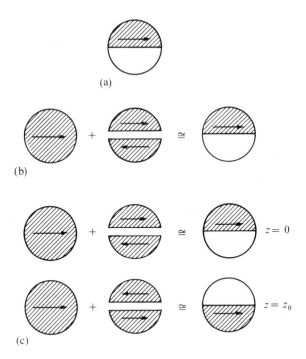

FIG. 9.41. (a) Semi-circular image assumed incident on bee rhabdom. (b) Construction of the image by a superposition of 1 and 2 *x*-polarized modes. (c) The change of polarization of a propagating 2 mode relative to the 1 mode. The image has completed its flip in a distance $z_0 = \pi/(\beta_1 - \beta_2)$ = the beat length, β_1 and β_2 being the phase velocities of the modes.

velocity (β_1, β_2). The situation is analogous to two sound waves of different frequencies. We know that constructive and destructive interference of the two sound waves can occur producing the well known beat phenomenon. Similarly, the modes on the rhabdom interfere or 'beat' with each other as they propagate. The way that the mode intensities change with respect to each other is illustrated in Fig. 9.41c. Basically, the electric vectors of mode 1 compared with mode 2 reverse 180° relative to one another in a distance $z_0 = \pi/(\beta_1 - \beta_2)$, producing the effect of rotating the image by 180°. This means that, as the two modes travel along the bee rhabdom, the image which is initially superimposed on rhabdomeres 1, 2, 3, 4 gradually 'flips' over to rhabdomeres 5, 6, 7, 8 in a distance $z_0 = \pi/(\beta_1 - \beta_2)$. Clearly, at the position z_0 where the image has completely flipped, rhabdomeres 5, 6, 7, 8 detect an image while rhabdomeres 1, 2, 3, 4 do not.

The degree to which the rhabdom can detect the image is determined by the stability of the image pattern. If the length of the rhabdom is much shorter than the distance, z_0, required for 180° image rotation, then the image is detected. However, if the rhabdom length is greater than z_0, the detection of the image is greatly reduced.

For a wavelength of 500 nm, $\beta_1 - \beta_2 \approx 0.036 \ \mu m^{-1}$ so that there are approximately 4·5 flips in a rhabdom of length 400 μm. The situation is depicted in Fig. 9.42 where the shaded area in the upper (lower) graph represents the light energy absorbed by rhabdomeres 1, 2, 3, 4 (5, 6, 7, 8), assuming a total of 90 per cent of the light energy is absorbed.

Close inspection of the shaded areas of Fig. 9.42 reveals that rhabdomeres 1–4 only absorb slightly more (10 per cent at most) than rhabdomeres 5–8. Thus the ambiguity caused by the 'beating' of the modes badly distorts or blurs the image detected by the rhabdom. Similar results follow for unpolarized light.

We can regard our conclusion, that image information is practically obliterated, as a theoretical upper limit on a rhabdom's ability to detect an image. This is because (a) the worker bee's dioptric apparatus is known to focus light 40 μm above the entrance to the rhabdom making it impossible to form an image as crisp as the one used for our analysis. If the bee rhabdom has difficulty detecting a well-defined image, it would have even greater difficulty detecting an initially blurred one. (b) Irregularities of the rhabdom tend to mix in other modes, which adds further ambiguity to the detection system.

Modes other than those of Fig. 9.39 exist for wavelengths less than 500 nm, but they are less important for the formation of the image of Fig. 9.41. Furthermore, these additional modes 'beat' rapidly and hence add further confusion to the image detection process.

We have considered a simple kind of image. Considering the dimensions of the bee's photoreceptor, a more complicated image seems to be of little

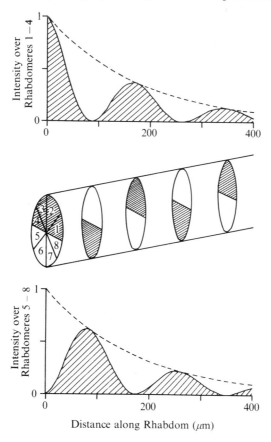

FIG. 9.42. The intensity of light, as a function of distance along the rhabdom. The shaded area in the upper (lower) graph represents the light energy absorbed by rhabdomeres 1–4 (5–8) as the image of Fig. 9.41a propagates down the rhabdom. The central drawing shows the state of this image at certain points along the rhabdom. Clearly all the rhabdomeres absorb light, so that the rhabdom cannot distinguish the image.

practical importance. In any case, our analysis suggests that the property of blurring the image is quite general.

9.4. Summary and conclusions

Rhabdoms (or rhabdomeres) are dielectric waveguides. Thus dielectric waveguide theory must be used to determine their absorption characteristics. Using dielectric waveguide theory we have been able to understand how the physical properties of rhabdoms (their size, shape, and index of refraction), effect their spectral, polarization, angular and absolute sensitivity. We summarize a few of our most significant conclusions.

9.4.1. *Spectral sensitivity* $S(\lambda)$

(1) The effect of confining photopigment to within a rhabdomere of a small cross-section is (a) to shift the visible absorption peak to lower wavelengths and (b) to increase the u.v. peak absorption relative to the visible. The smaller the diameter of the rhabdom the greater the effect (Fig. 9.15 and 9.16).

(2) We showed that there are three spectral receptor types in the Diptera eye consistent with the single cell measurements of Burkhardt (1962). The spectral types arise due to physical differences in the rhabdomeres, not due to different photopigments (Fig. 9.19). Rhabdomeres 1–6 are the green type, rhabdomere 7 is the blue type and rhabdomere 8 may be the yellow–green type. The enormous u.v. sensitivity of the blue receptor is due principally to the physical properties of the rhabdom rather than the sensitivity of the photopigment. Rhabdomere 8 is below the rhabdomere (number 7) which has an enormous u.v. absorption. Thus rhabdomere 7 acts as a u.v. colour filter reducing the u.v. sensitivity of rhabdomere 8 and altering its $S(\lambda)$ curve.

9.4.2. *Polarization sensitivity* S_{pol}

(1) The S_{pol} of a fly (Diptera) rhabdomere is dependent on the quantity γl where γ is the absorption per unit length along the rhabdom and l is the length of the rhabdom. The lower the γl of a rhabdomere the greater its S_{pol} and the lower its absolute sensitivity. The smaller the diameter of the rhabdomere the smaller γ (waveguide effects) and the larger the S_{pol}. The effect of placing one rhabdomere on top of another is to increase the S_{pol} of the lower rhabdomere. The longer the top rhabdomere the greater the effect.

(2) There are probably three classes of S_{pol} receptors in the fly. The least sensitive are rhabdomeres 1–6 because they are the longest and widest (i.e. γl is largest). Rhabdomere 7 is two-thirds as long and $\frac{1}{2}$ as wide as 1–6. It thus has a larger S_{pol}. Rhabdomere 8 is one-third the length of rhabdomeres 1–6 and lies below rhabdomere 7. It therefore has the largest S_{pol} sensitivity. Similarly the position of the 9th cell in the worker bee is consistent with its high S_{pol}.

(3) A salient characteristic of the open or unfused rhabdom is its dependence on the absolute sensitivity (via γl). Only when the absolute sensitivity of the cell becomes negligible does the S_{pol} become comparable to the dichroic sensitivity Δ of the photosensitive material within the rhabdomere. However, retinula cells of the Crustacea (layered) type rhabdom have a S_{pol} identically equal to the dichroic sensitivity Δ of the photopigment without sacrificing absolute sensitivity.

(4) The criterion for determing when a fused rhabdom (which is not layered) is an effective polarization analyser is presented. When the criterion is satisfied the unlayered fused rhabdom has the same S_{pol} as the Crustacea, i.e. $S_{pol} \equiv \Delta$. This occurs when the light power absorbed by the entire rhabdom due to light polarized parallel to the microvilli of the measured cell equals that due to light perpendicular to the microvilli.

9.4.3. *Angular sensitivity*

(1) The field of view of an ommatidium depends on the dioptric apparatus and the acceptance properties of the rhabdom. These acceptance properties must be found using dielectric waveguide theory. The visual field of the worker bee ommatidium $A(\theta)$ is equal to a product of the dioptric apparatus $A_D(\theta)$ and rhabdom's $A_R(\theta)$ angular sensitivities, i.e.

$$A(\theta) = A_D(\theta)A_R(\theta).$$

The field of view of the worker bee is limited approximately 50 per cent by its rhabdom.

There are many more examples of the application of dielectric waveguide theory to the analysis of insect photoreceptors. The use of dielectric waveguide theory has only just begun to unravel the influence that a rhabdom's physical properties has on its visual response. Much more experimental data and theoretical interpretation is necessary before the optical properties of the various insect photoreceptors are fully appreciated. Receptor optics is in its infancy.

10. Examination of the classical optics of ideal apposition and superposition eyes

P. CARRICABURU

ACCORDING to the classical work of Exner (1891) compound eyes can be classified into two groups: those with retinula cells directly in contact with the crystalline cones, and those with a clear space between crystalline cones and the retinula cells. The first were called by Exner 'apposition eyes', the second 'superposition eyes', here called 'eyes with crystalline threads' or 'eyes with crystalline tracts'. Exner postulated completely different dioptrics for the two kinds, and his theory has generally been accepted until the recent work of Carricaburu (1965); Allen and Bernard (1967); Horridge (1969); and Døving and Miller (1969). Subsequently, Exner's concept of superposition was revived by Kunze (1969); Seitz (1969); Horridge (1969, 1971); and Miller (1972) who observed eye glow and the focusing of rays in eyes with crystalline threads.

The following theory of the dioptrics of eyes with crystalline threads does not use Exner's superposition principle, but entirely explains the glow, the focusing of rays, and the effects of pigment migration during dark-adaptation. Following an outline of Exner's theory, evidence against it will be presented, and finally new concepts.

10.1. Exner's theory

According to Exner, the compound eye is composed of radially arranged optical systems, whose axes are concurrent. According to whether each optical system has a focus, or is like a telescope (afocal), we have to deal with an apposition or a superposition eye.

10.1.1. *Centred optical systems*

Centred optical systems have a radial symmetry around a straight line called the *optical axis*. Figure 10.1 shows such a system with foci. A ray from left to right parallel to the axis will pass through F', the image focus. Similarly, a ray through F, called the object-focus, will leave the system parallel to the

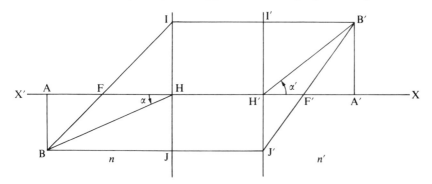

F IG. 10.1. Centred converging system, with different refractive indices, *n* and *n'*, on the two sides.

axis. There exist two planes perpendicular to the axis, called *principal planes*, which are the images of each other, with a linear magnification equal to *plus one*. Quantities $f = \overline{\text{HF}}$ and $f' = \overline{\text{H'F'}}$ are called the *focal lengths;* H and H' are the intersections of the principal planes with the axis. If $f < 0$ and $f' > 0$, the system is converging. A simple geometrical construction allows one to construct the image A'B' of an object AB. First, ray BJJ' is plotted in a direction parallel to the axis: it leaves the system along J'F'. Then, BF cuts principal plane H at I, and leaves the system in a direction parallel to the axis, along II'. The intersection of II' and JJ' gives the image B' of B. It can be shown that between the focal lengths there exists the following relation:

$$\frac{f}{n} = -\frac{f'}{n'}.$$

where *n* and *n'* are the indices of the object space and of the image space. Let us now consider Fig. 10.2: a point B at infinity not on the axis gives an image B' in the image focal plane. One can write:

$$\alpha = \overline{\text{HI}}/\overline{\text{FH}} \quad \text{and} \quad \alpha' = \overline{\text{H'I'}}/\overline{\text{H'F'}},$$

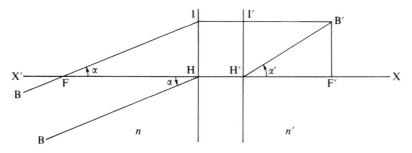

F IG. 10.2. Image B' of a point at infinity.

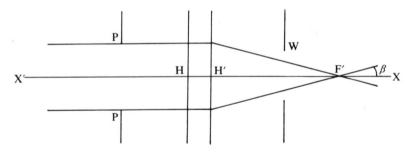

FIG. 10.3. Pupils P and window W.

from which

$$\alpha'/\alpha = -f/f' = n/n' \text{ when } \alpha \text{ is small.}$$

Suppose now a centred system with a point source at infinity on the axis X'X, limited by a stop P, called the *entrance pupil*. From Fig. 10.3, the pencil of rays leaving the system is a cone with vertex F' and half-angle β, called the *aperture angle*:

$$\beta = d/2f'.$$

where d is the diameter of the entrance pupil. With a point source at infinity, but not on the axis X'X, all the rays entering the entrance pupil will emerge from the system, if the angle between the axis and the direction of the source is not too large. For angles α greater than a certain value, some rays will be stopped by a diaphragm W, which is called the *window*. The maximum angle which allows all the rays to leave the system is called the *full-light field*, the angle which allows half of the rays to leave the system is the *half-light field*. For the sake of simplicity, we define *field* as the sum of the full-light field and the half-light fields on either side. The aperture angle and the field are easy to measure by experiment and play an outstanding role in the formation of images and of pseudoimages in the compound eye. With the window between the image principal plane H' and the focus F', rays from a point source at infinity converge through F' as illustrated (Fig. 10.4) but some are stopped

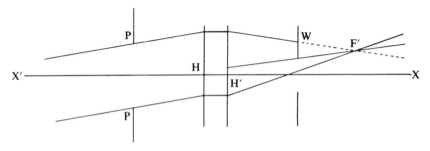

FIG. 10.4. Centred system with a window between H' and F', showing effect on average deviation.

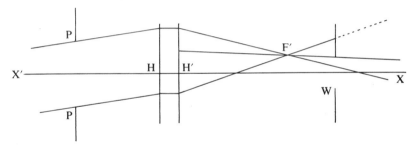

FIG. 10.5. Centred system with a window beyond F', again showing effect on the average ray direction.

by the window. The average ray of the pencil diverging from F' is no longer in the same direction as the average ray without the window: it is deviated further. On the other hand, if the window is located beyond the focus the mean ray is deviated less or even in the opposite direction (Fig. 10.5).

Let us now consider two converging systems such that the image focus of the first coincides with the object focus of the second (Fig. 10.6). An incident ray parallel to the axis now emerges parallel to the axis, and there is no focus, and so this is called an *afocal* (or *telescopic*) system. The incident ray F_1J_1 gives the ray $J_2'F_2$ from which, with small angles:

$$\frac{\alpha'}{\alpha} = \frac{f_1}{f_2'} = k$$

Notice that k is negative and a pencil of parallel incident rays emerge as parallel rays deviated in the direction of the incident pencil.

An incident pencil I_1I_1' parallel to the axis, limited by a pupil P, emerges along I_2I_2' (Fig. 10.7). From triangles $H_1'I_1'F_1'$ and $F_2H_2I_2$ we have:

$$\left|\frac{H_1I_1}{f_1'}\right| = \left|\frac{H_2'I_2'}{f_2}\right| \text{ so that } \left|\frac{f_2}{f_1'}\right| = \frac{H_2'I_2'}{H_1I_1}$$

The diameter of the emerging pencil is the diameter of the second lens, when that lens is the exit pupil. However, this is not generally the case, so

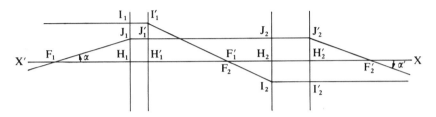

FIG. 10.6. Afocal system.

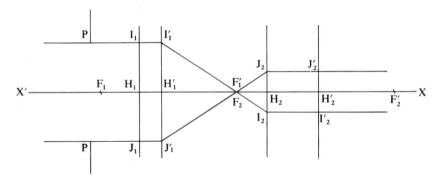

FIG. 10.7. Afocal system with pupil.

that the diameter of the emerging pencil is generally smaller than the diameter of the second lens. As previously, it is possible to define a window and a field of view.

10.1.2. *The compound eye*

Consider any eye composed of centred systems, M, N, etc., with foci and with axes pointing at 0 (Fig. 10.8). Rays from a source A on axis at infinity will emerge from each system as a conical pencil with half-angle β. By a computation similar to the one which is used for diopters, one sees that the mean rays of all pencils will pass through F' on X'X, such that:

$$\overline{MF'} = \overline{MO} \frac{n'}{n'-n},$$

where n is the refractive index of the substance outside, and n' the index inside the eye. Of course, only a few systems around M will send pencils, for the source A is outside the field of the others. Consequently, we have in F' a pseudo-image of A. The diameter of the pseudo-image can easily be computed:

$$d = |MF'| \times 2\beta$$

Such a set of centred systems corresponds to Exner's *apposition eye*.

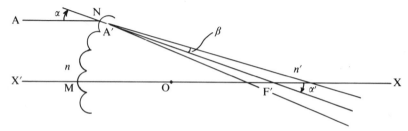

FIG. 10.8. Apposition eye.

For a set of afocal centred systems, with axes pointing at 0, incident parallel rays are transformed into emerging pencils of parallel rays: the axes of symmetry of those pencils will pass through A' (Fig. 10.9). One can easily show that:

$$\overline{H'_2A'}/\overline{OA'} = \alpha/\alpha' = \text{constant.}$$

The constant α/α' has been encountered in a previous equation. A' is a pseudo-image of A, with diameter approximately equal to that of the pencils emerging from the ommatidia. This is Exner's *superposition eye* which gives an erect pseudo-image of an object placed in front of it. According to Exner, it is possible to transform a superposition eye into an apposition eye by movement of pigment. However, the two sorts of eyes differ in the position of the pseudo-image, which in an apposition eye lies behind the centre of curvature of the eye, but between the centre and the cornea in a superposition eye.

10.1.3. *Structure of the ommatidia according to Matthiessen*

Exner postulated ommatidia to be converging systems, but could not find sufficient converging power in actual eyes, so he postulated the transparent media of the ommatidia to be non-homogeneous in refractive index. As discussed in Carricaburu (1967), Matthiessen considered a transparent cylinder with refractive index continuously variable from axis to periphery, with the variation of index given by the expression:

$$n = N_1\left(1 + \zeta \times \frac{b^2 - y^2}{b^2}\right) = N_m\left(1 - \frac{\zeta}{1+\zeta} \times \frac{y^2}{b^2}\right)$$

where N_1 is the index of the outer layer, N_m the index on the axis, b the radius of the cylinder and $\zeta = N_m - N_1$. A ray in a plane passing through the axis

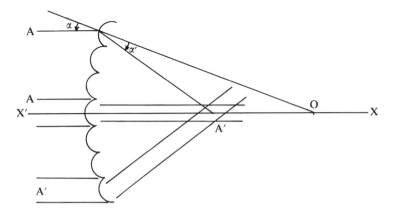

FIG. 10.9. Superposition eye.

describes a sinusoidal curve given by the following expression:

$$y = b_1 \cos \frac{\sqrt{(2\zeta)}}{b} x$$

where b_1 is the maximum distance between the ray and the axis. The wavelength of that sinusoid is:

$$L = \frac{2\pi b}{\sqrt{(2\zeta)}} \quad \text{so the ray crosses the axis every} \quad \frac{\pi b}{\sqrt{(2\zeta)}}$$

When the cylinder has a length $l/2$, a line parallel to the axis leaves the cylinder still parallel and the cylinder is afocal, so that:

$$f_i' = -f_2 = \frac{\pi b}{2\sqrt{(2\zeta)}}$$

In this simple system $n\alpha = -n_1\alpha'$, and the diameter of the pencil remains unchanged. If a superposition eye is made with the above Matthiessen systems, the pseudo-image A′ falls about midway between the dioptrics and the centre of the eye. From data for an actual ommatidium, it is possible to compute the necessary index variations, and rather high values are found. For example, for the ommatidium of a moth, 60 μm long and 20 μm in diameter, I calculate that there should exist, between the axis and the periphery, a difference of index as great as 0·14. The variations of index which I have so far found are insufficient to validate the theory of Matthiessen.

10.1.4. *The glow*

When one facet of a compound eye is illuminated, other facets become bright. According to Kunze (1972) the light rays entering the ommatidium, are scattered back from the plane of the erect image, and, by the principle of reversibility of light, leave the eye by other ommatidia. Kunze assumed that the above result would only happen in superposition eyes, but he simply did not consider alternative explanations.

According to Exner's Theory, superposition eyes have the following characteristics:

(1) A clear homogenous space between the crystalline cones and the rhabdoms.
(2) An afocal centred system, with an angular magnification equal to $c. -1$.
(3) A transparent media of the dioptric system with inhomogeneous refractive index which obeys Matthiessen's law or any similar law.
(4) The pseudo-image of a point must be small otherwise an erect image cannot be formed.
(5) Eye glow must be observable if there is a tapetum.

Exner and Matthiessen postulated that apposition eyes also had inhomogeneous cones. Consequently, condition (3) is necessary but not sufficient. In the same way, condition (5) is not sufficient, for the glow does not show that rays inside the eye are focused, as Horridge (1971) has clearly shown.

10.2. Consideration of Exner's theory

Each of the five conditions above will now be discussed in relation to Exner's theory.

10.2.1. *Homogeneity of the space between cones and rhabdoms*

Only eyes with a clear space between cones and rhabdoms can be superposition eyes (Exner 1891) but the space must be optically homogeneous as well as transparent, otherwise an erect image cannot be formed. Such homogeneity, however, is not usually found. For example, in Crustacea Macrura, ommatidia are composed of facets, crystalline cones, and tracts. The cones and tracts are quadrangular pyramids (Fig. 10.10) separated by a liquid layer, and the cones, tracts and liquid have different refractive indices. In most insects with 'superposition eyes' (fireflies, moths, etc.), the cones are continued as high index threads surrounded by low index tissues. In the firefly eye the space between cones and rhabdoms is not homogeneous (Fig. 10.11) so that the erect pseudo-image postulated by Exner could not be formed. That image has only been observed in fireflies in preparations of the cleaned cornea. After studying numerous eyes of nocturnal and other insects Horridge (1971) has introduced the term 'clear-zone eyes' and pointed out that clear-zone eyes are not necessarily 'superposition eyes'.

FIG. 10.10. Crayfish crystalline cone and tract.

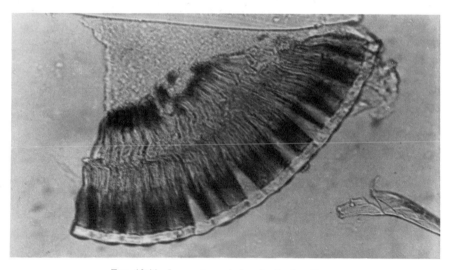

FIG. 10.11. *Lampyris* eye in longitudinal section.

10.2.2. *Ommatidia as afocal systems?*

By definition, a point source placed before an afocal system produces an image at infinity. With a microscope, however, we can observe objects only a few millimetres from the lens. An image at infinity could be observed with a mineralogical microscope fitted with a Bertrand's lens.

To examine the images, a *Lampyris* eye was cleaned out, leaving only the cornea. (In *Lampyris* the cones are outgrowths of the cornea as in *Limulus*). With a luminous cross *c*. 50 cm in front of the cornea each ommatidium yields an inverted real image of the cross seen in the interior of the cone, near its tip (Fig. 10.12). It appears that the system is as illustrated in Fig. 12.5, with a window located beyond the image focus, which does not agree with Exner's theory. In *Lampyris*, the ommatidia are *not* afocal systems otherwise this inverted image would not be seen. Other 'superposition eyes', for example, of the moth *Dysauxes* yield the same real inverted image. In other animals, the first image can be quite fuzzy, but afocal ommatidia have never been observed.

Cones of *Lampyris* have continuously variable refractive indices (Seitz 1968) but this alone is not a sufficient condition for a superposition eye. By tracing rays, Seitz found that the ommatidia were afocal, but the computation cannot be correct, because having shown that the cones are inhomogeneous, he used elementary formulae which hold only for homogeneous media. An exact computation, as shown long ago by Matthiessen (1886*b*), would require the exact gradient of refractive index, and the use of adequate formulae (Aoki 1966; Kapron 1970; McLeod 1954; Marchand 1970; Matthiessen 1886*a,b*;

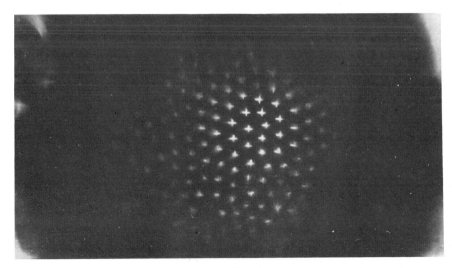

FIG. 10.12. Real images seen in the ommatidia of a *Lampyris* eye near the cone tip.

Schwarz 1885). The correct procedure is to trace rays by experiment, and not by computing.

Severe optical methods exist for determining the cardinal elements of a centred system. Each method isolates one or several pencils emerging from the lens, and visualizes the ray paths, e.g. by means of a knife blade (Foucault's method), by a pinhole (Harmann's method), by a grating (Ronchi's method), or by a half-silvered mirror (Twymann's interferometer). These methods are appropriate for an unknown system, or to control mass production in industry, but *never* does an optical engineer dismantle an unknown photographic lens, measure the radii and the indices, and compute the image. Computation is used in the design of new lenses and never to control actual ones. Let us recall that the 5 m mirror of the Mount Palomar telescope was controlled by Foucault's method, and not by computing. When studying an insect eye, we are in the position of an optical engineer studying an unknown lens, and must use appropriate methods, all the more so because we know the indices and the radii of curvature with very poor accuracy.

10.2.3. *The transparent media of the ommatidium must have continuously variable indices*

This condition is necessary but not sufficient. In several Crustacea and Insecta, the ommatidia are homogeneous (Carricaburu 1967), as in the crayfish (*Astacus fluviatilis*) illustrated in Fig. 10.13.

The interference microscope operates with two beams, one of which passes through the preparation to be studied, the other is a reference beam, passing

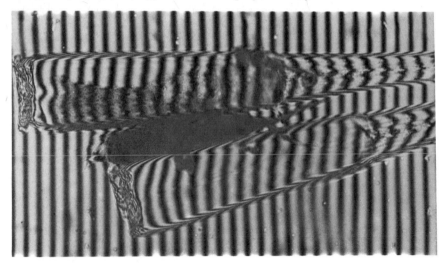

FIG. 10.13. Interferometry of Crayfish ommatidia.

through a wedge. If the preparation is homogeneous, with sides parallel, the fringes observed are straight and parallel. Retardation in the optical path causes a proportional shift by the fringes. If we measure the fringe shift caused by a body introduced into the observation beam, and know its thickness, we may compute its index. A prediction of the fringe shapes has been made for cones obeying Matthiessen's law (Carricaburu 1967). For a square pyramid of constant index, the shifted fringes remain straight and parallel (Fig. 10.13). *Astacus* has $n = 1.425$ constant for the crystalline 'cone' and 1.415 for the tract, and therefore the eye cannot give superposition images. In the crystalline cones of the moth *Dysauxes* (Fig. 10.14) the fringes are elliptical, which, for bodies of this shape, indicates a constant index (Fig. 10.15). In *Dysauxes* $n = 1.424$ for the cone and 1.393 for the thread.

In *Lampyris noctiluca* the index of refraction increases from the surface of the corneal cone to the interior, and from the cornea to the proximal end of the cone (Fig. 10.16 and 10.17). The index of the cornea is 1.510; the cones have indices equal to 1.532 (interior of the base), 1.525 (surface of the base), 1.540 (interior of the vertex). These differences are large but less than those predicted by Matthiessen.

10.2.4. *The erect pseudoimage*

Consider an ommatidium M, located beneath the axis X'X and illuminated by a point source A on its axis (Fig. 10.18). From the real image A', there emerges a pencil through 0. From adjoining ommatidia emerge pencils which meet around F'. Ommatidia further away, at N, admit no light from

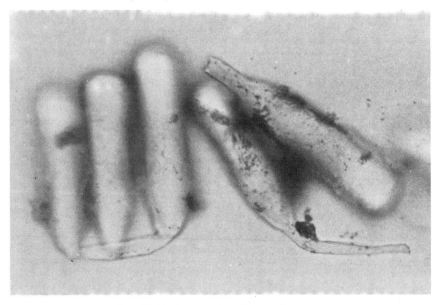

FIG. 10.14. *Dysauxes* (moth) crystalline cones.

A. Now place the whole under a microscope lens L focused on G. Rays from M first diverge from A′, pass through the lens and give a real image A″, located between the lens L and observing plane G′, then diverge beyond A″. In the plane of observation G′, one will see a spot S. Ommatidia near M give other spots, almost superimposed on S. Now, if the window is located

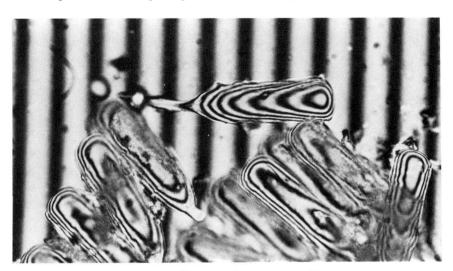

FIG. 10.15. Interferometry of *Dysauxes* crystalline cones.

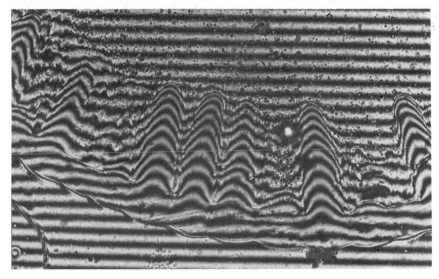

FIG. 10.16. Interferometry of *Lampyris* cones, parallel to cone axis.

beyond the image focus (Fig. 10.5), as in *Lampyris* and *Dysauxes*, the pencils emerge from the ommatidia as though from afocal systems. If one calculates pencils from various ommatidia, one can see that spots S produced from ommatidia by the source A are better superimposed for some locations of the windows. In this way, an eye that Exner would have defined as 'apposition'

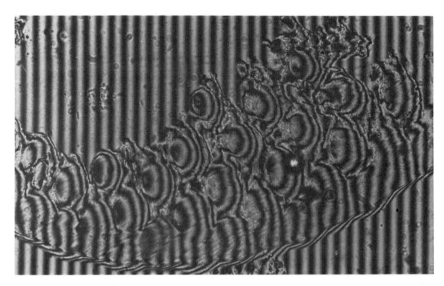

FIG. 10.17. Interferometry of *Lampyris* cones, oblique section.

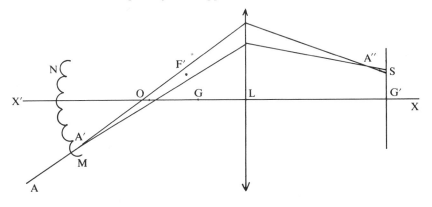

FIG. 10.18. Erect pseudo-image given by an apposition eye.

generates a pseudo-image S, which is on the opposite side of the eye from the source A.

Consequently, if A is moved from bottom to top, its pseudo-image seen *through the microscope* moves from top to bottom, as though the eye gives an erect image. The observation of an erect pseudo-image, therefore, is not proof of a superposition image, but only that the ommatidia have small fields. The same erect pseudo-image is observed in *Dysauxes* in which the ommatidia are certainly not afocal systems. The image is better if the window is located as in Fig. 10.5. When the fields are large numerous ommatidia contribute to the pseudo-image, which will be large and fuzzy, and possibly no longer perceived as an image. The pseudo-image in Fig. 10.19, produced by the eye which gave the inverted images in Fig. 10.12, is much larger than an ommatidium. In a perfect superposition eye, however, the pseudo-image should be no wider than an ommatidium, and observations of actual pseudo-images show that the optics are not ideal.

10.2.5. The glow

As will be proposed later, a point source A at infinity on axis, generates a real image near the cone tip and the thread (Fig. 10.21). The image, because of diffraction and aberrations, is a small spot. If A moves its image moves, until no ray from A enters the thread. But the rays are not necessarily absorbed; they can leave the ommatidium and diffuse into the clear zone. More precisely, they will be able to leave the eye by other ommatidia, and cause eye glow. The less the screening pigment extends around the cone tips (the more dark adapted the eye), the more important the glow will be. To sum up, when an ommatidium is illuminated by a pencil parallel to its axis, almost all the rays will enter the thread and not return. Oblique rays, however, will be able to leave the eye. If, like Kunze, we illuminate a single

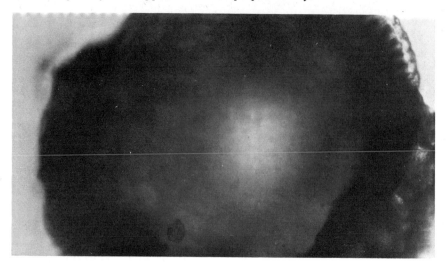

FIG. 10.19. Erect pseudo-image given by a *Lampyris* eye.

ommatidium by casting upon it a real image of a source, we illuminate it by means of an infinity of rays contained in a cone whose vertex is the facet and whose base is the illuminating lens. A large part of the rays will not enter the thread and will escape into the clear zone and out of the eye (Fig. 10.20). Therefore, this eye-glow experiment is not a proof of the existence of a superposition image, but is simply explained by the transparent space between the cones and the rhabdoms (Horridge 1971).

10.3. Crystalline tracts

In recent years a number of descriptions have appeared of crystalline tracts between the cone tip and the receptors in various clear zone eyes (Allen

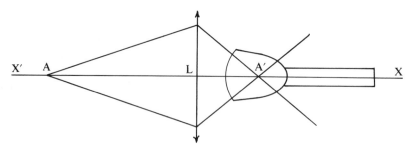

FIG. 10.20. Stray light and glow formation.

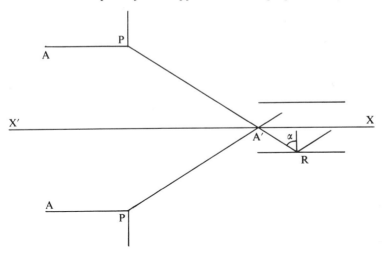

F I G. 10.21. Propagation of light by total reflexion in threads.

and Bernard 1967; Horridge 1968, 1969; Carricaburu 1969, 1972; Døving and Miller 1969). The general theory is that the dioptric system of the cornea brings rays from infinity on the axis to a real image near the entrance of the thread (Fig. 10.21). Then, if $n_i > n_e$, rays entering the thread within an angle θ to the axis suffer repeated total internal reflection if

$$\sin (\pi/2 - \theta) = n_e/n_i,$$

where n_i is the refractive index of the thread and n_e the refractive index of the medium around it. Any ray diverging from the axis by more than θ will leave this optical system and contribute to the glow. Inside the threads, rays can be propagated by successive total reflections, when the threads are large in comparison to the wavelength, or by modes when the threads are thin (Kapany 1967). To demonstrate this mechanism we must first prove the existence of high index threads immersed in a low index medium, then show that light is actually propagated by them.

10.3.1. *Refractive indices*

Direct measurement by interference microscopy shows that in the Crab *Carcinus maenas*, there exist high index threads (Carricaburu 1972). In the moth *Dysauxes*, we found for the threads a refractive index $n = 1\cdot393$; this value is much higher than that of the surrounding cytoplasm. Similar tracts are found in crayfish and lobsters.

10.3.2. *Propagation by threads or tracts*

In the firefly *Photuris*, Horridge (1968) sliced off the dark-adapted eye and observed the plane of section while illuminating the cornea. He observed

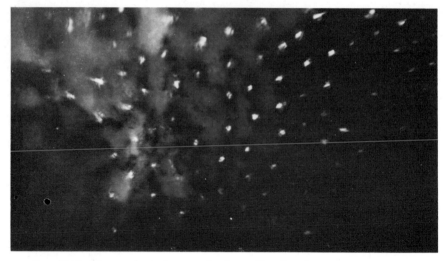

F I G. 10.22. Wave-modes in the cut ends of crystalline threads of the moth *Dysauxes*.

bright spots produced by light emerging from the cut ends of the threads.
The same experiments were made on *Dysauxes* (Fig. 10.22). Each bright spot
is the section of a thread. Their shapes are variable presumably because we
are seeing different wave modes (Kapany 1967). A similar phenomenon can
be observed in the crab *Carcinus* (Carricaburu 1972), and in the crayfish
Astacus. Because of the large diameter of the tract, there is no propagation
by modes in this case (Fig. 10.23). The eye of *Astacus* illustrated was cut

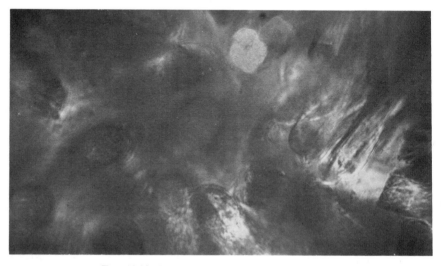

F I G. 10.23. Propagation of light in a Crayfish tract.

across the tracts at about the middle of their length, and illuminated by casting the real image of a point source on one facet. One single tract transmits light.

Propagation by threads allows one to explain the effect of movement of the screening pigment during adaptation. When light is totally reflected, if the external medium is perfectly transparent, all the energy is reflected. If the low index medium is absorbent, however, a part of the energy is absorbed. This is the phenomenon known as frustrated total reflection, covered by many authors (Hansen 1965; Harrick and Du Pre 1966; Lotsch 1968; Picht 1929; Tamir and Oliner 1969). A fibre immersed in a low index absorbent medium will transmit only a part of the incident energy. Horridge (1969) described the movement of the screening pigment surrounding the threads in the eye of the firefly *Photuris* as occurs also in Crustacea (Debaisieux 1944).

A model demonstrating frustrated total reflection is made of a glass rod placed along the axis of a PVC tube (Fig. 10.24). Between the rod and the

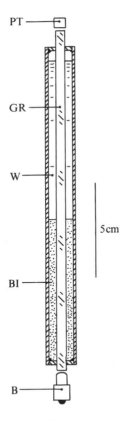

FIG. 10.24. Diagram of the thread model; PT, phototransistor; GR, glass rod; W, water; BI, black ink with gelatine; B, bulb.

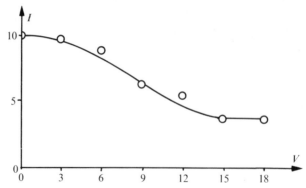

FIG. 10.25. Propagation of light in the model. Abscissa: volume of ink in millilitres. Ordinate: Intensity in phototransistor arbitrary units.

tube is poured a layer of black drawing ink (with added gelatine) then a layer of water. A bulb B illuminates one end of the rod, and a photo-transistor placed at the other end measures the luminous energy passing through. Increase of the length of ink reduces the intensity of the light emerging (Fig. 10.25).

10.4. Conclusion

(1) Anatomically, compound eyes are of two types, without a clear zone between cones and retinula cells (Exner's apposition eyes), and with one (Exner's superposition eyes). The ideal optical systems of these two kinds of eye are described and the mechanisms of the formation of inverted and erect pseudoimages are outlined in terms of ray theory.

(2) Most eyes with a clear zone have a crystalline tract which conducts light by successive total reflections (or modes) to the rhabdoms from the cones. For these to function well a real image must be cast upon the distal end or the tract, and the refractive index of the thread must be higher than that of the surrounding tissue. During light adaptation, the screening pigment surrounds the threads, and, by frustrated total reflection, absorbs a part of the light passing along the threads.

(3) Rays which escape or never enter the light guides, however, can cross the clear zone according to a variety of pathways determined by the dioptric system and distribution of pigment. Experiments on eye glow are valid only when they cannot be explained by scattered light, and direct observation of the optical system by following selected pencils of light is the most appropriate method of analysing it.

11. Optical mechanisms of clear-zone eyes

G. A. HORRIDGE

11.1. Introduction

A clear-zone compound eye is one with a wide clear region between the lens system of cornea or cone and the receptor layer of rhabdoms. Clear-zone eyes are characteristic of crepuscular insects but in skipper butterflies and a few groups of diurnal moths (e.g. Agaristidae) the zone is pigment-free in daylight. Relatively few clear-zone eyes have been analysed by the different techniques necessary to elucidate the optical mechanism and the detailed anatomical framework, complete with physiological or functional properties of the components, in both light- and dark-adapted states. The main types of ommatidia, and some of the principal optical mechanisms are now known in outline, but receptor recording in the great variety of clear-zone eyes is only just beginning. Only direct microelectrode recording is able to show the properties of the receptors, such as field of view and spectral sensitivity, which are of significance to the animal. The physiological measurements also provide the ultimate test of the optical theories, but as yet this approach is in its infancy. It is not possible, therefore, to present a complete functional description of the several types of clear-zone eyes; only the skipper butterflies have been examined by a range of available methods. Therefore I take the alternative approach of considering types of optical mechanisms and methods of analysing them, after a preliminary brief anatomical survey.

11.2. The anatomical variety of clear-zone eyes

On account of overlapping characteristics, clear-zone eyes are not easily placed in categories, but rather broadly they fall into the following four natural groups, of which the first two have retinula cell nuclei lying close to the rhabdom layer, and the other two have retinula cell bodies near to the cones at the ends of long retinula cell columns which reach to the receptor layer (Figs. 11.1 and 11.2). Besides the optical mechanisms, the most interesting feature of each type is the variety of mechanisms of adaptation to light. Light acting apparently on the receptors themselves stimulates cells and

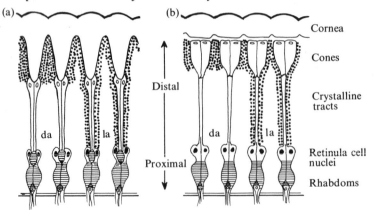

FIG. 11.1. Eyes with fixed tracts crossing the clear zone, dark-adapted on the left and light-adapted on the right within each figure. Cone cell nuclei are drawn as empty circles, retinula cell nuclei are drawn black. (a) With corneal cones, and crystalline tracts formed by extension of the cone cells, as in fireflies and related beetles. (b) With crystalline cones and tracts formed by retinula cell columns, as in skipper butterflies and some moths.

pigment to move so that light is attenuated just to the level which maintains that level of adaptation.

11.2.1. *Retinula-cells fixed and proximal*

(1) In families of beetles related to fireflies, notably Elateridae, Lampyridae and Lycidae, the cornea protrudes inwards as a cone in each ommatidium. A thin bundle formed by four cone cells, and called the

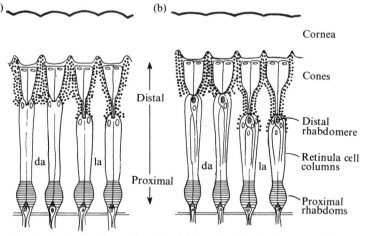

FIG. 11.2. Clear zone eyes with peripheral nuclei in retinula cell columns and, in the light adapted state, an extension of the cone cells in the form of a crystalline tract. (a) With only the main layer of proximal rhabdoms and basal retinula cells, as in Neuroptera, Megaloptera, Scarabaeoidea and most moths. (b) As in (a), but with a distal rhabdomere, as in large aquatic beetles (Gyrinidae, Hydrophilidae, and Dytiscidae) and some Carabidae.

crystalline tract, extends from the tip of each corneal cone to the rhabdom and retinula cell bodies, which lie entirely on the proximal side of the clear zone (Figure 11.1a). In some of these beetles, particularly the fireflies, pigment in the light-adapted state migrates from between the corneal cones and is distributed closely around the whole length of the crystalline tract. The pigment is inferred to act as a screen which prevents light from freely crossing the clear zone, and also as an attenuator of the light passing down the crystalline tract. As illustrated by Walcott in this volume, in the dark-adapted state the pigment withdraws between the corneal cones and presumably many rays then freely cross the clear zone. Some of the elaterid and cantharid beetles have little pigment and perhaps restricted powers of adaptation. Only two genera, *Phausius* (*Lampyris*) and *Photuris* (Lampyridae) have been investigated (Horridge 1969; Seitz 1969), and much anatomical and optical work remains to be done.

(2) Some moths (notably Bombycoidea) and skipper butterflies (Hesperiidae) have a very thin column of retinula cell extensions, which stretch out to the crystalline cone from cell bodies at the proximal side of the clear zone. The skipper butterflies are diurnal, and what little distal pigment they have always remains between the cones in the typical position of a dark-adapted moth eye. The bombycoid moths, on the other hand, have extensive pigment migration (Fig. 11.1b) which attenuates the light and also absorbs rays that are not restricted to the retinula cell columns. The only clear-zone eyes without anatomical light guides of some kind are those of the skipper butterflies. In most moths the fixed retinula cell column must act as a light guide in the light-adapted eye because it is then the only light path. As argued in Section 11.6.6, one can infer that in the dark-adapted eye the column also contributes by acting as a light guide although it may then carry an insignificant fraction of the total light reaching the receptors.

11.2.2. *Retinula cells mobile and distal*

(1) Many moths, some night-flying beetles, notably Scarabaeoidea, and many Neuroptera and Megaloptera examined, have clear-zone eyes in which a transparent column formed by the retinula cells crosses the clear zone in each ommatidium. In the dark-adapted state the nuclei of these cells lie close against the cone and the pigment migrates distally between the crystalline cones (Fig. 11.2a). Rays can then cross the clear zone freely from the broadly exposed tip of the cones, and the retinula-cell column can also act as a light guide. In some beetles and moths both paths of rays have been observed (see Section 11.4.1).

In the light-adapted state the retinula cells move away from the cone tip, and the peripheral cytoplasm of the crystalline cone is extended as a crystalline tract up to 100 μm long, as illustrated in Fig. 11.10c,d) and by Walcott in this volume. Distal pigment from between the cones migrates proximally to various extents in the numerous different families of neuropteroid insects with this type of eye: usually the pigment extends only as far as the level of the retinula-cell nuclei. The crystalline tract between the cone and the retinula-cell column in each ommatidium is then the only available light path to the receptors in the light-adapted eye.

(2) Eyes of beetles of the groups Carabidae, Dytiscidae, Gyrinidae, and Hydrophilidae differ in one important detail from other clear-zone eyes. One of the retinula cells has its rhabdomere situated at the distal end of the retinula-cell column close to the cone tip in the dark-adapted state (Fig. 11.2b). This distal rhabdomere is in a position to receive light that has entered by one fact only. In the dark-adapted eye the proximal rhabdomeres receive light that has entered by many facets but little light reaches them when the eye is light-adapted. In *Hydrophilus*, *Cybister*, and *Dytiscus*, the retinula-cell column is optically and histologically similar to the clear accessory cells that surround it, and the clear zone, composed only of these two cell types, is remarkably transparent and homogeneous. Ray tracing in a model eye of *Cybister* (Dytiscidae) (in which all optical constants had been measured by interference microscopy) showed that rays are bent by lens cylinders of the *cornea* (see Fig. 11.1b) but they are poorly focused upon the proximal receptor layer (see Fig. 11.21). The acceptance angle of these receptors in dark-adapted *Dytiscus* exceeds 30° (Horridge, Walcott, and Ioannides 1970). In carabid and some scarabaeid beetles, on the other hand, the retinula-cell columns are of obvious higher density than the surrounding cells, and in some species they have been observed to act as separate light guides in each ommatidium (Horridge 1971).

Before leaving this anatomical topic it is worth noting the great diversity of the proportions and shapes of the components of clear-zone eyes in the three orders where they are best known, Lepidoptera, Coleoptera, and Neuroptera. There seems to be all possible combinations of the features that one might suppose are relevant to the functioning of the eye. On the other hand, the anatomical requirements for a well-focused superposition eye are very stringent, as outlined below.

11.3. Terms relating to focusing of the eye

The typical dark-adapted clear-zone eye differs from the light-adapted eye, and from apposition compound eyes, in that *the eye as a whole has an aperture*. A parallel beam of light, incident on the eye, enters by the numerous facets forming this aperture, and the effect of the clear zone is to allow rays to be

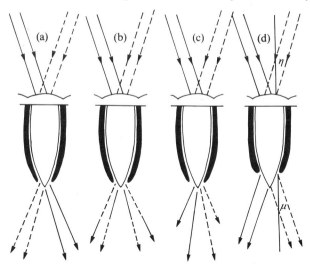

F I G. 11.3. The distribution of light from the cone tip in relation to the focusing of a clear zone eye. (a) In a single weak lens, as illustrated in Fig. 11.14a, rays that are well off-axis tend to emerge on the opposite side of the axis. One of the functions of the cone is to avoid this by exclusion of off-axis rays, by refraction at the tapering edge of the cone, and by other possible means summarized in this paper. (b) In an unfocused eye the distribution of light from the cone tip is independent of its direction of incidence on the facet; compare Fig. 11.4b. (c) In a partially focused eye, rays tend to spread from the cone tip into the same quadrant as that by which they enter the facet; compare Fig. 11.4c. (d) In a well-focused eye, parallel rays incident on the facet at an angle η to the axis emerge from the cone tip at the appropriate angle μ to the axis, as in Figs. 11.19 and 11.20.

redistributed between ommatidia within the eye. The term *focusing* of the eye refers to the redistribution across the clear zone towards the receptor layer of rays that are initially parallel when they strike the eye. The feature of interest, which determines the ray paths across the clear zone, is the angular distribution of rays emerging from a cone tip as a function of the direction from which rays strike its facet.

When rays emerge from the narrow proximal end of the cone with no reference to their direction of entry into the facet (Fig. 11.3b), the eye is defined as *unfocused*. This was the 'scattered light' eye of Horridge, Ninham, and Diesendorf (1972) summarized in Figs. 11.4a,b and 11.5. The unfocused eye provides a readily-defined model eye with minimum assumptions about ray paths.

When rays tend to emerge, on average, in the quadrant below that of their origin (Fig. 11.3c), the eye is defined as *partially focused*. Usually the rays diverge as they emerge from the cone tip even when incident rays are parallel (see Figs. 11.4c and 11.12).

To be *well-focused* an eye must fulfil two conditions (1) the rays emerge from the cone tip in a narrow distribution in the quadrant below that of their

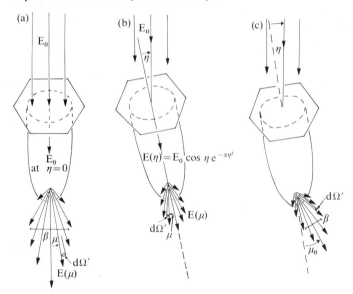

FIG. 11.4. Definitions for the mathematical treatment of ray distributions from the cone tip in unfocused and partially focused eyes. (a) Parallel rays on axis are distributed symmetrically in the clear zone. At an angle μ the intensity of the light emerging from the cone tip is

$$E(\mu) = f_0 \exp(-b\mu^2)\, d\Omega$$

where $f_0 = b/\pi E_0 \exp(-\alpha\eta^2)\cos \eta$. Symbols are defined on the diagrams. (b) When the facet is at an angle to the rays less light passes through, but in the unfocused eye its distribution into the clear zone remains symmetrical about the axis of the facet. (c) When some focusing is present the rays emerge into the clear zone with a distribution centred about an angle μ_0 that lies in the same plane as the incident rays, on the same side of the axis as the incident ray. In contrast with (a) now $E(\mu) = f_1 \exp[-b(\mu-\mu_0)^2]\, d\Omega$, where f_1 must be computed for each value of η.

origin (as in Fig. 11.3d), and (2) the ratio of angles μ/η (Fig. 11.4c) fits the dimensions of the clear zone (see Fig. 11.13), so that rays are focused on the receptor layer.

When rays emerge in the opposite quadrant, as they would tend on average to do for a single lens, the eye is negatively focused (Fig. 11.3a). An explanation for the existence of the cone at all is that it tends to prevent this situation, by a variety of mechanisms discussed in this paper.

11.4. Available optical mechanisms for clear-zone eyes

A variety of quite distinct optical mechanisms can be found in clear-zone eyes, some are remarkably subtle, all are elegant, and all require a marvellous perfection in growth to ensure that they function adequately, because the optical properties are apparently not adjustable after metamorphosis.

In the experimental analysis it is important to remember that a demonstration that one mechanism exists does not necessarily rule out other

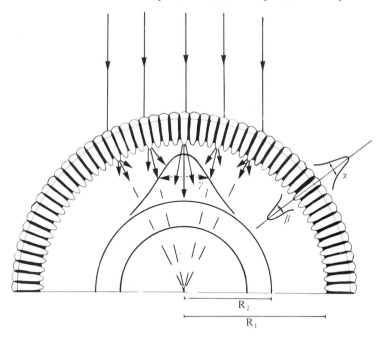

FIG. 11.5. Distribution of light over the clear zone in an unfocused eye. Rays emerging from the cone tips are distributed about the axis of the ommatidium. The resulting distribution of light on the receptor layer is wide. The widths of the distributions at the 50 per cent contour are as follows: α is the width of the η distribution (the admittance function of the facet), β is the width of the distribution of μ (the rays from the cone tip), γ is the width of the angular sensitivity of a receptor, all widths being taken at the 50 per cent contour.

mechanisms that might also be at work. In particular, there are many examples where some rays reach the receptors via light guides, while other rays freely cross the clear zone. Commonly one mechanism dominates in the light-adapted state and another in dim light, with a smooth transition between.

11.4.1. *Action of the cone extension or retinula-cell column as a light guide*

The idea of a light guide (see Fig. 11.10c,d) from cone to receptor in each ommatidium was originally suggested by Exner (1891, p. 134) for the eyes of some deep water crustaceans that had been described in detail by Claus (1879). In these animals the crystalline tracts are very long and obvious in living specimens. The idea was allowed to lapse for many years and was then revived by de Bruin and Crisp (1957) to explain why they found no change of acuity when pigment movements are caused by adaptation in the eyes of prawns and other crustaceans. The idea that the pigment around the light guide acts as a 'longitudinal pupil' was formulated by Kuiper (1962), who considered this the chief reason why the receptors lie at a distance from the cones in clear-zone eyes. Kuiper considered that the graded migration of pigment around the outside of the light guide is essential to protect sensitive receptors against overstimulation. In the first text-book account to stress the light guide as a possible mechanism, Horridge (1965, p. 1064) outlined the action of the pigment and light guide as an attenuator and considered that this theory agreed with the anatomy of nocturnal eyes as known at that time. At this stage there was no actual evidence of light-guide action or of the mechanism of attenuation by pigment, and no anatomical distinction between extensions of the cone and of retinula-cell columns as possible light guides.

Evidence for light-guide action across the clear zone in dark-adapted insects was not slow in appearing. In a detailed study of pigment migration during adaptation, Horridge (1968) showed that in the firefly an extension of the four cone cells crosses the clear zone in each ommatidium and is surrounded by pigment in the light-adapted state. When a dark-adapted firefly retina, which has been lightly fixed so that it is sufficiently firm to cut, is examined from behind, with the receptor layer removed, it is possible to focus upon the broken ends of the crystalline threads, where bright points of light can be seen when a light is placed in front of the eye (Horridge 1969). For parallel work on crustacea see Carricaburu (Chapter 10). Also, in several species of moths, it is directly observable that light entering the facets is channelled along the crystalline threads in the dark-adapted eye (Døving and Miller 1969). The distribution of pigment suggests that this is even more true of the light-adapted eye.

Subsequently, Horridge (1972) observed directly the light-guide action of the column of retinula cells in several beetles and in the moth *Ephestia*, but found that in other beetles, notably Dytiscidae and Hydrophilidae, that the short light guide of the light-adapted eye does not extend right across the clear zone. The moth *Ephestia* is peculiar in that the rhabdom in each ommatidium is carried up to the cone as a highly refractile thread within the retinula-cell column, forming one light guide within another in the dark-adapted state. In the light-adapted eye of *Ephestia*, the light guide formed by the rhabdom and retinula-cell column contains retinula-cell pigment *inside*, which is again unusual, while more distally the cone extends to form a typical neuropteroid crystalline tract that is closely surrounded by the pigment of the principal pigment cells (Horridge and Giddings 1971*b*).

Besides the direct observation of its function, one can infer that a tract or column crossing the clear zone is a light guide. Two arguments prevail, (1) when the refractive index of the tract or column is higher than that of surrounding cells and the diameter is greater than *ca*. 0·5 μm it will necessarily have the physical properties of a light guide; (2) a narrow tract or column surrounded by pigment in the light-adapted state must act as a light guide because it is the only path for the light. Then, in the dark-adapted state the tract or column necessarily acts even more efficiently as a light guide because the removal of the pigment from around it effectively reduces the attenuation of light within (see Section 11.6.6). Whether the fraction of light carried by this path is a significant part of the total light reaching the receptor is a different question.

11.4.2. *Unfocused light from the cone tip*

Five quite different types of observation suggest that light can spread from the cone tips across the clear zone: (1) Unfocused light is obvious in the examination of eye slices of noctural moths or beetles. (2) Tracing of rays through model systems of cornea plus cone quickly shows that not all rays are focused. (3) When the dark-adapted eye of nocturnal beetles, neuropterans and some moths is illuminated by a parallel beam the eyeshine rays diverge over a wide angle. (4) Local stimulation with a light guide shows that light can reach a receptor via many facets in *Dytiscus* (Horridge, Walcott, and Ioannides 1970). (5) In optomotor tests the dark-adapted eyes of nocturnal beetles with clear-zone eyes show relatively poor acuity (Meyer-Rochow, unpublished observations).

In the search for a way to treat unfocused light spreading from the cone tip, a theory was first developed with the minimum number of assumptions (Horridge, Ninham, and Diesendorf 1972). The direction of a ray crossing the clear zone was assumed to be a gaussian distribution from the cone tip about its axis (Fig. 11.4a). The total intensity from the cone tip spreading across the clear zone was assumed to be a simple function of the angle of incidence of the light on the outside of the facet (Fig. 11.4b). In the unfocused eye the main feature is that the direction of a ray crossing the clear zone is not correlated with its direction outside the eye, and the light crossing the clear zone has been described as 'scattered' from the cone tip.

Although this eye is completely unfocused, rays sum upon the receptor layer to increase sensitivity at the expense of acuity. For a diffuse source the sensitivity increase is proportional to R_1^2/R_2^2 where R_1 and R_2 are the radii from the centre of the eye to the cone tips and to the receptor layer. Therefore, the wider the clear zone the more the unfocused light increases sensitivity. To a parallel beam the sensitivity increases to a maximum when the receptor layer is at about half the radius of the eye, when typical values are introduced for other parameters. Further details are given in Section 11.5.1.

11.4.3. *An image formed by a pinhole at the cone tip*

In many clear-zone eyes there is a stage in adaptation when the pigment reaches just as far as the tip of the cone, which therefore must act as a pinhole. In some compound eyes examined the convex outer surface of the facet produces an inverted first image within the cone. With a pinhole at the cone tip this image must produce a second and erect image at whatever plane is

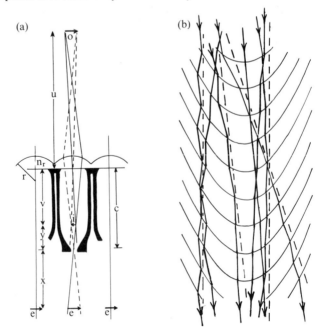

FIG. 11.6. Two of the three ways of forming a partially focused image without a lens cylinder. The third way is in Fig. 11.7g. (a) When the first image is within the cone, the pigment around the cone tip can form a pinhole so that an erect image is seen behind each facet in an eye slice (after Horridge 1971). (b) Layers in the corneal cone of the firefly are shaped like concentric paraboloids (as in Fig. 11.10a). Such a structure causes bending of rays towards the axis if alternate layers have slightly differing refractive index. Basically the mechanism is that at a discontinuity between layers rays which turn towards the axis stay within one layer for a longer path than rays which turn away from the axis. The dashed lines are straight to show that the rays curve (after Horridge 1959).

examined with a microscope focused in the clear zone (Fig. 11.6a). The object must be very bright because the pinhole has small aperture, and the erect image will be very fuzzy. Even so, the pinhole mechanism has not been excluded as an explanation of the observations of the poor erect images seen by Exner (1891) in firefly, moth, and melolonthid beetle eyes; by Eltringham (1919) in butterflies; by Kunze (1969) and Døving and Miller (1969) in various moth eyes. Pinhole images might explain the observations of erect images in eye slices, but do not provide a reasonable mechanism of vision for the nocturnal compound eye because in their formation the eye is partially light-adapted, and most of the light falling on the eye must be absorbed (see open triangles in Fig. 11.16).

11.4.4. *Image formation by the curvature of the corneal surfaces*

At least since the description of the bee's eye by Swammerdamm (1738) it has been evident that the outer convex surface of the corneal facet acts as a

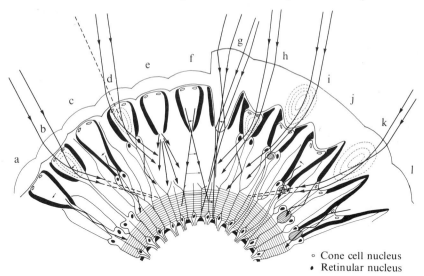

° Cone cell nucleus
• Retinular nucleus

FIG. 11.7. Clear zone eye mechanisms of 1972, with principal references. Retinula cell nuclei are solid black, cone cell nuclei open circles; pigment cells are solid black, inhomogeneous zones dashed, rhabdoms cross-hatched and light rays bear arrows. (a) Fixed light guide inferred in retinula cell column with crystalline cone; possibly some Bombycoid moths and *Bibasis*, Hesperiidae (Yagi and Koyama 1963). (b) Well-focused superposition image formed by inhomogeneous crystalline cone; Hesperiidae (Horridge, Giddings, and Stange 1972). (c) Typical neuropteroid retinula cell column with crystalline tract surrounded by pigment in light-adapted state; Neuroptera, many moths (Horridge and Giddings 1971a,b). (d) Dark-adapted state of (c) with poorly focused rays crossing the clear zone, and retinula cell column acting as a light guide (Horridge 1971). (e) Unfocused clear zone eye; model only (Horridge, Ninham, and Diesendorf 1972). (f) Pinhole formed by pigment round the cone tip; model only (Horridge 1971). (g) Erect image formed only by the outer and inner surfaces of the cornea; model only (Winthrop and Worthington 1966; Horridge 1969). (h) Rays crossing the clear zone are poorly focused by homogeneous cornea and inhomogeneous crystalline cone; otherwise as in (c) and (d); melolonthid beetles (Meyer-Rochow 1973). (i) Rays crossing the clear zone are poorly focused by thick inhomogeneous cornea. Note tiered retina with distal as well as proximal rhabdomere. The retinula cell column is not a light guide; aquatic beetles (Exner 1885; Meyer-Rochow 1972). (j) Light-adapted state of aquatic beetle eye in (i), with crystalline tract surrounded by pigment (Horridge and Giddings 1971a). (k) Rays crossing the clear zone are well-focused by inhomogeneous cone: inferred for firefly (Exner 1891; Seitz 1969). Cone cells form crystalline tract. (Horridge 1968). (l) Light-adapted state of (k), with pigment surrounding the crystalline tract; firefly (Horridge 1968).

lens, and many early microscopists saw a tiny inverted image behind each facet of the cleaned cornea. But I think Winthrop and Worthington (1966) were the first to point out that when the first image lies within a corneal cone then the curvature of the inside corneal surface, or the rounded tip of a corneal cone, can produce an erect second image deep within the eye. Ray diagrams are illustrated in Figs. 11.7g and 11.15a. It is possible that this is the principal mechanism of forming an erect image on the receptor layer in some cantharid, elaterid, lampyrid, and lycid beetles, and a contributory mechanism in many other groups, but no examples have been properly examined.

11.4.5. *Inhomogeneity of refractive index of the cornea*

The standard textbook account in this century has been that an erect image is formed upon the receptor layer by a lens cylinder in each corneal cone of the firefly, and that one combined image is formed. In the original account, Exner (1891) observed the combined erect image but was careful to propose the lens cylinder only as a hypothetical mechanism for its formation. He proposed a model lens cylinder twice as long as its focal length, forming an afocal system (Fig. 11.8b). We have seen that other possible explanations need to be examined, and that clear-zone eyes are very varied, but there is in fact very good experimental evidence in Exner's earlier work, for lens cylinders in the cornea. The earlier observations leading to Exner's general theory are worth bringing back into modern literature.

In 1876 Exner calculated the refractive index of the cornea of the water beetle *Hydrophilus* as 1·82 from the position of the first image. He had no explanation of this improbably high value. Having no means of measuring the refractive index of the crystalline cone he assumed it to be uniform with the cornea.

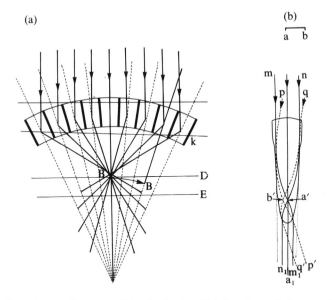

FIG. 11.8. Exner's own diagrams of the firefly, in which he observed the erect image and inferred the mechanism. (a) The erect image, BB′ and its formation by appropriate bending of the rays in the dioptric system k, were both observed. (b) The inferred mechanism of bending the rays by an inhomogeneous corneal cone and by a second refracting surface in the firefly. Compare Fig. 11.15a and b.

In 1885 Exner had used his new microrefractometer to measure refractive index of the crystalline cone of *Hydrophilus* as 1·557–1·561, with only slight changes after storing in ethanol for months. With the same instrument he showed the calculated value of 1·82 for the cornea to be erroneous, demonstrated the gradients of refractive index and realized that the non-homogeneous cornea of each ommatidium of *Hydrophilus* acts as a lens cylinder. In sections of the *Hydrophilus* cornea parallel to its surface (see Fig. 11.11b) he found $n = 1·545$ for the outer layer and 1·565 for the inner layer of the lens cylinder. Exner's experimental demonstration that in each ommatidium the first image is formed by the action of the lens cylinder within the thick cornea was as follows: (1) it still formed a lens when the curved inner surface was removed (the outer surface is already flat); (2) it forms the same image when immersed in a medium of much higher refractive index (barium mercury iodide, $n = 1·7783$) which would otherwise convert a convex lens to a diverging lens; (3) longitudinal sections of the cornea had the appearance of a cylinder viewed from the side, although both sides of the section were definitely flat. Exner found that the refractive index changed by a steady gradient with no cause visible in stained histological sections.

These results (Exner 1885) on the inhomogeneous refractive index of the thick cornea of the water beetle were not reproduced in Exner's great monograph of 1891, and have not appeared in the literature on this topic in the present century. In his introduction of the light guide theory and survey of superposition eyes, Kuiper (1952) was quite wrong in suggesting that it seems very doubtful if 'lens cylinder' properties will ever be found in a compound eye, and in saying that Exner did not have the tools to do the necessary experiments. It is true that Exner (1891) never showed how the lens cylinders of *Hydrophilus* act in the vision of this animal, and he was very careful in his statement never to say that he saw a lens cylinder in the corneal cone of the firefly; Exner always proposed his explanation of the combined erect image in the firefly as an *inferred* lens cylinder. Unfortunately it was the inferred story for the firefly, based only upon the observation of the remarkably perfect erect image, which came into the standard textbook literature, while the partially focused eye of the water-beetle, with its marvellous lens cylinders within the cornea, was forgotten. When I was working out the anatomy and process of light-adaptation in the firefly eye in relation to the optics (Horridge 1969), I was unaware that the observations on refractive index gradients had been made on the water-beetle and not on the firefly. Even worse, when working out the process of adaptation in the water-beetle eyes (Horridge and Giddings 1971a) and the field sizes of the dark-adapted *Dytiscus* eye (Horridge, Walcott, and Ioannides 1970), I was unaware of the peculiar optical properties of the thick cornea demonstrated by Exner in his preliminary paper.

In the corneal cones of the fireflies no lens cylinder was found in *Phausis*

(*Lampyris*) by Kuiper (1962) nor in *Photuris* by Horridge (1969) with crude methods, but Seitz (1969) demonstrated a gradient in the corneal cone of *Phausis* by interference microscopy. Although he traces only four rays through this inhomogeneous cone, Seitz states that two rays entering parallel to the axis emerge parallel to the axis, and that two rays parallel to each other but not to the axis emerge parallel to each other in a direction that could be compatible with the formation of an erect image. However, the ray paths traced by Seitz in the firefly cone are hardly curved at all (Fig. 11.9a) and it is not at all clear what would be the path of rays at a large angle to the axis. What is needed is the tracing of a large number of rays per facet and a continuation of the rays through the clear zone to see how they sum upon the receptor layer. It is still not known whether an image is formed on the receptor layer in the dark-adapted firefly eye, and the distributions of rays (α, β, and μ/η in Figs. 11.4 and 11.12) which describe the action of the eye, have yet to be measured in the firefly. A better example of the superposition principle is now provided by the skipper butterflies, but there the mechanism is different in that it depends on an inhomogeneous crystalline cone, and the cornea is thin and homogeneous (see Figs. 11.9c, 11.11a, and 11.20).

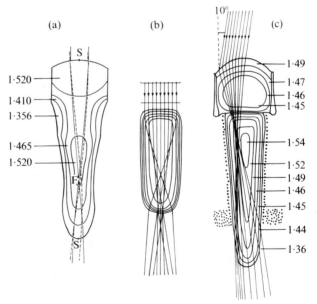

FIG. 11.9. Gradients of refractive index in the dioptric system and their effect in bending the rays. (a) The corneal cone of the firefly (from Seitz 1969) showing values of the refractive index and the paths of four rays. Note the refraction at the outer and inner surfaces. (b) Tracing of rays through a lens cylinder by modern computer methods shows that imperfect focusing of rays is the general rule. (Computation by Dr. Meggitt, A.N.U.). (c) The inhomogeneous crystalline cone of the skipper butterfly *Toxidia peroni*, which has a well-focused eye. (From material in Horridge, Giddings, and Stange 1972.)

The partially-focused beetle eyes are better known. The aquatic beetle *Cybister* (Dytiscidae) has a corneal lens cylinder corresponding to each ommatidium, although the very thick cornea has a smooth outer surface without separate facets. By tracing equally spaced rays incident on the dark-adapted eye at a range of angles up to 60°, Meyer-Rochow (1972*a*) showed: (1) the lens cylinder of the cornea acts as the first lens in the eye; (2) the curvature of the inner surface of the cornea and of the cone surface makes them significant refracting surfaces; (3) the light is partially focused upon the receptor layer; (4) the angular sensitivity curve of the receptors, as one can infer from ray tracing, is *c.* 30° wide at the 50 per cent contour in the fully dark-adapted eye. A similar study on another beetle is published in this volume (Meyer-Rochow 1972*b*).

The optical arrangement in the aquatic beetle eye is presumably an adaptation that is equally effective in air and water because the inhomogeneous cornea eliminates the need for a convex facet.

11.4.6. *Growth layers in the corneal cone*

Recently it has been found that insect cuticle has a molecular structure such that elongated chitin micelles are orientated parallel to the corneal surface in spirals which run through the cuticle from the inside towards the outside (reviewed by Neville and Cavaney (1969)). When cuticle is cut perpendicular to the surface and the edge examined under the electron microscope, one sees a series of layers formed by the alternate ends and sides of these micelles (see Meyer-Rochow, Chapter 12).

This pattern occurs in the corneal lens of all insects examined, and in horizontal sections through the cornea of the facet one sees a spiral pattern of layers under the electron microscope. In beetles with long corneal cones, the whole cone consists of hundreds of layers arranged like paraboloids fitting inside each other (Fig. 11.10a).

The refractive index of a mass of orientated fibres depends on their angle relative to the rays. The layers in the cone necessarily present to light passing into the eye a series of alternate refractive indices. The optical effect of this peculiar structure is not known in detail but preliminary ray tracing in two dimensions (Fig. 11.6b) suggested that the curved layers bend rays towards the axis, as in a lens cylinder (Horridge 1969). It is possible that the measurements of refractive index made with an interference microscope (Seitz 1969) include a contribution from these layers, but a proper optical treatment of them has never been made.

11.4.7. *Inhomogeneity of the refractive index of the crystalline cone*

Exner (1891) observed rather poor erect images behind slices of eyes with crystalline cones in several beetles and in fixed eyes of several moths. Apparently he did not examine the crystalline cones but he assumed that they

FIG. 11.10. Anatomical details. (a) Transverse section of a corneal cone of the firefly *Photuris* showing the layers formed by the alternate ends and sides of the spiral chitin micelles. (b) The same in longitudinal section. From Horridge (1969). This layered structure is typical of many insect facets and corneal cones. (c) Typical crystalline tract surrounded by pigment grains in the light-adapted eye of *Chrysopa* (Neuroptera). (d) The tract in transverse section at higher power.

are inhomogeneous in refractive index and that they act as lens cylinders because this was his only available mechanism for the formation of the erect image.

No further observations on this topic can be found until Kunze (1971) announced that the crystalline cones of *Ephestia* are non-homogeneous, with a greater optical density towards the centre. This provided one possible explanation of the erect image previously photographed behind each cone tip in *Ephestia* (Kunze 1969). Ray tracings by Horridge (1972) through the convex cornea and inhomogeneous cone of *Ephestia* suggest that the eye is poorly focused, but ray tracings in two dimensions are apparently not a reliable guide to the operation of this eye, and small errors in the measurement of refractive index and surface curvatures lead to large errors in tracing rays.

In further observations with an interference microscope we have now found that in the clear-zone eyes of several insects the crystalline cones have a refractive index which increases towards the centre of the cone. Examples include beetles and moths. In particular, the inhomogeneous refractive index of the cone in skipper butterflies bends rays so that a well-focused image (Fig. 11.11a) is formed on the receptor layer (Horridge, Giddings, and Stange 1972).

11.5. Mathematical models of light distribution from the cone tip

When a large number of equally spaced rays are drawn through the cornea and cone of an ommatidium for each direction (η) of light falling upon the facet, it is possible to plot the distribution of the angle (μ) at which light emerges from the tip of the cone as a function of its direction of origin outside the eye. In some cases this relation between μ and η can be directly measured (see Section 11.7.14), and clearly it defines the way in which the light behaves on entering the eye, and also how rays sum upon the receptor layer within the eye.

Based upon arbitrary distributions of μ that hopefully cover the range found in real eyes, the summation of rays on the receptor layer has been computed or derived analytically for three different cases as set out below. These models calculate the angular distribution of light on the receptor layer for a parallel beam stimulus. This is called the admittance-function of the eye for light, and has width γ at the 50 per cent contour. If the effectiveness of the light is independent of its direction of arrival upon the receptors, the admittance function is equal to the angular sensitivity of the receptors of width $\Delta\rho$ at the 50 per cent contour (see Section 11.6.5). The ideal models can then be related to measurements of optomotor or receptor acuity.

11.5.1. *The unfocused eye*

In the model of a completely unfocused eye (Fig. 11.5) the following assumptions are made: (1) The eye is symmetrical, of constant radius, the

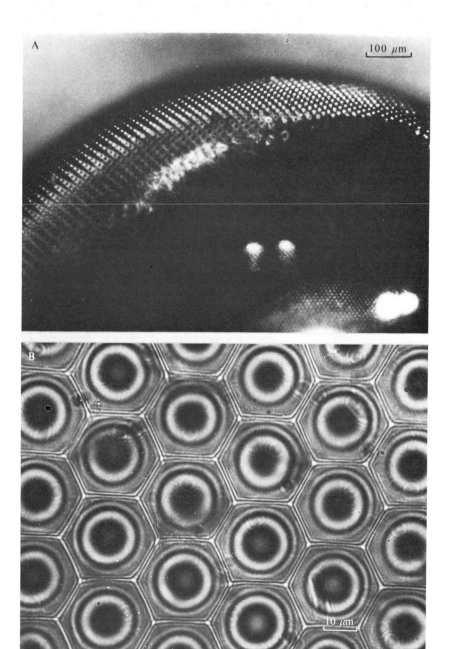

FIG. 11.11. Optical details. (a) The real erect image on the receptor layer in the skipper butterfly *Trapezites*. The two spots, formed by two distant point sources that subtend 8° at the eye, are seen through a cut in the top of the eye. The spots are about four receptors wide; the rhabdom ends appear as black dots around the image because they absorb light. From Horridge, Giddings, and Stange 1972. The image is formed mainly by the crystalline cone in skipper butterflies. (b) A parallel-sided tangential section of the cornea of the water beetle *Dytiscus* under an interference microscope in green light. The dark and light zones represent wavelength multiples of retardation, as caused by the inhomogeneous refractive index. These are the lens cylinders inferred by Exner (1886). Although possessing this marvellous optical device, the eye is poorly focused in several beetles examined. Photo by Meyer-Rochow.

clear zone is of constant width, cones and facets are similar, etc. (2) Less light enters a facet as its angle (η) relative to the axis is increased. The relation is assumed to be gaussian of width α at the 50 per cent contour. (3) The same light emerging from the cone tip is distributed symmetrically about the axis of the cone. Again the distribution is assumed to be gaussian (of width β at the 50 per cent contour).

Mathematical analysis of this very simple model eye (Horridge, Ninham, and Diesendorf 1972) shows that light is summed upon the receptor layer in the following way, where the symbols have the meaning shown in Fig. 11.5.

(1) The admittance function of the eye as a whole (closely related to angular sensitivity) has a width at half height given by:

$$\gamma^2 = \beta^2(R_1/R_2 - 1)^2 + \alpha^2 \qquad (11.1)$$

(2) The sensitivity of the eye to *diffuse* light depends on the width of the clear zone, and, other factors being equal,

$$E_{\text{peak}} = E_0 K R_1^2/R_2^2, \qquad (11.2)$$

where the constant K depends on the fraction of axially incident light that enters an ommatidium.

(3) For *parallel light* incident on the eye, there is an optimum width of the clear zone (for maximum sensitivity) when

$$R_1/R_2 = 1 + (\alpha^2/\beta^2) \qquad (11.3)$$

Since α is similar to β, the optimum is near $R_1/R_2 = 2/1$.

(4) The unfocused eye can be several times as sensitive for parallel light on axis as an apposition eye with similar losses of light in pigment around the cone.

(5) A large additional increase in sensitivity is now possible by increasing the cross-sectional area of the receptors, up to a hundredfold that in typical apposition eyes.

11.5.2. *The partially focused eye*

The essential feature of a focused clear-zone eye is that rays from the cone tip are distributed around an angle μ_0 which is not zero when the incident light is not axial, and that this angle μ_0 increases as η increases (measured as in Fig. 11.4c). Light now spreads over the clear zone, on average, towards a receptor which is looking in the direction of origin of the light outside the eye. A partially focused eye is illustrated in Fig. 11.12; a well-focused eye in Fig. 11.13.

Analysis of the geometry of the partially or well-focused eye (Diesendorf and Horridge 1973) brings out the following points:

(1) The three significant parameters in understanding the optics are the width β of the μ distribution at various values of η, the average value

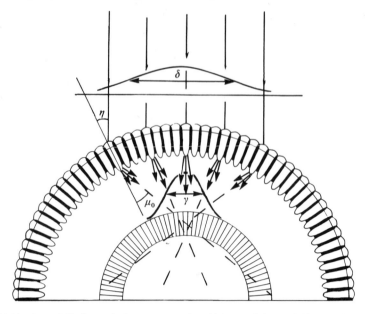

FIG. 11.12. A partially-focused clear-zone eye in which parallel rays falling on the eye pass through many facets and converge approximately on the receptor layer. Eyes of this type are less than perfect either because the rays converge towards the wrong place or because the distribution from the tip of each cone is too wide. A well-focused eye is so difficult to achieve for all rays falling on the eye from all directions that poorly focused clear-zone eyes must be the usual compromise when in the dark-adapted state most rays are not absorbed by pigment.

μ_0 for various values of η, and the fraction of light passing through the facet and cone for rays parallel to the axis. These quantities determine the sensitivity and the acuity, which are directly related. In turn, these angles and distributions are derived from the physical properties of the cornea, cone, and pigment.

(2) The acuity of the eye is improved when the β distribution is narrowed and when μ_0/η is appropriate for the geometry of the eye. Sharp focusing depends very critically on these parameters.

(3) The sensitivity depends on the excellence of the focus and the maximum intensity on the receptors is thereby increased as $1/\gamma^2$. The sensitivity also depends on the total aperture and is theoretically proportional to δ^2 (Fig. 11.12) if the relations in (2) still hold. There is a limit to the useful increase in aperture because rays that are far off axis are unlikely to be appropriately focused. The increase in sensitivity with focusing is calculated in Table 1.

(4) In a partially focused eye, the optics of the eye cannot be worked out from a measurement of the angular sensitivity, absolute sensitivity, or effectiveness of the different facets in admitting light, because these

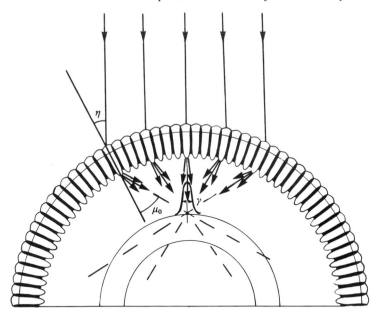

FIG. 11.13. A well-focused clear zone eye in which parallel rays falling on the eye pass through many facets to converge on a small region of the receptor layer. This is the situation demonstrated for the diurnal eye of skipper butterflies; it represents a modification of Exner's theory of the superposition image, with his perfect image replaced by a definable distribution of light, and some light absorbed by pigment around the cone. In such an eye the rhabdoms need to have a wide acceptance angle and to be isolated from each other to prevent spread of scattered light. These features are achieved in skippers by enclosing each rhabdom in a sheath of air spaces.

derived properties of the optics of the eye are the result of many possible combinations of the primary parameters set out in (1) above.

(5) Summation across the clear zone is compatible with the simultaneous propagation in light guides of a different fraction of the incident light.

The second model of the eye is only partially focused because rays emerge from the cone tip with a distribution of width β. However, for each cone they emerge from the tip in a distribution about a mean angle μ_0 (Fig. 11.4c). The equation for the Exner line (Fig. 11.19) is:

$$\mu_0 = \frac{R\eta}{1-R} \quad \text{where} \quad R = R_2/R_1 \tag{11.4}$$

When this relation is obeyed, the following conclusions stand (Diesendorf and Horridge 1973):

(1) The distribution of light on the receptor layer has a width at the 50 per cent contour of

$$\gamma = \beta/R\left\{1-R+\tfrac{1}{2}\left(\frac{R\alpha^2}{1-R}\right)\right\} \tag{11.5}$$

TABLE 11.1

Approximate values of the sensitivity to parallel rays eqn (11.3), and the admittance angle γ, for the unfocused eye and for the partially-focused eye, for different values of β when $\alpha = 15°$ and $R_1/R_2 = \frac{1}{2}$, where the symbols have the meanings shown in Figs. 11.4 and 11.5. This table illustrates that as β is reduced (ignoring the probable effect that more light incident parallel to the axis would then be absorbed in pigment) there is only a slow increase in sensitivity to parallel rays on axis, and a small narrowing of the field of view for the unfocused eye. For the focused eye, on the other hand, narrowing of the β distribution causes a parallel reduction of the visual fields and a large increase in sensitivity.

β	$\dfrac{\alpha}{\beta}$	Unfocused eye		Partially focused eye	
(degrees)		Model 1		Model 2	
		Parallel light sensitivity (eqn (11.3))	Acceptance angle (degrees) (eqn (11.1))	Parallel light sensitivity (eqn (11.6))	Acceptance angle (degrees) (eqn (11.5))
15	1	2	21·2	2	16·0
7·5	2	3·2	16·8	4	8·0
6	2·5	3·45	16·1	5	6·4
4	3·75	3·73	15·5	7·5	4·27
2·5	6	3·89	15·2	12	2·67
2	7·5	3·94	15·14	15	2·14
1·5	10	3·96	15·1	20	1·6

showing that γ depends directly on β and R_1/R_2 but only to a small extent on α.

(2) The gain in sensitivity to a parallel beam is

$$E_{\text{peak}}/E_0 = \frac{A}{2R(1-R)} \cdot \frac{\alpha}{\beta} \qquad (11.6)$$

showing that sensitivity is mainly governed by β, α, and A, where A is the fraction that is admitted of light that is incident parallel to the axis of a facet.

11.5.3. *The under- and over-focused eye*

In the third model of the eye, parallel rays that fall upon the outside of the eye are directed from the cone tips towards a focal point that is above or below the plane of the receptors. For a given value of η there is only one value of μ (i.e., the distribution of μ is assumed to be narrow), but the eye is either long-sighted or short-sighted (Diesendorf and Horridge 1973). This is a less satisfactory model than the previous one but it describes the summation on the receptor layer of those rays that would be brought to a focus at another level. The following conclusions emerge:

(1) The distribution of light on the receptor layer has a width

$$\gamma = \alpha(1 - R_3/R_2)/(1 - R_3/R_1) \qquad (11.7)$$

where R_3 is the radius to the focus, and other symbols as before. This implies that the acceptance angle depends directly on α and ranges from 0 when the eye is perfectly focused (actually it depends on β because the second model is then appropriate) to a value near to α when the eye is severely short- or long-sighted.

(2) The gain in sensitivity to a parallel beam by focusing is the ratio of the aperture to the area of the focused spot

$$E_{\text{peak}}/E_0 = A \cdot \frac{R_1^2}{R_2^2}\left(\frac{1-R_3/R_1}{1-R_3/R_2}\right)^2 \tag{11.8}$$

This again fluctuates wildly in the region where the eye is focused when $R_3 = R_2$, but model 2 is then appropriate. As in the other models, sensitivity depends on the fraction of axial light admitted and on the width of the clear zone.

The numerical consequences of convergence of rays on the receptor layer therefore depend on the choice of model. In real eyes and in ray tracing, both of these models of partial focusing have some relevance. The significant feature is that any situation is described by the parameters α, β, A, and the relation of μ_0 to η.

11.6. Physical consequences of the optical system

In Section 11.4 a variety of possible optical mechanisms have been mentioned as candidates in different clear-zone eyes. A closer look will now be taken at the physical consequences of some of these systems, as they relate in particular to the optical properties of the components of a single ommatidium. The particular topics below have been selected as bearing upon the interpretation of the experimental analysis to be discussed in Section 11.7.

11.6.1. *Divergence due to a single lens and the functions of the cone*

If we draw equally-spaced parallel rays incident on a convex surface they are brought to a focus, but beyond that they inevitably diverge (Fig. 11.14a). The typical insect cornea has a refractive index of *ca.* 1·5 and the radius of curvature of the facet is commonly similar to the facet diameter, *ca.* 10–30 μm. Therefore the focal plane is 40–100 μm behind the lens, and experimentally it has always been found within the cone in the few cases examined. From this it follows that parallel rays from any direction falling on the facet will produce a *divergent* bundle of rays in the clear zone beyond the cone *unless some other optical mechanism stands in the way*. Moreover μ_0/η will tend on average to be negative (Fig. 11.14c). In brief, six types of mechanism have so far been found which reduce this divergence of the rays and tend to make μ_0/η positive. Yet other mechanisms may achieve the same effect:

(1) a second curved corneal surface, as commonly occurs in beetles with a thick cornea, but rarely elsewhere (Fig. 11.7o,k);

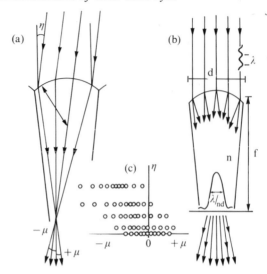

FIG. 11.14. Divergence of parallel rays. (a) Caused by a single lens. Parallel rays at a small angle η to the axis emerge with a distribution angle of μ, on both sides of the axis. As η increases, μ becomes more negative. (b) Caused by diffraction. A lens of small aperture converges parallel incident rays to an Airy disc in the focal plane. Imperfections of the optics widens this angular distribution. (c) The relation between the angle η to the axis of an incident ray, and the angle μ of the same ray after it has passed through a single lens. Note the drift to the left with increasing η. Compare Fig. 11.15c.

(2) a pinhole at the cone tip (see Figs. 11.6a, 11.7f), possibly more common than supposed when adaptation is just right;

(3) greater refractive index in the cone centre (see Fig. 11.9,a,c) occurs but not universally in moths and beetles;

(4) paraboloid layers of alternating refractive index in corneal cones (Fig. 11.6b);

(5) the shape of the cone itself acts as a converging mechanism if not surrounded by pigment (Fig. 11.16);

(6) a lens cylinder in the thickness of the cornea (Fig. 11.11b).

Qualitatively, these are mechanisms which influence the relation between η and μ for each ray away from that in Fig. 11.14c toward that in Fig. 11.15c. The more the value of μ/η is appropriate to the geometry of the clear zone, the better focused and the more sensitive is the eye. Natural selection has acted to generate focusing mechanisms and to perfect the action of those present in any given eye. By contrast, in apposition eyes the cone is always homogeneous so far as is known, and is presumed to act only as a spacing element, like the inside of a telescope.

11.6.3. *An afocal system*

One particular combination of two converging lenses in each ommatidium would allow reasonable focusing, on the receptor layer, of parallel rays

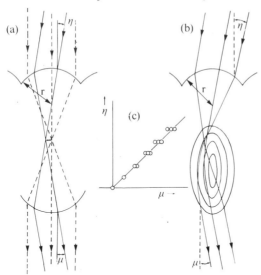

FIG. 11.15. Afocal systems, through which rays pass and remain parallel to each other, but not necessarily parallel to the axis. In each case an inverted first image is situated within a two lens system. (a) and (b) can occur together. (a) Two curved surfaces or converging lenses, placed so that their foci coincide. In an ommatidium the curved surfaces could be the corneal facet and the tip of the corneal cone, as in Fig. 11.7g or 11.9a. (b) A curved surface combined with a refractive index gradient, which could be situated in the cornea (Fig. 11.7i,k) or in the crystalline cone (Fig. 11.7b,d,h). (c) The relation between the angle η to the axis of an incident ray, and the angle μ of the same ray after it has passed through an afocal system as in a or b. For a given value of η, a scatter of μ values is caused by diffraction and imperfect optics.

falling on many facets. When the focus of the first curved surface is in the focal plane of a second converging lens, parallel rays at an angle η to the axis emerge through the second lens at an angle μ to the axis and they are still *parallel* (Fig. 11.15). The ratio μ/η is positive and its value depends on the ratio of the two focal lengths. As is clear by comparison of Figs. 11.5, 11.12, and 11.13, *any positive* average value of μ/η will produce a partially focused eye, and an appropriate average value of μ/η of about unity will produce a well-focused eye. Such a combination of two lenses has no focus for parallel rays, and is called afocal, but within the eye as a whole the effect would be to focus parallel rays falling upon different facets into a spot on the receptor layer. An afocal lens combination must be clearly distinguished from an unfocused system, referring to the whole eye.

This afocal system may be approached, if not perfectly achieved, in a compound eye by adding one of the following converging systems behind the outer curvature of the facet:

   (1) a second curved surface of the cornea, which must then be very thick so that the first image is within it (Fig. 11.7g and 11.15a);

   (2) an inhomogeneous cornea, approximating to a lens cylinder with

greatest refractive index on axis (Figs. 11.7i,k, 11.9a, 11.11b, and 11.15b);

(3) an inhomogeneous crystalline cone similarly approximating to a lens cylinder (Figs. 11.7b,d,h, and 11.9c);

(4) a pinhole, or system of layers in the cone (Figs. 11.6a, 11.7f, and 11.16);

(5) the shape of the narrow end of the crystalline cone has a weak but definite effect in eyes where the pigment is sufficiently retracted in the dark-adapted state (Fig. 11.16).

Present indications are that an attempt at an afocal system by a combination of these mechanisms produces a partially-focused dark-adapted eye in small Lepidoptera, Coleoptera, and Neuroptera that fly in dim light. However, skipper butterflies, some day-flying moths, e.g. Agaristidae, have an excellent afocal combination in daylight and so have many groups of large moths, elaterid and lampyrid beetles, and neuropterans when dark-adapted.

11.6.3. *Divergence of rays due to diffraction*

Parallel rays on axis falling on a perfect lens do not converge exactly to a point, but to a patch called the *Airy disc* (Fig. 11.14b). At the 50 per cent

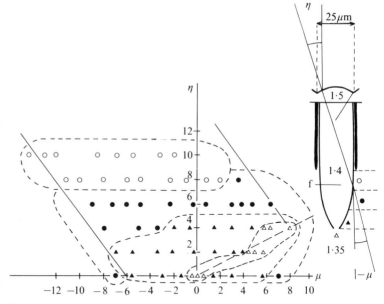

FIG. 11.16. Ray tracing through a typical convex facet and homogeneous cone with the pigment extending to four different limits shown by the symbols ○●▲△. Note the wide spread of the angle of emergence (μ) for a given angle of incidence relative to the axis (η). When the pigment creates a pinhole at the cone tip (open triangles) the value of μ/η is c. 2, which would produce an erect image as in Fig. 11.6a. A symmetrical distribution of rays (unfocused eye) is caused by an intermediate pigment position (solid triangles). In general, as the pigment withdraws from the cone tip, more rays pass through (dashed zones on both figures), and the acuity of the eye as a whole must decrease.

contour the Airy disc has an angular width of *c*. λ/nd radians (where λ = wavelength, n = refractive index of the medium in which the focus lies, and d = diameter of the lens). For an imperfect lens, parallel rays converge to a *circle of confusion*, which is always larger than the Airy disc. In a typical insect facet the divergence produced by diffraction alone is *c*. $1\cdot0$–$2\cdot0°$ and so it is important in only the most perfectly focused ommatidia.

Diffraction by a second aperture at the cone tip could become significant in a clear-zone eye, especially in cases where rays parallel to the axis are focused by the facet exactly upon a pinhole at the cone tip. This would reduce the width of the α distribution but increase the width of the β distribution, so that α/β would be reduced. A clear-zone eye constructed in this way would be inefficient, for sensitivity to a large source depends upon α^2/β^2.

11.6.4. *The migration of pigment around the cone tip*

As movement of the distal pigment toward the cornea during dark-adaptation leaves more of the cone tip exposed, progressively more rays are admitted and cross the clear zone freely (Fig. 11.16). This is the principal control of the aperture of the eye, as shown by the increase in size of the pseudopupil and of the patch of eyeshine. The important question is how the angular distribution of rays entering the clear zone changes as the pigment moves and more light enters. The result depends critically on the optics of the cone. In different types of clear-zone eyes the problems have hardly yet been tackled.

11.6.5. *Light summed upon the receptor layer and angular sensitivity*

The feature of the eye that is most important in vision is the angular sensitivity of the individual receptors. This property is measured electro-physiologically as the sensitivity of a single receptor to a distant point source that is moved around the eye. The angular sensitivity, however, depends on the distribution of rays upon the receptor layer. Light from the distant point source, i.e. parallel rays on the eye, passes through the facets and part of it is distributed from the cone tips upon the receptor layer. As the point source is moved round the eye, this distribution of light moves across the receptor layer and is the cause of the change in the recorded receptor response. Therefore the angular sensitivity of the receptor is the same angular function as the distribution of light on the receptor layer (Fig. 11.13), so long as all light that falls on a receptor is equally effective in stimulating it.

11.6.6. *Conditions under which light guides are effective*

A tract or column crossing the clear-zone must act as a light guide if its refractive index is greater than that of its surroundings. Light guide transmission for a cylindrical solid dielectric rod is governed by the parameter V, where

$$V = \frac{\pi d}{\lambda}\sqrt{(n_1^2 - n_2^2)}$$

and d = diameter of guide, λ = wavelength, and n_1 and n_2 are refractive index inside and outside respectively (Snyder, Chapter 9).

When this parameter V is greater than c. 15 the rod, column, or tract behaves according to ray theory, but when V is between 15 and 2 the guide transmits light as a number of modes of vibration; insignificant energy is transmitted when V is less than 1. When $V = 4$, the first four modes are transmitted and this is sufficient for the light guide to carry a large fraction of the incident light; more importantly, if an extension and thinning of the crystalline tract causes V to fall through the range from 6 to 1 then the ability of the light guide to operate falls dramatically. Control of V by change of refractive index is not yet known for light guides across a clear zone but occurs around the rhabdom in apposition eyes.

Let us take two typical cases, with $d = 4\ \mu m$ as for a crystalline tract and with $d = 10\ \mu m$ as for a retinula-cell column. Let $V = 4$, $\lambda = 0.5\ \mu m$ and let us assume that $n_1 + n_2 = 2.7$ (which is not a critical value). Then for the $4\ \mu m$ tract

$$\sqrt{(n_1^2 - n_2^2)} = \frac{1}{2\pi}$$

$$(n_1 + n_2)(n_1 - n_2) = \frac{1}{4\pi^2} \quad \text{so} \quad n_1 - n_2 = \frac{1}{10.8\pi^2} = 0.01$$

For the $10\ \mu m$ diameter column

$$\sqrt{(n_1^2 - \tfrac{2}{2})} = \frac{1}{5\pi} \quad \text{so} \quad n_1 - n_2 = \frac{1}{67.5\pi^2} = 0.0025$$

To carry the first four modes as a light guide a retinula cell column $10\ \mu m$ diameter must therefore have a refractive index which is 0.002–0.003 higher than that of the surrounding cells. This is readily achieved because a difference of 0.001 is caused by an excess of only 0.6 per cent of protein or 0.6 per cent of sodium chloride. On the other hand, a crystalline tract of $4\ \mu m$ diameter would require an excess refractive index of 0.01, which would need an additional 6 per cent of protein or equivalent material inside. This may be possible, but it is easy to see why most moths and beetles with clear-zone eyes have $10\ \mu m$ retinula-cell columns as light guides rather than $4\ \mu m$ crystalline tracts. Crystalline tracts, and the narrow retinula cell columns of some moths, contain a relatively high proportion of membrane which must confer a high refractive index on the structure on account of the high proportion of lipid. The typical crystalline tract is, in fact, an attenuating mechanism of the light-adapted eye. The movement of pigment grains closely around the crystalline tract, and the lengthening of the tract during light-adaptation in the neuropteran-type eye, are changes which occur under the influence of light and they reduce the amount of light reaching the receptors. The mechanism is that pigment grains *outside* the guide attenuate the light transmitted. With light to spare the tract can be narrow, which has the advantage of reducing the visual field of the ommatidium and improving acuity. Moreover, in bright light it need operate only with the first mode. When $\lambda = 0.5 \mu m$, $d = 2 \mu m$, the difference between n_1 and n_2 need be only 0.0025 for the crystalline tract plus pigment to carry the first mode. With these values the light energy is smoothly attenuated over a wide range that is limited only by the absorption coefficient of the pigment. From the measurements of Seitz (1969) the value of V for the crystalline tracts of the dark-adapted firefly is *ca.* 5, which means that about the first four modes are readily propagated. Therefore it must be even more effective in the transmission of light in the dark-adapted eye, when the pigment no longer surrounds it, than in the light-adapted eye.

11.6.7. *The aperture of rhabdom columns*

The shape of the rhabdom, its cross-sectional area, and the refractive index in relation to its surroundings necessarily have a large effect on the angular acceptance function (or field of view) of the rhabdom itself. There are several distinguishable factors.

(1) Light at an angle θ to the axis of the rhabdom is less effective by the factor cos θ. Compared with other effects, this is small.

(2) Where rhabdomeres are fused, as in all clear-zone eyes, the rhabdom size limits the 'grain size' for the subdivision of a superposition image if one is formed on the receptor layer, although no sufficiently sharp image is yet known for this effect to be significant. Large rhabdoms, and continuous sheets of rhabdomeres, are presumably a sign of a poor image.

(3) When the rhabdom acts as a light guide it necessarily has its own field of view, which can be calculated for a cylinder by the methods of Snyder (review in this volume). A columnar rhabdom with a small refractive index differential will have a narrow field of view, which must improve the acuity of each receptor. Examples where the rhabdom column is surrounded by pigment, as in some beetles, e.g. *Gyrinus* (Fig. 17b in Horridge 1971) are known.

(4) In other clear-zone eyes the rhabdoms occupy long closely-packed columns that are separated from each other by highly-developed trachea in which the spiral bars are specialized as reflecting plates. In skipper butterflies, where the light falling on the receptor layer has been examined in detail, the aperture of the whole retinula cell column is very wide, as indeed it has to be to catch the oblique rays from the patch of the eye admitting a parallel beam (Horridge, Giddings, and Stange 1972). This is evident from a perusal of Fig. 11.11a, in which each receptor column beneath the focused light is seen as a dark spot because it absorbs the light falling on it over a wide angle. In another

part of the same photograph, receptor columns which appear bright are inclined at least 30° to the observer, showing that light emerges from them (and by the principle of reversibility, enters) at a large angle. Recording from an individual receptor when the skipper eye is stimulated by a narrow beam upon the patch of cornea which admits light shows that light crossing the clear zone at a measured angle of *c*. 20° from the ommatidial axis nevertheless enters the rhabdom column.

The effect of a tracheal sheath around the rhabdom and associated cytoplasm is to make the whole receptor column into an *isolated* light guide of wide aperture and low loss. The light falling upon it enters, passes down and is partially absorbed. The light is then reflected by the equidistant parallel bars of the tracheae acting as quarter-wave-plate reflectors, and retraces its path through the column into the clear zone, finally being seen as eyeshine. The effects of the trachea are (1) to subdivide the image without scattered light and (2) to put the whole of the long absorbing receptor column into one optical plane, which is at the entrance to the receptor column. The tracheal sheaths are therefore an essential anatomical device of a well-focused clear-zone eye, in which the rhabdom layer must be long to ensure a high capture efficiency for photons and yet the eye aperture must not be limited by the fields of view of individual receptors.

11.6.8. *Increase in sensitivity caused by focusing*

In the simplified situation in which light is admitted through a circular patch of facets and focused upon an area of the receptor surface, the increase in sensitivity is proportional to the ratio of these two areas, or of the squares of the angles subtended by these areas at the centre of the eye. In a more realistic model, the fraction of a parallel beam admitted by each successive facet across the aperture of the eye is a gaussian distribution of angular width δ at the 50 per cent contour (Fig. 11.12), and the distribution of light on the receptor layer is another gaussian of width at the 50 per cent contour. So, for a constant loss of light *on axis* to pigment in each ommatidium, the increase in sensitivity is proportional to δ^2/γ^2. In a completely unfocused eye this ratio may reach only *c*. 2, but in the well-focused skipper butterfly can be 20–30. Optics which are sharply focused may, however, be compatible with admission of only a fraction of rays parallel to the optical axis of the ommatidium. This fraction is the height of the peak in the δ distribution, and has not been measured in any clear-zone eye.

Even without mathematical analysis, it is evident that selection should favour improved focusing, which simultaneously increases acuity and sensitivity. The question is why partially-focused eyes still exist when the skipper butterflies prove that excellent focusing is possible. One answer may be that sharp focusing is possible only when each ommatidium is stopped down, so that the increased sensitivity is not realized. A more likely reason is that sensitivity is at a premium in insects that are active in dim light, and sensitivity for objects that fill the field is increased with the size of the field of view of a receptor, so that the compromise between sensitivity and acuity is decided by the size of the objects of interest. Certainly there is a very wide range of clear-zone eyes with different combinations of pigment migration and cell movement on adaptation, but at present we are ignorant as to how these are correlated with illumination level during activity, with what objects are resolved, and with the ability to flit between different levels of illumination. Evidently the skipper eye is not an adaptation for seeing gross outlines of trees at night but for flying through woody undergrowth at medium levels of illumination during the day.

11.7. Methods available for analysis of clear-zone eyes

Examination of the methods available for experimental analysis, and the limits of interpretation that can be placed on the results, immediately throws into prominence the dependence of theories of mechanism on a few experimental procedures. In all branches of science it is necessary to invent new analytical techniques to pursue mechanisms in progressively greater detail. We have a historical process of progressive discovery based on this interaction between observations and the techniques that make them possible. Secondly, analysis is tentative because we build up a model system, concerned with what we think the real system to be like, but we cannot always show that our model corresponds to the real system because alternative and equally satisfactory models can be subsequently invented. This distinction between possible causes and necessary causes applies equally to quantitative as to qualitative analysis. Thirdly, where two or more mechanisms exist side by side, experiments may be designed to illustrate one, and conclusions may be drawn, with no attention to other mechanisms that are active at the same time. In the following list of methods of attack on the insect eye, these three epistemological considerations reappear again and again.

11.7.1. *Search for superposition images behind eye slices*

The driving force behind Exner's theory of superposition was undoubtedly his observation of the erect image behind the cleaned cornea of a firefly. He photographed this and correctly showed it to be formed by convergence of rays from many facets. The achievement was remarkable and the only major objection to the argument based on the observation is that all clear-zone eyes were for many years assumed to work in the same way. It was never stressed that the firefly differs from most insects in having corneal cones, that despite considerable efforts on Exner's part he could not show a good image in other clear-zone eyes, and that several types of eye have structures that may serve as light guides across the clear zone.

A fuzzy erect image was seen by Exner (1891) behind eye slices of moths and night-flying beetles, which are the two main groups of clear-zone eyes with crystalline cones. This observation implies that these eyes are focused to some extent. Exner never showed even the firefly eye to be well-focused, because (1) he did not show that the image lies on the receptor layer, (2) he observed only rays within the aperture of the observing microscope, and (3) he worked with the cleaned cornea, without cone cells and pigment cells, and made no experiments on the normal eye.

The erect image on the receptor layer in skipper butterfly eyes has an excellent resolution of a few degrees. Receptor fields have an angular width of 5–7° at the 50 per cent contour and the superposition image is the only mechanism of vision (Horridge, Giddings, and Stange 1972).

11.7.2. *Observation of rays behind single cones*

With a lamp placed in front of a cleaned cornea, Exner used the focusing system of the microscope to follow the rays from several facets and show that

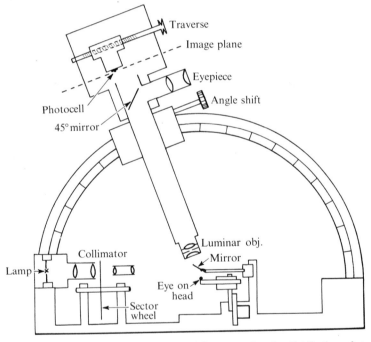

FIG. 11.17. The verstaile device that is essential for measuring the distribution of rays that diverge from a small region. This is used in the measurement of the divergence of eyeshine and of the relation between μ and η. The essential feature is that a microscope of narrow aperture can be swung around the object at any angle. The eye is observed through a half-silvered mirror at any angle to the direction of the incident rays. The sector wheel is to chop the light for the AC amplification of the photocell output (from Horridge, Giddings, and Stange 1972).

they converge when parallel rays fall on the outside of the eye. Kuiper (1962) described the same experiment on several crustaceans with clear-zone eyes but found no convergence. Limitations of the early work are that the microscope axis is fixed and its aperture does not distinguish the direction of the light. A better arrangement is to mount a narrow aperture microscope on a turntable (Fig. 11.17) and use it to measure the angular distribution of light. So far the distribution of μ for various values of η (as in Fig. 11.15c) has not been measured for an actual eye, although that is the vital relation for any clear-zone eye. However, using various fixed values of μ the corresponding values of η have been measured for a skipper butterfly (Horridge, Giddings, and Stange 1972). The results (Fig. 11.19) agreed rather well with the theoretical values required by the geometry of the eye (Fig. 11.20).

11.7.3. *Observation of an erect image behind single cones*

A requirement of a well-focused eye is that each cone generates a small erect image if viewed in isolation, as photographed in *Ephestia* by Kunze 1959). These images show that the eye must be at least partially focused but not that rays converge to a well-focused combined image in the right place. In all work with eye-slices, particularly when cut at the level of the cone tip, the question of whether the images are modified by structures further back in the clear zone has to be separately examined.

A general comment on the observation of images in eye slices is that a very bright source is required to see them, and when found they are remarkably fuzzy. Apart from those behind the firefly cornea, the images seen so far are not very promising as mechanisms of vision. If formed by the pinhole mechanism (Section 11.4.3) they could be an accidental product at a particular stage of light adaptation with no increase in sensitivity. We lack the comparative measurements of absolute sensitivity which are needed to show that focusing is in fact effective despite the loss of light in shielding pigment.

11.7.4. *Observation of light guides in eye slices*

By looking into the back of eye slices, Horridge (1968, 1969) observed threads filled with light crossing the clear zone in the firefly *Photuris*, and Doving and Miller (1969) photographed the same in several moths. Subsequently Horridge (1971) described retinula cell columns acting as light guides in a typical scarab beetle, *Repsimus*, and the moth *Ephestia*. The objection to this approach is that no quantitative estimate is available for the fraction of light that passes by this route, so it is not known how significant it is compared with the rays freely crossing the clear zone outside the light guides. In aquatic beetles and skipper butterflies the clear zone is remarkably homogeneous and direct observation suggests that the retinula cell columns have no action as light guides.

11.7.5. *Measurement of refractive indices in cornea and cone*

Exner observed in the water beetle cornea, and supposed for the firefly corneal cone, that the optical mechanism required to produce focusing consisted of an appropriate gradient of refractive index. A lens cylinder similar to that in the water beetle, could produce the erect image by an afocal combination as in Fig. 11.8b, although actual tracings of rays in the beetle lens cylinder (Fig. 11.9b) now shows that focusing there is poor. Elsewhere, refractive index gradients appropriate at least for partial focusing have been found in the corneal cone of the firefly, and in the crystalline cone of many moths and Neuroptera, the skipper butterflies and in several beetles.

Refractive index measurements by modern interference microscopy make possible ray tracings which show how the optics of the cornea and cone actually function, neglecting diffraction effects. The width of the facet and cone is sufficient (more than ten wavelengths) for ray optics to apply. However, refractive index measurements are approximate because (1) the thickness of the necessary frozen sections is hard to measure, (2) the anatomical location of the values of refractive index is very important in ray tracing but hard to define from frozen sections. The subsequent ray tracing can be done easily only in two dimensions, but the full three-dimensional situations have not yet been computed. Measurements of the beautiful lens cylinders discovered by Exner in the water beetle cornea have now been used for tracing rays (Meyer-Rochow 1973). This work shows that it is not surprising that the image is fuzzy and the acceptance angle c. $30°$ wide, because aberration is severe for off-axis rays. A lens cylinder designed to focus rays parallel to the axis cannot focus rays incident at another angle (Fletcher, Murphy, and Young 1954).

11.7.6. *Tracing rays to the receptor layer*

On a scale diagram of the eye, in which all values of refractive index have been entered in zones, it is possible to trace into the eye samples of rays falling on the cornea. When incident rays are numerous, equally spaced, and parallel, the 'closeness together' of rays in any region inside the eye is a measure of the intensity. In this way one can define (in two dimensions) how well the eye is focused and can make an informed guess at the angular sensitivity curve of the receptors (Fig. 11.21).

11.7.7. *Measurement of eyeshine*

Convenient measurements of eyeshine can be made on an intact clear-zone eye, but unfortunately not all of them shine, even when dark-adapted. Exner (1891) realized that in a perfectly focused eye, the eyeshine rays should be directed straight back in the direction of the parallel beam on the eye, but the only observations he made were on moths in which he had to explain the observed divergence of eyeshine by assuming that the eyes were not perfectly

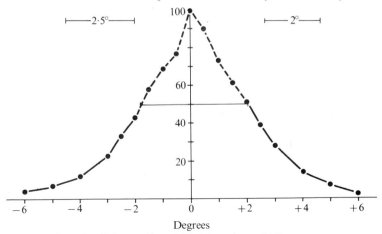

F IG. 11.18. The intensity of the eyeshine of the skipper butterfly *Taractrocera papyria* as seen from different angles to the incident beam. The total light from the surface is integrated within an observing aperture of 2·5° (from Horridge, Giddings, and Stange 1972).

focused. Exner realized that the tapetal reflecting layer is not at the same level as the receptor layer, but made no quantitative analysis.

The size of the eyeshine patch shows the area of the eye which emits light in a given direction. By the principle of reversibility of light this patch must also admit light of a parallel beam to some *destination* within the eye (not necessarily focused). The size of this patch fixes the maximum aperture of each facet. Usually the patch admitting or emitting light is 16–40 facets diameter and subtends 25–35° at the centre of the eye. During light-adaptation of the eye, the size of the patch decreases smoothly. It is supposed, but not yet demonstrated by instant fixation techniques, that this reduction in aperture of the whole eye occurs as pigment migrates along the narrowing portion of the cone.

The divergence of the eyeshine shows how well the eye is focused. When the source is a parallel beam on the eye the angular distribution of eyeshine is approximately a gaussian curve. For a well-focused eye with receptor columns of wide aperture this curve should be about twice as wide as the angular distribution of light on the receptor layer which is the angular sensitivity of the receptors. In skipper eyes, the patch admitting light subtends *c.* 35° on the eye but the divergence of the eyeshine is only *c.* 8° wide at the 50 per cent contour (Fig. 11.18). These two measurements alone tell us that the intact eye is well-focused without even looking inside, and when measured the receptor acceptance angle turned out to be *c.* 5°.

11.7.8. *Receptor acceptance angle and acuity to stripes*

One way of measuring the performance of an eye is to measure the angular sensitivity of the receptors directly by microelectrode recording as a distant point source is moved round the eye. This is the most important measurement which determines the acuity and contrast transfer (see article by Wehner, Chapter 5). The results allow tests of optical theories but do not of themselves suggest mechanisms.

Another approach to a maximum value for the angular sensitivity of the receptors is to measure the acuity towards *equally-spaced* stripes of differing width by the optomotor response in a large drum. If the fields of individual receptors are equal and approximately gaussian, the angular repeat period for the narrowest stripes seen is approximately equal to the acceptance angle of the receptors. This is because stripes of this width cause intensity modulation in the receptors that is approximately at the threshold for intensity discrimination.

11.7.9. *The relative effectiveness of facets within the patch admitting light*

By recording from one receptor intracellularly and moving a small light guide from facet to facet it is possible to show that many facets admit light. The method, however, does not provide appropriate data because a light guide is a divergent source.

With parallel light that is axial to the receptor containing the electrode, a narrow slit can be moved across the eye, and the sensitivity can be measured at each position of the slit. This stimulus is convenient because the axial facet cannot then be missed. The results, however, do not reveal whether there is a special sensitivity of the axial facet greater than that of its six immediate neighbours, which would test whether a light-guide is present and favours one facet. A pinhole stimulus is harder to manipulate than a slit, and impossible with short-lived cells, but would give more exact information about the relative amounts of light from a parallel beam which enters the eye by different facets and which then reaches the receptor that lies on the axis of the beam. The slit method showed that in skipper butterflies the effectiveness of facets in the patch which admits parallel light to a receptor rises like a cone almost linearly from zero at the edge of the patch to a peak at the centre (Horridge, Giddings, and Stange 1972). However, this method tells nothing of the light which never reaches the receptor that is recorded from, and therefore this experiment alone does not show how well the eye is focused or by what mechanism it functions.

11.7.10. *Shining a narrow beam on the eye*

Kunze (1970) showed that when a narrow parallel beam is directed upon the dark-adapted eye of a moth (*Ephestia*) one sees eyeshine from the whole area which normally shines when the eye is illuminated. The result is consistent with any theory of the clear-zone eye in which some light freely crosses the clear zone, and is therefore not evidence for a focused eye. This experiment can also work in eyes where some light travels down light guides, and so it is not evidence against them, although it eliminates eyes with a complete separation of ommatidia.

Measurement of the angular sensitivity towards a narrow parallel beam shining on one facet only would show how well focused are the facets near the centre, or alternatively at the edge, of the patch which admits parallel rays. This is one experiment which would distinguish between reasons for poor focusing in a clear-zone eye. Alternatively, the use of a narrow parallel beam that could be focused on each facet in turn and seek for the axis of each while recordings are made from a single receptor, would also make possible the measurement of whether one facet in particular is extra sensitive, as it should be if a light-guide mechanism exists in addition to summation across the clear zone.

11.7.11. *Optomotor stimulation through a single facet*

Kunze (1970) introduced the idea of focusing a pattern of moving stripes upon a single facet by means of a microscope objective, and observed optomotor responses in the same direction as the movement of the stimulus. Not sufficient detail has been published for one to decide whether the stimulus situation provides an unambiguous test of a well-focused superposition eye. From Kunze's account it appears that the result is proof of some degree of focusing as in Fig. 11.3c, which is already known from the observation of erect images in *Ephestia*.

11.7.12. *Response to a parallel beam versus diffuse light*

As recently as 1971 I thought that a useful experiment was the measurement of the ratio of the sensitivity of a receptor to a parallel beam on axis and to diffuse light from all directions equally. The idea was to measure the contribution from the rest of the eye via the clear zone (Horridge 1971). Analysis subsequently showed that the ratio of the light reaching the receptor for the two stimulus situations is no more than a measure of the acceptance angle for the receptor (Horridge, Ninham, and Diesendorf 1972).

11.7.13. *An object poked into the eye*

Hoglund (1966) introduced the idea of bypassing the screening effect of the visual pigment by stimulating directly into the clear zone via a light guide poked into the eye. Although he was limited to the ERG in his measurements, and recorded a total receptor response, he was able to show that the receptors are as sensitive in the light-adapted eye as they are when dark-adapted. He estimated that the distal pigment reduces the sensitivity of the eye to *c.* 1/1000 of that of the dark-adapted state. It would clearly be a refinement to repeat these experiments by intracellular recording, but at present I do not see how the results would help to distinguish between optical mechanisms.

When a light guide is poked through the back of a clear-zone eye, rays from it emerge from the facets. If these rays are parallel, one can infer that the tip of the light guide is at the focus of a well-focused eye (Horridge, Giddings, and Stange 1972).

Similarly, if a needle inserted into the clear zone is seen from outside the eye, one can safely infer that the μ distribution is narrow for a given value of η and that the cornea and cone can act as an afocal combination. This, however, is exactly the same as can be inferred from the erect image behind an eye slice, and tells little about the distribution of light on the receptor layer, about the fraction of light passing down light guides, or about the overall acuity and sensitivity of the receptors.

11.7.14. *Observation of rays through the cornea and cone*

For rays freely crossing the clear zone the relevant measurement is the μ distribution of light from the cone tip as a function of the angle η at which light strikes the facet relative to the axis. The distribution of μ can be narrow as a result of two effects: (1) better approximation to an afocal system (Fig. 11.15c); (2) absorption of unwanted rays in screening pigment.

As illustrated in Figs. 11.19 and 11.20, η has been measured directly for various values of μ in the eye of a skipper butterfly, and the relation between the two angles fits very well the geometry of this eye. It is only because the skipper eye is well focused, however, that the measurement can be made this way round. If the μ distribution is more than a few degrees wide for parallel light falling on the facet then the distribution of μ must be measured directly.

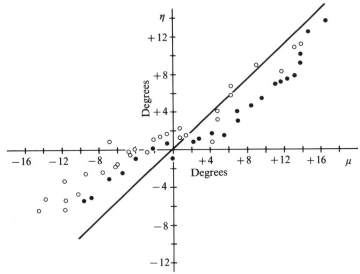

FIG. 11.19. Values of η plotted for different values of μ in the skipper butterfly *Toxidia peroni* (see Fig. 11.20). Filled and open circles are from different eye slices. The line shows the relation that would give a perfect focus (Exner line). (From Horridge, Giddings, and Stange 1972).

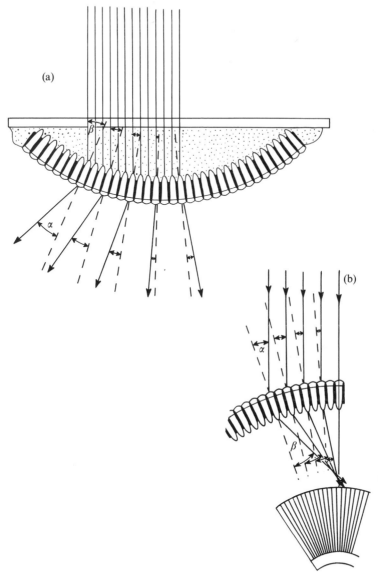

FIG. 11.20. Direct measurement of the direction in which rays pass through the cornea and cone. (a) Parallel rays shine upon the back of an eye slice that is mounted on a cover slip. The light enters each cone tip at a different angle μ that is determined by the curvature of the eye. These angles, relative to the axis of each cone are calculated from the position on the eye. The distribution of η of the ray emerging from each facet is measured directly by means of a narrow aperture (2·5°) microscope which can be rotated with the eye at its centre. Measurements are plotted in Fig. 11.19. (b) From the measured relation between η and μ, rays are now drawn inside the eye showing how a parallel beam falling on the eye is focused. This example is a scale model of the eye of the skipper butterfly *Toxidia peroni* (from Horridge, Giddings, and Stange 1972).

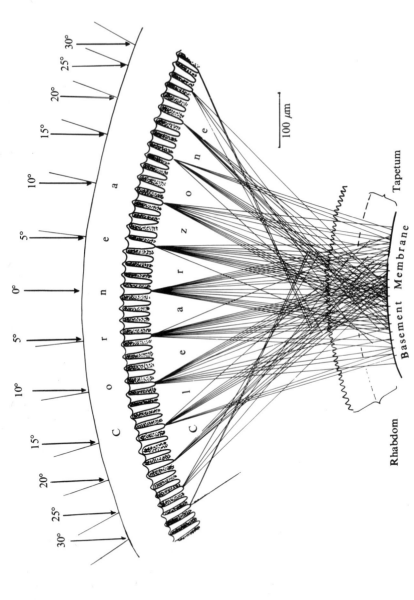

FIG. 11.21. Ray tracing across the clear zone in the beetle *Anoplognathus* (Scarabaeidae) to illustrate a partially focused eye. Parallel rays incident on the eye strike the ommatidia at different angles to their axes. Careful tracing of regularly spaced rays within each corneal lens cylinder and cone, as in Fig. 11.9, reveals the distribution of rays from the cone tip for each ommatidium. These rays are then entered on a scale drawing of the eye. The accuracy depends upon the measurement of refractive indices and dimensions of the dioptric system, and the result can be checked by direct recording of the angular sensitivity of the receptors (after Meyer-Rochow 1972).

A much easier way of giving an upper limit for the distribution of μ and for determining its relation to η in a well-focused eye is to measure the size of the illuminated spot on the receptor layer when parallel light falls on the eye. Then from the size of the patch which admits a parallel beam into the eye one can draw rays covering this spot from each of the cones which admit light. However, this is possible only in a well-focused eye because an eye can be imperfectly focused either because the μ distribution is wide or because μ/η has the wrong value.

An educated guess about the relation between μ and η can be obtained by tracing approximately fifteen equally spaced parallel rays through the facet and cone for each of approximately ten directions of incidence at equal intervals of 1° to the axis (Fig. 11.9c) and continuing the rays to the receptors (Fig. 11.21). The rays arriving upon a receptor from each direction are then counted and an angular acceptance curve plotted. Experience with ray tracing in *Ephestia* (Horridge 1972) and skipper butterfly cones, suggests that the eyes are in fact better focused than the ray tracing in two dimensions suggests, but to-date diagrams of this type provide the only information we have of the ray paths within partially focused eyes.

11.7.15. *Measurement of the distribution of light on the receptor layer*

The distribution of light upon the receptor layer can be examined directly while a parallel beam shines on the eye. This can be done by slicing off an edge of the eye and looking in through the clear zone. By taking two parallel beams at a small angle to each other two distributions are thrown on the receptor layer. The observer may then readily make judgements whether two lights at this angle are resolved by the optical system (Fig. 11.11a).

By accurate measurement it should be possible to seek for a discrepancy between the measured angular distribution of light on the receptor layer (Fig. 11.13) and the angular sensitivity of a typical receptor. If the latter is more sharply peaked then one must seek additional optical mechanisms, of which the most likely are restricted aperture of the rhabdom column and additional light passing down a light guide in each ommatidium.

11.8. Conclusion

11.8.1. *Summary of experimental findings*

However theories are bandied about, the experimental observations upon which they are based are always worth consideration, if only as antique inspiration for the better employment of modern equipment. The following is a summary of the main observations.

LARGE WATER BEETLES, *Dytiscus*, *Hydrophilus*, and *Cybister* are structurally similar.

Lens cylinder in thick cornea (Exner 1885) *Hydrophilus*; (Meyer-Rochow 1973) *Cybister*

Fuzzy erect image (Exner 1891) *Hydrophilus*

Ray tracing showing poor focus on receptor layer (Meyer-Rochow 1973) *Cybister*

Crystalline tract in light-adapted eye of *Dytiscus* (Schultze 1868; Grenacher 1879; Horridge 1969)

Extensive movement of retinula cells on adaptation. *Dytiscus, Hydrophilus* (Walcott 1969; Horridge and Giddings 1971)

Crystalline cone of secondary importance (Exner 1885) *Hydrophilus* (Meyer-Rochow 1973) *Cybister*

Acceptance angle of receptors very wide, *Dytiscus* (Horridge, Walcott, and Ioannides 1970)

Distal rhabdomere must be a separate receptor system (Grenacher 1879; Horridge 1969)

FIREFLIES, *Phausis* (= *Lampyris*) and *Photuris* are structurally similar.

Erect image behind cleaned cornea (Exner 1891) *Phausis*, confirmed by others (Tinbergen 1966)

Lens cylinder inferred in corneal cone (Exner 1891)

Erect image attributed to curved inner-surface of corneal cone (Winthrop and Worthington 1966)

Peculiar optics attributed to layers in corneal cone. Fixed extensions of the cone cells act as light guides in dark-adapted eye. Pigment surrounds light guides in light-adapted eye (Horridge 1968, 1969) *Photuris*.

Corneal cone inhomogeneous (Seitz 1969) *Phausis*

Acuity of dark-adapted eye is still unknown.

MOTHS

Fuzzy erect images behind slices of fixed eyes in a few species (Exner 1891; Døving and Miller 1969)

Erect images behind individual cones (Kunze 1969) *Ephestia*

Inhomogeneous crystalline cone in *Ephestia* (Kunze 1971)

Some focusing inferred in *Ephestia* (Kunze 1969, 1970)

Light guides observable in dark-adapted eyes (Døving and Miller 1969; Horridge 1971)

Pigment migration around retinula cell columns (Exner 1891; Umbach 1934; Tuurala 1954; Yagi and Koyama 1963)

Movement of retinula cell columns (Umbach 1934; Tuurala 1954; Yagi and Koyama 1963; Horridge and Giddings 1971*b*)

Formation of crystalline tract in light-adapted eye (Umbach 1934; Day 1941; Tuurala 1954; Horridge and Giddings 1971*b*)

Rhabdom of *Ephestia* extends from cone across clear zone as light-guide (Horridge 1971)

Most large moths and especially neuropterans have well-focused dark-adapted eyes (Horridge, unpublished observations)

SKIPPER BUTTERFLIES

Assumed to be superposition eyes (Tuurala 1954)

Narrow field of view of individual receptors (Døving and Miller 1969) *Epargyreus*.
Proved to have a well-focused erect image on the receptor layer *in situ;* mechanism lies in lens cylinder of crystalline cone (Horridge, Giddings, and Strange 1972) many species.

OTHERS

Retinula cell columns function as light guides in some beetle eyes (Horridge 1971) *Carabidae Scarabaeidae*

Eyeshine divergence as a measure of degree of focusing (Horridge, Giddings, and Stange 1972)

Extensive movement of retinula cells on adaptation. *Scarabaeidae* (Horridge and Giddings 1971*a*)

Differences of acuity and sensitivity in LA and DA eyes. *Scarabaeidae* (Meyer-Rochow 1973)

Circadian rhythm of adaptational state. *Scarabaeidae* (Meyer-Rochow 1973)

Change in shape of cone upon adaptation could be a mechanism of accommodation. *Gyrinidae*. (Horridge, unpublished).

It is important to note that the three fundamental relationships in a clear-zone eye are: the way μ depends on η; what fraction A of light parallel to the axis of a facet is effective in stimulation; and what fraction of the light which crosses the clear zone travels in light guides. None of these features has been measured in a partially focused eye by any of the methods so far employed for the study of compound eyes. Also, measurements of absolute sensitivity of receptors and intact eyes are as yet almost totally lacking.

11.8.2. *A figure of merit for clear-zone eyes*

Engineers sometimes bring together the important parameters of a complex structure to form an overall estimate of its performance. An example of a figure of merit is the *rating* of a sailing boat which depends mainly on immersed length and area of sails. This enables a comparison between boats, and predicts suitable handicaps. Similarly we can judge an engineering structure in terms of strength per unit weight or cost per unit of effectiveness. Competition between designs of engines, bridges, aeroplanes, or anything else depends first on this overall effectiveness. Diversity persists, however, because different solutions of the compromises are optimal under different circumstances, for example yachts of a given rating differ in performance under different conditions, and because extinction has not yet caught up with the less effective designs. Apparently this is true also for clear-zone eyes. A figure of merit could include the absolute sensitivity, the dynamic range over which the eye can usefully operate, and the acuity at different parts of this range. Quantitatively, this data is yet to be gathered.

11.8.3. *An evolutionary cul-de-sac*

Experience in ray tracing through various types of cone (Figs. 11.6, 11.9, 11.14, 11.15, 11.16, 11.19) suggests that the clear zone evolved because increased sensitivity was a selective advantage, but that all of the varied mechanisms run into problems of acuity when a wide dynamic range is also required. The intensity ratio of 10^7 between sunlight and moonlight probably explains why the crystalline tract persists in most clear-zone eyes. Possibly there exists no other optical solution for a compound eye with sharp focus and wide aperture if it must have a wide dynamic range.

11.8.4. *Lessons*

The acceptance of the superposition theory for all clear-zone eyes, and in particular the acceptance of the crystalline cone as a lens cylinder which would produce a well-focused eye in all dark-adapted moth or beetle eyes, illustrate very well the way in which a theory becomes established, discordant details are forgotten, text books copy the one from the other, and the original experimental data are replaced by a comprehensive explanation which doesn't agree with the actual details of the animals themselves.

Progressively better descriptions of clear-zone compound eyes, particularly of the changes in anatomy on dark- and light-adaptation, have illustrated an ever-increasing diversity of components and mechanisms (Fig. 11.7). There has also been the necessity of keeping separate the data on each type of eye, and of not arguing from one species to another. Thirdly, there has been a tendency by all workers to take evidence for one optical system as being evidence against other systems which are later shown to operate at the same time. Finally, it has taken some time to clarify what measurements ought to be made to explain the optical mechanism of just the known types of clear-zone eye, and these experiments will bring to light further wonders at present unsuspected.

12. The dioptric system in beetle compound eyes

V. B. MEYER-ROCHOW

12.1. Introduction

The morphology of the cornea and cone provides the basis for one classification of compound eyes. The cornea and cone are more significant, however, as the optical components. Measurement of their curvatures and refractive indices leads to an understanding as to how light passes through them, and helps to explain the optical properties of eye slices or whole eyes. Manual or computerized ray tracing can be compared with the performance of the whole structure. Further, the examination of optical properties such as refractive index, transmission, birefringence, and dispersion, leads to a further analysis of the physical and chemical basis of these properties. There are many aspects, therefore, of the path of the light before it reaches the receptors.

Beetles are very suitable objects for the investigation of insect eyes on account of the wide variety, covering most of the known types of compound eye. Yet no review of coleopteran eyes is available since the comprehensive work of Kirchhoffer (1908). The eyes can be totally absent as in some cave dwelling forms such as *Lathrobium coecum*, and *Trechus bilimeki*, or highly reduced as, for example, in the myrmecophilic *Claviger testaceus* (Jörschke 1914). The total number of facets varies from a few in *Bryaxis* to almost 30 000 in *Necrophorus americana*, and there are numerous species in which eyes of individuals of the opposite sex differ in shape, size and number of facets, e.g. *Machaerites* and *Phausis* (= *Lampyris*) (Kahmann 1947).

The eyes are usually situated on either side of the head and are separated from each other, except that in some male Lampyridae they meet above and below the head, and in male Lymexilidae they fuse on top of the head (Britton 1970). The compound eye is generally round, oval, or kidney-shaped. The two eyes sit on short lateral cuticular projections in *Minemodes laticeps* and eyes are partially divided by a lobe, called a canthus, in a number of families including Passalidae, Geotrupidae and others. In Gyrinidae and in some Lucanidae the division into an upper and lower eye is complete. A median ocellus occurs on the frons of some Dermestidae (e.g. *Anthrenus*) and there are 2 ocelli on the head between the compound eyes of Staphylinidae–Omaliinae; otherwise ocelli are absent (Britton 1970).

Cones and receptor layer are separated by a clear zone devoid of pigment in Dytiscidae, Hydrophilidae, Scarabaeidae, Elateridae, Cantharidae, Lampyridae, and some others, and are rather sharply distinguished anatomically from apposition eyes, where no clear zone is found e.g. Coccinellidae, Chrysomelidae, Cerambycidae, Staphylinidae, Corylophidae, and Curculionidae. Usually insects with clear-zone eyes are mainly nocturnal and those with apposition eyes diurnal. Anatomically there are hardly any intermediates between the two types of eye. Among clear-zone eyes we expect to find within the Coleoptera all degrees of partial or perfect focusing of rays across the clear zone, but unfortunately it is not yet known whether the fireflies, or any other beetles, have eyes that are well-focused when the eye is intact. Fixed crystalline tracts occur in clear zone eyes of some beetles, notably those with corneal cones, such as Lycidae, Elateridae, Cantharidae and Lampyridae, but the majority of clear zone eyes in Coleoptera appear to be of the neuropteran type with a crystalline tract extending from the cone only in the light-adapted state. Mechanisms of adaptation are very varied and not yet known sufficiently widely to be classified with confidence.

12.2. The cornea

12.2.1. *Morphology*
The diameter of the corneal facets usually ranges between 25 and 35 μm, even for species of beetle that vary considerably in body length (Meyer-Rochow 1972). There seems to be a lower limit in ommatidial size, and facets of smaller dimensions do not appear to exist. The reason advanced for this fact is the compromise between diffraction produced by tiny apertures and number of facets required for excellent vision (Barlow 1952). Exceptionally large facets are found in certain longicorn beetles, e.g. *Cerambyx heros* (Cerambycidae) with corneal diameters of nearly 100 μm (Kahmann 1947), whereas tiny facets of only 9 μm in diameter have been found in *Corylophodes sp.* (Meyer-Rochow unpublished observations). The cornea varies in thickness from only 4 per cent of the total length of the ommatidium (Agee and Elder 1970; Butler, Roppel, and Zeigler 1970) to as much as one-fifth of the ommatidial length (Meyer-Rochow 1972). Deane (1932) distinguishes eyes with facets which have a strong external convex curvature, called 'uvate' (Fig. 12.1a,b), from those with facets which have a small individual convex curvature, called 'favate' (Fig. 12.1d), and those that have no individual facet curvature at all called 'glacial' (Fig. 12.1c). Uvate facets seem to be more readily found in small forms with eyes of only a few facets, whereas glacial facets tend to be developed more strongly in aquatic, nocturnal, and large beetles.

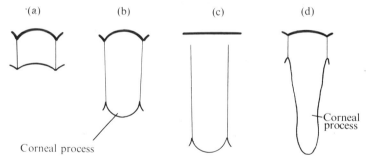

FIG. 12.1. Four main types of corneal lenses in the compound eyes of beetles. (a) convex-concave, e.g. young *Attagenus megatoma* (Butler, Roppel, and Zeigler 1970); (b) bi-convex, e.g. *Saperda carcharias* (Grenacher 1879); (c) plano-convex, e.g. *Dytiscus marginalis* (Günther 1912); (d) exocone, e.g. *Phausis* (*Lampyris*) *splendidula* (Schultze 1868).

Interfacetal hairs, distributed randomly over the whole eye, are present mainly in some small beetles, e.g. *Sartallus signatus*. As the hairs project over the eye's surface they must cast shadows. Neither the consequences of this effect nor the function of the hairs is fully understood. The hairs may indeed have nothing to do with vision at all and most likely are mechanoreceptors related to cleaning reflexes.

Corneal nipples so far have not been reported in any beetle species. From an extensive description of corneal nipples in many insects Bernhard, Gemne and Sällström (1970) suggest that beetles did not have the genetic background to develop these structures.

On the proximal (inner) side of the cornea the curvature can be either concave (Fig. 12.1a), e.g. *Anthonomus grandis* (Agee and Elder 1970) and *Attagenus megatoma* (Butler, Roppel, and Zeigler 1970), more or less convex (Fig. 12.1b,c) or cone-shaped (Fig. 12.1d). Thus morphologically one can distinguish convex–concave (Fig. 12.1a), biconvex, and plano-convex types of lenses from facets with a corneal process (Fig. 12.1d). The latter type is often referred to as 'exocone' when the inner process is very long, but all intermediates exist. Morphological changes of the cornea correlated with dark and light adaptation have not been seen yet. Changes of the corneal morphology with age have, however, been reported: the cornea of *Attagenus megatoma* is of convex–concave type shortly after the beetle's final molt, but within nine days increases in thickness and becomes bi-convex (Butler, Roppel, and Zeigler 1970).

12.2.2. *Constitution of the cornea*

In 1931 Nowikoff found that the cornea of Lepidoptera consists of three layers which stain differently in borax carmin–Mallory and iron–haemotoxylin (Heidenhain stain). The outer (distal) layer stains bright red with Mallory and black in Heidenhain. The middle layer stains dark blue with

Mallory and is hardly stained at all in Heidenhain. The inner layer stains light blue with Heidenhain and pink with eosin. Sannasi (1970), working on the beetle *Photinus*, finds the same three layers and calls them (i) the epicuticle, (ii) the hyaline exocutive, and (iii) the mesocuticle; the latter producing the processus corneae where it is present. The inner anilin blue-positive layer, which is called endocuticle in typical arthropod body cuticle, is absent in the cornea. According to Locke (1966) the cornea can be considered as the direct continuation of the body cuticle without the innermost layer. The morphogenesis of the lens cuticle seems to have some features in common with the events taking place in the morphogenesis of body cuticle (Delachambre 1971). Pinocytotic vesicles and cell processes projecting into the lens cuticle are commonly seen in developing cornea as well as body cuticle.

The cornea of all insect eyes examined has the layered arrangement that was first analysed by Bouligand (1965) in crustacean body cuticle. The layers consist of superimposed planes parallel to the surface of the cuticle (each 5–10 nm thick), in which the axial direction of the parallel anisometric components (sometimes called 'micelles') in any one plane is set at a slight angle to the directions of neighbouring planes (Neville, Thomas, and Zelazny 1969). The whole system has a progressively-rotating direction of planes, which, when cut obliquely, appear as layers, and as a special case they form a large double spiral pattern in sections of the inner corneal process parallel to the surface of the eye (Fig. 12.2a,b). The arrangement of layers is not a series of concentric ellipsoids, as illustrated by Exner (1891), but is a series of paraboloids displaced along the radial axis (Horridge 1969). The pitch of the spiral varies but is *c.* 1 μm along the axis where it is greatest, and is less towards the edge of the corneal cone process,

(a)   (b)

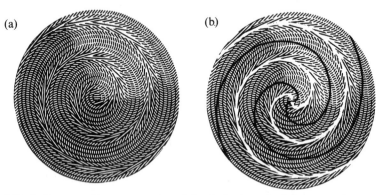

FIG. 12.2. A diagrammatic representation of the derivation of spiral patterns in transverse sections of corneal cuticle (from Bouligand 1965). (a) The chitin microfibrils are arranged parallel to each other forming layers, and the direction of the fibrils changes from one plane to the next. (b) Because of the paraboloid fitting of the layers in the cornea, transverse sections produce the well-known spiral pattern.

indicating that either the angle of displacement between the peripheral planes is larger than that of the central planes or that the central planes are spaced wider than the peripheral ones. A definite optical significance for these spirals has not been found in eyes but it is possible that they are able to bend rays that strike them at an angle, and therefore that they may create optical path differences indistinguishable from those caused by non-homogeneous refractive index. The spirals are not growth layers and have no known relation to a circadian rhythm. Many spirals are formed in a single day as the cornea is laid down.

Growth layers, formed by alternate bands of spirally structured cuticle and unstructured cuticle, as found in locust body cuticle, are not known in eyes. Anisotropic crystals, which cause reflection at interfaces between layers in the body cuticle of some scarabaeid beetles (Caveney 1971), are not known in eyes. There is evidence, however, for at least two types of loading of corneal cuticle with crystals or other material which increases the refractive index: (1) in the outer layers of the cornea one frequently finds bands of differing refractive index, although no special development which causes interference colours has been described from Coleoptera; (2) where the refractive index is nonhomogeneous, values on the optical axis may reach $n = 1.8$, which implies loading of the chitin by solid substances as yet unanalysed.

So far as is known, the cornea is composed of the same kind of chitin micelles, embedded in protein, as in more typical cuticle. A colourless, elastic protein, called resilin has been found in the lens cuticle of the firefly (Sannasi 1970), and a variety of hydrocarbons, wax esters, free fatty acids, and sterols are found covering the outer surface of terrestrial insects (Arnold, Blomquist, and Jackson 1969). Chitin, a polysaccharide, ordinarily accounts for 25–50 per cent of the dry weight of the cuticle (Lafon 1943 in Foster and Webber 1959). It may be regarded as a derivative of cellulose, in which the C_2-hydroxyl groups have been replaced by acetamido residues and, indeed, chitin resembles cellulose in many of its properties. Chitin consists of long unbranched chains of 2-acetamido-2-deoxy-D-glucose residues with a screw axis and β-D-(1,4)-glycosidic linkages (Foster and Webber 1959). As with cellulose, equal numbers of chains probably run in opposite directions, but unlike cellulose adjacent chains in the same plane are oppositely directed. Hydrogen bonds are thought to be involved between C=O···HN groups of adjacent acetamido side-chains and C=O···HO linkages are involved to form connected piles of chitin chains within the crystalline regions (Foster and Webber 1959). It is likely that the now classical analysis by Carlson will have to be revised by renewed x-ray crystallography. Chitin, although insoluble in concentrated KOH, undergoes partial de-*N*-acetylation, yielding chitosan. With iodine–potassium iodide, chitosan exhibits a brown colour which becomes red–violet with sulphuric acid.

12.2.3. *Optical properties*

To function as an efficient window for the ommatidium, the corneal lens has to be highly transparent and preferably colourless. In some other insect orders coloured facets have been reported (briefly reviewed by Friederichs (1931) and Bernard (1971)). Among Coleoptera the tiger beetle *Cicindela* shows modified corneal lenses in that the amount of light entering the ommatidium is reduced and restricted to a central 'pupil'. A greyish ring, 3–4 μm thick, thought to contain finely dispersed black pigment, is found in each ommatidium at a radius from the centre of *c*. 9 μm (Friederichs 1931). As *Cicindela* is exclusively active in bright sunlight the presence of a diaphragm in the facet is easily understood. In some dung-beetles, e.g. *Onthophagus*, the colour of the central facet area and the interfacetal edges differ from each other, especially when viewed with light of shorter wavelengths, e.g. 408 nm. These observations indicate different absorption characteristics within parts of one and the same facet (Meyer-Rochow unpublished observations). However, microspectrophotometry of the central part of the cornea (15 μm fresh transverse section) shows that transmission is here constant throughout the visible spectrum (Fig. 12.3a). Absorption is least in the longer wavelengths and increases towards the ultraviolet, but nowhere in the visible spectrum does the absorption exceed 30 per cent.

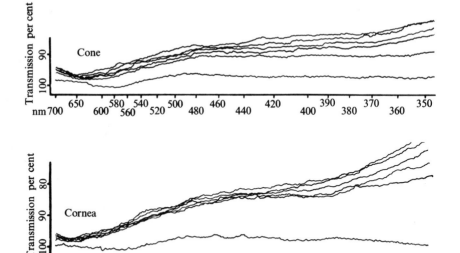

FIG. 12.3. Microspectrophotometry along the optical axis of the dung-beetle *Onthophagus africanus*. Transmission in the cone and cornea is best at longer wavelengths and decreases towards the u.v., but never falls below 70 per cent in the visible spectrum. The isolated lower trace is the reference beam.

Similar results have been obtained for the dragonfly cornea plus cone (Kolb, Autrum, and Eguchi 1969).

The refractive index of the cornea can be measured with an interference microscope (for methods see Seitz (1968); Meyer-Rochow, (1973)). Two principal arrangements have been found. Beetles so far investigated with convex–concave and biconvex lenses, e.g. *Creophilus erythrocephalus* exhibit an optically homogeneous structure, sometimes with a slightly smaller refractive index between the facets (Meyer-Rochow 1972). Plano-convex corneae may show a homogeneous arrangement also, e.g. *Anoplognathus sp.* (Fig. 12.4a) or belong to the second group, e.g. *Cybister sp.* and other water beetles (Fig. 12.4b) (Exner 1885; Meyer-Rochow 1973), in which the optically-inhomogeneous cornea has the form of a lens cylinder. The characteristic feature of such a lens cylinder is a radial gradient of refractive index, in which the highest value is along the optical axis. A similar lens cylinder has been found in the corneal cone of the firefly (Seitz 1969) (Fig. 12.4c). The difference of refractive index between centre and periphery, in lens cylinders so far investigated (Exner 1885; Seitz 1969; Horridge, Giddings, and Stange 1972; Meyer-Rochow 1973) lies in the range 0·1–0·2.

12.2.3.1. *Birefringence.* Stockhammer (1956) was the first to report birefringence of the insect cornea, and speculated about its function in the perception of polarized light. It has long been known that organized cuticle is optically anisoptropic, and exhibits positive birefringence in relation to the principal axis of the structural elements parallel to the surface (Castle (1935). Thus it resembles cellulose in which the maximal difference between n_α and n_0 is 0·06, where n_α is the refractive index for the extraordinary ray and n_0 is the refractive index for the ordinary ray (Frey-Wyssling and Blank 1948).

(a)	(b)	(c)

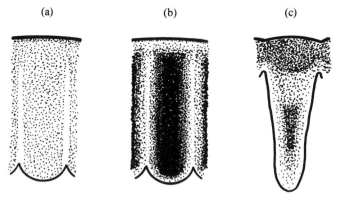

FIG. 12.4. Types of plano-convex corneal lenses in Coleoptera: (a) in Scarabaeidae the cornea is homogeneous, (b) in large aquatic beetles, e.g. *Hydrophilus* there is a lens cylinder formed by the high central and the lower perpheral refractive index. (c) The corneal cone of the firefly also is an inhomogeneous cylinder structure.

The cornea of insects is a negatively birefringent uniaxial crystal in which the optical axis runs parallel to the axis of the ommatidium (Seitz 1969). The same author gives the value $-0{\cdot}0012$ for the birefringence of the fly cornea. The discrepancy between this result and the positively-birefringent polysaccharide micelles with their optical spindles perpendicular to the ommatidial axis is explained by the change of axis. In addition, Wiener (1912) in Frey-Wyssling, and Blank (1948) showed that form-birefringence will always be negative in a system of lamellae with single thicknesses very much smaller than the wavelength, no matter what the individual birefringence of the micelles may be. As each of the superimposed micellar unit planes in the cuticle of the cornea has a thickness of 5–10 nm, the requirements of Wiener's formulation are met and a negative form birefringence relative to the ommatidial axis is no surprise.

However, birefringence of the cornea is very weak when viewed along the axis of the ommatidium and no functional significance for it has been found in beetles. The oriented layers of chitin micelles cause plane-polarized light passing along the axis of the spiral to be rotated in polarization plane and if the pitch is *c.* 150 μm circularly polarized incident light is reflected if its direction is against that of the spiral. The general impression, from what little is known, is that the *beetle* cornea is designed to avoid these interference effects or absorption or similar impediments to transmission that could be caused by its fine structure.

Other interference effects, inferred to arise from other kinds of layers, e.g. refractive index steps, have been reported in jumping spiders (Land 1969) and the fly *Condylostylus* (Bernhard 1971), which seem to have a functional significance.

12.3. The cone

12.3.1. *Morphology*

A remarkable feature of the eyes of Coleoptera is their diversity, and they include many examples of all four types of cone (Fig. 12.5).

Cones of the 'acone' type are found in Staphylinidae (Hesse 1901; Kirchhoffer 1908) and Tenebrionidae (Eckert 1968). They are characterized by large centrally-situated nuclei and the absence of a large optically-dense central area. The four cone cells contain rather clear protoplasmic material with a few cell inclusions and exhibit an arrangement typical for fluids surrounded by thin membranes (Fig. 12.5a) (Wolf 1968). Acone cones are optically homogeneous and have a relatively low refractive index of *ca.* 1·36 (Meyer-Rochow 1972).

'Eucone' eyes occur in Adephaga and Scarabaeidae (Grenacher 1879) and are defined by the dense cone-shaped core, secreted intracellularly towards the axis during development (Fig. 12.5b). In cross-section the edges of the four cone cells meet at right angles in the middle signifying the solid nature

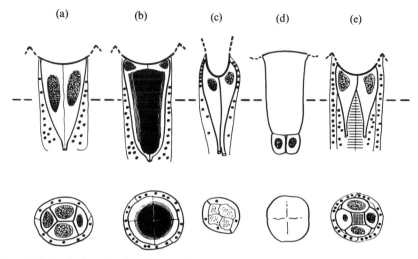

FIG. 12.5. Semischematic drawings of the main types of crystalline cones found in beetle compound eyes (not to scale): (a) acone, e.g. *Saperda carcharias* (Grenacher 1879); (b) eucone, e.g. *Melolontha vulgaris* (Grenacher 1879); (c) exocone, e.g. *Phausis splendidula* (Schultze 1868); (d) pseudocone, e.g. *Anthonomus grandis* (Agee and Elder 1970); (e) super-acone, e.g. *Corylophodes sp.* (Meyer-Rochow unpublished observations). The upper row represents longitudinal sections, the lower, transverse sections in the plane of the dotted line.

of the inner core. At the periphery the four cells contain multiple concentric layers of endoplasmic reticulum, glycogen granules, mitochondria, and a thin layer of protoplasmic material containing microtubules. The refractive index of the structure is relatively high for the dense central region and drops steeply in the outer more cellular margin of the cone.

'Pseudocone' eyes are unusual in beetles, but have been reported to occur in Dascilliformia and Malacodermata (Meixner (1935) in Paulus (1972)). In the pseudocone eye each cone cell produces a clear gelatinous substance which is secreted extracellularly. The pseudocone is situated between the cornea and the cone cells (Fig. 12.5d). No interferometrical studies have been carried out on this structure in beetle eyes, but examinations of the fly pseudocone show that the structure is optically homogeneous with a low refractive index of 1·337 (Seitz 1968).

Corneal cones, the 'exocone' type, are best known from the Lampyridae, but occur also in related families such as Elateridae, Lycidae and Cantharidae (Horridge and Giddings 1971). The corneal cone is a cone-shaped proximal projection of the cornea formed by the mesocuticle. Its refractive index is high in the centre (*ca.* 1·5) and low at the periphery (*ca.* 1·35) with a continuous radial gradient between (Seitz 1969). The *Semper* cells are situated proximal to this corneal cone (Fig. 12.5c). They are small and form only a bundle of four narrow extensions as far as the receptor layer, each containing microtubules and vesicles (Horridge 1969). The bundle of cone cell roots has a slightly higher refractive index than the surrounding clear-zone (Seitz 1969).

Another type of cone cell arrangement seems to be present in Corylophidae, where the four Semper-cells diverge proximally to leave a central space through which retinula cells achieve direct contact with the cornea (Meyer-Rochow, unpublished observations) (Fig. 12.5e). Such an arrangement may simply be the consequence of the lack of space, as the total length of Corylophidae often is less than 1 mm.

Anatomically it is possible to distinguish subgroups, and transitional forms between acone and eucone type, e.g. hourglass shaped cones of eucone type have been reported by Kirchhoffer (1908) to occur in *Scarabaeus variculosus* and similar shaped cones, but of acone type, have been illustrated by Eckert (1968) for *Tenebrio* and Meyer-Rochow (1972) for *Creophilus*.

12.3.2. *Structure and optical properties*

The solid central core in cones of the eucone type is not fully analysed, but the association with glycogen (Perrelet 1970) and mitochondria, with an apparent synthetic zone around the outside suggests that a glycoprotein is secreted. Kim (1964) has found mucopolysaccharides and proteins in *Pieris* cones, and Barra (1971) mentions material rich in sulphydryl (SH) and disulphide (SS) groups in cones of Collembola.

Where cone cells with patterns of glycogen granules are found, the latter have been claimed on dubious grounds to be of optical significance in the perception of polarized light (Skrzipek 1971). Light absorption is thought, according to this author, to vary with the amount of glycogen granules present in the cells. Differences in density of glycogen granules between the four cone cells are rather common, and have also been found in sections through the proximal end of dung beetle cones, but the functional significance, if any, is unknown.

Occasionally cones are found which do not consist of the usual four cells, but of three, five, or other numbers with no apparent functional consequence. In any case cones of the eucone type are of extraordinary transparency with little absorption over the whole visible spectrum (Fig. 12.3b). The higher refractive index in the centre of some cones of the eucone type could be achieved by a higher concentration of solids as was suggested for the cornea.

12.4. **Ray tracing**

Knowing the inner and outer radii of curvature, and the refractive index of the corneal lens, one can calculate the path of a bundle of parallel rays by applying the thick lens formula

$$F = n(n-1)(r_2 - r_1) + d(n-1)^2, \quad \text{where } F = F_1 + F_2 - d(F_1 - F_2) \quad (12.1)$$

and $r_{1,2}$ = radii of curvature, n = refractive index, and d = thickness of the lens (Longhurst 1967). The result however, will only be correct if the dioptric structures are optically homogenous. The paths of rays within such a structure can then easily be visualised by geometrical ray tracing through a two-dimensional model. Homogeneous structures allow ray tracing simply by the application of Snell's law of refraction

$$\frac{n_1}{n_2} = \frac{\sin \beta}{\sin \alpha} \qquad (12.2)$$

When rays pass through a dioptric system in which horizontal layers have different refractive indices ray tracings may be made simply by applying at each interface Snell's law of refraction, as done by Varela and Wiitanen (1970).

A lens cylinder is more difficult to work with, for the focus and path of rays depend in a complex way upon the radial distribution of refractive index, but here again ray tracing illustrates in principle the distribution of light within the eye.

According to the superposition theory of Exner (1891), light entering the dioptric apparatus would be focused within the lens cylinder and would leave the cone at an angle and direction such that another sharp focus is formed at the receptor level. To test whether rays do this, large drawings of a longitudinal section of the dioptric apparatus are required. The distribution of refractive index, measured in transverse and longitudinal sections with an interference microscope, are added to the drawing. For simplification, layers of equal refractive index are drawn. A bundle of twenty parallel equally spaced rays is drawn to strike the eye at regular intervals across this facet. The procedure is repeated at various angles to the ommatidial axis (Fig. 12.6a,b), and, to represent what happens at neighbouring facets when rays leave the model, the rays are drawn entering the model at the corresponding place and angle on the other side.

An additional complication, where the lens cyclinder runs along the optical axis is that a ray travelling parallel to the axis is in fact bent although it stays within one layer. As a result of the continuous transition in refractive index, a ray entering such a 'layer' parallel to its adjacent layers (the inner one of which has a higher refractive index and the outer a lower) would be bent towards the optical axis where the higher refractive index lies. A good approximation of the radius of curvature for such a ray is given by

$$R = \frac{\bar{n}}{\sin \alpha \dfrac{n_2 - n_1}{\Delta r}} \qquad (12.3)$$

(see Meggitt and Meyer-Rochow, Chapter 13).

It is very important to calculate R for each ray and for each 'layer' unless

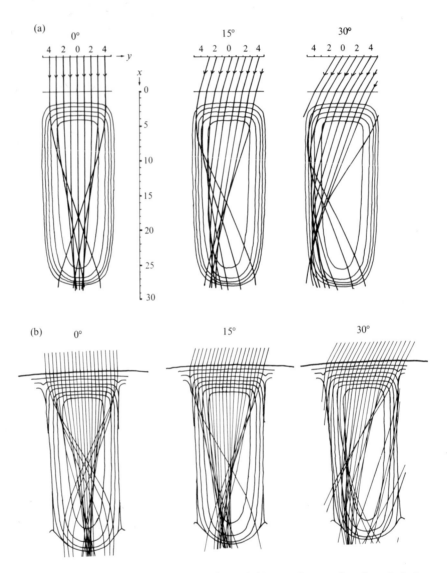

FIG. 12.6. Examples of (a) computer ray-tracing and (b) manual ray-tracing, through the lens cylinder of the cornea of *Cybister fimbriolatus* in which the difference between central and peripheral refractive index is 0·15 (after Meyer-Rochow 1973). Rays are drawn to strike the cornea at regular intervals, meeting the surface at angles of 0°, 15°, and 30° to the axis. There is a focal region, but no sharp focus, and peripheral rays have the shortest focal length (caustic curve phenomenon). Note that rays are unlikely to be found accurately at a point on the receptor layer by any structure of this type.

the ray crosses the 'layer' perpendicularly. Ray diagrams of cornea and cone lens cylinders, traced as described, are shown in Figs. 12.6b, and 12.7. The diagrams show clearly that rays entering the lens cylinder peripherally are focused more strongly than rays which enter the structure closer to its axis, forming a caustic curve. When the whole eye is considered, another caustic curve occurs and can be visualized by large scale ray diagrams (Fig. 12.7).

Ray tracings can also predict the acceptance angle of the whole eye, and this can be done in relation to its state of adaptation if the position of the screening pigment is considered. Rays entering the clear-zone are plotted in relation to the incident angle, thus giving an idea of the amount of light which reaches a receptor from each direction outside the eye. The angle corresponding to the 50 per cent level of the number of rays is defined as the admittance angle of the dioptric system. In the light-adapted state light has to travel down the crystalline tract, which is surrounded by screening pigment. In ray tracing the light-adapted eye, therefore, one counts only those rays that leave the cone through an aperture corresponding to the diameter of the crystalline tract, and at an angle allowing the light to be totally reflected in the light guide.

From ray diagrams of the dioptric systems of the water beetle *Cybister* (Meyer-Rochow 1973), the Christmas beetle *Anoplognathus* and dung beetles (Meyer-Rochow unpublished observations) it has become evident that rays entering by a large patch of facets are only partially focussed at the photoreceptor level in the dark-adapted state. Partial focusing is achieved by the lens cylinders of cornea or cone, which have a steep peripheral gradient of refractive index. Sensitivity is greatly increased, but at the expense of acuity because there is no sharp focus. This 'partially focused' type of clear-zone eye has been treated mathematically by Diesendorf and Horridge (1973). The fuzzy erect image that can be seen behind eye slices of some clear-zone eyes is caused by the lens cylinders, but is not necessarily in the plane of the receptors. The lens cylinders simply provide a convenient way of making a partially focused eye and it seems that the erect image is less significant for vision than is the increase in sensitivity by summation of light.

A lens cylinder can theoretically produce a sharp focus for incident rays parallel to the axis. This has been shown by Fletcher, Murphy, and Young (1954), who give for the required axially symmetrical distribution of refractive index the formula

$$N_{(r)} = N_0 \, \text{sech}(\pi r/2F). \tag{12.4}$$

This ideal distribution, compared with one measured from interference micrographs, is shown in Fig. 12.8, which demonstrates that a sharper focus would be obtained in a single facet if the peripheral decrease of refractive index were less steep than that found experimentally. The required difference

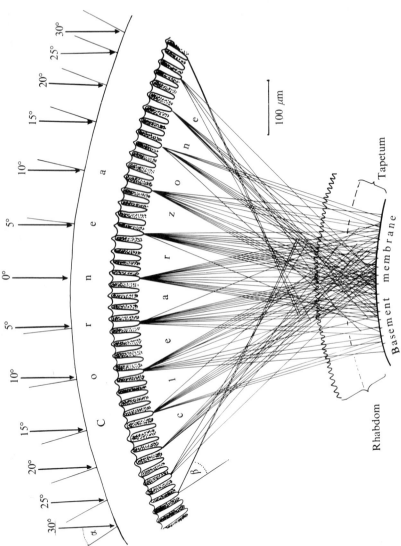

FIG. 12.7. Longitudinal section through the eye of *Anoplognathus pallidicollis*, showing convergence of rays which will increase sensitivity without necessarily forming a sharp image. This eye has a homogeneous cornea and a lens cylinder in the crystalline cone.

Refractive index

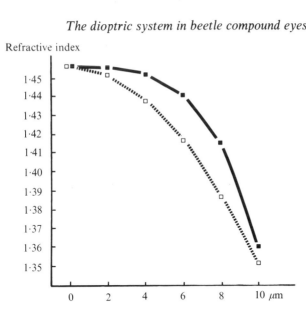

FIG. 12.8. The radial distribution of refractive index in the crystalline cone of *Repsimus sp.*, measured experimentally (solid line). The dotted line shows the distribution in an ideally focused lens cylinder after Fletcher, Murphy, and Young (1954). The experimentally-obtained values show a steeper peripheral fall in refractive index than would be required for an ideal focus.

between central and peripheral refractive index for a given length of focus (50 μm) and a given diameter of the lens cylinder (20 μm) to produce a fairly sharp image is $1.03n_p$ where n_p = refractive index of peripheral region, calculated after eqn (12.4). This value is in good agreement with experimentally obtained values. However, we still await the calculation of the refractive index distribution that produces a sharp focus *on the receptor layer* for rays that are incident *over a range of angles to the axis.*

ACKNOWLEDGEMENT

The author wishes to thank Dr. I. Meinertzhagen for helpful discussion and constructive criticism, and Professor Horridge for encouragement and attention to this chapter.

13. Two calculations on optically non-homogeneous lenses

S. MEGGITT and V. B. MEYER-ROCHOW

13.1. Computer ray tracing

A straightforward repetitive calculation such as ray tracing is very suitable for a rapid three-dimensional computer analysis. A computer program could also more easily compensate for absorption of rays, reflection, dispersion, and other optical phenomena than manual procedures, if accurate data are available.

The first computer ray tracing through the dioptric system of an insect eye was carried out by Varela and Wiitanen (1970). They did not deal with a lens cylinder, however, but with the cornea of the honey bee which does not show a radial gradient of refractive index.

Lens cylinders, which are found in the dioptric systems of some nocturnal insects, are more difficult to deal with. In order to make a comparison with manual ray tracing a program was devised at the Australian National University.

In the specific application to the eye of a beetle, it has been assumed that the refractive material is contained in a hexagonal prism, representing the ommatidium, or in a circular cylinder representing the cone. The X_3 axis lies along the centre line of this prism, and the light enters through the surface $X_3 = 0$. The refractive index n is assumed to be axially symmetric, so that it depends only on X_3 and $R = \sqrt{(X_1^2 + X_2^2)}$, the radial distance from the axis (Fig. 13.1).

The first stage in the program is to store the values of n at mesh points, equally spaced in X_3 and in R, covering the whole structure. By reading in these values from cards, punched with experimentally measured values of n, it is possible to deal with a wide variety of axially symmetric eyes (although use has also been made of mathematical functions to generate n).

Then wherever a value of n, or of its gradient,

$$\text{grad } n = (\partial n/\partial X_1, \partial n/\partial X_2, \partial n/\partial X_3),$$

is required at some point in the program, a subroutine is used to calculate it by interpolation from the nearest mesh points.

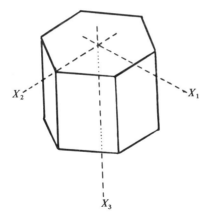

FIG. 13.1. For the corneal lens cylinder the refractive material is assumed to be contained in hexagonal prisms, while the refractive index n is axially symmetric solely depending on X_3 and $R = \sqrt{(X_1^2 + X_2^2)}$, the radial distance from the axis.

Each ray is traced through the medium by updating two vectors: one giving the current position $\mathbf{X} = (X_1, X_2, X_3)$ and the other (a unit vector) giving the ray direction $\mathbf{A} = (A_1, A_2, A_3)$. Here the components A_1, A_2, A_3 are the cosines of the angles which the ray makes with the three axes. In order to start a ray, the initial position $(X_1, X_2, 0)$ and direction are read in. It is necessary to provide error exits in the program if $A_1^2 + A_2^2 + A_3^2 = 0$ (in which case the initial direction is not defined), or if $A_3 \leqslant 0$ (when the ray is not travelling inwards), or if the initial position is outside the allowable values for X_1 and X_2 in the prism. It is also necessary to normalize the vector $\mathbf{A}$ if the values read in do not satisfy $A_1^2 + A_2^2 + A_3^2 = 1$.

The ray enters the refractive medium through the surface $X_3 = 0$ where it undergoes an initial refraction. The ray direction after refraction at such an isolated surface may be calculated as follows. Setting $\mathbf{A}$ = unit vector in direction of initial ray; $\mathbf{B}$ = unit vector in direction of final ray; $\mathbf{U}$ = unit vector normal to surface; n_1 = initial refractive index; n_2 = final refractive index.

Compute
$$\alpha = \mathbf{A} \cdot \mathbf{U} = A_1 U_1 + A_2 U_2 + A_3 U_3$$
$$\beta = 1 + (\alpha^2 - 1)n_1^2/n_2^2.$$

When β is positive, the ray undergoes refraction, and the final direction is:
$$\mathbf{B} = (n_1/n_2)\mathbf{A} + [\text{sgn}(\alpha) \cdot \sqrt{\beta} - \alpha n_1/n_2]\mathbf{U}.$$

(Note if $\beta < 0$ then the ray is reflected and
$$\mathbf{B} = \mathbf{A} - 2\alpha\mathbf{U},$$
a case which cannot arise here because $n_1 < n_2$).

To continue the ray through a medium in which n varies continuously, it is necessary to integrate the vector differential equation

$$\frac{d}{ds}\left(n\,\frac{dX}{ds}\right) = \text{grad } n,$$

where s denotes the distance measured along the ray from an arbitrary point. For numerical work this is conveniently expressed as a set of six first-order differential equations,

$$dX_1/ds = A_1, \qquad dX_2/ds = A_2, \qquad dX_3/ds = A_3,$$

$$d(nA_1)/ds = \partial n/\partial X_1, \qquad d(nA_2)/ds = \partial n/\partial X_2, \qquad d(nA_3)/ds = \partial n/\partial X_3.$$

In the program, these are integrated numerically using a standard fourth-order Runge–Kutta type method (see for example Kopal (1955)). If this method is considered applied to the vector system of differential equations

$$dY/dt = f(Y)$$

starting at the point Y with step length h the calculations for one step are

$$K_1 = hf(Y_0)$$
$$K_2 = hf(Y_0 + \tfrac{1}{2}K_1)$$
$$K_3 = hf(Y_0 + \tfrac{1}{2}K_2)$$
$$K_4 = hf(Y_0 + K_3)$$

The estimate of the value Y_1 at the next point is then

$$Y_1 \approx Y_0 + \tfrac{1}{6}(K_1 + 2K_2 + 2K_3 + K_4).$$

At the end of each step, after printing out position and direction, it is necessary to test whether the ray still remains within the structure. In the applications of the present program to the cornea, the latter has been considered to be made up from a hexagonal mosaic of facets. So a ray leaving through one face would immediately enter another hexagon of identical structure. This is dealt with in the program by adjusting the co-ordinates X_1, X_2 to place the ray in the corresponding place in the original hexagon. In this way the ray is continued to the final value of X_3. In the case of a cone, surrounded by pigment, rays leaving the cylinder through its curved side, would be terminated.

Having outlined the basic principles of the program it should be possible to make it fit the requirements of any particular case of insect dioptric system. If further information is needed, requests should be addressed to the authors.

13.2. Calculation of radius of curvature of rays in a gradient of refractive index

In ray diagrams rays can be drawn as straight lines only when they pass for short distances through zones of different optical density. If long distances are covered within a layer of one refractive index the ray will be bent towards the region of higher index, which is usually towards the axis of the structure.

Consider a series of homogeneous layers, as described, in which the refractive index n has constant values in the different layers. Then by Snell's law, as shown in Fig. 13.2a.

$$n_1 \sin \alpha_1 = n_2 \sin \alpha_2 = n_3 \sin \alpha_3$$

In the limiting case in which n varies continuously with y (= distance from the central axis of the ommatidium)

$$n \sin \alpha = k, \tag{13.1}$$

where k is constant for a particular ray but varies from one ray to another. Set

$$y' = \frac{dy}{dx} \quad \text{and} \quad y'' = \frac{d^2 y}{dx^2},$$

calculated along the ray path. According to Fig. 13.2b

$$n/k = \operatorname{cosec} \alpha = \frac{\sqrt{(dx^2 + dy^2)}}{dx} = \sqrt{(1 + y'^2)},$$

and

$$y'^2 = \frac{n^2}{k^2} - 1. \tag{13.2}$$

Given n in terms of y this differential equation integrates to give the path. The radius of curvature is

$$R = (1 + y'^2)^{3/2} / y''. \tag{13.3}$$

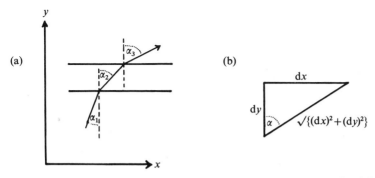

Fig. 13.2. (a) Path of light through layers of progressively changing refractive index, following Snell's law of refraction. x = central ommatidial axis; y = distance from central ommatidial axis. (b) Calculation of the change in path length caused by refraction of a light ray through an angle α.

From above eqn (13.2) $1+y'^2 = n^2/k^2$ is differentiated to give

$$2\frac{dy}{dx}\cdot\frac{d^2y}{dx^2} = \frac{1}{k^2}\cdot 2n\frac{dn}{dy}\cdot\frac{dy}{dx} \quad \text{as} \quad n^2\{y(x)\},$$

which becomes,

$$y'' = \frac{n}{k^2}\frac{dn}{dy}. \tag{13.4}$$

Hence:

$$R = \frac{n^2}{k\,dn/dy}$$

and with eqn (13.1) we obtain

$$R = \frac{n}{\sin\alpha\,dn/dy} \tag{13.5}$$

An example of the path within one layer of refraction can be demonstrated with the following assumption: $n = A+By$, where A and B are constants and $dn/dy = B$. A corresponds to the axial refractive index. Equation (13.2) thus becomes

$$y'^2 = (A+By)^2/k^2 - 1. \tag{13.6}$$

Taking the square root and solving for dx, eqn (13.6) becomes

$$\frac{dy}{\sqrt{[(A+By)^2/k^2-1]}} = dx. \tag{13.6a}$$

This is a standard integral and integrates to

$$\frac{k}{B}\,\text{arc cosh}\,\frac{A+By}{k} = x + \text{const.} \tag{13.7}$$

The constant is 0, if $y' = 0$ at $x = 0$, for

$$y'(x) = \frac{k\,\sinh\{B/k(x+\text{const})\}B/k}{B} \tag{13.8a}$$

By simple algebraic procedure eqn (13.7) becomes

$$y = \frac{k}{B}\cosh\frac{B}{k}x - \frac{A}{B}. \tag{13.8}$$

If $y = y_0$ at $x = 0$, where $y' = 0$, then

$$k = A + By_0 \tag{13.9}$$

which for $|B| > 0$, is illustrated in Fig. 13.3 This inserted in eqn (13.8) changes the equation to

$$y = \left(\frac{A}{B}+y_0\right)\cosh\frac{Bx}{A+By_0} - \frac{A}{B}. \tag{13.10}$$

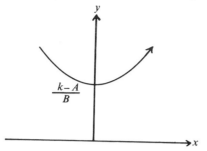

FIG. 13.3. Schematic path of a ray in the x–y-plane (as in Fig. 13.2) where A and B are constants and k is constant for each ray.

For the particular case of a ray travelling between the two layers $n_1 = 1\cdot616$ and $n_2 = 1\cdot670$ (Fig. 13.4), $\mathrm{d}n/\mathrm{d}y$ becomes approximately

$$\frac{1\cdot616 - 1\cdot670}{2\ \mu\mathrm{m}} = -0\cdot027, \quad \text{with} \quad \bar{n} = 1\cdot643.$$

As $\sin\alpha = 1$, R of eqn (13.5) becomes

$$R = \frac{1\cdot643}{0\cdot027}\ \mu\mathrm{m} = 60\cdot8\ \mu\mathrm{m}.$$

The path of this ray is illustrated in Fig. 13.4.

Leaving the pathway of a ray within one layer of refractive index and turning to a ray that traverses several layers, conditions become more complicated. Assuming:

$n = \sqrt{(A - By^2)}$, because such an equation has a similar graph to that

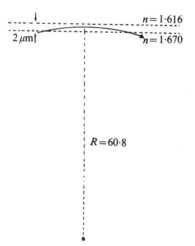

FIG. 13.4. The radius of curvature of a ray path calculated according to eqn 13.5 along one layer of assumed 'homogeneous' refractive index.

shown in Fig. 13.3, y'^2 becomes

$$y'^2 = \frac{A}{k^2} - 1 - \frac{B}{k^2} y^2. \tag{13.11}$$

From this equation as in eqns (13.6 and 13.6a) it follows that

$$\frac{dy}{\sqrt{\left(\frac{A}{k^2} - 1 - \frac{B}{k^2} y^2\right)}} = dx, \tag{13.11a}$$

which in its integrated form becomes

$$-\frac{k}{\sqrt{B}} \text{ arc cos} \sqrt{\left(\frac{B}{A-k^2}\right)} y = x + \text{const.} \tag{13.12}$$

As in eqn (13.7) $c = 0$ if $y' = 0$ at $x = 0$.
Therefore

$$y = \sqrt{\left(\frac{A-k^2}{B}\right)} \cos \frac{\sqrt{B}}{k} x. \tag{13.13}$$

As in eqn (13.8) k is the value of $n = $ at $y' = 0$, thus if $y = y_0$ at $x = 0$ and $k^2 = A - By_0^2$

$$y = y_0 \cos \sqrt{\left(\frac{B}{A-By_0^2}\right)} x. \tag{13.14}$$

From this formula it is clear that with increasing y_0, i.e. distance from the centre, the curve will be steeper and cut the x-axis earlier than with y_0 almost 0.
As

$$\sqrt{\left(\frac{1}{A/B - y_0{}^2}\right)}$$

cannot become negative, y_0 approaches the value $\sqrt{(A/B)}$. Representative rays illustrating these points are shown in Fig. 12.6 of the article by Meyer-Rochow (Chapter 12).

14. Theoretical and experimental analysis of visual acuity in insects

J. PALKA and R. B. PINTER

14.1. Introduction

THE study of the visual acuity of the compound eyes of insects has had a complex history during the past ten years. In 1962, Burtt and Catton first published experiments indicating that locusts could resolve test patterns with a stripe spacing up to ten times finer than would be predicted from the diffraction limitations imposed by the diameter of single facets. If correct, these results would demand a reconsideration of the optics of apposition eyes, and Burtt and Catton proposed a theory of 'deep diffraction images' to account for the anomalous results.

However, a subtle difficulty with these original measurements was soon pointed out by Palka (1965) and by Barlow (1965). Burtt and Catton have tested large units in the ventral nerve cord of locusts (termed Descending Contralateral Movement Detectors, or DCMDs, by Rowell (1971)) with a rectangular bar grating moved abruptly behind a square aperture in a black screen. It is easy to verify for oneself that as black stripes emerge from or disappear behind the edges of the aperture, detectable events usually described as edge flicker occur *independently of the subject's ability to resolve the individual bars of the grating*. This edge effect can be abolished in humans by arranging for the window to be defocused when the eye is focused in the plane of a theoretically unresolvable grating. Insects have fixed focus eyes, but the anomalous responses of locust neurons can be abolished by two other manipulations on the window edge: by rotating the window so that the edges of the window and the grating are not aligned, and by brightening the window edge.

Thus, not only was low spatial frequency information shown to be inherent in the test situation used by Burtt and Catton, but when it was removed the responses indicating anomalous resolution also disappeared—i.e. the evidence was strong that the animal in fact utilizes this information in cases when it is unable to resolve the grating itself. Responses to patterns which are resolvable theoretically are only slightly affected by changing edge conditions.

Unfortunately, Burtt and Catton did not attempt to repeat or explain these experiments. Instead they devised an ingenious method of stimulus presentation which avoids the edge flicker phenomenon. They prepared a radial grating, consisting of black bars of constant width arranged like the spokes of a wheel. This pattern was rotated behind an annular aperture so that only the periphery, where the bars were nearly parallel, could be seen.

Much to our surprise, Burtt and Catton obtained the same results with their radial patterns as with the original rectangular grating test, even though those original results had been shown to depend on specific aperture boundary conditions and these were no longer present. Finding this situation no less anomalous than the original result, we undertook to repeat the Burtt and Catton radial-grating experiment. We have been unable to confirm their results—thus far not a single animal has yielded response curves like those of Burtt and Catton's major study of 1969, or those reported by Northrop (Chapter 17).

This is a difficult state of affairs, because we cannot claim that we have demonstrated any specific artifact in the results from Burtt and Catton's laboratory—we simply have not obtained the same results. We therefore present an analysis of technical difficulties inherent in acuity measurements, a demonstration of a weak mechanical artifact which appears as the threshold for the visual response is approached, and our own experimental results which are entirely compatible with diffractive limitations in single ommatidia and which do not extend at all into the anomalous region.

14.2. Theory and methods

In any optical imaging system composed of a lens and aperture, the light intensity at a point in the image plane depends not only on the intensity at the corresponding point in the object plane, but also on the intensity at all other points in the object plane. Thus an object intensity distribution is transformed into a different but related image intensity distribution. The relationship between these two distributions may be calculated by means of the incoherent optical transfer function (OTF).

14.2.1. *Determination of the OTF*

The OTF is a direct function of object spatial frequency, and multiplying the object spatial frequency spectrum by the OTF yields the image frequency spectrum. The image intensity distribution is obtained by taking the inverse Fourier transform of the image spatial frequency spectrum. It is known that all simple lens and aperture combinations have OTFs which pass low spatial frequencies and attenuate high ones, eventually falling to and remaining at zero above some spatial frequency which depends directly on the aperture diameter in the case of simple circular apertures. The physical basis for this is Fraunhofer scalar diffraction theory, which has been assumed in all the calculations in the present study.

It should be pointed out that any image processing system, not only optical ones, can be characterized by an OTF; diffraction theory is not a necessary condition. Even physiological processing systems, such as lateral inhibition, can be treated this way. Cornsweet (1970) has presented a general discussion of the physiological implications of the OTF, together with several examples from the experimental literature.

To provide an unambiguous measurement of any system OTF, spatial frequency by spatial frequency, it is necessary to avoid stimulus patterns having very broad spatial frequency spectra. To do otherwise is directly analogous to measuring a colour spectral sensitivity curve in a receptor using a white light source and no narrow band filters.

As an example, small spots, even though they contain relatively more high spatial frequency components than large spots, have a very broad spatial frequency spectrum, and are therefore unsuitable for measuring an OTF. An infinite rectangular sinusoidal grating (in some cases simulated by placing a finite grating on a large drum) contains only a single spatial frequency, the frequency of the sinusoid, and is therefore an ideal object for measuring the OTF. Under certain conditions an infinite rectangular bar grating is suitable.

A radial bar grating consisting of an annular segment of a spoked wheel divided into alternating black and white bars of equal angular width is suitable for measuring at least the upper limit of resolution of an OTF. This was the solution adopted by Burtt and Catton (1969). However, a closer examination of the spatial frequency spectrum of this pattern reveals the difficulty of eliminating low-frequency artifacts caused by dimensional errors in the fabrication of the pattern. An anecdotal observation can serve to illustrate the problem.

In the summer of 1966 one of us (J. P.) visited Professor Burtt's laboratory, and together with another observer examined the appearance of patterns then in use, at distances which rendered the bars of the radial grating unre-solvable. When the patterns were moved, both observers were able to report not only the occurrence of rotation of the unresolved pattern, but also its direction and approximate angular extent, by watching the movement of indistinct gradations in the mean brightness of different regions of the pattern, i.e. by watching low spatial frequency noise in the pattern. This observation was subsequently confirmed by other subjects (Burtt, personal communica-tion). The initial reports of Burtt and Catton on acuity measurements using radial gratings were based on these particular patterns. Recognizing that their original patterns were flawed, Burtt and Catton then prepared a high quality, though of course not perfect, master pattern and measured its physical characteristics.

For a rigourous analysis of the spatial frequency spectrum of a radial grating pattern, it is insufficient to consider only the angular component; the entire two-dimensional spatial frequency spectrum must be examined. To obtain this spectrum, the two-dimensional Fourier transform of the pattern has been calculated for an error-free pattern and for a pattern containing dimensional errors on the order of those reported by Burtt and Catton (1969). Then the relative intensities of the signal at the periodicity of the bars of the error-free pattern and the noise, or low-frequency artifact, at the periodicity of the error were calculated. On a simple intensity ratio basis, the spatial signal-to-noise ratio for a narrowing of two opposite black stripes each by just 10 min of arc (determined approximately from the harmonic error table given by Burtt and Catton (1969)) is less than ten (Feder 1971). In other words, the low-frequency noise intensity is more than 10 per cent of the intended, high-frequency signal intensity.

The OTF of an optical aperture falls to zero at some value of spatial frequency, and remains zero above that value. For an ommatidial lenslet, as will be shown in greater detail below, this value is very close to 1 cycle deg^{-1}. When Burtt and Catton's high precision pattern is set at the purported upper limit of resolution (20 min of arc per cycle of pattern measured from the cornea), the spatial frequency spectrum of the pattern has intensities within the non-zero portions of the OTF of at most 10^{-36} relative units. However, at this same pattern distance, the above error, being of much lower spatial frequency, has an intensity of 17×10^{-4} relative units. The ommatidial aperture acts as an extremely steep filter for spatial frequencies, and in this particular geometrical configuration produces a signal-to-noise ratio of c. 10^{-33}, or, to put it more simply, selects very strongly in favour of the low-frequency noise components of the test pattern.

Having estimated the actual intensity of the low-frequency error in Burtt and Catton's best pattern, and having recognized the effectiveness of the aperture in filtering out the intended signal while preserving the unintended noise, we must ask the empirical question, is the visual system capable of detecting the noise? What is required is an independent measure of the ability of the ventral cord fibres to respond to patterns of very low contrast. Figure 14.1 shows some previously unpublished measurements of the response to 4·8° rectangular bar gratings (spatial frequency 0·2 cycles deg^{-1}) of diminishing contrast. The contrast is defined as

$$M = (I_{max} - I_{min})/(I_{max} + I_{min}).$$

A response was still obtained with a contrast of 2·9 per cent. Our estimate of the noise in Burtt and Catton's pattern was 10 per cent of the contrast of the error-free pattern. Of course, we do not know the actual noise spectrum of the pattern, but these calculations indicate the very strong likelihood that the best master pattern available to Burtt and Catton still contained low-frequency noise detectable by the animal.

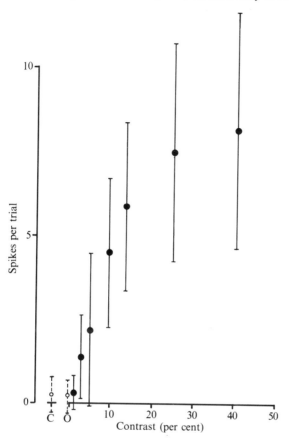

F IG. 14.1. Response of locust DCMD to movement of parallel gratings of spatial wavelength 4·8° and contrasts varying from 1·4 to 40 per cent; movement through 0·5° at 4 deg s^{-1}. Blank control (no motion) at C, active control (motion of pattern carrier with pattern removed) at O. Means and standard deviations of ten trials are shown. Experiment of 10 August 1964.

It is also important to point out that the physical measurements presented by Burtt and Catton refer to a master pattern, but this pattern was never used in their actual experiments. Instead, they used a photographic reduction whose quality was never measured. In the reduction process the pattern's spatial properties might have been altered through aberrations in the optics of the camera and enlarger, or through imperfect alignment of the optics with the pattern, the negative and the photographic paper. The pattern's intensity properties might have been altered through uneven illumination of the master or negative or uneven development of the negative or final print. Unevenness in exposure or development of the presumably high contrast negative and print emulsions might also have led to dimensional changes.

For example, 'bleeding' of black lines on high-contrast materials is a well-known source of frustration to photographers.

14.2.2. *Test patterns and their presentation*

For our own study, four radial grating patterns were scribed on Layout Scribing Glass (Type 133-12, Optical Gaging Products, Rochester, N.Y.) in a high precision jig borer (SIPÉ) at the Applied Physics Laboratory, University of Washington. These patterns have 45, 20, 10, and 5 stripe cycles each composed of a transparent and an opaque stripe of equal angular width. The upper limit on the number of stripe cycles is dictated primarily by the increasing probability of error in scribing with increasing number of stripe cycles. After cleaning and inspection, these glass masters were transferred to Kodak Photoplast Plates (emulsion coated methyl methacrylate, 0·060 in. thickness) by contact printing. These photocopies were then machined to fit the rotating apparatus, and further inspected under a microscope for ragged edges, pinholes, nicks, and other imperfections. Dimensions of all patterns were: annulus of stripes, 6·00 in. o.d., 2·50 in. i.d., and an opaque ring surrounding the annulus 0·25 in. wide, giving a total diameter of 6·50 in.

Dimensional tolerances for the glass masters were smaller than $\pm 0·001$ deg ($\pm 0·060$ min) on the angular dimensions, and $\pm 0·0001$ in. on displacements. It is estimated that the dimensional stability of the Kodak Photoplast Plates under our laboratory conditions is of the same order.

The ratio of densities between the opaque and transparent stripes is $3·75 \times 10^3$, yielding a contrast of 99·95 per cent. Illumination of the patterns is accomplished by placing them in the round window of an illumination integrating box, where the twenty-four 1·73 W lamps which provide the illumination of the box are arranged in radial symmetry and out of view of the preparation. The lamps are supplied with 14·4 V from heavy copper bus bars wired to a regulated power supply. The voltage on each lamp is the same within ± 100 mV, keeping illumination non-uniformity caused by differences in applied voltage to a calculated 0·005 per cent. The light intensity through the transparent stripes is 1200 lm m^{-2}, measured with a standardized and spectrally calibrated Gamma 2020 photometer (Gamma Scientific, Inc., San Diego, Calif.). Special attention has been paid to coating uniformly the interior of the integrating box with a high-reflectance white matte paint, and all visible exterior surfaces with a similar non-reflecting black. Within the field seen through the transparent stripes there are no apparent shadows cast by the lamps or the pattern rotation shaft, which is out of view behind the pattern and is coupled at the back of the box to the servo-drive motor. This shaft has been carefully designed with precision ball bearings at both ends, pattern and drive, to eliminate lateral displacements during rotation, and nevertheless allow pattern changes to be made with reasonable speed.

The annulus of stripes is defined entirely by the high precision pattern, the round window of the integrating box covering only one-half the outer 0·25 in. opaque ring on the pattern itself. A cone-shaped hood on the front of the box avoids projecting any light from the pattern to the darkened room where it might be visible from the animal's position. The entire setup and room are carefully inspected for light artifacts during each experiment.

14.2.3. *Experimental procedure*

Our recording arrangement has been simple, but sufficiently automated to make it feasible to collect large amounts of data in each experiment. As in previous studies, the locust (*Schistocerca gregaria* male, obtained through the courtesy of Dr. John Phillips, University of British Columbia) is mounted ventral side up, the ventral nerve cord exposed between the first and second thoracic ganglia, and one of the connectives picked up on a silver wire hook. Vaseline expressed from a syringe serves to insulate the recording site and prevent desiccation; the indifferent electrode is a silver wire inserted under the cuticle of the abdomen. We have generally not cut the connectives. The ipsilateral eye and the three ocelli are covered with thick black paint. The contralateral eye views the pattern, usually *c*. 45° posterior to the animal's transverse plane since this is the most responsive part of the DCMDs receptive field. The head is fixed in a natural posture with wax or glue, and for standardization the whole preparation is tilted so as to make the eye stripes vertical. No anaesthesia is used, but we often chill the animal enough to made the mounting procedure easier.

After conventional a.c.-coupled amplification the desired spikes, consistently the largest seen in our recordings (Fig. 14.2), are picked out by a window circuit whose levels are monitored throughout the experiment. The window output pulses are simultaneously led to counters and histogramming units; the counters are gated on for the length of pattern rotation, the histogrammers include both a pre- and a post-stimulus time period. All timing is done by a digital programmer. In all of the experiments reported here, pre-stimulus time was 400 ms, rotation time 300 ms, and post-stimulus time 580 ms.

The drive motor for the pattern ($\frac{1}{8}$ h.p. with a servo speed control) is engaged by electromagnetic clutches, one for clockwise and another for counterclockwise rotation. The clutches are operated through reed relays driven by the digital programmer. Braking is accomplished by a teflon friction collar. Pattern rotation rate has been $360°\text{ s}^{-1}$ except for a few experiments conducted at $720°\text{ s}^{-1}$. Displacement is linear with time.

Our routine protocol has been to collect thirty-two responses for each stimulus condition, with an inter-stimulus interval of 30 s. Each complete measurement therefore takes 16 min. During this time, and even more during the several hours required to complete each experiment, the amount of non-stimulus-associated spiking waxes and wanes considerably. To take

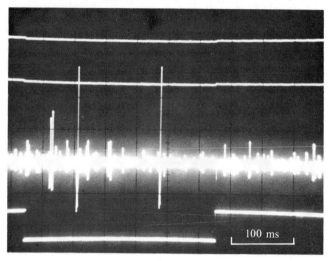

F IG. 14.2. Response of DCMD to single rotation of 3·0° radial grating through *c*. 100° at 360 deg sec⁻¹. Upper marker traces monitor the levels of the window circuit used to select desired spikes; lower marker trace indicates time of rotation. The two large spikes falling in the window are from the DCMD; the two smaller ones are from auditory interneurons.

account of this we have sampled the spontaneous activity at 30 s intervals interleaved with the stimulus presentations. Two counters and two histogrammers are used to accumulate spikes occurring in the presence of pattern rotation and, with identical timing, in the absence of rotation but in full view of the pattern. A three-channel flip-flop circuit alternates stimulus and non-stimulus conditions and routes the window output pulses accordingly.

The stimulus apparatus and the locust are mounted on two separate carts. Nevertheless, there is a weak mechanical artifact associated with movement of the pattern; whether it is due to airborne sound or some other mechanical stimulus is not known. To evaluate its effect we have measured the animal's spike output (from the same visual neuron—auditory units are easily separated out by the window circuit) during rotation of a pattern which cannot be seen because the illumination is turned off and a black paper mask is placed in front of the pattern.

14.3. Results

Our basic result is easily stated—we obtain responses exclusively from the range of pattern spatial wavelengths which our calculations show to be resolvable theoretically by the optical apparatus of a single ommatidium. We have never obtained any responses in the anomalous region of Burtt and Catton except those which are entirely attributable to a weak mechanical artifact. More importantly, our entire response curve is shifted by a factor of three to four towards the longer spatial wavelengths.

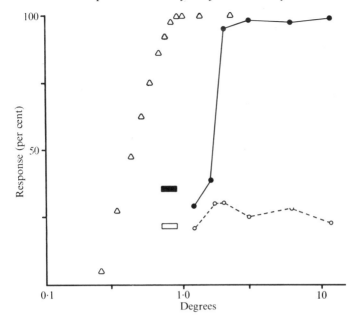

FIG. 14.3. Mean responses of six locusts to rotation of radial gratings of spatial wavelength from 1·26° to 11·6°. Solid circles and line, responses; open circles and dashed line, blank controls; solid rectangle, active control; open rectangle, blank control on active control (see text for full definitions). Open triangles show data from Burtt and Catton (1969). Responses in our experiments fall to control levels at spatial wavelengths at which Burtt and Catton still obtain spikes on 100 per cent of trials.

Figure 14.3 compares our data with those of Burtt and Catton. The solid circles and curve represent the pooled results of six experiments conducted between 25 October and 30 November 1972; only one response curve obtained during this period has been excluded, because the level of response was low and the level of spontaneous activity high. The results are plotted according to the convention of Burtt and Catton, as the proportion of trials in which one or more spikes occurred during the time of stimulus presentation. The lower, dashed curve and open circles represent non-stimulus-associated, 'spontaneous' activity occurring, as described above, during 300 ms time samples each taken midway between two consecutive stimulus presentations. These constitute blank controls. The solid rectangle is an active control, during which a pattern was rotated as usual but could not be seen because its illumination was turned off and a black paper mask was placed in front of it. The open rectangle is the blank control corresponding to the active control. The mean response to the active control is not significantly different from the response to illuminated patterns of 1·2° and 1·5°. It is consistently greater than activity during blank control periods, which we interpret simply as evidence of a weak, non-visual, presumably mechanical artifact associated with the rotation of a pattern, and becoming equally apparent whether the pattern is not seen, or is seen but is not resolved.

The open triangles of Fig. 14.3 are taken from Fig. 6 of Burtt and Catton's major study of 1969. There is a clear and major difference between the two response curves, which can be summarized by saying that, on the average, we obtain no visual responses at spatial wavelengths of 1·5° and less, while Burtt and Catton still obtain a response on 100 per cent of trials even at 0·9°. They do not give actual scores for their blank controls, but state them to be between 5 and 10 per cent. Establishing a single spatial wavelength as the threshold of the response is, of course, impossible in both sets of results.

The shape of these response curves is determined by the choice of plotting only the proportion of trials in which at least one spike occurred, but ignoring the number of spikes per response. If actual spike counts are plotted, the progressively increasing magnitude of the response with increasing pattern wavelength becomes apparent. Figure 14.4 shows these more detailed results for a representative experiment (2 November 1972), and compares them with results obtained by Northrop and with the calculated OTF for locust ommatidia. The basic findings can again be described in terms of the location of the response curve with respect to spatial wavelength: our responses first become distinguishable from controls slightly above 1·5°, which is near the maximum value tested by Northrop.

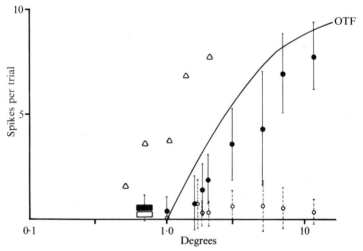

F IG. 14.4. Single experiment (2 November 1972) testing responses of DCMD to radial gratings of spatial wavelength 1·0° to 11·6°. Solid circles, responses; open circles, blank controls; solid line, optical transfer function of 30 μm aperture assuming 500 nm incoherent illumination of test grating; solid rectangle, active control; open rectangle, blank control on active control. Open triangles show data from Northrop (1971) normalized to the maximum response of this experiment. Our response has fallen to control values at spatial wavelengths at which Northrop obtained approximately half-maximal responses. Means and standard deviations of thirty-two trials are shown.

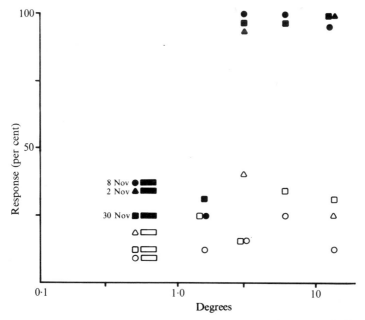

F ɪ G. 14.5. Results of three experiments in which animal-to-pattern distance was held constant and variation in spatial wavelength was obtained entirely by substitution of patterns containing different numbers of stripe cycles. Total patterns subtended 19°. Solid figures, responses; open figures, blank controls; solid rectangles, active controls; open rectangles, blank controls on active controls.

The theoretical OTF reaches zero at 1.0 deg cycle^{-1}, and, when plotted on semi-logarithmic co-ordinates, approaches saturation above $c.$ 20 deg cycle^{-1}. When comparing the theoretical curve with experimental results the ordinate scaling is arbitrary because (i) the response of the DCMD does not saturate in the range of pattern wavelengths we have used, and (ii) even if it did, this might be because of neural rather than optical reasons. But for illustrative purposes, we have adopted the convention that the maximum value of the OTF (unity) is equivalent to ten spikes per response, and the minimum value (zero) is equivalent to 0 spikes per response. Our responses first rise above control values somewhere between 1.5 and $2.0°$, well to the right of the foot of the theoretical curve. All of the values presented by Northrop lie far to the left of the theoretical curve.

The pattern wavelengths of Figs. 14.3 and 14.4 have been obtained by the combined use of patterns with different numbers of pattern cycles and different animal-to-pattern distances, with the obvious result that the total angular subtense of the test patterns has varied. But in the range of distances we have employed this has very little, if any, effect. Figure 14.5 shows the results of the three separate experiments in which the distance of the patterns was constant at 18 in., and variation of spatial wavelength was obtained

TABLE 14.1

Orientation—	45° posterior		0°		45° anterior	
	Active control	1·26°	Active control	1·26°	Active control	1·26°
Stimulus	0·406 ±0·979	0·437 ±0·669	0·406 ±0·797	0·437 ±0·619	0·281 ±0·523	0·562 ±1·162
Blank Control	0·156 ±0·447	0·094 ±0·390	0·094 ±0·296	0·562 ±1·795	0·156 ±0·574	0·343 ±0·653

entirely by substituting patterns. The data from these three animals are in close agreement with each other, and with the composite curve of Fig. 14.3.

There are some differences in the way Burtt and Catton, Northrop and we have prepared our animals. We have checked the possible effects of most of these, and found that they do not influence the results. For example, we have oriented our animals so that the pattern is positioned equatorially and 45° posterior to the transverse plane of the eye, because this is the most responsive part of the receptive field of the DCMD, both for stationary (Palka 1967) and moving (Rowell 1971) targets. We presume that the other authors placed their patterns directly to the side of the animal, as was done in earlier studies (e.g. Palka 1965). But, as Table 14.1 shows, we obtained no responses to 1·26° patterns no matter what the relative orientations of animal and pattern were over a wide range.

Burtt and Catton used CO_2 anaesthesia in preparing their animals, Northrop used no anaesthetic, and we chilled the animals used earlier in our series of experiments. But the individuals from which the data of Tables 14.1 and 14.2 were obtained, and one of the individuals on which Fig. 14.5 is based, were not chilled or anaesthetized in any way. Again, this factor had no effect on the results.

Both Burtt and Catton and Northrop cut the two connectives of the ventral nerve cord just posterior to the recording site. We have not done this,

TABLE 14.2

	Connectives intact			Connectives cut			
	Active control	1·26°	3·0°	Active control	1·26°	3·0°	3·0°
Stimulus	0·500 ±0·950	0·219 ±0·420	1·812 ±0·821	0·125 ±0·336	0·219 ±0·491	2·500 ±1·107	1·531 ±0·842
Blank Control	0·125 ±0·421	0·250 ±0·440	0·156 ±0·448	0·156 ±0·369	0·656 ±1·285	0·125 ±0·336	0·250 ±0·508

because it has been our impression that the responses of maximally intact animals are more consistent for longer periods of time. But in order to explicitly test the possibility that low-level responses are somehow released and made more prominent by cutting the connectives, we have compared the responses of one animal (experiment of 1 December 1972) before and after cutting the connectives (Table 14.2). And again, neither the responses to theoretically-unresolvable patterns (1·26°) nor to resolved patterns (3·0°) were affected.

All investigators seem to have used paint to cover the ocelli and the unwanted compound eye, except Northrop who used wax. The animal represented in Table 14.2 had only soft wax at room temperature smeared over the ocelli and one eye, and the absence of an organic solvent on the cuticle did not enhance the response to theoretically-unresolvable patterns.

To sum up, we have found no differences in the methods of preparation of the animal which in any way alter the basic results.

14.4. Discussion

The empirical results of this investigation clearly conflict with those obtained by Burtt and Catton (1969) and Northrop (Chapter 17). At the present time no single factor has been directly demonstrated to account for this discrepancy. We interpret the data of Fig. 14.5 and Tables 14.1 and 14.2 as strong evidence that the performance of our locusts was not depressed by some peculiarity of our way of making and mounting the physiological preparation. We may note, parenthetically, that our animals gave consistent results for many hours. We test them with 1·26° patterns both very early and late in each recording session, and not in one instance since the start of the experimental series in June 1972, have we obtained any response to these patterns above the mechanical artifact, let alone the 100 per cent response reported by the other authors at pattern wavelengths even well below our standard 1·26°.

On the other hand, we propose that there is a reasonable likelihood that the patterns used in Burtt and Catton's laboratory, while carefully prepared, are not quite good enough. Given the DCMDs demonstrated ability to respond to low spatial frequency events of very low contrast, we feel that our estimate of 10 per cent low-frequency noise in their best master pattern, coupled with the fact that photocopies of unknown quality rather than the master were used in their actual experiments, cast reasonable doubt on the validity of their data. Our own patterns are made to tolerances one hundred times closer than Burtt and Catton's master, and it is possible that this accounts for the difference in the experimental results. Yet, we continue to find it puzzling that not only the toe, but the location of the entire response curve, should be so different.

In view of the difficulty of making patterns of adequate precision, and of the complex geometry of radial gratings, we would like to discuss in greater detail our view of the limitations on visual performance imposed by the optics of apposition eyes.

The best performance to be expected of a single ommatidium in a compound eye has generally been expressed in terms of the Rayleigh criterion, given in angular terms as $\theta_0 = 1\cdot 22\, \lambda/l$. Now, this criterion derives historically from astronomy, where it was developed as a measure of the ability of a telescope to show to the human eye that two nearby stars were indeed separate point sources of light. At the angular separation specified by the Rayleigh criterion the image still shows appreciable contrast: there is a 19 per cent drop in intensity between the peaks of the two adjacent Airy discs in the image. It is the performance of the human eye, the inability of most observers to detect the two peaks as distinct, that made the Rayleigh criterion such a useful measure of the imaging capabilities of optical systems.

However, this rather arbitrary criterion cannot be applied directly when a system is being tested with extended, incoherent, nonmonochromatic objects such as the grating patterns used in studies of insect visual acuity. Furthermore, the criterion is only a single number. It is far more useful to have a description of the performance of an optical system that allows the calculation of the intensity distribution in the image plane for any arbitrary object pattern. One may regard the generalized Fraunhofer diffraction theory in the following incoherent Fourier optics form as the most elementary theoretical model of ommatidial optics capable of generalization to objects other than point sources.

In a Fourier optics model of the ommatidium, we consider the cornea and crystalline cone as an 'optical black box'. The object (an incoherently-illuminated stimulus pattern) is transduced via this optical black box into the image, the latter having its intensity variations lower in contrast than those of the object. These contrast attenuations are a function of the spatial frequency of the object. A formal statement of the optical transduction in terms of spatial frequency is then

$$G_i(f_x, f_y) = H(f_x, f_y)G_0(f_x, f_y) \qquad (14.1)$$

where $G_i(f_x, f_y)$ is the image intensity spatial frequency spectrum as a function of spatial frequencies f_x and f_y cycles m^{-1}; $G_0(f_x, f_y)$ is the object intensity spatial frequency spectrum; and $H(f_x, f_y)$ is the optical transfer function (OTF) of the optical black box.

The object intensity spatial frequency spectrum can be obtained by a two-dimensional Fourier transform of the object intensity distribution, and the image intensity distribution can be obtained from an inverse two-dimensional Fourier transform of the image intensity spatial frequency spectrum (making appropriate normalizations to mean intensities if necessary). The limitations of monochromaticity and Fraunhofer diffraction are present, but the relationship of eqn (14.1) is valid for any physically realizable object intensity distribution. For an optical black box consisting of a lens and circular aperture the OTF is given by (Goodman 1968)

$$H(\rho) = \frac{2}{\pi}\left[\cos^{-1}\left(\frac{\rho}{2\rho_0}\right) - \frac{\rho}{2\rho_0}\sqrt{\left\{1-\left(\frac{\rho}{2\rho_0}\right)^2\right\}}\right], \rho \leqslant 2\rho_0 \quad (14.2)$$
$$= 0 \qquad\qquad\qquad\qquad\qquad\qquad\qquad\qquad , \text{otherwise}$$

where ρ is radial spatial frequency, $\rho = \sqrt{\{f_x^2 + f_y^2\}}$, in cycles m^{-1}; l is the diameter of the aperture in m; d_i is the distance from aperture to image plane in m; λ is the wavelength of the monochromatic illuminating intensity; and where ρ_0, one-half the cut-off spatial frequency, is given by $\rho_0 = l/2\lambda d_i$. Since the optical black box has radial symmetry, there is no angular dependence. Values of the above parameters for the locust ommatidium (Horridge 1966) which yield the upper bound on resolution in spatial frequency are $l = 30 \times 10^{-6}$ m; $d_i = 60 \times 10^{-6}$ m; and $\lambda = 515 \times 10^{-9}$ m, and thus the OTF remains zero above $0\cdot97 \times 10^6$ cycles m^{-1}, corresponding to an angular resolution limit of c. 1 deg. The minimum resolved spatial wavelength would be increased for broad-band non-monochromatic sources of the type used in this investigation. The use of Fraunhofer diffraction theory for the dimensions involved here is perhaps somewhat questionable, but the complexity of complete vector calculations of the intensity patterns at the entrance of the rhabdom for our extended incoherent non-monochromatic objects is beyond the scope of the present investigation. Such a result would further require calculation of the resulting modal patterns in the rhabdom (e.g. Snyder and Pask 1972a,b).

It is of interest to compare the above calculation with the traditional Rayleigh two point resolution criterion. In this case, $D = 1\cdot22\,\lambda d_i/l$, where D is resolution in m in the image plane, and thus the minimum angular resolution is $1\cdot2$ deg. This numerical value is very close to the value at which the OTF first reaches zero. Therefore, the Rayleigh limit formula is, in fact, a convenient way of predicting the limiting performance of an optical system tested even with extended grating objects. It can be regarded as a reasonable estimate of one point on the optical transfer function.

But the full OTF gives us far more information. In particular, it emphasizes the action of a lens or aperture as a filter of spatial frequencies, attenuating high frequencies more than low ones and, unlike RC electrical filters, cutting off absolutely above $2\rho_0$. It is this information which, for example, allowed us to calculate the signal-to-noise ratio in the image of an imperfect object grating (see Theory and Methods earlier). Similarly, it is this viewpoint that leads us to reject as entirely inappropriate the use of single discs, which contain very broad spatial frequency spectra, in the evaluation of visual acuity as has been done by Northrop.

In considering the properties of test objects, a question arises concerning the use of a single number spatial wavelength to describe the ideal radial grating pattern's spectrum when it is in fact a complicated function of spatial frequency. This single number for a given pattern and position, λ_s, is the angle subtended at the cornea of the animal by one complete stripe cycle at the outer edge of the pattern, in keeping with earlier definitions. For spatial frequency the inverse of this number is used, $f = 1/\lambda_s$. Our calculations show that the spatial frequency spectrum of this pattern does correspond to the conventional single number description in that the single number is a kind of least lower bound on intensity as a function of spatial frequency of the pattern. Up to a spatial frequency of 1·3 f the intensity of the spectrum remains very small (10^{-11} per cent of the maximum intensity), at which point the intensity has become 1 per cent of the maximum. Maximum intensity is attained at a spatial frequency of 1·64 f. This behaviour of the spectrum is due to the rapid rise of the Bessel function $J_n(x)$ as one approaches $x = n$, and the fact that the radial component of the spectrum is given by

$$ R(\rho) = \frac{1}{(2\pi\rho)^2} \int_{0\cdot637(2\pi\rho)}^{2\pi\rho} x J_n(x) \, dx, \text{ for an } n \text{ stripe cycle pattern,} $$

where ρ is the radial component of spatial frequency. Thus the single number description of the spectrum is very close to the point on the spatial frequency spectrum where intensity first becomes sensible.

The fact that the pattern must be moved to be detected by the locust DCMD must be considered. It is the co-ordinates of the motion which determine the pertinent co-ordinates for calculating the spatial frequency spectrum of the pattern. The radial symmetry of the pattern has been used in our calculations because the pattern is rotated about its centre in these experiments.

The time dependence of the response of the retinula cells (Pinter 1972) and all subsequent neuronal elements leading to the DCMD response also enter into detection of pattern motion, and must be considered in a complete theory. But if the pattern is not resolved in any way spatially, it cannot be detected by temporal processing. This, of course, in no way denies the existence or utility of neuronal theories explaining spatial frequency response broadening (Northrop 1971) or spatial frequency response sharpening (Harth and Pertile 1971).

A rather different viewpoint, pioneered by Götz (1964), is based on the observation that the angular acceptance curve of retinula cells is often approximately Gaussian in shape. Presumably, the OTF combines with other factors, such as the dimensions of the rhabdomeres, to produce this shape. The angular acceptance curve, in turn, can be used to predict the response of the retinula cells to various spatially-distributed stimuli, including rectangular gratings. These calculations successfully predict the optomotor responses of flies (Götz 1964) and locusts (Thorson 1966) as a function of pattern spatial wavelength, and have been tested directly by means of intra-cellular recording in retinula cells of locusts (Tunstall and Horridge 1967). The latter authors found general agreement with Götz's equation, but observed two deviations which they interpreted as probably causally related. The angular acceptance curves were more peaked than Gaussian curves, and the responses to stripes of spatial wavelength below 5° were greater than predicted. Nevertheless, their responses fell to zero between 2° and 1°, and they were unable to find any evidence of responses in Burtt and Catton's anomalous region.

In sum, the optical transfer function constitutes a model of the optics of apposition eyes, in the sense that it predicts the optically controlled portions of the response characteristics of neurons in the visual system. It is of interest to inquire whether the contrast attenuations predicted by the OTF are dominant in the response of retinula cells, or whether other factors, such as the dimensions of the rhabdomeres, are so powerful that the OTF is actually a very weak predictor of retinula cell responses. An answer to this question for *Calliphora* has been provided by Kuiper (1966), who has shown that there is excellent agreement between the shape of the angular acceptance curves of single retinula cells and the shape of the Airy disc, derivable from the OTF, for lenses of the appropriate size. On the other hand, the agreement for *Musca* was not so good. For the worker bee, Snyder (1972) has shown that the angular acceptance curve is the product of the acceptance curve of the optical apparatus determined by the diffraction pattern in the focal plane, and the acceptance curve of the rhabdom. In the locust, the Airy disc calcu-lated from our fit of the OTF to experimental data is circa one-fourth the width of the angular acceptance curve in the light-adapted state. (Horridge 1966). It would appear, then, that at least in some insects the OTF is a model that has a direct bearing not only on the first stage of the visual system, the lens and crystalline cone, but also on the output of the entire first neutral layer, the retina.

Part IV

Electrophysiology of the optic lobe

15. The function of the lamina ganglionaris

S. B. LAUGHLIN

15.1. Introduction

A major problem in sensory physiology is establishing the neural mechanisms
by which a sophisticated visual system abstracts information on contrast,
movement, direction, and spectral composition from a spatial array of
receptors that are each capable of responding in a single mode. The optic
lobes of insects with apposition compound eyes are favourable systems for
study for the following reasons. First, the insects investigated show all the
discriminatory capacities of a well-developed visual system. Secondly, the
responses of the primary photoreceptors, the retinula cells, can be accurately
characterized because they can be recorded intracellularly. Thirdly, these
receptors have well-defined narrow fields of view that can be defined precisely
relative to each other from the geometry of the eye according to the classical
mosaic theory. This means that any stimulus configuration can be interpreted
in terms of patterns of individual receptor output. Finally, the successive
sites of integration, the optic neuropiles, are arranged in the sequence of
lamina, medulla, lobula/lobula plate, and protocerebrum. Furthermore, the
neuropiles of the lamina and medulla are arranged and connected in parallel
morphological units that reflect the visual axes of the receptors.

The lamina is the first optic neuropile and receives as its input the axons
of the retinula cells. It is, therefore, the site of the initial processes of visual
integration. Any information on its function is extremely important for
interpreting results obtained at a higher level in the system. For example, a
mathematical analysis of the output of the motion-detection system in fly or
locust has led to the proposition of detailed models of movement detection
in mathematical terms (Reichardt 1961; Thorson 1966). Extracellular
recordings from movement detectors in the medulla, lobula, protocerebrum,
and ventral nerve cord have tested the validity of these models and generated
new models (see Mimura (Chapter 19), Kaiser (Chapter 16), and Kien
(Chapter 18)). This model building depends on the black-box approach,
varying input conditions and observing the modification in output. However, as long as the lamina remains as an additional black box in series with

the higher order unit under investigation the true function of that unit cannot be derived in terms of its actual input. This is as true for motion detectors as it is for other less intensively investigated processes such as colour coding and the processes of light/dark adaptation. In this article I shall attempt to review our present knowledge of the physiology and anatomy of the lamina and suggest the possible functional significance of these findings.

15.2. The anatomy of the lamina

The first.comprehensive study of the cells of the lamina was made by Cajal and Sánchez (1915). Two types of lamina were described in detail, the first in a number of flies and the second in the bee *Apis mellifera*, with supplementary observations from other insect groups including the Odonata. The work was based on Golgi impregnated preparations and although limited conclusions were reached on the topographical relationships between neurons, it has a lasting significance as a descriptive catalogue of a large number of cell types and as a demonstration of the subtlety and precision with which these are recognized. These studies, together with similar early investigations have been conveniently summarized in Chapter 19 of Bullock and Horridge (1965), and comprehensively evaluated and amplified in two recent studies (Strausfeld 1970a; Strausfeld and Blest 1970). Subsequent work has been almost wholly confined to Muscoid flies, using either electron microscopy (Trujillo-Cenóz 1965; Boschek 1971) or reduced silver and Golgi methods (Braitenberg 1967; Strausfeld 1971a), so that we now have for the lamina of these species the most comprehensive description of cell type and connectivity in any known neuropile. The following general account of lamina anatomy draws heavily on these studies and covers not only muscid flies but Odonata, Hymenoptera, and Orthoptera as well. Although not complete in detail, the picture (Fig. 15.1) provides an anatomical background for the electrophysiology.

The lamina lies as a curved plate beneath the retina and consists of a ganglion layer containing the somata of monopolar cells overlying the external plexiform layer which is the synaptic region. The external plexiform layer is divided conspicuously into many cylindrical lamina cartridges, which are the basic functional units of the lamina. There is one lamina cartridge for each ommatidium and the axons of those retinula cells with the same field of view converge upon a single cartridge in *Musca* (Braitenberg 1967; Kirschfeld 1967) and in a number of insects with fused rhabdom ommatidia, including locust, bee, and dragonfly (Horridge and Meinertzhagen 1970). In addition Horridge and Meinertzhagen (1970) and Strausfeld (1971b) show that the axons leaving a single lamina cartridge project through the external chiasma to a single medullary cartridge. This means that each ommatidial field of view is projected by a non-divergent pathway to the medulla.

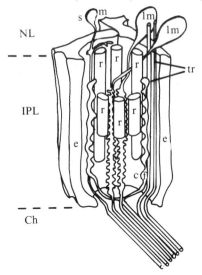

FIG. 15.1. A diagrammatic representation of a generalized single lamina cartridge showing, in simplified form, the organization of the principal neurons described in the text: NL = nuclear layer, IPL = inner plexiform layer, Ch = first optic chiasma; e = epithelial pigment cell; sm = small monopolar cell; r = retinula cell axon terminal; tr = throughgoing retinula cell axon; lm = large monopolar cell; cf = centrifugal basket fibre.

15.2.1. *Neurons in the lamina*

Associated with each lamina cartridge we find short retinula axon terminals, long retinula axons, a number of types of monopolar neuron, and a number of types of medulla centrifugal fibre terminals. The relationships between these components in the cartridge have been thoroughly investigated in the fly and the limited investigations of other insect orders allow the following generalizations to be made about their involvement in the lamina cartridges.

15.2.1.1. *Retinula cell axons.*   There are eight retinula cells in each ommatidium in Odonata (Horridge 1969; Eguchi 1971) and flies (Boschek 1971) and nine in drone bee (Perrelet 1970) and worker bee (Menzel and Picker in the press). Six of their axons form a cylinder in the lamina cartridge, associated with the two large monopolar cells, upon which they repeatedly synapse. The remaining long retinula axons pass through the cartridge and continue to the medulla. In the fly the long retinula axons do not synapse in the lamina (Boschek 1971), while in the bee limited synaptic interaction probably occurs (see Strausfeld (1970b)). There is no anatomical evidence for electrical or chemical synapses between retinula axons of the fly (Boschek 1971) or worker bee (Varela 1970).

15.2.1.2. *Monopolar cells.*   The principal centripetal components of the lamina have in the ganglion cell layer a cell body which sends a single axon through the cartridge, where it sends out lateral processes, and goes via the chiasma to the medulla, where it terminates in a medulla cartridge. There are two types, large and small, distinguished by Cajal and Sánchez (1915) by the size of the somata and the extent of lateral dendrites. A small number of large monopolar cell (LMC) axons lie at the centre of the cartridge, close to the retinula axon terminals. In flies there are two LMCs and apparently each LMC is postsynaptic to all six retinula axons (Trujillo-Cenóz 1965). LMCs terminate at different levels in the medulla cartridge and cannot be assumed to be functionally equivalent. Small monopolar cells have a small number of lateral processes, both within and beyond their own cartridge, that are restricted to certain strata of the lamina neuropile. In fly, Strausfeld (1971a) has deduced from their position and the shape of their dendritic fields that they can only synapse with one or two retinula axons. In addition he has shown that there are three small monopolar neurons in each cartridge, two of which have dendritic fields confined to their own cartridge and one of which sends a collateral to LMCs in two adjacent cartridges and may not receive a synaptic input from its own cartridge (Strausfeld and Braitenberg 1970; Strausfeld 1971a). Small monopolar cells similar in type to those studied in fly have been described by Cajal and Sánchez (1915) in worker bee and Odonata but their position and distribution in the cartridge is not known.

15.2.1.3. *Centrifugal basket fibres.*   These are called centrifugal on anatomical grounds only, because each has a cell body in the medulla cortex with a long axon across the external chiasma to the lamina and a short axon into the medulla cartridge. The lamina terminal was described by Cajal and Sánchez (1915) as the distinctive nervous bag, a dense system of dendrites that envelopes the cartridge. Trujillo-Cenóz and Melamed (1970) have identified positively the lamina terminals in flies from Golgi preparations that have been resectioned and examined electron microscopically. In bee (Varela 1970) and *Musca* (Boschek 1971) and several flies (Trujillo-Cenóz and Melamed 1970) the lamina terminals are associated with retinula axon terminals but opinion is divided as to whether they are functionally centrifugal and therefore presynaptic to retinula axons (Trujillo-Cenóz and Melamed) or are postsynaptic (Boschek). In addition, the medulla terminal is closely associated with the LMC terminals from their own lamina cartridge and they could therefore form an angle-preserving feed-back or feed-forward loop within their own cartridge unit.

15.2.1.4. *Epithelial glial cells.*   The major glial elements separating lamina cartridges are particularly obvious in fly where Boschek (1971) has shown that each epithelial glial cell borders three cartridges in a regular lattice arrangement. In worker bee the glial elements are more tenuous but still appear to enclose tightly each cartridge (Varela 1970). These cells may play an important role in the physiological operation of the cartridge as they are linked to retinula axon terminals by capitate projections (Trujillo-Cenóz 1965) and appear to receive a synaptic input from retinula axons and centrifugal fibres (Boschek 1971).

15.2.1.5. *Wide field elements*. There are a number of other cell types associated with the lamina whose dendrites appear to spread over a number of cartridges. The most notable are amacrine cells (Cajal and Sánchez 1915; Strausfeld 1971*a*) which appear to contact three or four cartridges. There are also several types of centrifugal fibre with small fields extending over a few cartridges and several distinct types of wide field tangential elements that run across the lamina for considerable distances. These elements have not been assigned a role in the connectivity of the lamina but it is important to bear in mind that they must have a physiological function.

15.3. The physiology of the lamina

The physiological analysis of an integrative system proceeds at several levels that are not necessarily hierarchical. The responses of units are characterized, i.e. action potentials, graded potentials, frequency distribution, or waveform of response etc. and the response is correlated with the appropriate input, e.g. intensity of illumination, velocity of movement etc. Stimulus parameters are carefully manipulated while the input to the unit is correlated with the output so that the integrative transforms on information taking place at the level of that unit can be assessed. The mechanics of integration can then be examined at the cellular level and interpreted in terms of the cell's excitable properties and its inferred connectivity. Finally the unit can be identified positively anatomically and its structure and connectivity correlated with function. At the present time no sensory system or single stage of a sensory system has been analysed completely at all these levels. However, it is possible that some of the more accessible components of the lamina may be analysed to this degree in the near future.

Until recently (Arnett 1971), all investigations of the lamina and external chiasma using extracellular recording techniques have failed to detect spike activity (Burtt and Catton 1956; Ishikawa 1962; Horridge *et al.* 1965; Bishop and Keehn 1967). For this reason the analysis of lamina function has had to wait upon the perfection of intracellular microelectrodes capable of obtaining stable recordings from axons with diameters of $< 5~\mu$m (Shaw 1968; Autrum *et al.* 1970). This account of lamina physiology draws heavily upon the studies of Shaw on *Locusta* (1968), Scholes on *Musca* (1969), Zettler and Järvilehto on *Calliphora* (1971, 1972) and my own work on Odonata (Laughlin, 1973).

15.3.1. *Extracellular properties of the lamina*

The lamina compartment has a standing negative potential of -30 to -100 mV (Burtt and Catton 1956; Mote 1970*a*; Zettler and Järvilehto 1971). Extracellular recordings from the lamina detect a positive d.c. potential in response to illumination of the retina. This is almost certainly due to the lamina acting as a sink for current from the component of receptor potential that is conducted down the retinula axons (see later), for the potential has the same waveform as the receptor potential and is insensitive to the synaptic inhibitor, carbon dioxide (Leutscher-Hazelhoff and Kuiper 1964). However, the magnitude of this positive potential does not appear to be uniform over the whole lamina and depends very much on the recording site. Thus in dragonfly lamina and worker bee lamina (Menzel, personal communication) micropipettes detect positive potentials of between 5 and 20 mV, depending on their position. These always show the waveform of the receptor potential and have an angular sensitivity that is related to the maximum response amplitude. Thus the small responses have a broad angular sensitivity and the larger responses have a narrow sensitivity function. Similar observations have been made by Scholes (1969) and Mote (1970*a,b*) has carried out a number of experiments on similar positive potentials in fly, from which he concluded that the large responses reflected a dominant input from a small number of ommatidia with a lesser contribution from ommatidia over a large field. This is in agreement with the typical angular sensitivity functions for large responses, an example of which is illustrated in Fig. 15.2, which show a sharp peak on axis, reflecting an input from one ommatidium and a

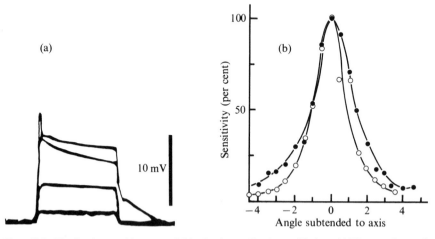

F IG. 15.2. The lamina positive potential in the dragonfly *Anax gibbulosa*. (a) The waveform of the response to 1 s axial flashes of increasing log intensity from $-3 \cdot 0$ to $0 \cdot 0$. (b) The angular sensitivity function of the lamina positive potential (—●—) compared to that of a retinula cell (—○—).

broad shoulder showing a contribution from ommatidia over a wide field. The positive potentials recorded in worker bee lamina are insensitive to the plane of polarization of incident light, indicating that they represent a summed response, originating from several retinula cells (Menzel, personal communication). These findings can be explained on the hypothesis that each cartridge is separated from its neighbours by a high resistance barrier, the epithelial glial cells. The cartridge acts as a sink for current from the retinula axon terminals and depolarizes to a degree that depends predominantly on the size of its own retinula input, and to a lesser degree on the depolarization of neighbouring cartridges. This hypothesis is consistent with the observation that damaged monopolar cells pick up a positive waveform that resembles the receptor potential (Shaw 1968; Autrum *et al.* 1970). If cartridge depolarization occurs it could be extremely important in regulating the input to the cartridge from retinula axons, because it would subtract from the positive retinula cell input signal at the axon terminals, thus acting as an electrotonic negative feedback.

15.3.2. *The input from retinula cell axons*

The response of retinula cells to light is a depolarization that increases with increasing light intensity and becomes bi-phasic at high light intensity (lower line in Fig. 15.3). The receptor potential is not encoded as action potential frequency for transmission to the lamina but is transmitted in its

Monopolar

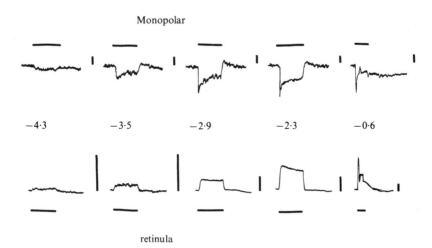

$-4\cdot3$ $-3\cdot5$ $-2\cdot9$ $-2\cdot3$ $-0\cdot6$

retinula

FIG. 15.3. A comparison between the waveforms and amplitudes of retinula cell and LMC responses to the same axial point source at a number of different stimulus intensities, from $-4\cdot3$ to $-0\cdot6$ log units. In each case the horizontal bars indicate the duration of a flash of 500 ms and the vertical bar to the right of each waveform is a scale of 10 mV. Responses were obtained from intracellular recordings in the lamina and retina of the Australian Corduliid dragonfly, *Hemicordulia tau*.

original form by passive electrotonic conduction down the retinula axon. The evidence for this is as follows:

(1) Action potentials are rarely recorded intracellularly from retinula cells in the retina (Shaw 1969; Baumann 1968).

(2) Action potentials from units that give a sustained discharge in response to illumination and have the same receptive field as retinula cells have never been found in the lamina or the medulla.

(3) Intracellular recordings from retinula axons show that their response is identical to that recorded from receptors except that it shows a slight temporal smoothing that is to be expected from passive conduction down a cable structure (Ioannides and Walcott 1971; Järvilehto and Zettler 1970).

(4) The structure and response of receptors in the dragonfly ocellus is similar to that found in retinula cells. Application of tetrodotoxin to the ocellus in concentrations that would block completely any action potentials did not modify the hyperpolarizing response, recorded intracellularly from second order neurons in the ocellar nerve (Chappell and Dowling 1972).

(5) In comparison, the photoreceptor axons of the barnacle ocellus are over 1 cm in length. The receptor potential is passively conducted down the axon and is only attenuated by a factor of three. Thus graded potentials can be conducted over greater distances than was previously thought due to the high specific membrane resistance of these axons (Shaw 1972).

At present all physiological data confirms the anatomical finding that axons with the same field of view converge upon a single cartridge. In Odonata and *Calliphora* (see later) monopolar cells receive an excitatory input corresponding to the field of view of a single retinula cell. In addition, Scholes (1969) performed a series of elegant experiments on large positive potentials recorded with high-resistance micropipettes in the lamina of *Musca*. These potentials have the same fields of view and waveform as retinula cells but were shown to be the summed response from all six retinula cells in the pseudopupil. This experiment gives indisputable proof that in fly the inputs from six retinula cells with the same fields of view converge upon a single cartridge. The narrow fields of view and the large amplitude of these potentials suggest that these were intracellular responses from axons and that the axons must be coupled electrically. Unfortunately no other criteria of an intracellular locus were given and it can be postulated, with just as much uncertainty, that the responses were extracellular recordings from a well insulated cartridge.

The sensitivity of retinula cell soma responses has been compared to that of lamina positive potentials using sinusoidal stimuli, in *Calliphora* (Smola and Gemperlein 1972; Gemperlein and Smola 1972). The positive potentials resemble those recorded by Scholes (1969) but there is no direct evidence that they are intracellular recordings from retinula axons. These positive potentials are more sensitive and are more efficient at transmitting modulation at low intensity. It is claimed that this demonstrates a coupling of axon responses in the lamina which improves the signal-to-noise ratio. This coupling, however, could result equally well from an extracellular recording site. The study does emphasize that summation of retinula axon pathways in the lamina upon the next stage in transmission will reduce the signal-to-noise ratio and amplify the signal thus increasing the sensitivity of the visual system. However, the summation may not take place at the level of the axons themselves, as these investigations suggest, but could arise by convergence of inputs, from uncoupled axons, upon a single large monopolar cell.

Coupling of retinula cell axons in Diptera could be expected on comparative grounds for it is the only possible functional equalivent of the retinula cell coupling described in drone bee and *Locusta* (Shaw 1969). However, a direct attempt to demonstrate axon coupling in *Calliphora*, by recording simultaneously from two lamina loci gave a negative result (Rehbronn 1972). The wide angular sensitivity functions of the positive potentials used for this study suggest that the responses are definitely extracellular in origin and do not represent intracellular recordings from retinula axons as asserted.

The organization of retinula cell axons in a single lamina cartridge is apparently adapted to amplify the retinula cell output and improve the signal-to-noise ratio. This would go some way to explain the means by which the visual system can reliably detect movement when it is represented at the receptors by a signal of <0.5 mV (Scholes and Reichardt 1969). In addition to the response summation, discussed above, two other adaptations may be important. Firstly, the retinula axons, acting as passive cables, function as low pass filters (Shaw 1972) but it should be noted that Smola and Gemperlein's (1972) results fail to show this. Secondly, the repeated synapsing of single retinula axon terminals upon a large monopolar cell would not only amplify the receptor signals but would maintain synaptic noise at a high and constant level.

15.3.3. *Large monopolar cells*

Intracellular recording and dye injection techniques have shown that in the fly (Autrum, Zettler, and Järvilehto 1970; Zettler and Järvilehto 1971; Arnett 1972) and Odonata (Laughlin unpublished results) LMCs respond to retinal stimulation with graded hyperpolarizations. The waveform and response characteristics are similar in both orders and resemble the largest amplitude responses recorded by Shaw (1968) in locust lamina, and Menzel (personal communication) in worker bee.

15.3.3.1. *Response characteristics.*
A successful penetration of a LMC is signalled by the appearance of up to 5 mV of a distinctive low-frequency

noise and an indeterminate resting potential. This noise is presumed to be synaptic in origin, for its frequency increases and its amplitude decreases with higher intensities of stimulation (see Fig. 15.3). Presumably the hyper-polarizing response is generated by the summation of these ipsps. In *Locusta* the discrete hyperpolarizing potentials are more pronounced than in fly and dragonfly, and in many cases their summed tonic effect is small. This may well be a recording artefact, but Shaw (1968) took the frequency of these discrete potentials as a measure of cell response. As he referred all responses to a calibration of response frequency against axial light intensity, this is a valid stimulus parameter. Caution must be used, however, in interpreting the absence of discrete potentials at high light intensity (see later).

The amplitude and waveform of a LMC response to an axial point light source is intensity dependent. This is illustrated in Fig. 15.3. At threshold, the response is monophasic and uneven in amplitude. As the intensity is increased the response becomes triphasic with an initial 'on' transient peak, a sustained plateau and a positive 'off' transient. The amplitude of both the

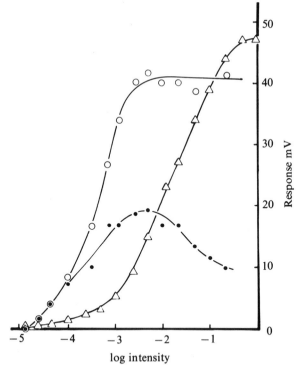

FIG. 15.4. Response–intensity curves for the peak (—O—) and plateau (—●—) amplitudes of the monopolar cell responses compared with the curve for a retinula cell (—△—). In each case the response amplitude is plotted against the intensity of the axial light source. These curves are derived from two full series of responses, of which some have been used in Fig. 15.3.

peak and the plateau of the response rapidly rises to a maximum so that the dynamic range between threshold and saturation corresponds to 1·0–2·0 log units of intensity in Odonata, *c*. 2·0 log units in *Calliphora* (Zettler and Järvilehto 1971) and *c*. 1·0 log units in *Phaenicia* (Diptera) (Arnett 1972). A typical response/intensity curve for a dragonfly LMC is shown in Fig. 15.4. As the light intensity is increased above saturation the amplitude of the 'on' peak remains constant but the amplitude of the plateau decreases and changes during the course of the stimulus. This has been observed in Odonata, in *Calliphora*, in worker bee, and in locust. In damaged units the plateau potential may become positive and it has been suggested that this is due to summation with the lamina positive potential (Shaw 1968; Autrum, Zettler, and Järvilehto 1970). This surmise is probably correct, but in more stable recordings from units giving large responses the changes in plateau potential bear no temporal relationship to the lamina positive potential (see Fig. 15.3). This means that the plateau response could well be modulated by neurons synapsing upon monopolar cells or retinula cell axons.

15.3.4. *Integrative properties of large monopolar cells*
The LMCs constitute a major centripetal pathway between the retina and the medulla. It can be expected, *a priori*, that their output is not just a repetition of the retinula input but is transformed in a number of different ways. These transformations can be unmasked by comparing the monopolar cell output under different stimulus conditions to their retinula input, as determined anatomically and physiologically. The significance of these transformations can then be assessed in terms of the major processes of visual integration.

In order to measure the sensitivity of a LMC to changes in stimulus parameters one must have a measure of the cell's responsiveness in terms of a standard input. This essentially amounts to calibrating the cell. At the present level of analysis it is best to calibrate the cell with respect to the intensity of a light on axis (the point of maximum sensitivity in the visual field). This is the measure used by most investigators on retinula cells and has been used by Shaw (1968), Zettler and Järvilehto (1971, 1972), and by myself on studies of LMCs. It is important to position the calibrating light source exactly on the axis in order to minimise lateral interactions, and to standardise them so that the responses to different stimuli and of different individual neurons may be compared.

15.3.4.1. *The major input.*
The hyperpolarizing response has a latency, relative to the retinula cell response, of 1 ms in *Calliphora*, 2 ms in Odonata, and 5 ms in *Locusta*. In addition it has been shown in all three cases that the angular sensitivity function of LMCs is as narrow or narrower than the angular sensitivity of individual retinula cells. This is illustrated for the

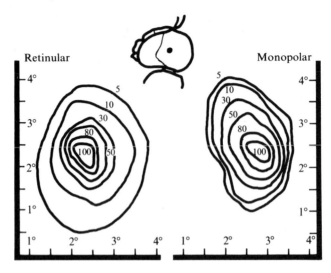

FIG. 15.5. The angular sensitivity of a retinula cell and of a LMC in the dragonfly, *Hemicordulia tau*. Contour lines of isopercent sensitivity were derived from measurements of sensitivity obtained from a point source positioned at nodes on a half degree grid, represented by the two sets of axes. The inset shows the region of the eye from which both sets of data were obtained.

dragonfly in Fig. 15.5. These two pieces of evidence together show that the primary input to large monopolar cells comes from retinula axons with the same fields of view. The anatomical data suggests that the inputs from at least three and possibly all six retinula cell axons converge upon a single large monopolar cell (Trujillo-Cenóz 1965). In order to test this electrophysiologically either polarized-light sensitivity or spectral sensitivity must be examined. If the LMC sensitivity is the product of two or more retinula cell sensitivities then convergence has taken place. The converse, however, is not necessarily true for it could demonstrate the selective convergence of retinula cell axons with the same sensitivity characteristics. At present the evidence on this point is inconclusive. Shaw (1968), in locust, has reported the instance of a single hyperpolarizing unit that was sensitive to the plane of polarization but most units were insensitive. In addition, Shaw found in several cases that the amplitudes of the discrete hyperpolarizations that he observed in a single unit fell into two distinct classes. He correctly interpreted this as demonstrating the occurrence of two spatially separated synaptic sites. It cannot be concluded, however, that the occurrence of two amplitude classes demonstrates convergence, because repeated synapses by a single retinula cell axon upon one large monopolar cell could produce the same effect.

15.3.4.2. *Retinula–monopolar transfer*. It is apparent from the latency determinations that the monopolar cell is hyperpolarized in response to a pre-synaptic depolarization of the retinula cell axon. In a dark-adapted animal the large monopolar cells are extremely sensitive to light falling on the retina. In Fig. 15.4 the dark-adapted response–intensity curves of a LMC and a retinula cell, measured with the same axial light source are compared. It is clear that a small amplitude retinula cell depolarization generates a much larger LMC response. In other words there is an amplification as well as inversion of signal in the transfer between retinula cell and monopolar cell, together with a change in waveform. This amplification was first observed and discussed by Shaw (1968) who demonstrated in locust that individual receptor quantum bumps generate discrete hyperpolarizations of at least five times their own amplitude. Järvilehto and Zettler (1971) have demonstrated an eightfold amplification of response using sinusoidal stimuli in the frequency range of 10–100 Hz., in *Calliphora*. When pre- and post-synaptic responses to square wave stimuli of constant intensity were compared, no direct amplification of peak or plateau could be seen, but a twofold amplification was apparent if the rates of potential change for the rising phases were compared. Zettler and Järvilehto (1971) concluded that the rate of rise of the retinula response generated the rate of rise of the monopolar response, and that the transfer of signal from retinula cell to LMC had the properties of a frequency-dependent amplifier. They emphasized that the dynamic characteristics of transfer are important for describing LMC responsiveness. Furthermore, they suggested that the rate of rise is the best parameter for quantifying the monopolar response, because it has the same dynamic range over 6 log units as the receptors, whereas the LMC response amplitude has a narrow dynamic range (2 log units). While it is certainly true that the rate of response rise is useful for analysis of monopolar cell function, this does not imply that the monopolar cell response sends information to the medulla encoded as the rate of response rise. There is no evidence from quantitative analyses of the optomotor response to suggest that the differential of the retinula input is the significant input for movement perception. Furthermore, there is no compelling *a priori* objection on functional grounds against second-order neurons having a narrower dynamic range than their receptors. In fact, in vertebrate retina (*Necturus*) the receptors have a dynamic range of 4 log units while the second-order horizontal cells and bipolar cells have a dynamic range of 3 and 1 log units, respectively (Werblin 1971). In view of the clear-cut amplitude amplification found in Odonata it would seem better to assume that the presynaptic retinula potential generates the post-synaptic LMC potential, the magnitude of which transmits information to the medulla. The absence in *Calliphora* of a clear-cut amplitude gain to square-wave stimuli shows that there can exist different frequency–response characteristics for excitatory and inhibitory inputs from those in the dragon-fly.

Amplification could be the result of axonal convergence together with a large number of retinula–LMC synapses. In the dark-adapted animal the gain in the lamina aids the detection of weak stimuli and the system appears to be adapted for maximizing the signal-to-noise ratio. Amplification also effectively increases the responses caused by movement of objects across the visual fields of retinula cells and so may increase acuity (Northrop personal communication). There is at present no clear-cut evidence on the behaviour of LMCs in the light-adapted state, so it is impossible to assess whether amplification occurs throughout the intensity range of the insect.

15.3.4.3. *Transmission to the medulla.*   Action potentials have not been recorded from LMCs and all investigations using extracellular electrodes have failed to demonstrate a large class of units with a tonic discharge inhibited by light similar to those found in the ocellar nerve (Ruck 1964; Chappell and Dowling 1972). Zettler and Järvilehto (1971) have recorded from LMC axons at different levels in the medulla and shown that the graded hyperpolarization is conducted down the axon. They found no clear-cut response decrement, but this could have been obscured by a scatter of potential amplitudes due to the instability of recording. If conduction is passive a small amplitude decrement requires a high specific membrane resistance, which would in turn severely attenuate the high-frequency components of the response. The 'on' transient is not attenuated, however, and it seems possible that an active membrane process reinforces the graded hyperpolarization as it is conducted to the lamina, although evidence for this is not conclusive.

15.3.4.4. *Synaptic mechanisms.*   On account of the inversion of the signal the retinula–LMC synapse must be chemical, although the nature of the transmitter and the post-synaptic changes induced are obscure. Shaw (1968) found that the input resistance of locust lamina hyperpolarizing units decreased on stimulation and that the response had a reversal potential 25 mV negative to the resting potential. This suggests that the response results from a potassium or chloride ion conductance increase, similar to that found in ipsps. In vertebrate cones the hyperpolarizations are generated by a decrease in sodium ion conductance (Baylor and O'Brien 1971). Zettler and Järvilehto (1971) studied the effect of injecting constant current into a LMC, and found that depolarizing current decreased the response amplitude whereas hyperpolarizing currents accentuated it. They attributed this phenomenon to a presumed non-decremental conduction mechanism but it could also be interpreted as showing a resistance increase on stimulation that is responsible for response generation, as in vertebrate cones. The difficulties associated with passing constant current through high resistance microelectrodes make this point difficult to clarify.

15.3.4.5. *Inhibition.*   If LMCs are to have an integrative function, it seems likely that various inhibitory inputs will act upon their response. A good example of this is the lateral inhibition found by Zettler and Järvilehto (1972) in *Calliphora*. They have demonstrated the inhibitory effect by comparing the angular sensitivity of LMCs to that of retinula cells. The width of the LMC angular sensitivity function at the 5 per cent level is 3° as opposed to a corresponding value of 11° for the receptors. The narrowing of the visual field must result from inhibitory inputs from adjacent cartridges and has an important functional role in increasing the acuity of the visual system. It has been pointed out above that the triphasic waveform of the LMC response could result from inhibitory inputs and it is worth bearing in mind that this type of waveform is typical of lateral inhibitory networks.

Immediately after a high-intensity stimulus is turned off the receptor potential decays exponentially to its resting value over a period of several seconds. This receptor afterpotential should be large enough to generate a response in the LMCs. However, Shaw (1968) found that discrete hyper-polarizing potentials could not be evoked from locust lamina units by low-intensity stimuli delivered during the receptor afterpotential and suggested that the units were desensitized during this period. The component generated by the afterpotential is clearly seen in the response of dragonfly LMCs, however, so desensitization must here be small in effect.

Inhibitory inputs may act to reduce the sensitivity of LMCs during the process of light adaptation, for in the dark-adapted state they utilize only a small part of the receptors' dynamic range. However, it has also been suggested by Strausfeld (1971a) that in flies the LMCs represent a scoptopic system and that their function is taken over by the small monopolar cells at higher light intensities. The behaviour of LMCs at progressively higher light intensities together with the properties of light-adapted receptors must be thoroughly investigated before the role of the LMCs in the visual system can be assessed.

15.3.5. *Unit responses from the lamina*

Several investigators have recorded both extracellularly and intracellularly from other units in the lamina and chiasma. Unfortunately direct anatomical identification of these units has not been possible so that one can only conjecture about their role in the visual pathway.

15.3.5.1. *'On–Off' and 'Sustaining' units.* In the first chiasma of the fly *Phaenicia* Arnett (1972) has studied extracellular activity using tungsten microelectrodes. 'On–Off' units give a brief spike discharge at the beginning and after the cessation of a stimulus anywhere within the homogeneous visual fields. The angular sensitivity of the visual field, measured as the intensity required to produce a constant (threshold) response, is an elliptical Gaussian function with half-widths along the major and minor axes of 5·5° and 4·4°, respectively.

In response to a sustained spot stimulus the 'Sustaining' units give a tonic discharge which adapts slightly in an intensity dependant manner that is reminiscent of the adaptation of the receptor potential of retinula cells. The receptive field is more complex, consisting of an elliptical 'on' region flanked along the medio-lateral axis by two circular 'off' regions. Spot stimuli within the 'on' region produce the tonic discharge that is characteristic of this unit and the 'on' region has a Gaussian angular sensitivity function of half-width 2·2° in the medio-lateral plane and 2·6° in the dorso-ventral plane. The 'off' regions have a diameter of 4° and stimuli falling in these regions produce an 'off' discharge. The stimulation of the adjacent 'off' regions inhibits discharge due to a stimulation of the 'on' region. This lateral inhibitory effect is responsible for the narrowing of the 'on' region angular sensitivity in the medio-lateral plane, for the angular sensitivity of the retinula cells is 2·3°.

Both types of unit appear to receive an input from the retinula cells that terminate in the lamina, for they have the same spectral sensitivities as retinula cells 1–6 and are insensitive to the plane of polarization of light (McCann and Arnett 1972). Arnett suggests that 'On–Off' and 'Sustained' units are the two large monopolar cells, but he readily acknowledges that this view is incompatible with the intracellular recordings that he has made from large monopolar cells. It is not possible to identify the cell types giving the spike responses nor even to assess whether they are second order or not, for there is no accurate measurement of response latency and the recording locus is not marked. Arnett claims that they are centripetal for he cut the first chiasma and still obtained unit responses. However, the chiasma is less than 300 μm long and separates two curved surfaces, so it would seem impossible, without direct histological examination of the lesion, to be certain that a part of the outer layer of the medulla was not left intact and attached to the lamina. Both units have small receptive fields corresponding to single receptors and they could be either centrifugal basket fibres or small monopolar cells. In order to gain a better understanding of their function in the absence of direct anatomical identification it would be interesting to correlate the temporal and spatial properties of their responses with changes of sensitivity in large monopolar cells.

15.3.5.2. *Multimodal units and tonic light units.* These have only been recorded extracellularly from sites in the lamina and first chiasma of the lubber grasshopper, *Romalea*, by Northrop and Guignon (1970). The recording locus was positively identified using a radiofrequency lesion technique. Multimodal units have an erratic steady state discharge in both the light and the dark, and respond with a short burst of spikes at either 'on' or 'off'. They also respond to movement of a contrast in any region of the visual field although continued movement in the same small region leads to adaptation. Multimodal units derive their name from the fact that acoustic or tactile stimuli also evoke a short discharge. Tonic light units have little or no dark discharge but respond to illumination of the visual field with a sustained discharge. Both units are found throughout the medulla and have large receptive fields. The multimodal units are certainly centrifugal and their existence emphasizes that feedback onto the lamina from higher-order neurons may have an important function in modulating the response of second-order neurons in the visual pathway.

15.3.5.3. *Units in Drosophila.* In *Drosophila* Alawi and Pak (1971) have recorded intracellular responses which consist of an initial graded spike-like depolarization followed by a sustained small amplitude depolarization. Injection with procion yellow dye showed that the responses came from sites in the lamina, but no clear-cut correlation between response and cell type could be made. Again it is impossible to assess the functional significance of these findings.

15.4. Conclusions

It should now be apparent that our knowledge of lamina function is rudimentary and many investigators disagree over the interpretation of their results. Disagreement results from the lack of clear-cut evidence, as a result of the enormous technical difficulties associated with obtaining any records at all from the diminutive lamina neurons. I will conclude by briefly summarizing the properties of the lamina that I feel are incontestably established and alluding briefly to their function.

(1) Retinula-cell axons transmit visual information to the lamina as graded depolarizations, which are generated by passive conduction of the receptor potential down the axons to their terminals.

(2) Large monopolar cells (LMC) are the only centripetal component of the lamina with response identified. They transmit visual information to the medulla as graded hyperpolarizations.

(3) LMC responses are generated by synaptic inputs from retinula axons terminating in the lamina. All axons terminating in one lamina cartridge have the same field of view and the pathway to the medulla through large monopolar cells maintains acuity.

(4) In *Calliphora* at least, large monopolar cells are laterally inhibited by adjacent cartridges so that acuity is not only maintained but improved.

(5) In the dark-adapted animal the visual signal is amplified when it is transferred from retinula axon to large monopolar cells. The anatomical organization and physiological properties of this pathway show several adaptations that improve the signal-to-noise, ratio and this system functions partly to increase the reliability and sensitivity of low level detection.

(6) Unidentified units, encoding visual information as action potential

frequency, have been found in the chiasma. They reflect the parallel arrangement of visual pathways through the lamina and show lateral inhibition.

(7) In transferring visual information from the receptors to the medulla, the lamina improves the acuity of the system and increases the reliability close to the visual threshold. It also receives a substantial centrifugal input and is ideally situated to control the gain of the visual system in response to the ambient light intensity.

ACKNOWLEDGEMENT

I would like to thank Dr. Ian Meinertzhagen for correcting many of my erroneous anatomical notions and Dr. Randolf Menzel for stimulating discussions on lamina function.

16. The relationship between visual movement detection* and colour vision in insects

W. KAISER

16.1. Introduction

T H E presentation of the topic may strike some readers as unusual: it follows in its broad outlines the historical sequence of events which have contributed to the present state of our knowledge. It is worthwhile to consider occasion-ally the historical development of a topic because it provides us with the understanding that new scientific knowledge is obtained through the continuing development of new research methods and new points of view and that both of these extend beyond the scope of any individual researcher. A historical survey can also lead to the realization that a great deal of successful modern scientific work has its foundations in ideas formulated many years earlier—we should therefore not neglect the study of older literature.

The responses of insects which will be discussed in this chapter are elicited when the entire surroundings of the animal, or a large part thereof, move relative to one (or both) eye(s) of the insect. Such reactions were reported for the first time by Radl in 1903. He established that different insects, when they were rotated together with their substrate in a stationary environment, tried to counteract the rotation by making walking movements. Radl was able to demonstrate that these rotation-induced movements of the animals were elicited via visual pathways, and he interpreted them as compensatory movements. von Uexkuell (see von Buddenbrock 1952, p. 88) and von Buddenbrock and Moller-Racke (1953) probed into the biological signifi-cance of these compensatory movements. They came to the conclusion that

---

* The term 'movement detection' is used here in a broader sense than that employed by von Buddenbrock (1952), who differentiated between reactions caused by displacement of the entire retinal image of the surroundings and reactions in response to the movement of individual objects relative to a stationary background ('true' movement detection). According to our present knowledge, however, we can say that the basic mechanisms underlying the two reactions are not totally independent of one another. It thus appears meaningful to extend the range of the term.

these reflex movements serve to hold the retinal image of the surroundings stationary, in order to make it easier for the animals to perceive moving objects when the surroundings themselves are also in motion (see also Collett and King, Chapter 20 this volume). This concept also included the special theoretical consideration of reflex course corrections in passively-transported flying and swimming animals discussed by Dijkgraaf (1953). He believed that the function of these compensatory movements was to perform course corrections. These movements, elicited by image displacement on the retina, can, without doubt, aid in course stabilization. This supposition gains strength from the fact that flying insects react extremely sensitively to movement of their surroundings.

16.2. Experiments with striped rotating drums made out of pigment-papers

Schlieper (1926) coined the term 'optomotor reaction' and was the first to use this reflex reaction to study colour vision in those arthropods from which conditioned behavioural responses could be obtained only with great difficulty, when at all. The animals, resting on a firm substratum, were placed in a rotating cylinder which consisted of strips of pigment-papers pasted vertically onto a grey background. The brightness of the grey stripes was changed in small steps during the course of the experiment. The first experimental animal was the decapod crab *Hippolyte*, an animal whose ability to differentiate between colours had been inferred from experiments involving adaptive colour-change reactions. Schlieper observed, to his astonishment, that the crab completely ceased to react when the brightness of the grey stripes attained a particular value. In other words, the animals reacted as if they were colour-blind. Nonetheless, Schlieper did not immediately draw the conclusion that *Hippolyte* was colour-blind; instead he proceeded (1927) to perform control experiments by testing the reactions of bees whose ability to perceive colour had by then been established beyond all doubt (von Frisch 1914; Kuehn 1924). The bees *also* reacted in the colour–grey rotating drum as if they were colour-blind; the optomotor responses were extinguished within a narrow range of grey tones. The conclusion arrived at by Schlieper was that colour differences did not play a role in the optomotor response. Nevertheless, optomotor behaviour appeared to be an ideal instrument with which to study the brightness values of different colours for the animals (Schlieper 1928). Using this method, Schlieper discovered, for example, that different genera of butterflies had markedly different sensitivities for red. All of the insect species that Schlieper studied, including the butterflies, reacted similarly to the bees—they were colour-blind with respect to the optomotor response.

Schlieper (1927) explained this seeming colour-blindness by invoking '...various central circuitry of the reflex arcs ...'. von Buddenbrock (1929) came to a similar conclusion in his review of Schlieper's results.

It is of interest to note that vertebrates *show* colour specific reactions in response to moving, equally bright colour–grey patterns (Birukow 1950—see also Reuter and Virtanen 1972; Musolff 1955; Thomas 1955; Trincker and Berndt 1957; Himstedt 1972). The results of the older experiments, in which pigment-papers were used, have gained in consequence through the recent studies of Himstedt, who used spectral lights.

An observation made by Ilse (1932) appeared to throw new light on Schlieper's results: butterflies definitely known to be able to perceive colour made colour-specific approaches towards coloured targets resting on a grey background *only* when there was a difference in brightness for the animals between the target and its grey background.

von Buddenbrock and Friedrich (1933)—inspired by this result—developed a *two*-colour striped rotating drum consisting of combinations of two different pigment-papers. The basic idea behind their method was that the stronger colour contrast (in comparison with the colour–grey drum) should elicit optomotor reactions. If this proved to be the case, then the optomotor response would–in contrast to Schlieper's opinion—indeed be suitable for use in the study of colour vision.

von Buddenbrock and Friedrich used in their drum a pattern of two different pigment-papers, each of which, when previously combined with the one and the same grey tone, had caused extinction of the optomotor response. The pigment-papers chosen in this way apparently were of equal subjective brightness for the animals.

The expectations of the authors were fulfilled: the *two*-colour striped rotating drum elicited an optomotor reaction from their experimental animal *Carcinus maenas*. It was thus possible to establish the presence of colour vision in *Carcinus* and to hope that the newly-developed experimental method would provide insights about the colour sense of many arthropods.

In the following years, the two-colour pigment-paper drum was used to test a large number of insects for their colour sense (Schlegtendahl 1934; Rokohl 1942; von Buddenbrock and Moller-Racke 1952; Moller-Racke 1952; Schoene 1953; Resch 1954).

Schlegtendahl and Moller-Racke found, during the course of their optomotor experiments, insects which were colour-blind and others which were not. Insects such as bees, bumblebees and butterflies, all of which were known to perceive colour, displayed colour-specific optomotor responses. However, these reactions in response to colours of equal brightness were relatively weak in comparison to reactions to patterns with brightness differences (Moller-Racke).

16.3. Spectral characteristics of the photoreceptors in the compound eye

The extremely rapid development of new electrophysiological and optical techniques in the late 1950s and early 1960s brought for the first time much knowledge about the basic mechanisms underlying insect colour vision. Burkhardt (1962-review) and Autrum and von Zwehl (1964) were able to measure the spectral sensitivities of individual retinula cells in flies and bees, respectively. Langer (1967-review) determined the spectral extinction of single rhabdomeres in the eye of the fly *Calliphora*.

Flies possess at least two receptor types, differentiated on the basis of their spectral characteristics; bees have at least three different receptor types (see also Menzel, this volume).

16.4. Behavioural experiments with spectral lights

Daumer (1956), using spectral lights and trained bees, studied the ability of these animals to differentiate between wavelengths within certain regions of the spectrum, including the ultraviolet (u.v.). Another important aspect of this work was his investigation into the ability of bees to distinguish colour mixtures whose spectral composition was precisely defined. Daumer's results were subsequently found to be in accordance with the receptor properties of bee visual cells as determined by Autrum and von Zwehl.

The experiments of Burkhardt and Langer, on *flies*, had established the presence of the peripheral prerequisites for colour vision: several types of receptors with different spectral sensitivities. However, it was still unknown whether flies had the necessary central nervous mechanisms required for colour vision. That this vital piece of information was missing was mainly due to the fact that flies can be conditioned only with extreme difficulty.

What was more obvious than to proceed from the foundations established by von Buddenbrock and Friedrich and by Moller-Racke and develop a rotating drum which permitted the use of spectral lights, including u.v.? There was the distinct hope that insects in such a drum would demonstrate stronger colour-specific reactions than those that could be elicited by the broad reflection spectrum characteristic of pigment-papers. These conditions were met by the apparatus (Fig. 16.1) which I used to test the fleshfly *Phormia regina* (Kaiser 1968). The torque exerted by the animals during fixed flight in response to rotation of the striped cylinder was used as a measure of the reaction. Kunze (1961) had developed this method which enabled extremely precise measurement of the strength of the optomotor reactions.

To my great astonishment, *Phormia* behaved nevertheless as if it were colour-blind: at all wavelength combinations which were tested the optomotor reaction disappeared at a precisely defined intensity ratio between the two coloured lights. This, despite the fact that the animals reacted sensitively to intensity contrasts. The intensity of the stripes in the foreground needed to match the brightness of the constant background (matching intensity) is precisely delineated: even small departures from this intensity can evoke a clear reaction. It is thus possible to obtain highly-accurate spectral sensitivity values from the matching intensities. The ordinate values

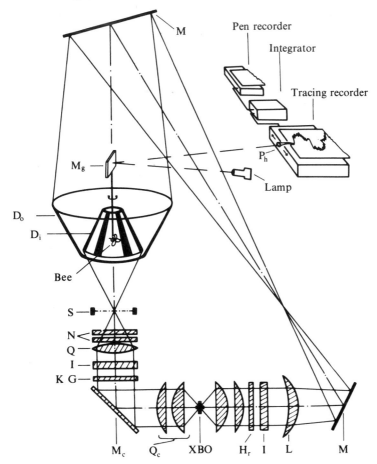

Pen recorder

Integrator

Tracing recorder

Lamp

F<small>IG</small>. 16.1. Schematic experimental arrangement of the two-colour rotating drum used to investigate the optomotor behaviour of bees and flies. D_0 outer wall of drum; D_i inner slotted drum; H_r heat reflecting filter; I interference filter; KG heat absorbing filter; L lens; M mirror; M_c cold mirror; M_g galvanometer mirror; N neutral density filter; P_h photocell; Q quartz lens; Q_c quartz condenser; S shutter; XBO xenon arc. Right-hand beam: standard beam (background light); left-hand beam: test beam. The experimental animal is rigidly connected to the mechanical torque meter. A tracing recorder registers the torque exerted by the flying animal. From KAISER (1968), with alterations.

for the points on the spectral sensitivity curve (*Phormia*, light adapted, Fig. 16.2) are therefore extremely precise; unfortunately the number of wavelengths which could be tested was relatively small.

The results of the studies on the visual cells of flies made it unlikely that *Phormia* is colour-blind. Thus another hypothesis gained in plausibility, namely that the optomotor system of the fly is independent of its colour-vision system; only brightness contrasts play a role in the former. This

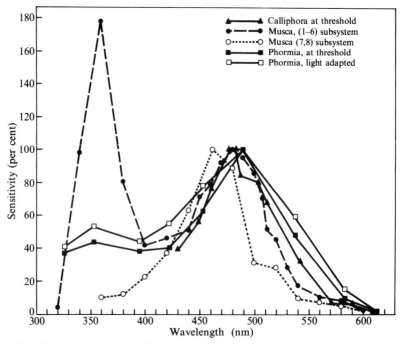

FIG. 16.2. Spectral sensitivity of flies as revealed in studies of their optomotor behaviour. The spectral sensitivity of *Musca* (1–6) was also determined at threshold and is thus to be compared with the curves for *Calliphora* (Schneider 1956) and *Phormia*, at threshold. The *Musca* (1–6) subsystem was studied using a spatial pattern wavelength of 90° (for the spectral region 380–750 nm) and 10·6° (for the spectral region below 380 nm). The (7, 8) subsystem was selectively stimulated with a 3° pattern; the animals then turned in the opposite direction to the pattern (negative reactions). The spatial pattern wavelength in the experiments on *Phormia* was 45°. The curve for light-adapted *Phormia* was determined in the two-colour striped drum using a light intensity which was 8×10^3 times higher than the average intensity at threshold. The curves for *Musca* are from Eckert (1971), those for *Phormia* are from Kaiser (1968).

explanation is also the one chosen in earlier times by Schlieper (1927) and von Buddenbrock (1929), in their attempts to interpret the significance of the absence of colour-specific optomotor reactions in the colour–grey drum.

16.5. The contribution to the optomotor response made by the two receptor systems of the fly

The question then arose: what kind of circuitry could form the basis of such a separation between the optomotor system and the colour vision system? The simplest imaginable situation would be that only one of the two receptor systems, namely the retinula cell system 1–6, controlled the optomotor reaction. The spectral sensitivity of the retinula cells 1–6 is identical (Langer 1967-review; McCann and Arnett 1972).

Support for this hypothesis came from several sources: firstly, the absence of colour-specific reactions in *Phormia* in the rotating-drum experiments; secondly, the spectral sensitivity curve which these experiments yielded is similar to the spectral extinction curve for the 1–6 system which Langer obtained; and finally, the results of Bishop (1969), who had investigated the spectral sensitivity and the discharge patterns of directionally-specific motion-detecting neurons in the optic lobe of flies. It is very probable that these neurons play an important role in the optomotor system (Bishop and Keehn 1967).

Kirschfeld and Reichardt (1970), in experiments concerning the optomotor behaviour of the fly *Musca*, were able to demonstrate, however, that the retinula cell system 7, 8 *also* participates in the optomotor response of these animals: the system 1–6 and the system 7, 8 *simultaneously* deliver their contributions to the optomotor reaction, when the pattern consists of broad stripes of sufficient intensity. The subsystem 7, 8 proved to be sensitive to the *E*-vector direction of linearly-polarized light; the subsystem 1–6 was not.

Eckert (1971), also working with *Musca*, was able to measure the spectral sensitivity of the two subsystems by selective stimulation of the receptor systems (Fig. 16.2). He studied the behaviour of the animals in a rotating drum during fixed flight. The spectral sensitivities were in at least qualitative agreement with the spectral extinction curves of the two systems which Langer had obtained previously using microspectrophotometric methods. Another important aspect of Eckert's work was his experimental proof that the difference in *absolute* sensitivity between the two receptor systems was of the approximate order of magnitude which had been predicted in earlier experiments (Kirschfeld and Franceschini 1968; Kirschfeld 1971): the 1–6 receptor system was 25–30 times more sensitive than the 7, 8 system (the predicted factor was 24–48).

The results obtained by Eckert were essentially corroborated by McCann and Arnett (1972), who recorded from directionally-specific motion-detecting interneurons in the optic lobe of different fly species (Fig. 16.3). The spectral sensitivity curves of the neurons which McCann and Arnett obtained are in general agreement with those which the behavioural experiments of Eckert yielded. There are differences however, with respect to the values for u.v. sensitivity. For reasons which are unknown, it was possible only with *Musca* to achieve activation of the neurons by selective stimulation of the 7, 8 receptor system. Even in *Musca*, approximately only 50 per cent of the neurons responded to stimulation of the 7, 8 system—a fact which indicates that two different neuron types were investigated. Through the use of patterns consisting of stripes of medium width during selective adaptation with long-wavelength light, McCann and Arnett could demonstrate that the two receptor subsystems, 1–6 and 7, 8, normally participate simultaneously in the optomotor response. They thus confirmed the results of Kirschfeld and Reichardt (1970). In presenting the results of their selective adaptation

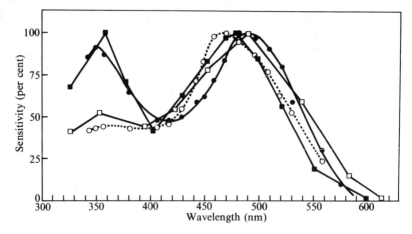

FIG. 16.3. Spectral sensitivity of flies. ■—■: in cells 1–6 of *Calliphora*. This curve is in agreement with the one found most frequently by Burkhardt (1962). ●—●: directionally-specific movement-sensitive neurones in *Musca*, *Calliphora* and *Phaenicia*. This curve was obtained during stimulation with a pattern of 20° spatial wavelength and is supposed to reflect the input of the 1–6 receptor system. O···O: the same movement-sensitive neurones of *Musca*; curve obtained during stimulation with a pattern of 3·6° (in this situation, the input is supposed to come only from the 7, 8 receptor system). □—□: spectral sensitivity of *Phormia*, obtained from colour-matching experiments (behavioural studies) in the two-colour striped drum; pattern spatial wavelength 45°, animals adapted to the light intensity in the drum ('light-adapted'). The curves for *Musca*, *Calliphora*, and *Phaenicia* are from McCann and Arnett (1972); for *Phormia* from Kaiser (1968).

experiments, McCann and Arnett unfortunately give only a few sensitivity values from the visible range of the spectrum (see their Fig. 9). The interesting question, how selective adaptation affects the sensitivities in the near-u.v., thus still remains unanswered.

The authors present an important result as follows: 'In all cases the influence of the (1–6) subsystem greatly dominated, and it was only under very special conditions that the influence of the (7, 8) subsystem could be observed.' This result is in agreement with Bishop (1969).

In a very elegant manner, Kirschfeld (1972) was subsequently able to provide the first direct proof of the participation of the 1–6 subsystem in the optomotor system: by means of successive illumination of two rhabdomeres of the 1–6 type within *one* ommatidium, he was able to elicit optomotor reactions in *Musca*. Up till now, Kirschfeld's experiments using stimulation of single visual cells have been restricted to the 1–6 system. It is to be hoped that analyses of this sort will eventually provide definitive information not only about the size of the contribution made by the 7, 8 receptor system to the optomotor response, but also about possible interactions between the 1–6 and 7, 8 subsystems.

The advances in our knowledge about the functional structure of the optomotor system of the fly outlined in the preceding paragraphs bring the results of my experiments on *Phormia*, performed in the two-colour rotating drum, into new focus. If one assumes that *Phormia* possesses a visual system whose structure and function is not in principle different from that of *Musca* then both receptor systems must have been in action during the two-colour experiments: these experiments were performed using a pattern of broad stripes and the light intensities were greater, by a factor of 8×10^3 and $1 \cdot 6 \times 10^4$, than the average intensities at which the spectral sensitivity at threshold (receptor system 1–6) was measured (Fig. 16.2: *Phormia*, at threshold). The participation of the 7, 8 system at the higher intensities is probably reflected by the higher values in blue and u.v. (in comparison with the curve at threshold). The low u.v. sensitivity of *Phormia* (*cf. Musca*, 1–6 subsystem, i.e. at threshold) is striking. *Phormia*'s sensitivity in the visible region of the spectrum, in contrast, is in seemingly good agreement with that of *Musca* and *Calliphora* (Fig. 16.2).

Despite the fact that both of *Phormia*'s receptor systems were participating in the reaction in the two-colour drum, no colour-specific reactions were found: for all wavelength combinations tested, the response disappeared at a sharply defined ratio of intensities of the two colours, i.e. the response curves passed through zero ('zero-point'; *cf.* also Fig. 16.5). The fact that the response curves of the two-colour experiments passed through zero is of very special significance—not only does it mean that there is no transfer of colour information via the optomotor system, it also restricts to a major extent the range of possible models dealing with the peripheral circuitry of the optomotor system (Kirschfeld 1973). The existence of a 'zero-point', even though both receptor subsystems are in action, makes it likely that signals from both receptor systems go to *one* summation station. A correlator (Hassenstein and Reichardt 1953; Reichardt 1957; Hassenstein 1958; Reichardt and Varju 1959) then receives the summed signals.

The experimental results described here show that both receptor systems participate in the optomotor system of the fly. On the other hand, the hypothesis outlined above, which postulates a separation between the colour-vision system and the optomotor system, is not yet proved. It is still an open question whether flies can actually see colour. The critical test of the hypothesis must therefore be made on an insect which is definitely known to be able to differentiate between different colours. Only the bee satisfies this condition (von Frisch 1914–1915; Daumer 1956; Menzel 1967; von Helversen 1972). In the bee it should also be easier to recognize the participation of the individual receptor types than is the case for the fly, since there are clear-cut differences between the spectral sensitivities of the bee's retinula cells (Autrum and von Zwehl 1964; Autrum 1968).

The results of Schlegtendahl and Moller-Racke (see earlier) may appear to the reader to make a test of the hypothesis on bees redundant, since these authors demonstrated the presence of colour-specific optomotor reactions in bees in the two-colour pigment-paper drum. Recent experiments on the optomotor response of flies (*Musca:* Fermi and Reichardt 1963; *Phormia:* Kaiser 1968) and bees (Kunze 1961) made it clear that even small brightness contrasts could elicit very distinct reactions from these animals. I considered it therefore essential to check whether the previously observed 'colour-specific' optomotor responses of the bees were in fact responses to small brightness differences. Such a precise test would be permitted by the use of a pattern composed of various monochromatic colours, because the intensity of spectral light can be graduated in much finer steps than the brightness of pigment-papers.

16.6. Optomotor response studies using spectral lights : bees

The first extracellular recordings from interneurons in the optic lobe of the bee were made by Kaiser and Bishop (1970), who found four types of directionally-specific movement-sensitive interneurons, having different preferred directions of motion. Their investigations made it likely that these neurons belong to the optomotor system. It thus seemed worthwhile to look at the reaction of these fibres in response to stimulation with a moving two-colour striped pattern. Parallel experiments on the optomotor *behaviour* of the animals under corresponding conditions were begun. The principle underlying the optical setup in my electrophysiological experiments on bees is exactly the same as that developed for, and employed in, the experiments on *Phormia:* stripes illuminated by one monochromatic beam move in front of the homogeneous background of a different colour (Fig. 16.4). It is thus possible, for example, to present the bee with a pattern consisting of ultra-violet stripes on a yellow background. The result of these experiments is surprising (Kaiser 1972*a,b*)—the neurons cease to react to movement of the pattern when the brightness contrast between the two colours is eliminated (Fig. 16.5). The matching intensity is once again very sharply defined. The reciprocals of the relative numbers of quanta required to achieve matching between foreground and background yield a very precise spectral sensitivity curve. The similarity of this curve to the green-receptor curve (Autrum and von Zwehl 1964) cannot be overlooked (Fig. 16.6). The green-receptor curve represents the results from intracellular recordings made from retinula cells of worker bees. It is thus clear that the input to these interneurons comes mainly from the green-receptors. The sensitivity of these neurons in the short wavelength region of the spectrum is also in agreement with this conclusion, because the green-receptor of the bee—as was already demonstrated by Goldsmith (1960)—is also sensitive to ultraviolet light. It is still an open question whether this reaction of the neurons includes a small contribution from the blue and/or u.v. receptors which have been found in the bee.

Naturally, the entire optomotor system does not necessarily have to

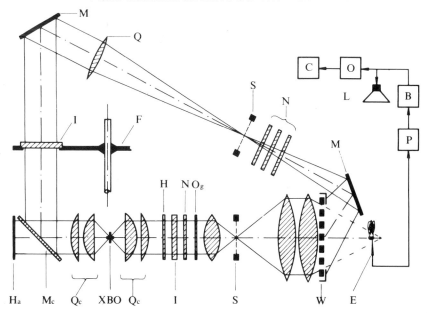

FIG. 16.4. Apparatus used to investigate the reactions of the bee's directionally selective movement-sensitive inter-neurons to two-colour patterns. The receptive fields of the neurons are contralateral and monocular; they extend over the whole eye. The preferred direction of the neurons is back → front. There is also a small ipsilateral input with a front → back preferred direction. B bandpass filter; C electronic counter; E extracellular recording microelectrode; F filter wheel; H heat reflecting filter; H_a heat-absorbing plate; I interference filter; L loudspeaker; M surface mirror; M_c cold mirror; N neutral density filter (quartz); O oscilloscope; O_g opaque glass; P preamplifier; Q quartz lens; Q_c quartz condenser; S shutter; W pattern wheel: a large wheel with slots and spokes of equal breadth, the bee could only see a circular area at the periphery of the wheel extending over 20° of the visual field of the animal's right eye; XBO xenon arc.

behave exactly like the neurons. Other cells with a strong input from different (blue and/or u.v.) receptors may also participate in the system. The way in which the whole system reacts can only be clarified by behavioural experiments. These were performed using the apparatus shown in Fig. 16.1 (Kaiser and Liske 1972, 1974): the bees react sensitively to brightness contrasts; colour contrasts, on the other hand—even between colours which are exceptionally well differentiated in training experiments—apparently play no role (Fig. 16.7).

The response curves (response as a function of brightness contrast) pass through zero (minimum response value ≈ 0). This result applied to all the wavelength combinations used. An example of a critical test of the 'zero-point' is provided by Fig. 16.8. The wavelength combination 394/509 nm is also extremely well differentiated in training experiments. Nevertheless, the

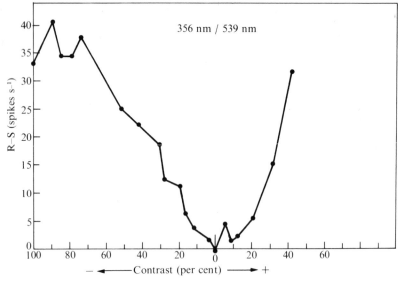

FIG. 16.5. Response (R–S) of a directionally-specific movement-sensitive neurone of the bee as a function of the brightness contrast of a two-colour pattern (wavelength combination as indicated). R number of spikes counted during the 10 s period of pattern movement; S number of spikes counted during the immediately preceding 10 s period of spontaneous activity. It is assumed that minimum responses occur only then, when both the foreground and background have the same subjective brightness for the animal (contrast 0 per cent, 'zero point'). The contrast is calculated by $(a-b)/(a+b)$, where a is the intensity of the foreground corresponding to the particular response and b is the foreground intensity corresponding to the minimum response. Negative contrast: the foreground appears darker than the background, for the animal; positive contrast: foreground brighter than the background.

reaction to this combination disappeared at a particular intensity ('contrast = 0 per cent') and remained at zero during the entire experiment, despite the fact that the bee reacted persistently to strong contrasts. The contrast sensitivity of the whole animal is probably greater than that of single interneurons. This assumption arises from the fact that more than one neuron (at least two monocular movement-sensitive neurons) participates in the optomotor response; this could result in an improvement in the signal-to-noise ratio. (In the behavioural experiments the pattern was located in the frontal visual field of *both* eyes; it was offered only monocularly in the electrophysiological experiments.) The assumption appears to be confirmed by the first results of a comparison between the gradients of the response curves obtained from the whole animal and from the neurons.

The very precisely determined spectral sensitivity of the optomotor behaviour of the bee is shown in Fig. 16.9: once again there is an amazing similarity to the green-receptor curve. This result, obtained under conditions

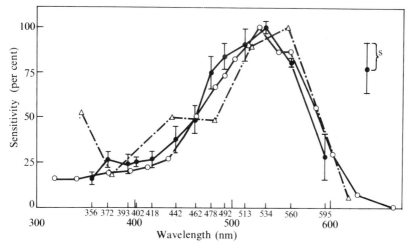

FɪG. 16.6. Spectral sensitivity of the bee's motion detectors: △—·—△: determined by measuring neurone response latencies after coloured flashes from a stationary light source (Bishop 1970). ●—●: average of four curves from four neurons of four different animals studied in the author's colour-matching apparatus (see Fig. 16.4). O—O: for comparison, intracellularly measured green-receptor sensitivity (Autrum and von Zwehl 1964).

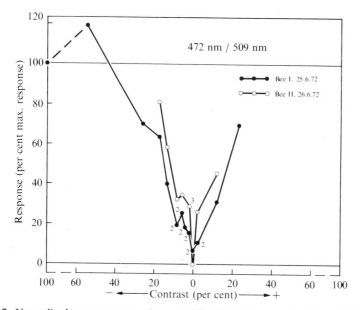

FɪG. 16.7. Normalized average response (torque) of fixed flying bees as a function of the brightness contrast of a two-colour pattern. For the calculation of contrast, see legend, Fig. 16.5. Wavelength combination as indicated. The two curves represent data taken from two animals. The numbers next to the points indicate the number of individual measurements contributing to an average response value.

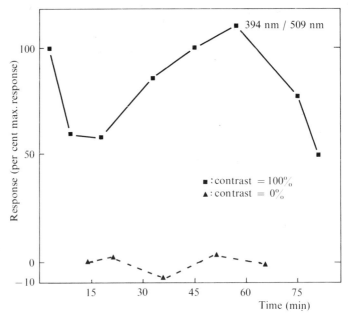

F IG. 16.8. Optomotor response of one bee at the matching intensity (contrast 0 per cent) as a function of time after the start of the experiment, compared to the control response (contrast 100 per cent). The control response values were normalized with respect to the value of the first control response (100 per cent). The response values at the matching intensity were normalized to the preceding control response (i.e. each control value was given the value 100 per cent). This normalization compensated for fluctuations in the animal's inclination to react—the fluctuations are evident in the control response curve. Measurements (not shown) were also made at contrast values other than those shown on the graph.

very similar to those in the electrophysiological experiments, indicates that all the motion-detecting neurons which participate in the optomotor system have identical visual input.

The extremely close similarity between the green-receptor curve and the spectral sensitivity of the optomotor system shown in our experiments could be attributed to the fact that in both of our experimental series, we stimulated a portion of the eye which contains only green-receptors. It is known from the work of Gribakin (1969, 1972) that there is a non-uniform distribution of the different receptor types over the eye. Gribakin divides the bee's eye into a ventral portion where there are six green and two u.v. receptors per ommatidium and a dorsal portion in which each ommatidium contains four green, two blue, and two u.v. receptors. The function and connections of the ninth cell are not clear, but it definitely does not belong to the green receptors.

In each of our series of experiments, we stimulated the central zone of the eye. One would in principle, therefore, expect a contribution from all three receptor types, with that of the more frequent green receptor dominating.

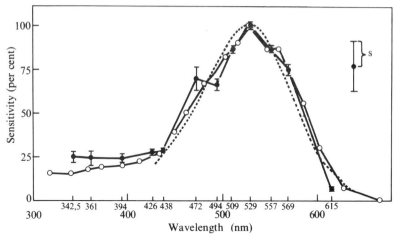

FIG. 16.9. Spectral sensitivity shown by the bees in their optomotor behaviour (●—●). The sensitivity is calculated from the inverse of the relative numbers of quanta needed to achieve matching (minimum response) in the two-colour drum. O—O: spectral sensitivity of the bee's green-receptor; • • • •: the Dartnall nomogram for a visual pigment whose λ_{max} = 530 nm. (The latter curves are from Autrum and von Zwehl 1964.)

The extensive agreement between our sensitivity curves and the green-receptor curve thus implies that—under our experimental conditions—the contribution coming from other receptor types is either very small or non-existent.

If it should turn out that the motion detectors which have been discovered to date in the bee only receive input from the green receptors, then an astounding analogy to the situation in the mexican ground squirrel will have been revealed. The visual system of this mammal, which possesses a pure-cone retina, was investigated by Michael (1968). He recorded from single units in the optic nerve and found that the movement-sensitive neurons were connected to only one type of receptor, namely the green receptor. The blue receptor exerted no influence on the movement-sensitive cells. Colour vision has been demonstrated in the mexican ground squirrel (Jacobs and Yolton 1971), but no experiments have been performed to date on the behaviour of movement-sensitive neurons or of the whole animal in response to stimulation with a two-colour striped pattern.

The lack of colour-specific reactions in the optomotor behaviour of the bee implies that the optomotor system of this insect is separate from the colour-vision system. Both systems, howev r, share at least one receptor system—the green-receptor system. The colour-vision system has in addition at least two other types of receptors at its disposal—the blue and the u.v. receptors, which actually play a dominant role in training experiments (Fig. 16.10).

There are indications that the separation between the colour-vision system and the motion-detecting system is a general principle among the insects. Experiments involving an aspect of insect behaviour seemingly far removed from the optomotor response also provide evidence for the validity of this principle.

Tinbergen, Meeuse, Boerema, and Varossieau (1942) (see also Collett and King, Chapter 20) studied the courtship behaviour of certain butterflies. Animals known to perceive colour behave as if they are colour-blind during courtship: only the brightness contrast and mode of movement of dummies are essential for the elicitation of the pursuing flight from the males. This observation shows that this first step in courtship is controlled by a movement-detection system which in its turn is 'colour-blind'.

The results of our experiments confirmed the supposition referred to above: the bees, in the pigment-paper drums of Moller-Racke and Schlegten-dahl, had in fact reacted, not to colour differences, but rather to small differences in brightness which arose from imperfect matching of brightness between the papers. As Fig. 16.7 shows, the optomotor behaviour of bees is very sensitive to brightness contrasts.

That Schlieper failed to find 'colour-specific' optomotor reactions in his colour–grey drum is probably due to the fact that grey pigment-papers can be graduated more finely with respect to their brightness.

Some of the experiments performed most recently on bees make a very important contribution to the problem of the participation of different receptors in the optomotor system of these animals. Kirschfeld (1973) investigated the optomotor behaviour of fixed *walking* bees in response to stimulation with a moving 'pattern' consisting of polarizing films having various polarization directions. Such 'polarization patterns' elicited optomotor reactions only when illuminated from behind with light of short

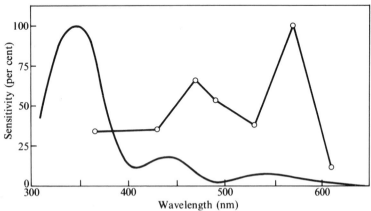

FIG. 16.10. Examples of different spectral sensitivies of bees as shown in different behavioural situations and under different conditions. ———: conditioned feeding behaviour (von Helversen 1972); O—O: phototactic behaviour (Sander 1933).

wavelengths. Long-wavelength light was ineffective; it could, however, elicit reactions when an intensity pattern was present. Kirschfeld concluded, on the basis of his results, that *in additon to* the green receptor system (which is polarization insensitive), a polarization-sensitive blue and/or u.v. receptor system particpates in the optomotor response. This conclusion seems thoroughly compelling, but on the other hand it appears to conflict with our results. Kirschfeld's results are supported by the work of Bishop (1970) who, on the basis of experiments on the directionally-specific motion-sensitive neurons (see earlier), came to the conclusion that all three receptor types of the bee are involved in the optomotor system. Bishop obtained the spectral sensitivity of the motion detectors by latency measurements after stimulation with light flashes from a stationary light source (Fig. 16.6). The difference between the results of Bishop and Kirschfeld on the one hand and those obtained by us (Kaiser, Kaiser and Liske) remains unexplainable at this time. For the time being we can only offer a choice of likely explanations.

Possibility A. The blue and u.v. receptors only come into play in the optomotor system at higher light intensities, in spite of the fact that threshold for the receptors themselves is already reached at low intensities (Thomas and Autrum 1965). Kirschfeld and Bishop would have then experimented at higher intensities than we had used. This hypothesis will be tested by carrying out experiments at higher intensities.

It is not to be expected, however, that *colour-specific reactions* would indeed appear at higher intensities. The intensities which we used in our behavioural experiments were with certainty even greater than those which Daumer (1956) employed in his training experiments (Daumer, personal communication). Daumer had found that his bees could distinguish exceptionally well between colours.

At this point, mention must be made of a strange phenomenon which is revealed upon careful reading of the literature concerning the spectral sensitivity of the bee. There are very obvious differences apparent in the various spectral sensitivity curves, even for one and the same behavioural response. For example, Sander (1933) determined the spectral sensitivity shown by bees in their phototactic behaviour (Fig. 16.10). The curve he obtained indicated that the green and blue receptors were active; the contribution made by the u.v. receptors was at the very best small. Bertholf (1931) and Heintz (1959)—even though they also studied the phototaxis of the bee—found a very high u.v. sensitivity relative to the sensitivity in the green range; the blue receptor, in contrast, appeared to make no contribution.

von Helversen (1972) also found a very high u.v. sensitivity when studying the conditioned feeding behaviour of bees (Fig. 16.10), whilst Thomas and Autrum (1965), who studied the conditioned feeding behaviour of more or less dark-adapted bees, obtained a spectral sensitivity curve, which was similar to Sander's, but had three peaks. In light-adapted animals Thomas and Autrum also obtained a curve with three peaks, but with the dominant maximum in the blue region.

The physiological mechanism responsible for all these various results is still only very poorly understood.

Possibility B. The green receptors may exert an inhibitory influence on the participation of the blue and u.v. receptors in the optomotor system. Since, in all our experiments, our constant background light was green, the green receptors were stimulated continuously. This objection can be refuted by the results of Menzel (1973). Menzel determined the spectral sensitivity of the movement-sensitive neurons of the bee by using coloured stripes on a black background as stimulus. A response–intensity curve was determined for each wavelength used. The spectral sensitivity curve which he derived from the response–intensity curves is also very similar to the green-receptor curve. In addition, the individual response–intensity curves do not show wavelength-specific differences.

Possibility C. The spectral sensitivity is dependent on which region of they eye is investigated. Gribakin, as mentioned above, provides some evidence favouring this view. However, Kirschfeld and Kaiser and Liske performed their experiments using approximately the *same* region of the bee's visual field—the medial frontal fields of both eyes. Thus it is clear that regional differences in receptor distribution cannot account for the different results.

Possibility D. The circuitry underlying the optomotor walking response is different from that for the optomotor flight reaction. This does not apply, however, to flies: evidence is available for the activity of *both* receptor systems in flying animals. The critical experiment is to test the optomotor walking reaction of fixed bees using a two-colour striped pattern as stimulus.

Possibility E. A contribution from the blue and u.v. receptors is present, but is so small in comparison to that of the green receptors, that it only becomes apparent in the overall spectral sensitivity under very special conditions.

Possible experiments to test this hypothesis include: (1) measurement of the spectral sensitivity during adaptation with light of long and short wavelengths. Experiments of this kind performed to date have not indicated a change in sensitivity, but they must be continued further; (2) increasing the blue and u.v. sensitivity of the bees by conditioning them with a mixture of these colours; (3) recording from the movement-sensitive neurons during stimulation with polarization patterns and short-wavelength light, i.e. under the conditions employed by Kirschfeld (1973), and thereby measuring the spectral sensitivity at different intensity levels.

The results of these critical experiments are eagerly awaited, since they should provide us with information about which types of receptors participate in the optomotor response and the nature of their involvement.

The question concerning the relationship between motion detection and colour vision in insects, implied in the title of this chapter can, however, already be answered for the bee, and in all probability for other insects capable of perceiving colour: there are obviously two *separate* systems responsible for colour vision and motion detection.

ACKNOWLEDGEMENT

The optomotor response studies with bees, using spectral lights, were supported by the Deutsche Forschungsgemeinschaft.

I would like to thank Miss E. Kuhn for technical assistance, and for the preparation of the figures. I am also very grateful to my wife, Jana Steiner-Kaiser, for her constant co-operation and valuable discussions throughout the work, as well as for the translation of the manuscript.

17. Information processing in the insect compound eye

R. B. NORTHROP

17.1. Introduction

In 1959, Lettvin, Maturana, McCulloch, and Pitts published the results of a unique study in which frog ganglion cell responses were examined when the eye was stimulated with a variety of simple visual objects presented on the inside surface of a matte grey hemisphere concentric with the eye. In their study, the frog retina was treated as a 'black box' with about 10^6 parallel inputs from receptor cells, and single outputs selected at random from $c.$ 0·5 million ganglion cells. They discovered that ganglion cell responses could be categorized into four major groups or 'operators' which detect or respond maximally to particular aspects of the visual object, such as local sharp edges and contrast, curvature of the edge of a dark object, movement of edges, and local dimming produced by movement or general darkening.

The great advance made in this work was the attempt to find the natural or best stimulus for each type of unit by exploration of its pattern-filtering properties. An influence upon studies of arthropod visual units is apparent from a comparison of early 'on-', 'off-', and 'on-off' responses (Burtt and Catton 1960) with the later work on *Podophthalmus* (Waterman, Wiersma, and Bush 1964a,b,c) and on the locust (Horridge *et al.* 1965). In the later work it is clear that the authors struggled with the problem of finding natural stimulus situations for unknown units, and at the same time tried to characterize them quantitatively with some measurable feature of the stimulus. Waterman *et al.* found evidence for afferent responses caused by simple components of visual patterns, such as those observed in the frog. They found several examples of *Podophthalmus* optic tract units which were active when small dark objects in a light field were presented to the eye. These units were less responsive to larger dark objects in a light field, and target shape was not a factor in the response. All receptive fields (RFs) were relatively large (30–180°), and some had irregular boundaries. Vector movement units responded maximally for certain values of target speed and direction. Other on-off, on-sustaining, and novelty units were noted. Evidence suggested that movement sensitivity could involve a release of sustaining units from inhibition. Visual information obtained by one eye is apparently relayed to the optic ganglia of the opposite eye, and some units transmit visual information in both directions, i.e. afferent *or* efferent impulses. The crab's optic ganglia also receive mechanoreceptor information, as recorded by Horridge *et al.* (1965) in the locust

Studies by Burtt and Catton (1960), Ishikawa (1962), Schiff (1963), and Dingle and Fox (1966), of single-unit responses in the optic lobes of locusts, silkworm moths, mantis shrimp, and crickets, respectively, made use of simple on and off of general illumination for stimulation. Predictably, units were found which responded to on, off, on and off, as well as sustaining units responding to continuous light or dark. Elementary operations on object form and motion were described by *Locusta* by Horridge *et al.* (1965), who did not investigate differential form sensitivity, e.g. as between spot sizes, but outlined the responses of some twenty classes of units. Many of these were still described only by their on-off behaviour, and form and/or motion-sensing properties were not identified for all, which makes it difficult to fit them into categories described by other workers. The most consistent type of unit found in these and other studies in Orthopteran, Lepidopteran, and Dipteran visual systems, in spite of the variety of stimuli used, has been the directionally-sensitive or vector response unit, discussed in Section 17.2.2.

17.2. Categories of units

The primary difficulty in making a comprehensive survey of insect optic lobe units is that no two researchers have used the same stimuli or investigated the same properties. Hence, in many instances there is available a detailed but still incomplete list of unit properties which makes uncertain their comparison between insect orders. Yet in spite of this difficulty there emerge broad patterns which may be used to define units.

The classification used in this paper is shown in Fig. 17.1. It is similar to a binary-decision tree. To avoid 'fuzzy' criteria, the decision boundaries are based on the presence or absence of a particular property in a unit rather than a point on a graded scale. The first branch in the classifier is based on an observation by Northrop and Guignon (1970) that some optic lobe units are form-discriminating and some are not. The ones that do not sense form have been found in general to have monocular RFs of limited extent. These include the on-type and tonic L-units in *Romalea* which have narrow fields as small as *c.* $10° \times 30°$, and net dimming (D) units which did not have definable RFs. Horridge *et al.* (1965) reported on units with RFs as small as $7°$ in *Locusta* (type BC), net dimming units with $15–25°$ RFs (type BD), and net dimming units with $60°$ graded RFs (type BE)

The form-discriminating units in the optic lobes respond maximally for particular object shape. This form preference is generally manifested as a preference for either spots or linear boundaries. Units responding preferentially to spots include Multimodal units in *Romalea* and locusts, and the

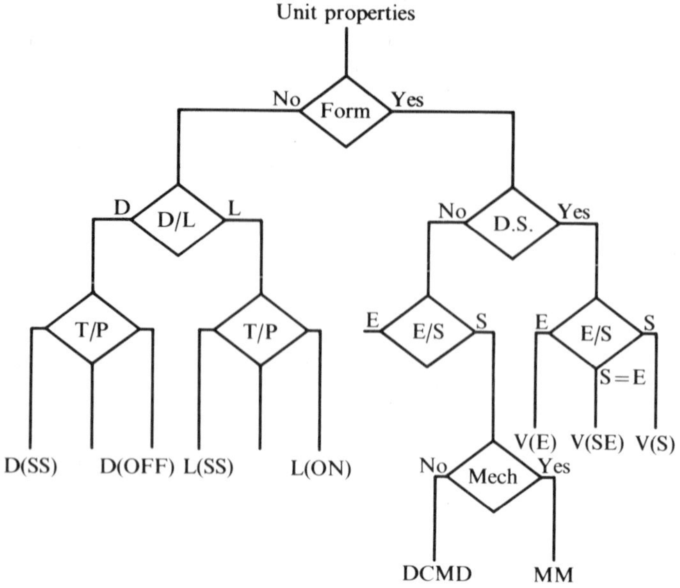

Unit properties

FIG. 17.1. A classification tree for insect visual units. The diamonds are decision points. As an example of classification, a DCMD unit is form-sensing, not directionally sensitive, prefers spots, and is not sensitive to other stimulus modalities. It also has a number of other distinctive characteristics such as its on/off response and its spatio-temporal habituation properties, which have not been included in the tree. Abbreviations are: MM = multimodal; DCMD = descending contralateral movement detector; DS = directional sensitivity; V(E) = vector edge; V(S) = vector spot; V(SE) = vector spot and edge; D = dimming or dark; L = light; (SS) = sustaining or tonic response. A rectilinear *versus* rotational vector distinction has been omitted from the figure. In the situation where there is a graded mixture of response types (e.g. ON response and tonic firing in the light in L units) the decision blocks have three outputs; $T > P$, $T = P, T < P$.

descending contralateral movement detector neurons (DCMD) in *Romalea* and locusts (Rowell 1971*a*). Edge-specific responses have been obtained in a number of species from vector (directionally-sensitive) units, and also from the static edge detectors in the fly (type I reported by McCann and Dill 1969). Only vector units, by definition, sense the direction of object motion. Classification of vector units has been made more difficult by the discovery in the lobula of *Romalea* of examples which respond equally well to spots or stripes (Northrop unpublished observations). Guignon (1974) has found a purely spot-sensing vector unit in the ipsilateral circumesophageal connective of *Romalea* which failed to respond to stripes with the preferred vector.

In the classification scheme of Fig. 17.1, no attempt has been made to include branch points based on units' on/off responses. These are important properties which can be used as an additional verification of most classes of unit in terms of their instantaneous frequency.

In identifying five classes of optic lobe unit in *Romalea*, Northrop and Guignon (1970) found that a unit's class could be determined with a great deal of certainty from a detailed inspection of the instantaneous frequency profile in response to on and off of general illumination. For example, Multimodal units gave on and off bursts, fired irregularly in steady-state light (LSS) or dark (DSS), sometimes faster in LSS. Vector edge units gave a very characteristic off burst, no on response, and fired more regularly in DSS although at about the LSS rate. Light units gave varying degrees of on response, no off response, and were usually silent in DSS. Net dimming units had on/off behaviour complimentary to the L-units. However, to determine identity of an unmodulatable and weakly sensitive unit other tests had to be used.

Units found more recently in *Romalea* optic lobes (Northrop and Guignon unpublished observations) include DCMD and Vector spot-sensing units. The former have on/off behaviour similar to Multimodal units, except they fire more slowly in LSS and DSS. The latter give a short burst at on, followed by a period of reduced firing, and otherwise are similar to Vector edge units.

It is evident that the on/off responses can provide a shortcut to identification of units, but only after unique behaviour has been defined in terms of form and motion responses.

It is significant that form-sensing units in insect optic lobes require object motion; even the type-I static edge sensors of McCann and Dill lose response a few seconds after presentation of the stationary object. It is possible that form-detection is an integral part of motion detection, as suggested by the model for vector unit behaviour shown in Fig. 17.11, although there is no reason why form-discrimination could not be separate from and precede the directional-sensing operation.

In the following sections, detailed descriptions of the properties of four major categories of optic lobe unit are given: DCMD, vector, multimodal, and non-form-specific units. Most attention has been given to the very thoroughly studied DCMD system because of its enigmatic role in the insect central nervous system and the fact that it possesses anomalous spatial resolution (Burtt and Catton 1962, 1966, 1969) and spot size preference (Palka 1967). The anomalous resolution (which has been measured down to a stripe period of $0.3°$) probably is the result of neural interactions in the optic ganglia because there is no evidence that optical diffraction images, which have been clearly shown to exist in model systems and dissected eyes, in fact occur in the intact locust eye (Horridge 1971).

17.2.1. *Descending movement detector (DCMD) neurons and anomalous resolution*

17.2.1.1. *Introduction.* Visually-elicited responses from the giant fibres in the Orthopteran ventral nerve cord (VNC) have been examined by a number

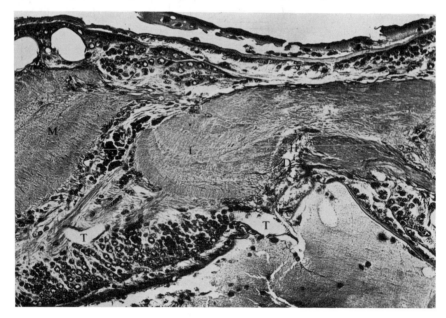

FIG. 17.2. A DCMD Unit Recording Site in the ipsilateral optic lobe of *Romalea*. The RF-lesioned DCMD unit fibre is one of a group of three silver-stained axons at D in the coronal section. L = lobula; M = medulla; T = trachea; H = hole caused by electrocoagulation.

of workers. In this section, we will focus our attention on the locust's DCMD neuron. This is a purely-visual monocular unit characterized by its large spike amplitude which may be traced electrophysiologically from the lobula region in the ipsilateral optic lobe through the tritocerebrum, across and down the contralateral oesophageal connective to the VNC where it proceeds as far as the third thoracic ganglion. A radio-frequency lesioned DCMD unit in the ipselateral optic lobe of *Romalea microptera* is shown in Fig. 17.2. The writer was able to record simultaneously from these units in the optic lobe with microelectrodes and in the contralateral VNC (between the first and second thoracic ganglia) with hook electrodes. The transmission is apparently without synapse, and has a delay of four to five ms.

A comprehensive 'Characterization and Review' of DCMD neuron properties has been given by Rowell (1971*a*), including habituation, dishabituation, and arousal, and DCMD-unit responses to the directions of object motion (Horn and Rowell 1968; Rowell and Horn 1968), and the output function of the DCMD system (Rowell 1971*b*). The 'type CF' unit described by Horridge *et al.* (1965) in the optic lobes of *Locusta* may also be a DCMD unit.

17.2.1.2. *Anomalous resolution.* One of the most interesting properties of the DCMD neuron was described by Burtt and Catton (1962) who showed that the DCMD unit would give a significant response to moving, rectilinear, black–white stripes with periods down to *ca.* 0·3°. This figure for stripe angular period at threshold has been questioned on the grounds that it exceeds the Nyquist limit for spatial sampling by roughly a factor of 6·67 in the vertical plane of the locust eye, and also because it is beyond the 'resolving power' of the apertures of the individual ommatidia, as determined by classical optical theory (Rayleigh criterion). The anomalous resolution of the DCMD system is also contradicted by the measured directional sensitivity function (DSF) of individual retinula cells. Hypotheses were advanced claiming that DCMD unit anomalous resolution was in fact due to imperfections in the pattern (spatial subharmonics) (McCann and MacGinitie 1965), or was due to an 'edge effect' (between stripes and their window) which extended the spatial high-frequency response to an anomalous value (Palka 1965).

Burtt and Catton (1966, 1969) countered the criticisms of their earlier experiments by using a precision, rotating 'wheel' pattern, viewed through

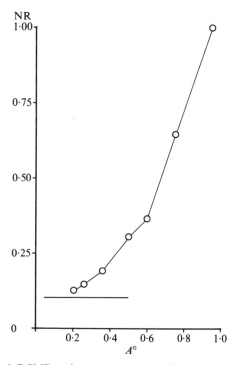

FIG. 17.3. Normalized DCMD unit responses to rotating checkerboard pattern viewed through an annular window. Other stimulus parameters are the same as in Fig. 17.5. Confidence (C) that data points are significant above the blank mean: $A = 0.205°$, $C > 95\%$; $A = 0.26°$, $C > 90\%$; $A = 0.36°$, $C > 97.5\%$.

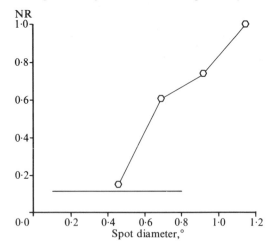

FIG. 17.4. Normalized DCMD unit responses to a single jittered black spot inside a large Grey hemisphere. 1 min counting intervals were used. The confidence that the DCMD unit responded to the jitter of the 0·46° diameter spot is better than 99·5% (one-sided t-test, 9° freedom).

an annular mask. This object had no relative edge motion, and was claimed to be sufficiently precise to render spatial subharmonics negligible. The period of the stripes at the edge of the wheel at limiting resolution was again found to be *ca.* 0·3°.

The writer has also verified the 0·3° period limit of resolution in DCMD units in *Romalea* and *Schistocerca*. In addition to a precision, radial striped pattern, a checkerboard pattern (also rotated in the annular window used for the radial stripes), and a single jittered black spot object inside a 50 cm diameter fibreglass hemisphere with a matte grey surface were used to test anomalous resolution. The radial stripes and the jittery spot tests typically gave threshold stripe periods of 0·35–0·45° and threshold spot diameters of 0·35–0·45°, respectively, for significant DCMD responses during object motion. The one-sided Student's t-test was used to test the hypothesis that a response sample mean was significantly greater than the blank mean. Results are given in the figure captions, as per cent confidence that the response mean was greater than the blank mean for critical values.

Figure 17.3 shows the response of a *Schistocerca* DCMD unit to the rotated checkerboard pattern. A 30 s interstimulus interval was used. The basic repeat distance of the checkerboard was 0·1824 cm; it was viewed through an annulus (i.d. = 8·2 cm, o.d. = 14·5 cm) in a grey board. Typical threshold periods ranged from 0·2–0·3° for this object, which was by far the most effective stimulus used. The results of a jittery spot test are shown in Fig. 17.4. In this experiment, a spot, which is a powerful stimulus for DCMD units, was manually jittered by a hand-held magnet in an erratic path on the

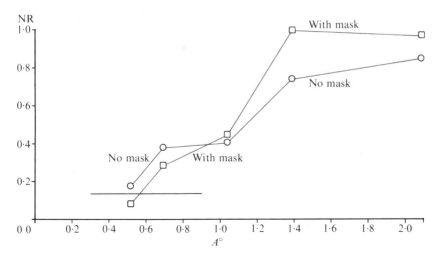

FIG. 17.5. DCMD unit responses to a rotating radial pattern. Rotation was 90° alternating clockwise (CW) then counterclockwise (CCW) at 0·643° sec^{-1} with a 30 s i.s.i. Each data point represents the normalized average of the responses to ten alternate CW and CCW movements. Circles: responses to pattern viewed through annular window. Squares: responses to pattern viewed through an eight-spoke mask over the annular window. Horizontal bar: mean of blank runs for fifty samples of the same durations as when the pattern was moved. Confidence that the data points are significant above the blank mean (one-sided t-test): With mask, $A = 0·695°$, $C = 99·5\%$; without mask, $A = 0·52°$, $75\% < C < 80\%$.

inside surface of the hemisphere. Because of spatial habituation, the object's path and speed was constantly varied by the experimenter to maximize the audible DCMD unit firing rate. The experimenter effectively closes a feedback loop in this test. This procedure was found to yield surprisingly consistent results because of the closed-loop conditions and the long stimulus presentation time (one-minute).

DCMD-unit resolution to a precision radial pattern was examined to study the 'edge effect' upon the response threshold. Data points are normalized cumulative spike counts during pattern rotations. Background was estimated by counting spikes before each rotation in the same dwell time as the rotation. Figure 17.5 shows a locust DCMD unit's responses to radial pattern motion with and without a black sector mask which overlay the annular window. The mask had eight equally-spaced 10°-wide spokes which reduced visible pattern area by 22·2 per cent.

The significant result illustrated in Fig. 17.5 is that the mask acts to *increase* the DCMD unit responses for pattern periods over 1°, but causes poorer resolution for stripe periods less than 1°. I have interpreted this as evidence that the edge effects are not associated with the basic operation of the DCMD unit near threshold, but do augment DCMD responses when the object's fundamental spatial frequency is below the Nyquist cut-off frequency for the eye. That is, the object, and the window edges can be unambiguously resolved as such and this information can be relayed to higher processing centres.

When a striped pattern is moved before the eye, the intensity contrast in locust retinula cells agrees approximately with that predicted by assuming a Gaussian model for the angular sensitivity (Tunstall and Horridge 1967). Intensity contrast is defined as:

$$C = \frac{I_{max} - I_{min}}{I_{max} + I_{min}} = \frac{\Delta I_{pk}}{\bar{I}},$$

where I_{max} = maximum 'effective light intensity' in the receptor giving rise to a retinula cell depolarization, V_{max}. I_{min} = minimum effective light intensity in the receptor giving rise to a retinula cell depolarization, V_{min}. $\bar{I}$ = mean effective light intensity (time average). ΔI_{pk} = peak change in effective light intensity above or below caused by motion of object.

Measured values of receptor depolarization were converted to effective intensity values using the retinula cell's intensity–depolarization curve $V = f(I)$.

Tunstall and Horridge's experimental results show a higher value of contrast C for small stripe periods than predicted by the Gaussian DSF model. They gave the hypothesis that this departure from the predicted curve was due to the fact that the measured locust DSF has a sharper peak than a Gaussian DSF with the same half-angle $\Delta\rho$. This hypothesis is in agreement with incoherent Fourier optical theory, and also serves to illustrate the point that the Gaussian DSF model is perhaps more of a mathematical convenience than a 'best-fit' to measured data.

The data given by Tunstall and Horridge in their Fig. 10 show that the intensity contrast C is $c.\ 0.035$ for a light-adapted eye when the stripe period is $c.\ 2°$. Extrapolation of this curve shows that C would be negligibly small for the DCMD threshold of $0.3°$!

The problem which must be resolved is how can these very small changes in I (and V) cause a DCMD response? It is clear that the neural events which lead to DCMD firing must involve the processing of more than one retinula-cell output. In the locust the rhabdomeres of an ommatidium are fused, and the eight retinula cells have the same field, i.e. they look in the same direction. All eight retinula axons project in a bundle to a single cartridge of the lamina; two of these axons run right through the cartridge to the medulla (Horridge and Meinertzhagen 1970). Whether all six short axons converge upon any one of the lamina ganglion cells of a cartridge is not known from physiological data, but the recordings by Shaw (1968) certainly suggest a convergence in some degree upon the largest lamina monopolar cells.

In a multiplicative signal processing (MSP) model developed by the writer it is shown that the parallel transmission of information by axons of one ommatidium, and its linear summation by a second-order lamina neuron, can result in improved spatial resolution. The linear summation of receptor potentials is assumed to be followed by an exponential transfer characteristic in the coupling to a subsequent neuron in the medulla. It is not necessary to invoke any theory of optical superposition or diffraction images to explain anomalous resolution. The MSP model uses purely neural interactions. In any event, anomalous resolution is seen to be only one of many interesting properties of the DCMD system, the most important of which are summarized later.

17.2.1.3. *Object size optimality.*   Palka (1967) showed, using luminous stationary stimuli to the locust eye that increasing the size of a dimmed spot increased the DCMD unit response up to a certain point, and then a further increase in area reduced it. He found that the weaker the stimulus intensity, the larger the optimal area, also, that stimulation has simultaneous excitatory and inhibitory effects which depend on the stimulus intensity. The inhibitory effect decays more slowly following a stimulus. This is directly illustrated by the fact that an ongoing response can be attenuated by a second weak stimulus, even if the latter is subthreshold.

Using *Romalea*, I have found that size optimality for black disc objects has a very broad maximum, and is variable between animals. Even so, we may say that a 5° spot given a controlled motion will generally elicit more DCMD spikes than a 0·5° or 50° spot given the same motion. We have found that the unit responds to the motion of white spots on a grey background approximately one-half as strongly as it does to black spots on the same field, the contrast being about equal in both cases. Discs, squares, or triangles of the same area are equally effective stimuli in the optimal size range, but no systematic quantitative study of the effectiveness of object shapes has been made.

My experiments have shown that a 20° black disc moved in the grey object hemisphere is a far more potent stimulus for DCMD units than a 9 × 180° black stripe pivoted at the poles of the hemisphere. The stripe will elicit a small burst when initially moved, and the unit will ignore it thereafter unless given a rest of over a minute. The spot will continue to cause large bursts as long as it is moved over a fresh part of the eye. Repeated movement of the spot in one area leads to a local spatial habituation, a well-known property of the DCMD system.

17.2.1.4. *Inhibition from multiple objects.*   Palka (1967) found that dimming of a large stationary spot will terminate or prevent the response of a DCMD unit to that from a small object just previously dimmed. The writer has found a parallel phenomenon in the study of DCMD behaviour in *Romalea*. The rapid slewing of a 9 × 180° black stripe in the display hemisphere would cancel or strongly attenuate the unit's response to a jittered 5° black disc in the centre of the same hemisphere. The stripe motion was coincident with, or just after the spot motion for this inhibition to be observed.

Rowell (1971*a*) has shown that DCMD responses to the motion of normally effective objects are suppressed by whole-field movements of the eye. The effect is graded, however, and fades for low angular velocities of eye motion. It appears that this effect lasts only for the duration of the whole-field or large object motion. In examining this effect, Rowell used a complex and relatively undefined background which included a portion of the laboratory and the pen recorder used to move the object. This object and its surrounding environment defy a quantitative description in terms of spatial frequencies, and probably contained linear contrasting boundaries which could activate the inhibitory mechanism of the unit as the eye was scanned past them. The quantitative relationships between inhibitory object size, velocity, and position in relation to the stimulus object parameters remain to be worked out.

17.2.1.5. Directional preferences.   Burtt and Catton (1966, 1969) observed that DCMD units often show a preference for either clockwise or counter-clockwise rotation of the radially-striped disc object used to test spatial resolution. This was especially noted near the threshold of resolution. Rowell (1971*a*) has observed that preferences for directions of linear motion of an object also exist. Rowell and Horn (1967) found that directional preference may vary over the receptive field, which includes the entire contralateral eye.

A most unusual property sometimes exhibited by the DCMD system, described by Horn and Rowell (1967), is called *primacy*. Primacy behaviour is defined as the unit showing a response preference for the *first* direction of object movement which excited it after a long rest interval, given alternate object displacements back and forth along the same path. Horn and Rowell have proposed a 'two-cell feeder model' to describe primacy which makes use of a long-lasting reciprocal inhibition of two directionally-sensitive (vector) units which have opposite preferred directions and which both act excitatorily on the output neuron. Because of the long time constant of inhibition, the first vector operator to be excited in its preferred direction produces a large response and at the same time inhibits the second vector operator, which, when stimulated in its preferred direction produces less output due to the long-lasting inhibition from the first one.

Another 'three-cell feeder model' for primacy proposed by Horn and Rowell uses two vector operators with opposite preferred directions, and also a non-directional movement sensor, all three of which excite the DCMD output unit. In this model, primacy results from habituation in the vector units' responses to stimuli in their preferred directions, and habituation in the general movement-sensing 'C' neuron. In an alternating sequence of object motions, the C unit habituates faster than vector units 'A' or 'B'. Horn and Rowell showed that this habituation can lead to either maintained directional sensitivity, primacy, primacy and directional sensitivity, or neither effect, depending on the magnitudes of the basic responses of cells A, B, and C, given equal rates of habituation for A, B, and C.

One difficulty in accepting the models of Horn and Rowell lies in their use of vector operators as components in the DCMD system. The writer has found (see Section 17.2.2) that vector unit directional sensitivity exists for objects with periods which exceed the Nyquist period for spatial sampling in the eye. The directional preferences of DCMD units have been noted for striped objects with periods between the Nyquist limit and threshold (Burtt and Catton 1966), where, presumably, ordinary vector units are incapable of responding selectively to the direction of object motion.

Because directional preferences in DCMD unit responses appear to be randomly-oriented, inconsistent, and often poorly-defined, it is safe to conclude that these units do not code or transmit information about the direction of object motion (this is the job of the vector units). Their primary function appears to be to provide information about the novel movement of small, convex dark objects anywhere in the entire, monocular receptive field of the unit.

17.2.1.6. *Purpose of the DCMD system.*   The efferent connections to and the purpose of the DCMD system have yet to be clarified. Rowell (1971*b*) has examined DCMD responses in fixed and free-roaming locusts with chronically-implanted electrodes and has found several factors which affect the over-all state of activity of the DCMD system. There was no direct correlation between the behaviour of free-moving animals and DCMD unit activity, and a relationship is also lacking between activity on metathoracic nerve IIIC and the DCMD units (Rowell 1971*a*). Rowell also has shown that the flight motor output is unaffected by DCMD unit activity. He formed the hypothesis that DCMD firing can initiate a shift of control in the thoracic ganglia, shifting this system from local, 'housekeeping' tasks to following inputs sent in by descending command fibres.

17.2.2. *Vector units*

17.2.2.1. *Introduction.*   Units which respond maximally for visual objects moving in a preferred direction have been found in a number of vertebrate and arthropod visual systems. In this discussion of the properties of these units, we will focus our attention on Vector responses found in Othopteran, Dipteran, and Lepidopteran optic lobes and nerves.

17.2.2.2. *Vector units in grasshoppers and locusts.*   Northrop and Guignon (1970) have recorded from single Vector edge units in the optic lobe medulla in *Romalea microptera*. These form-specific neurons fire at the same rate in steady-state light and darkness, although firing in the latter is more regular. There is a sharp burst at off, followed by a relatively long period of reduced firing after which the unit recovers to its steady-state rate. There is no effect at on, unless on occurs during the period of suppression following off. In this case, the unit accelerates abruptly to the steady-state rate. This class of unit is monocular and its RF involves the entire ipsilateral eye. Mechanical stimulation of legs and body has a very slight accelerating effect. The unit fires more rapidly than its base rate when a contrasting long straight, boundary is moved in the preferred direction (usually anterad). The basic unit rate is suppressed if the dark half of the boundary is anterior and the edge is moved in the anti-preferred direction. No such suppression occurs if the white half is anterior during this motion. Unit acceleration during preferred motion is independent of light–dark orientation, and only depends on direction and speed of motion (hence it is Vector). These units respond to single stripes (white on black or black on white) moving with the appropriate Vector. When multiple stripes are used, the Vector response adapts with continuous motion, and the suppression effect is noted for anti-preferred motion. It, too, adapts. Vector units are extremely sensitive to very small incremental displacements of linear contrasting edges, of $\frac{1}{3}°$ or less. A small abrupt movement of the edge often evokes a higher-rate burst than a more gross motion. Any slight, over-all dimming also evokes an abrupt burst. Vector edge units exhibit form discrimination. They do not respond to moving, small spots, and give only a slight response to a large spot moved in the preferred direction.

The preferred directions of the ipsilateral, medulla Vector edge units in *Romalea* range from anterio-dorsad to anterio-ventrad, and the response of a unit is distributed with an approximately cosine characteristic around its preferred direction. Vector edge units respond over a very wide range of object speeds; there is no sharp optimum. After an initial adaptation, Vector edge units maintain response to continual motion of stripes in the preferred direction, or in the case of the back-and-forth motion of a single $10° \times 180°$ stripe, actually show an increase in response as the motion continues in a $\pm 90°$ arc (high firing for preferred motion, deep suppression for anti-preferred motion).

The *Romalea* Vector edge units studied by Northrop and Guignon (1970) are probably most closely represented in *Locusta* by the 'class BG' units described by Horridge, *et al.* (1965).

The writer has found another distinct class of Vector unit in the optic lobes of *Romalea*. These are called Vector spot units. All such units which have been lesioned were found to be in the lobula or among the giant fibres running from the lobula to the photocerebrum. These units are identical to the Vector edge units found in the medulla of *Romalea* with the following exceptions.

(1) They are more strongly (bilaterally) mechanosensitive than Vector edge units. (2) They give a short, sharp bust at on, followed by a relatively long period of reduced firing, unlike their medullar counterparts which are practically insensitive to on. (3) They respond well, in a Vector sense, to contrasting, small spots (as well as to stripes) given motion in the preferred direction. There is virtually no adaptation to the spot response, which persists throughout the preferred motion. Anti-preferred motion of spots gives suppression of background firing.

17.2.2.3. *Projections of Vector units in the locust, and their spatial acuity.* Studies of Vector edge units in the medulla of *Romalea* and in the third cervical nerve of *Schistocerca gregaria* lead to the conclusion that these units do not possess anomalous spatial resolution as do DCMD units. In fact, they appear to function in a mode where motion acuity is traded off for form-discrimination of objects, because they can sense displacements of an edge or stripe in their preferred directions as small as 10 min.

Thorson (1966*a,b*) has made detailed measurements of the spatial and temporal parameters governing locust optomotor response in the form of isometric neck torque. Thorson's (1966*a*) Fig. 9 shows, for two animals, that pattern periods smaller than *ca.* 3° produce negligible neck torque in yaw. The cut-off for roll was closer to 4° in his Fig. 8. In order to examine the neurophysiological basis for these optomotor responses and to examine the spatial resolution of the head movements system, Northrop, Burtt, and Catton* recorded from the cervical nerves innervating neck muscles involved in the locust's head-turning reflex; details are given later.

17.2.2.4. *Methods and materials.* Adult locusts (*Schistocerca gregaria*, Forskal) of either sex were used in the experiments without preliminary anaesthetization. After removing the wings, legs, and antennae, animals were fastened securely to a perspex platform, dorsal side down, using Cenco tackiwax. The animal's left eye had an unobstructed view for about a hemisphere of solid angle. The right eye and the right and median ocelli were covered with opaque black wax. The animal mounting platform was attached to a universal ball joint, and could be pivoted to align the axes of the eye with the object, as desired.

Recordings were made from the ipsi- or contra-lateral third cervical nerves (N_3) which originate from the subesophageal ganglion and innervate the posterior neck muscles (Albrecht 1953). The throat of the locust was exposed by tilting the head dorsally *c.* 40° and securing it firmly with tacki-wax. The soft membrane covering the ventral neck was carefully removed with fine iridectomy scissors.

* The work described in this section was performed in collaboration with Drs. E. T. Burtt and W. T. Catton at the University of Newcastle-Upon-Tyne during the 1970–71 academic year. Support is gratefully acknowledged from the U.S. Air Force Office of Scientific Research Grant AFOSR 68-1539C, and from The Science Research Council (U.K.).

Fine platinum wire hook electrodes were used. A nerve was first lifted free of haemolymph on the electrode, then gently surrounded with a mixture of white vaseline and paraffin oil extruded from a hypodermic syringe. The nerve was then cut distal to the recording electrode. In all preparations, the VNC was cut bilaterally at the recording site. The vaseline–oil insulation generally permitted stable recording for at least 48 h, and in one exceptional case, for as long as four days.

Keithley model 103 amplifiers were used with 100–3 kHz bandwidth and gains of 10^3. Spike trains and cue record were stored on magnetic tape using a Hewlett-Packard 2900C instrumentation recorder. Single units were isolated with a precision pulse-height window. The discriminated pulses (window output) were counted by an electronically-gated digital counter which had a counting dwell time adjusted to slightly exceed the duration of stimulus motion. In any experimental run using a known constant object velocity, the counter dwell time was of course held constant; the value was the same as in the blank runs (with no object motion) used to sample spontaneous unit activity. The duration and extent of object motion were selected to give 'maximum response' for a given type of unit. These variables were optimized in an approximate manner, the main interest being the investigation of how object size affects unit response.

The N_3 Vector (motor) units were stimulated with relatively-large black-and-white striped objects viewed through various square windows having widths which were integral and non-integral multiples of the angular stripe period A. Square windows were also rotated with respect to the stripe axis to test edge effects on N_3 vector unit resolution. The striped patterns were carefully constructed and found to have no detectable subharmonics by the pinhole test described later.

Rectangular striped patterns were mounted on light rigid perspex sheets which were, in turn, firmly attached to the penholder on the cross-drive of a rectilinear, potentiometric chart recorder (Kelvin Servoscribe). This recorder had a maximum slewing speed of c. 200 mm s^{-1}. Its loop gain was adjusted to give critical damping with the increased mass of the object attached, to prevent overshoots. The recorder drive signal was obtained from a precision, triggerable ramp generator. The extent of object motion was usually restricted to an integral number of stripe periods.

In order to check for the existence of possible spatial subharmonics in the various periodic objects used, a simple spatial low-pass filter was made by putting a minute hole in a piece of aluminium foil. As the observer backed away from the moving pattern looking at it through the pinhole, a progressive blurring due to attenuation and loss of high spatial frequencies was noted, until only uniform grey was seen. All of the periodic test objects used in our resolution experiments passed this 'pinhole' test for uniform fadeout.

Patterns were illuminated from a 12 V 48 W lamp. This lamp was used at 15 V, giving an intense white light which was adjusted just to cover the test object. Illumination measured using a Type 48A Holophane Lumeter was 5000 lx, and was essentially constant over the object area.

Data was presented graphically in normalized form. The maximum, total response (generally obtained for the closest object) for five (or ten) trials was divided into the total responses for all other five (or ten) trial sets for a given experimental run.

The one-sided Student's t-test was used to test the hypothesis that a response sample mean was significantly greater than the blank mean. Results have been expressed as per cent confidence that the response mean was greater than the blank mean, or as probabilities that the hypothesis is wrong. Note that $C > 95$ per cent is equivalent to $P < 0.05$ in discussing significance, etc.

17.2.2.5. *Spatial resolution of units in the third cervical nerve* (N_3). Recordings from N_3 with both VNC connectives intact and both N_3s intact showed a predominance of responses elicited when body-surface mechano-receptors or the aerodynamic hairs on the head were stimulated. For example, giant fibres in N_3, giving spikes in the order of -4 to -8 mV, responded vigorously to a puff of air on the wind-sensing hairs on the ipsilateral side, and less strongly to the same stimulus directed to the contralateral side of the head. These and other interfering N_3 responses were eliminated by cutting the VNC bilaterally. and covering the aerodynamic hair bed surfaces with tackiwax during the mounting procedure.

Once these steps were taken to suppress N_3 mechano- and aero-dynamic responses, a definite hierarchy of purely visual responses was revealed. The lowest amplitude N_3 units (typically $-400 \mu V$ peaks), called tonic As, were non-visual; that is, their rate was constant regardless of the state of illumination or object motion. There were probably two or three tonic As active in each N_3. All units larger than the tonic As in N_3 were found to be directionally sensitive Vector units responding maximally to long linear contrasting boundaries moved in the preferred direction. These units are presumably motor units innervating the dorsal posterior neck muscles (Albrecht 1953), and no doubt are involved in visually-elicited head movements.

As a rule, we found that the larger the peak amplitude of the N_3 vector unit, the coarser was its spatial resolution and the more velocity dependent it was. D-units as large as -10 mV pk. were observed in a few fresh N_3 preparations. These large-amplitude vector units ceased to respond 0.5–1.0 h after the preparation was started. On the other hand, the tonic A-units, and the two next largest vector classes, the B- and C-units, typically maintained a consistent response for about 24 hours.

All ipsilateral N_3 vector units studied were found to respond maximally to posterio-dorsal (PD) stripe motion with an approximately cosine characteristic. Contralateral N_3 C-units were found to have an anterio-ventrad (AV) preferred direction, acting in a complimentary fashion to the ipsilateral

N_3 C-unit response. Other contralateral N_3 vector units were not examined in detail in this study, so we can not state whether they, too, have an AV preference for moving stripes.

Stripe resolution for a typical complementary pair of N_3 C-units is shown in Figs. 17.6a,b. The threshold for motion resolution for preferred stripe motion was found to be a smaller object angular period than the point where

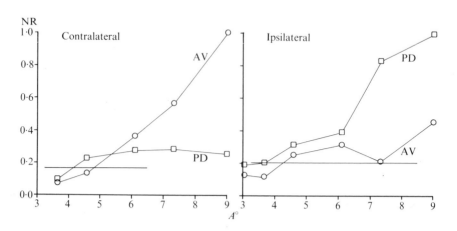

FIG. 17.6. Resolution of contralateral and ipsilateral cervical N_3 C-units. AV and PD rectilinear stripe motion was viewed in a square-window six-stripe periods on a side and 0° tilt. Note the strong vector response of the contralateral unit to AV stripe motion in (a) while the corresponding unit in the ipsilateral N_3 has a PD preferred direction in (b). Data points are normalized average responses to five trials each. A 30 s i.s.i. was used. The horizontal lines show normalized blank means. Spatial resolution for movement detection fails at $A_t = 4.5°$, and AV/PD discrimination appears to fail at $A_d = 6°$ for both ipsi- and contra-lateral C-units in this preparation. ○, PD motion; □, AV motion.

response to AV equalled that for PD stripe motion; that is, the system lost the ability to discriminate AV from PD motion, but could tell that motion occurred. We designated the object angular period where *directional* discrimination is lost, A_d, and that angular period where an ambiguous response is no longer detectable above chance firing, A_t. In both Figs. 17.6a and b, A_d is c. 6°; A_t is harder to estimate, but examination of the results for two preparations show it to be c. 4.5° for both the ipsi- and contra-lateral case.

Estimation of A_t and A_d where the data points approached the blank mean value asymptotically was found to be made easier at small values of A by plotting the data on logarithmic graph paper (Fig. 17.7). A_t for the ipsilateral N_3 C-unit is estimated to be c. 4.5°; A_d is c. 5.5°. A phasic D-unit from the same N_3 had $A_d = A_t \approx 9.0°$ for the same object parameters. Note that there was no N_3 D-unit response to AV stripes, or any spontaneous firing of this unit, which is typical of the class.

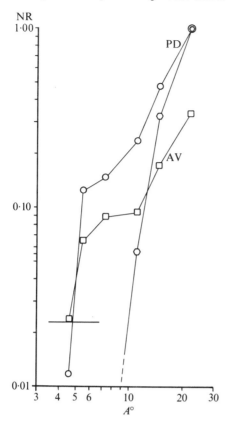

FIG. 17.7. log–log plot of normalized ipsilateral N_3 resolution of moving rectilinear stripes. Stripes were viewed in a square window 2·5 periods on a side with 0° tilt. Circles: C-unit responses to PD stripe motion. Squares: C-unit responses to AV stripe motion. Hexagons: D-unit responses to PD stripe motion (D refers to size of unit, see text). The D-unit had zero blank mean and did not respond at all to AV stripe motion at any velocity. Horizontal line: normalized blank mean for C-unit. Stripes were moved two periods in 1·4 s with a 20 s i.s.i.

B-units from another ipsilateral preparation gave the response shown in Fig. 17.8. B-units in general had a higher spontaneous rate than C-units, and appeared to be more sluggish in their response. $A_d \approx 7·0$ and A_t appears to be *c.* 1·05° by a lengthy extrapolation. The anomalous rise in response as the object angular period approaches A_t was noted in several other N_3 B- and C-unit runs. The reason for this phenomenon is not known.

12.2.2.6. *Window edge effects on N_3 C-units.* Because edge effects with rectangular stripes have been observed to cause enhanced responses for DCMD units (Palka 1965), we examined the vector C-unit responses as a function of stripe window tilt to see if a similar response enhancement was taking place. Figure 17.9 shows data representative of our findings. Clearly,

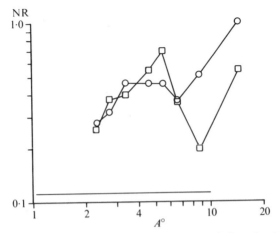

FIG. 17.8. log–log plot of normalized ipsilateral N_3 resolution of stripes. Circles: B-unit responses to PD stripe vector. Squares: B-unit responses to AV stripe vector. Horizontal line: Blank mean of normalized B-unit. Note the odd peak in the response at $A \approx 5.5°$ following loss of PD/AV discrimination at $A_d \approx 7°$.

C-unit responses are enhanced by window edges being parallel to the stripes. (Stripe periods are well above the Nyquist limit.) This edge effect causes an increase in vector resolution, A_d, of almost $1°$.

Figure 17.10 shows spike records from vector units in the VNC and the third cervical nerves. While the vector discrimination threshold, A_d, is well

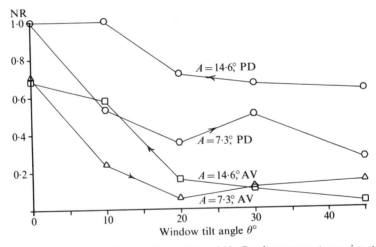

FIG. 17.9. Effect of window tilt angle θ on ipsilateral N_3 C-unit responses to moving stripes. Stripes were viewed in a square-window 3·25 periods on a side. Circles: normalized responses to PD vector, $A = 14.6°$. Squares: As circles, but AV vector, Hexagons: Normalized responses to PD vector for $A = 7.3°$ in same window. Triangles: As hexagons, but AV vector. Normalized blank mean was 0·049. Stripes moved 2 periods in 1·4 s with a 20 s i.s.i.

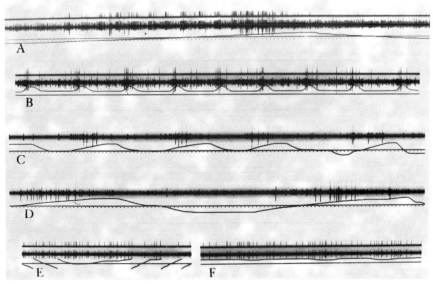

FIG. 17.10. Vector responses from the VNC and third cervical nerves in *Schistocerca*. Stimulus in each case was a 9° black stripe moved by hand inside the surface of a fibreglass display hemisphere. The hemisphere was painted gray, and the stripe was pivoted at the 'poles' of the hemisphere. (A) A contralateral VNC vector unit with PD preferred direction, recorded between the suboesophageal ganglion and the first thoracic ganglion (T1). Top trace is pulse-height window output, middle trace is raw data, bottom traces are a reference with 100 ms markers and a stripe position cue trace. Upward deflection of the cue is PD motion. Full span of the cue trace is *c.* 180° arc. Note modulations in the vector response due to stepwise stripe motion. (B). An ipselateral VNC vector unit responding maximally to a dorsad stripe motion. Trace arrangement as in (A) but time markers were omitted. The stripe was swept over the dorsal half hemisphere. (C) An ipselateral third cervical nerve D-unit responding to fast, dorsad sweeps of the stripe over about one-half the hemisphere. Note the C-unit firing which accompanies the giant unit response. (D) The same unit and preparation as in (C) but PD vector. Maximum unit sensitivity appears to be in the dorsal posterior quadrant of the eye. (E) and (F) Same preparation as in (C) but −750 μV C-unit is being selected by the pulse-height window. Its response is to very small PD movement of the 9° stripe. In (E) the jumps in the cue trace are 1°. The same scale is used in (F) showing that the unit detects movements in the preferred direction of 0·1° or less.

above the Nyquist limit for stripes, some vector units can resolve minute motions of a large object of less than 0·1°. This is shown in Fig. 17.10E and F.

17.2.2.7. *Discussion.* The N_3 vector units are of higher order than the medullar vector units reported by Northrop and Guignon (1970). The most important difference in these two classes of Vector units is that the N_3 units respond to aerodynamic organ and mechanical stimulation as well as to the visual input of moving edges. The aerodynamic response is a very strong one, and one has but to blow at the face of a locust (held in a position so that it can use its wings) to observe, in addition to the expected flight reflex, a dynamic positioning of the head and antennae. In certain individuals, this may be observed as a distinct bobbling of the head during the period air is directed at the head.

The N_3 vector C-units also give bursts at off and on of general illumination, while medullar vector units respond only to off with a burst, and do not sense on. Medullar vector units exhibit suppression of their average rate of firing when stripes are moved in the anti-preferred direction. Anti-preferred object motion elicits a weaker, *positive* response from N_3 C-units than does preferred object motion, and little suppression is evident.

We have noted a number of cases where N_3 B- and C-unit responses lost their vector discrimination at a stripe angular period well above the value where response is significant above noise; that is to say, the units respond at low A values but fail to give directional information. As the DCMD-unit resolution is *c.* $0.3°$ at cut-off, it is evident that information about novel movement is available in the optic lobe in the region between loss of N_3 vector response, and no N_3 response to movement.

17.2.2.8. *Vector units in diptera.*   Bishop and Keehn (1967) recorded from optic lobe units in *Musca domestica, Eucalliphora lilaea,* and *Calliphora phoenicia.* Directionally-selective and directionally-non-selective motion sensitive units were found in the medulla–lobulla region of all three species. Both types of units responded to on and off of general illumination; at on, there was a burst of activity which decayed to background; at off, there was a burst followed by a period of reduced firing which then returned to background. The visual fields of these units were *ca.* 2π steradians. Preferred directions were described by Bishop and Keehn for one class of fly vector unit in terms of three orthogonal vectors, motion along which would produce excitation: Right → left, up → down, and near → far (along body axis). Contrary motions produced inhibition of firing. The other class of true vector unit described by Bishop and Keehn responded to object-motion vectors which were the negatives of the first set. The authors correlated their recorded neural responses with fly optomotor responses to the same type of striped objects, as measured by McCann *et al.* (1965), Fermi and Reichardt (1963), and Götz (1965), including the optomotor reversal behaviour at a $3.6°$ stripe period in *Musca.*

A more-detailed quantitative study of fly optic-lobe unit responses for *Calliphora* and *Musca* was made by Bishop *et al.* (1968). The type II units which were recorded in the lobula–lobula plate region were broken into two major subcategories depending on whether they had entire contralateral (IIa) or ipsilateral (IIb), monocular visual fields. Preferred directions were listed as: IIA_1, 'Approximately horizontal, periphery to inward'; IIa_2, 'Vertical, downward'; IIa_3, 'Horizontal, inward to periphery'; IIa_4, 'Vertical, upward'; IIb_1, 'Horizontal, inward'; IIb_2, 'Horizontal, outward'. Among the salient features of fly vector-unit behaviour noted by Bishop *et al.*, was that 'short periods of motion in alternate directions caused an enhancement of both stimulation and inhibition as compared to their corresponding steady–state velocity responses'. This phenomenon has also been observed by the writer in *Romalea* vector-edge units, and by Northrop, Burtt, and Catton (unpublished observations) in locust cervical nerve-3 preparations.

McCann and Dill (1969) extended the quantitative study of fly optic lobe units. They defined twenty-one categories of form and/or motion sensing units of which the class I and II units will be mentioned here. Their class II units included eight members consisting of two groups of four for contra-lateral and ipsilateral fields. Four major orthogonal vectors were defined for the two fields. The eight, type II (vector) units had in addition to the pre-viously-described vector behaviour, the ability to sense the 'form' of a stationery object (a 62° disc with two half-stripes in it). This meant that the units would fire more rapidly when abruptly presented with the object than when presented with a disc of the same average intensity but with no stripes. In a form-sensing vector IIa$_1$ unit (such as shown in McCann and Dill's Fig. 6c), the form-sensing response component lasts *ca.* 2·3 s, dying out roughly exponentially toward the background rate which the disc alone would produce. Type I units observed by McCann and Dill also sensed form on a transient basis, but had relatively narrow fields (10–40° diam.), and did not respond to the direction of motion.

Mimura (1971) has shown, using a moving 1·15° bright spot object as a stimulus, that vector units in the lobula of *Boettcherisca peregrina* do not appear to have linear preferred directions, but instead are keyed to respond maximally to a clockwise or counterclockwise (circular) motion of the spot through the binocular field of the animal. Linear motion of the spot in any portion of the entire field also gives an apparent vector response, but leads to inconsistencies because, for example, a L→R spot motion in the ventral central field will give a strong response, while a L→R motion in the dorsal field will give an anti-preferred response. Only circular motion (in the case cited, CCW) produces continuous firing for 270° [*cf.* Mimura (1970), Fig. 1]. Mimura also found the more conventional (linear) vector response in vector units recorded in the medulla.

Mimura speculates that the rotational vector units may be used by the fly for fly roll control in flight. This seems a very reasonable hypothesis, and should lead to some interesting studies on flight-control mechanisms in flies.

None of the workers cited above have tested vector units for form dis-crimination; that is, there may be an optimum object shape and size to elicit maximum responses. For example, how would the rotatary vector units of Mimura respond to a wheel pattern, such as used by Burtt and Catton (1966) to examine locust visual activity?

17.2.2.9. *Vector units in lepidoptera.* Classic vector responses from 'fast dark units' have been recorded from the lobula–medulla region of the optic lobes of the hawk moth, *Sphinx ligustri*, by Collett and Blest (1966). Moving stripes with a period of 22° were used as a stimulus. Preferred directions for either eye were predominantly anteriad or posteriad; some dorsad and ventrad preferred directions were noted too. Units were truly binocular, the trend being that a preferred direction for the ipsilateral eye would be 180° from that for the contralateral eye, although a few units were observed to have the same vectors for both eyes.

Collett (1971) reported on further studies on hawk moth vector units, giving attention to responses recorded in the right anterior optic tract. These are claimed to be output cells from the optic lobe units. The optic tract vector units responded only to *anteriad* object motion, giving roughly the same output to black or white spots, bars, or edges, both when the leading edge was dark and when it was light. Receptive fields were plotted using a 2° spot on the display hemisphere, and were found to be *c*. 30° in diameter with fuzzy boundaries, although some were reported as large as 90° in diameter. They were all situated in the antero-ventral quadrant of the eye.

Collett (1971) investigated the effects of relative motion of a 30° period grating and a stimulus bar on optic tract vector unit responses. Anteriad motion of the grating had the effect of reducing vector unit response to anteriad bar motion in the ipsilateral eye. Backward motion (anti-preferred) of the grating often enhanced the response to bar motion anteriad. Collett states that 'with no stimulation at the field centre, a surround grating moving in the null direction often suppressed the resting rate'. This is expected, because the field of these vector units is very broad, and there may be sufficient form discrimination in the units so that responses to gratings' edges are significantly stronger than to the 2° spot used to map the field. Collett found that forward motion of a grating presented to the contralateral eye enhanced the response to bar movement in the preferred direction (anteriad to ipsilateral eye), particularly at low or subthreshold bar speeds. Posteriad motion of the grating suppressed the response to the bar. Collett found further that there was evidence for a variable 'gain' in a postulated velocity control system for visual tracking which was input dependent. He considered the motor output of the variable-gain controller to be vector unit, tracking neurons'.

Swihart (1968) has recorded several types of single units in the optic lobes of the butterfly *Heliconius erato adanis*. In the medulla, he found vector responses that lay predominantly along the anterior-posterior axis, and a few along the dorsal-ventral axis. Some units 'preferred' bright objects rather than dark objects. Swihart did not give a detailed description of the objects he used, except an observation that a black, 3×5 in. file card at *c*. 10 ft was an effective stimulus for one vector unit when moved 1–2 in. (This card subtended $1 \cdot 4° \times 2 \cdot 3°$ at 10 ft.) This is clearly a vector response to a spot, or a relatively high spatial frequency object, providing there was no additional lower spatial frequency vector input, such as edges associated with a stick or arm used to move the card.

17.2.3. *Multimodal units in insect optic lobes*

These units are probably the largest and most frequently encountered system in the optic lobes of insects when recording with extracellular, metal microelectrodes. They have been described in detail in *Romalea* by Northrop and Guignon (1970), in *Locusta* by Horridge *et al* (1965) (Type CE), and in moths by Blest and Collett (1965).

In *Romalea* these binocular units fire erratically in steady-state light (LSS) or darkness (DSS). The LSS rate is usually slightly greater than the DSS rate. There is always a burst at on and off of general illumination (sometimes the off burst is larger). Multimodal units respond with varying degrees of adaptation to the motion of black *or* white spots moved anywhere on the matte grey inside surface of the object hemisphere. Maximum response is evoked when spot motion is abrupt and changes direction often. Continuous smooth motion in one direction does not give a sustained response. A small (3·3°) spot (disc, square, or triangle) gives a greater response than larger ones given the same motion. Repeated jittery spot motion in one area causes the response to adapt, but moving the spot to a new area and jittering it yields a refreshed response which, too, adapts. A moving stripe elicits a burst only at the onset of motion. Direction of motion of spots and stripes is not important. The RFs apparently cover both eyes. Multimodal units also fire in a burst when any part of the body surface is gently stroked with a small paintbrush. This mechanosensory input is generally rapidly adapting, although some units show a good deal of latency to brush stroking and a persistent response. Most multimodal units also respond to acoustic stimuli.

Multimodal unit visual responses are qualitatively similar to those of DCMD units. At one time the writer suspected that DCMD units drove the multimodal visual component; however, simultaneous recording of a DCMD unit in the VNC and an ipsilateral multimodal unit in the optic lobe, under conditions of threshold stimulation, has shown that the multimodal can respond with one or two spikes while the DCMD is silent. In some preparations, this situation was reversed, the DCMD unit fired while the multimodal did not respond. Both systems respond to visual stimulation with approximately the same latencies. These features suggest that, rather than the DCMD unit driving the multimodal, both systems are driven in parallel from a common visual processing system.

The threshold spatial resolution of the multimodal system has not yet been measured, but it would not be surprising if it did turn out to have anomalous resolution like the DCMD system. Both systems exhibit the same form-preference for jittery spots (Northrop and Guignon unpublished observations), and have the same properties of spatial habituation, in addition to the similarities described above.

Horridge *et al.* (1965) found that the latency to sound stimuli is less (20 ms) than to light (90 ms) in *Locusta* optic lobe multimodal units. In *Romalea*, Northrop and Guignon found that the acoustic response latency was *c.* 60 ms. Mechanical stimuli produced responses that usually adapted more slowly than did visual responses to a moving spot in *Romalea*.

The effect of superimposing two stimuli (spot jitter and mechanical stimulation of body surface and legs with a crystal probe) in *Romalea* multimodal units was close to simple summation in all but one case in which the stimuli appeared to potentiate each other, causing a response about twice as strong as the summed response. Horridge *et al.* observed that simultaneous sound and light stimuli to *Locusta* produced increased responses which were less than those for the sum of the separate stimuli.

Lesioning of 30 multimodal units in *Romalea* has shown them to be located on the proximal, distal, and lateral surfaces of the medulla (8), in the outer chiasma (3), in the lamina (2), in the volume of the medulla (5), in the lobula (7), and in the inner chiasma (5). The DCMD units, which respond visually very much like the multimodal system, have been shown by RF lesioning never to occur *in* the medulla, but rather to lie between it and the lobula, and in the tracts running from the lobula centrally.

17.2.4. *Non-form specific optic lobe units*

The non-form specific units are, broadly speaking, Light (L) units and Dimming (D) units. They are characterized by receptive fields (RFs) which are in general smaller and better defined than those of the form-specific units, namely DCMDs, multimodals, vector-edge, vector (spot), and vector (motor) units. The relevant stimulus is average light flux in the RF. Responses can be tonic, rate-sensitive, or some combination thereof.

Horridge *et al.* (1965) have called *Locusta* tonic L-units type AD, their complement, tonic D-units, type AF. On-sensing L-units included Horridge *et al.*'s types AC (low sensitivity, monocular RF) and BC (with 'small receptive fields down to 7 degrees ...'). The tonic L and on units found in *Romalea* by Northrop and Guignon (1970) were found to have a wide range of thresholds and receptive fields. RFs may include the entire ipsilateral eye, or be in an ovoid region of as small as 20°.

A small number of tonic L-units in *Romalea*, having a marked on response, also exhibited an on-gated mechano-response, i.e. mechanical stimulation following the on burst would elicit another smaller burst superimposed on the tonic, LSS response. The on-gated mechano-response was not observed in DSS, or in LSS more than several seconds following on.

In *Romalea* net dimming units are not silenced by microspot illumination of the eyes, but any general brightening of the ipsilateral eye will turn them off. They are evidently ipsilateral monocular units, and are mechanoreceptor insensitive. Some D-units are clearly of a tonic nature; they fire slowly in

LSS and their rate accelerates smoothly at off. Others are off-types, firing a burst at dimming or off, and remaining silent otherwise. D-units are not sensitive to any form of visual object, except when the object is large enough to cause a significant reduction (or increase) of the light flux on the entire eye.

The *Locusta* counterpart to the off-sensing D-unit is probably type BD of Horridge *et al.*, who claim a 15–25° field for this unit. Northrop and Guignon (1970) were never able to define definite RFs for D-units in *Romalea*. The reason for this may lie in a difference in experimental methods. D-unit RFs were sought using a *collimated* microbeam (illuminating as few as seven facets) in DSS and listening for a slowing the units' firing to indicate an influence inside the RF. This method must be contrasted to that of Horridge *et al.*, who evidently positioned a point source of light at various points over the surface of the eye, and switched it off to test D-unit response. They obtained a marked positional effect for some D-units, defining RFs.

On several occasions in *Romalea*, small RF on-type L-units gave an apparent motion-sensitive behaviour with a black spot in the grey display hemisphere. As the spot was moved, the unit would fire. This was soon found to be due to the spot *leaving* the RF of the L-unit, causing a response to the increase in light flux to the RF. There was no optimum spot size (up to 50° diam. tested); response became stronger with increasing spot size. White spots also produced the same effect when moved into the RF of the L-unit. The same effects were noted for broad (9–10°) black or white stripes. Direction of motion did not seem to be important for any of the above L-unit responses; in fact, switching a disc of brighter light on, or removing a disc shadow had the same effect as the physical movement of the paper objects.

17.3. Spatial frequency analysis in insect vision: future applications

17.3.1. *Introduction*

Horridge *et al.* (1965) have pointed out that the classification of units recorded in insect optic lobes necessarily must depend on the tests used, and these in turn depend on the imagination and technical resources of the investigators. Procedures used in studies of the behaviour of visual units have evolved from simple tests using on and off of general illumination, through RF mapping with spots of light, to tests of form and directional motion discrimination with stripes, spots, rectangles, etc. A new viewpoint from which to examine insect visual units is described below.

17.3.2. *Spatial frequency domain description of objects*

With the advent of coherent light from lasers, holography, and a general interest in optical information processing, it was found that there are many advantages in describing coherent and incoherent optical systems in the spatial frequency domain using Fourier transforms (Cutrona *et al.* 1960). Spatial frequency domain operations were used at first in holography and optics for the design of filters and other signal processing operations. They were then extended to schemes of optical signal detection analogous to those already in use with conventional time-variable electrical signals (Vander Lugt 1964). Soon it became apparent that two-dimensional frequency-domain operations on signals and noise were not restricted to optical systems. Visual objects could be represented in mosaic form numerically, and operations by digital computer could effectively duplicate the spatial filtering operations formerly done optically. Such numerical filtering has been used in the design of the visual system of a robot (Sutro and Kilmer 1969).

17.3.3. Neural spatial filter models

More recently, it has been recognized that many of the properties of form-discrimination, movement detection, and directional sensitivity of visual units in vertebrates can be described as being the result of *neural* spatial filtering operations, that is, the organization, the signal transmission, and processing properties of interneurons in the visual nervous system are assumed to enable them to perform as elementary spatial filters. Fukushima (1969) illustrated this type of system mathematically and showed the types of neural spatial filters required to enhance boundaries and detect the orientation and ends of lines. Fukushima did not consider the filtering operations in the spatial frequency domain because of the built-in non-linearity of the nervous system—i.e. it is impossible to represent negative signal values as negative spike frequencies (in time). Even though input intensities to the visual system are always positive, it is theoretically possible to obtain negative values at certain stages in the assumed filtering operations. These of course cannot be represented by pulse frequency coding, but may be accepted in an electrotonically-coupled system.

Other investigations of linear spatial filter models and their possible role in vertebrate vision have been described by Niemann (1970) and von Seelen (1970) for the case of static objects. Korn and von Seelen (1972) and Zsagar (1972) considered the properties of spatial filters having time-dependent interconnections and input objects which change in time. They have shown mathematically that under such transient conditions, the spatial filtering characteristics of the system change. Such modelling work is essential to understanding mechanisms of information processing and feature extraction in the cortex, and is directly applicable to insect vision.

It is also possible that simpler, less mathematically-complex models, based on the work of Zorkoczy (1966) may better describe the basis for form- and movement-sensing units in the insect's optic lobes. Zorkoczy's modelling scheme is based on automata theory and logical operations rather than linear filtering, hence is simpler to describe analytically. A *Romalea* vector edge unit system is shown in Fig. 17.11 represented in a modified Zorkoczy

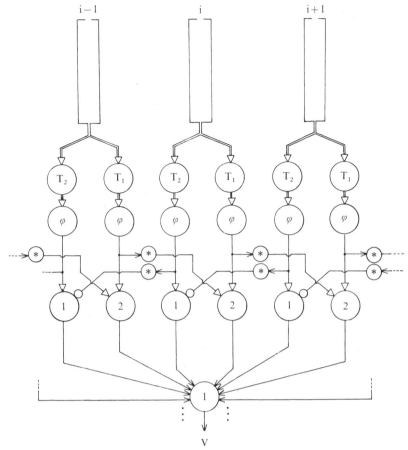

F IG . 17.11. Model of a *Romalea* vector edge detector. Edge detection is accomplished by a
correct sequential stimulation of a majority of receptors in columns of N ideal receptors
(Zorkoczy 1965). Each receptor produces a periodic train of impulses when illuminated and is
silent in the dark. Every receptor in a column is connected to an on detector (T1) and to an off
detector (T2). The T1 produces one output pulse at the time the first input pulse occurs (following a pause of one period or more). The T2 produces an output pulse one period following the
cessation of its input. At least $\phi < N$ pulses must occur simultaneously at the outputs of the
T1 or T2 systems for a 'threshold neuron', ϕ to fire. Hence a linear moving object must be
aligned nearly parallel to the receptor columns to enable the threshold neurons to fire for a
contrast change over a column. Moving convex objects will not produce threshold neuron
outputs as too few receptors in the columns will be simultaneously stimulated. Directional
preference in the model lies from left to right for a long edge parallel to the receptor columns,
regardless of whether dark precedes light, or *vice versa*. The neurons labelled '1' require one
excitatory input to fire. If an inhibitory input (through the knob synapses) occurs with the
excitatory input, or alone, there is no output. A neuron labelled '2' requires two simultaneous
inputs to fire. It is in essence an AND gate. The output V requires one or more inputs to fire.
It is analogous to a logical OR gate. The circles with asterisks represent delays of the clock
period T, so that an edge moving with velocity a/T in the preferred direction will cause the '2'
neurons, hence the V output, to fire as long as motion continues. There are assumed to be c. 90
columns of receptors over the entire surface of one eye, spaced $a = 2°$ apart. Off of general
illumination will produce an output pulse from the model, on produces no output. If the model
is relaxed to include temporal summation and decay in the threshold-logic 'neurons', and the
ϕ-neurons are assumed to be spontaneously active, then it is seen that edge antipreferred
motion will suppress the V output if the light side precedes the dark side of the edge. If the dark
side precedes the light, then no suppression will occur. This behaviour is in keeping with observations on edge polarity in vector edge units by Northrop and Guignon (1970).

form. Details of the operation of this model are given in the figure caption.

Certainly, other models may be created to describe the basis for form- and directionally-sensitive units. Models are fun to devise and give much insight into the neurobiology of the system because their behaviour must be reconciled with known neuroanatomy and all aspects of the system's input–output behaviour. Resolution of the modeller's dilemma (which one lies closest to the biological system?) depends on either being able to open the biological 'black box' to gain access to other signal points in its component interneurons, or to devise new quantitative tests to characterize more exactly the system's input–output characteristics. A case to support the latter approach is advanced below.

17.3.4. *Spatial frequency analysis*

One way to characterize form-discriminating optic lobe units is by their two-dimensional spatial frequency response. This is basically an extension of one-dimensional linear systems analysis to a two-dimensional, obviously non-linear system. The commonly-used black–white striped object viewed in a rectangular black mask or window has a two-dimensional Fourier spatial frequency spectrum given by the relation:

$$F(u, v) = \frac{zh}{2} \operatorname{sinc}\left(\frac{vh}{2\pi}\right) \sum_{\substack{n=-\infty \\ \text{odd}}}^{\infty} \operatorname{sinc}\left(\frac{n}{2}\right) \operatorname{sinc}\left[\frac{z\left(u - \dfrac{n2\pi}{A}\right)}{2\pi}\right]$$

Where z is the window width; h is the height; A is the stripe period; u is the spatial frequency in x; v is the spatial frequency in y; and

$$\operatorname{sinc}(x) = \sin(\pi x)/\pi x.$$

This is a complicated surface to visualize in two-dimensional, u, v space. It lies mostly along the u-axis because there is a little high frequency energy in v due to the length of the stripes in y *versus* their period A. The spectrum is repeated with decreasing amplitude along the u-axis with centres at $u = 0$, $u = \pm 2\pi/A$, $u = \pm 6\pi/A$, etc., because of the periodicity of the stripes at spatial frequency $2\pi/A$ radians. This type of object is obviously too complex in terms of its harmonics and energy distribution to be used in quantitative testing of spatial frequency response. In this respect, the rectilinear striped disc objects used by McCann and Dill (1969), and the radially-striped pattern used by Burtt and Catton (1966, 1969) are even more complex.

The approach needed is that used by Campbell (1968) in his innovative studies of spatial frequency response in the neural part of the human visual system. Using coherent light to generate interference patterns directly on the retina, Campbell presented subjects with one-dimensional sinusoidal intensity distributions to study how contrast perception is attenuated as a function of spatial frequency. The major result of this work was to show that the retina–brain system is more low-pass than the dioptrics of the eye to spatial frequencies.

Similar sinusoidal intensity patterns with negligible harmonic energy might be used to probe the spatial resolution of insect visual units. They could be generated on a large cathode ray tube (CRT) in one or two dimensions, taking into account the non-linearity in the phosphor's light-emitting properties with electron-beam energy. Another method of generating sinusoidal stimulus objects could use lasers as Campbell did, and a back-projection screen. It should be noted that all such patterns are bounded by the finite extent of the display area, and hence in the spatial frequency domain the object spectrum will not be pure impulse functions. This boundary effect causes the sine spectrum components to have a shape which is the shape of the window's spectrum. If the window dimensions are very large compared to the period of the sine pattern, then the window spectrum will be very low-pass, and the spectrum of the windowed sine wave will approach delta functions in frequency, centred at $u = \pm 2\pi/A$ spatial radians m^{-1}. The spectra of any sinusoidal objects necessarily must have a zero-frequency component because of the 'bias' illumination which must be supplied to the object in order to realize non-negative, sinusoidal intensity distributions. This zero-frequency component will also have to be considered in measuring spatial frequency response of a unit because of system non-linearity.

Bandwidth-limited single objects with circular symmetry may also be used to examine quantitatively the spatial resolution of spot-sensing units. This type of object can be generated by Fraunhoffer diffraction using coherent light, or can be computer-synthesized (with intensity correction) on a CRT. For example, the Fourier spectrum of the intensity pattern known as the Airy disc (from diffraction at a circular aperture) is a conical surface with a flared base (Papoulis 1968). Other objects having nearly ideal low-pass ('pill-box'), bandpass ('ring'), and high-pass ('well') spectra can also be generated and used to study optic lobe unit responses in the spatial frequency domain. Presentation of single objects of defined spatial frequency composition is analogous to pulse testing in electronic communications circuits. A unit's maximum response to a particular size and shape of object could serve as quantitative evidence for a spatial matched filter mechanism in insect vision.

17.3.5. *Conclusion*

Campbell (1969) has reviewed the evidence for the role of spatial frequency analysis in vertebrate vision and has shown how such measurements can be made. We suggest that neural spatial filters may operate in insect vision, and that the sinusoidal gratings in one and two dimensions and singular objects with defined spatial frequency spectra appear to be very suitable for use in quantitative studies of form-discrimination and spatial resolution in these systems. Because most form-discriminating units in the insect optic lobes require object motion for a response to be elicited, the speed and direction used in presenting these objects to the eye would have to be held constant while their spatial frequency content is varied. If it is desired to hold constant contrast changes per unit time over the photoreceptor array for a sinusoidal grating, then object speed must be made proportional to the period A.

Physical realization of objects with defined spatial frequencies is technically difficult, but it is felt that their use will lead to a new appreciation of how insects process and use visual information.

17.4. Summary

An examination of the information-processing properties of visual units in three orders of insects (Orthoptera, Diptera, and Lepidoptera) has revealed that they possess certain common features of behaviour which may be used to categorize them. These features include: (1) Form sensitivity or no form sensitivity. (2) Form sensitivity which can be subdivided into preference for spots *versus* edges. (3) A general sensitivity to motion *versus* a sensitivity to form moved in a preferred direction (vector response). Vector responses generally have broad optima in direction, shape, and speed of the object. (4) Ability to respond to other stimulus modalities. (5) If non-form sensitive, then responding maximally to a general on and/or LSS or a general off and/or DSS (L and D units).

A number of unique combinations of these properties have been found, and have served as the basis for naming units. It is expected that as new quantitative techniques such as spatial frequency analysis are applied to the study of insect vision, other unsuspected properties will be discovered which will lead to a more refined classification system for units and a deeper understanding of the neural mechanisms underlying visual feature extraction.

Many enigmatic properties of optic lobe units demand further study. In particular, what is the neural basis for directional sensitivity, anomalous resolution, and form-discrimination? Do neural spatial filters exist in the insect visual system? Does multiplicative processing occur as we have suggested to increase contrast sensitivity, and does it act as Reichardt (1969) has proposed to enable vector sensitivity? How is optic-lobe unit information used by the animal—what role do multimodal and DCMD systems have?

The answers to these problems will be found by neurobiologists who are willing to extend classical methods to include disciplines such as computer science, Fourier optics, spatial filtering, and modern electronics. The use of computers to cross-correlate and extract transfer functions and PST-diagrams (Gerstein and Perkel 1972) showing the relationships between two simultaneously-recorded units, and laser-generated stimulus patterns are two examples of the interdisciplinary effort which ought to be made to further our understanding of insect vision.

18. Motion detection in locusts and grasshoppers

J. KIEN

ALTHOUGH there has been much anatomical, electrophysiological, and behavioural study of central nervous and reflex function in insects it is still unknown how visual neurons are connected to analyse motion. I will discuss data obtained only from locusts and grasshoppers; movement perception in other insects will be reviewed elsewhere in this volume. Motion detection is of several known types: (1) perception of motion regardless of direction—neurons in this class respond to movements in all directions; (2) analysis of direction of motion—neurons respond to movements in specific directions only, usually rotation about the body axes. The animal must be able to distinguish its own movements from movements in its environment. Neurons responding to the animal's own movement will respond to motion of the whole visual field, whereas neurons responding to external movement will respond to motion of only small parts of the visual field. The directional information gained from a comparison of activity in these different neuron types would allow the animal to compensate for its own movements and also allow head stabilization against a visual background.

18.1. Direction-insensitive responses to movement

Neurons which are sensitive to movements in any direction have been found in locust and grasshopper optic lobes and brain (Burtt and Catton 1959, 1960; Horridge, Scholes, Shaw, and Tunstall 1965; Northrop and Guignon 1970). These neurons respond best to rapid movements of small contrasting objects but they usually habituate quickly. Some respond well to increase or decrease of light intensity (on–off, Burtt and Catton 1960), and do not discriminate between light intensity changes and movements. In some cases, after habituation to a particular movement, the neuron will respond to the movement elsewhere in the visual field. Neurons of this type have been called novelty units (Horridge *et al.* 1965). They have whole eye fields and are often binocular. They may respond also to auditory and tactile stimulation, and there may be a complex relationship between the different modalities. They are found in the medulla (Northrop and Guignon 1970; Horridge *et al.* 1965), lobula and protocerebrum (Horridge *et al.* 1965). Similar units, classified as 'jittery' movement fibres, have been found in crayfish (Wiersma and Yamaguchi 1967) and butterflies (Swihart 1968).

The destination and function of these fibres is quite unknown but their apparent novelty function leads to some speculation. If a movement in an insect's visual field is regarded as a change in environment any new movement irrespective of direction will cause an equal environmental change which diminishes as the movement is continued or repeated. Such a movement may then cause further change only when some parameter is altered. In this case, a change in direction may cause a response but it is novelty, not direction, that is the causal parameter. A unit which responds to a change in the environment may have alerting (warning) function or increase the insect's general excitability and activity level (arousal) (see Wiersma and Oberjat 1968; Wiersma and Fiore 1971). Arousal may serve to decrease reflex latency and increase information transmission in the insect nervous system as has been found in vertebrate nervous systems (e.g. rabbit, Fuster and Uyeda 1962).

Although the destinations of the novelty units of the brain and optic lobes are unknown, neurons which respond to movement in any direction have been found in the locust ventral cord (Catton and Chakroborty 1969). These may be related. These units also respond to light-on, or -off, or both. Some are binocular but the majority are contralateral with homogeneous fields covering the whole eye. Illumination changes on a single facet produce several spikes in a single unit and they habituate very quickly to even this tiny stimulus. They respond to only the first of two light flashes each on a single facet (even up to thirty facets apart) with an interstimulus interval of several seconds (Kien, unpublished observations).

Behavioural responses to this stimulation can be observed. Flashing light guides which cover only single facets can evoke large head movements in a locust left in the dark. Even when the head is fixed, appropriate potentials can reliably be recorded in the neck muscles. Sudden illumination by a moving light that may normally cause directional head following will, if the animal has been in the dark, cause large random leg, body and head movements (Kien, unpublished observations). The complete neuronal basis of these responses is not clear but the neurons in the ventral cord appear to play a part in initiating the muscle potentials seen before a movement. These potentials may function to improve efficiency and speed of contraction when movement is required (Kammer 1970).

A third type of non-directional movement detector is the descending movement detector (DMD) neuron reviewed recently by Rowell (1971). The field includes the whole eye and the properties vary in different species. These neurons respond to dimming of small areas and to moving, and preferably dark, objects. The receptive field plotted in locust and cricket (Palka 1967, 1969) shows a grading of sensitivity to a stationary dimming pulse with maximum sensitivity in the centre of the eye. The smallest reliable response to a dimming pulse requires stimulation of about seven facets but responses can sometimes be obtained by stimulation of fewer facets. These units show general post-excitatory depression and a local habituation of the area stimulated. Inputs to this detector are unknown but Rowell (1971) has suggested that the inputs lie in the lamina and pass through a highly labile synapse before reaching the DMD. An important feature of the DMD is that it responds to small movements in the visual field, while movements of the animal cause large movements of the whole visual field. At low speeds of head movement the unit is quite unresponsive to the movements of the visual field and at high speeds the activity of the unit is noticeably inhibited (Palka 1969; Rowell 1971). This property allows the neuron to detect any small external change in the animal's environment. As no behavioural correlate of activity in these neurons has been found it is possible that the DMD may be command fibres priming thoracic ganglia for faster processing of cerebral inputs (Rowell 1971).

Thus, there is what may be called a direction-insensitive motion detection system in the locust, which may prime the thoracic ganglia for information processing and may prepare the muscles for faster initiation of contraction. The system consists of (a) a command fibre (DMD) which acts on thoracic ganglia and which is responsive to small external movements even while the animal is moving and (b) neurons in the optic lobe which respond to any movement or change in the visual environment and send their information to neurons in the ventral cord. These prepare the muscles for faster initiation of contraction. Information on the direction and speed of the movement does not travel through this system so the muscles can only be prepared for action. Instructions for specific contractions must come from elsewhere.

18.2. Direction-detecting neurons

Most research on these neurons has aimed at finding the precise circuitry that allows a neuron to respond to movement in one direction but not the reverse. Relatively few of these neurons have been found in locusts or grasshoppers and their properties have not been explored extensively. In comparison there are many studies of such neurons in fly (Bishop and Keehn 1967; Bishop, Keehn, and McCann 1968; McCann and Dill 1968; McCann and Foster 1971; Mimura 1971), butterfly and moth (Blest and Collett 1965a,b; Swihart 1968; Collett 1970, 1971), bee (Bishop 1970; Kaiser and Bishop 1970), and beetle (Frantsevich 1970; Frantsevich and Makrushov 1970).

The direction detecting neurons in locusts and grasshoppers have been found in the medulla (Horridge *et al* 1965; Northrop and Guignon 1970) and in the lobula (Horridge *et al.* 1965). They are similar in all respects to the selective direction detecting units found in other insects (except Mimura 1971).

The smallest receptive field found for this type of neuron is 50° for some medullary cells (Horridge *et al.* 1965). Their rate of firing is increased by movement in the preferred direction. No plots of firing rates against speed of movement have been made but it has been reported that they follow a large range of velocities with no fractionation. However, some may saturate at very slow velocities. Motion in the direction opposite to the preferred, i.e. the null direction, causes a decrease from the spontaneous firing rate.

The optimal stimuli are not known precisely for these neurons but they respond to movement of black and white stripes, a single black stripe, and a single black–white edge. The black must be anterior (Northrop and Guignon 1970), a finding whose importance will be discussed later. Units in the grasshopper anterior medulla will respond to a moving spot, preferably a large one (Northrop and Guignon 1970) and similar units in the locust anterior medulla respond to virtual movement of two lights. The cells with large visual fields do not respond well to moving spots but prefer a large moving pattern.

The anatomy of these neurons is not known, but from their physiological properties it is suggested that some of the tangential fibres in the medulla may be these motion detectors. There has been no study on acridids similar to the two electrode mapping studies on moth (Collett 1970, 1971) and fly (McCann and Dill 1968; McCann and Foster 1971) which would reveal possible extent and connections of these fibres. The studies on moth and fly indicate the following scheme as a possible general connection pattern in insect visual system. The first direction detecting cells occur in the distal medulla. The precise inputs to these cells have not been found electro-physiologically; they may come from the lamina, the medulla, or both. As the information reaches a directional cell it undergoes some transformation, as there are cells with different responsiveness to different velocities. The first directionally-selective cells in the directional pathway have smaller visual fields. Information, at least partially coded for direction and velocity, passes to cells with larger receptive fields. These may be in the medulla or lobula. Optic lobe output units lie in the lobula. They probably receive inputs from only few wide-field cells and also may receive inhibitory inputs from wide-field cells with the opposite preferred direction. Fibres from the lobula pass to the medial protocerebrum on both sides. Here there are binocular fibres which pass to the medulla or lobula, (either ipsi- or contra-lateral) and to the ventral cord. These binocular cells respond as if they sum inputs from monocular cells with appropriate directions, e.g. backwards across one eye and forwards across the other. Their function is complex but it appears they may mediate a directional surround effect on the optic tract neurons, which may track a moving object while the animal itself is moving (Collett 1972). Another system allowing the animal to distinguish its own movements from the external movements may involve the different types of direction detectors that have been found. In the praying mantis there are two types of direction detectors, some responding to small movements but insensitive to large movements, others only sensitive to large movements (Butenandt 1972). In the fly both unidirectional and integrative cells can be found in the lobula (Mimura 1971). The small field cells would provide information about small movements in the visual field while the integrative type responds best to large movements such as those generated by the animal's own movement. Collett (1971) has proposed that asymmetrical inhibitory connections between optic lobe output cells may provide a means for the moth to perceive clearly external forward motion while flying, which produces backward movement of the visual environment across both eyes.

18.3. Direction-detecting output from the brain

A directional behavioural sign that is easily observed is the optomotor response. This is the movement the animal makes after a movement of the visual field so that the velocity of the field movement across the eye is reduced. This is similar to the optokinetic movements of the eyes of crustaceans (Horridge and Sandeman 1964) and vertebrates (Collewijn 1969). In the locust the optomotor response consists of head turning about the longitudinal axis of the body and also in the horizontal plane (Thorson 1965). For the insect to turn precisely in response to a visual movement, there must be interneurons transmitting directional information to the ganglia which innervate the appropriate muscles. Although there have been thorough searches (Horridge *et al.* 1965; Catton and Chakaborty 1969), they have failed to find any such neurons in the ventral cord either in the neck connectives or below the prothoracic ganglion. Direction detecting neurons have, however, been found in the circumoesophageal connectives, i.e. at the input to the suboesophageal ganglion (Kien 1974b; Northrop, in this volume). This ganglion innervates some of the muscles that turn the head (Shepheard, in the press). Perhaps in the locust only the neck is controlled by descending motion sensitive units and turning of the body is initiated in other ways.

Quantitative study reveals that the behaviour of the motion-sensitive units of the circumoesophageal connectives parallels the optomotor response, and it is likely that these are the output fibres of the optomotor system (Fig. 18.1). They appear to correspond to the binocular neurons of the moth and fly brain (Collett 1971; McCann and Foster 1971). Large moving objects are the preferred stimuli. The neurons do not respond well to small moving lights and prefer movements of black objects against a white background. All units respond well to a large range of constant velocities and frequencies of oscillation. Some units prefer higher speeds and frequencies of oscillation, up to 7·5 Hz. Others have their optimum in the slower range, below 0·25 Hz. The ability to follow higher frequencies is the only difference between the

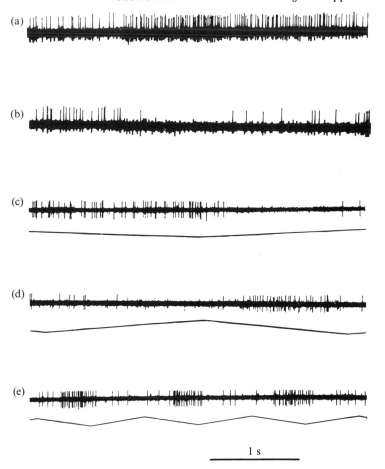

FIG. 18.1. An optomotor output fibre in the locust circumoesophageal connective. Movement in the preferred direction (forward over the contralateral eye) increases firing above the spontaneous rate (a). Movement in the opposite or null direction (b) causes a decrease in the firing. These fibres respond in a typical fashion to oscillation of a striped pattern in the preferred and null direction (c–e) and show no habituation. Lower trace shows movement of the pattern: (c) 0·01 Hz oscillation, an incomplete cycle is shown, (d) 0·25 Hz, (e) 1·0 Hz. In all cases amplitude of the oscillation was 15° of arc.

response of these neurons and the optomotor response, which begins to fail at *c.* 0·75 Hz.

A second type of directionally sensitive neuron can be found in a nearby region of the circumoesophageal connectives (Fig. 18.2). These units are much larger than the optomotor fibres. Preliminary testing implies that there are only a few of these fibres in each connective (two–four) and although they receive inputs from both eyes, only contralateral stimulation reveals a

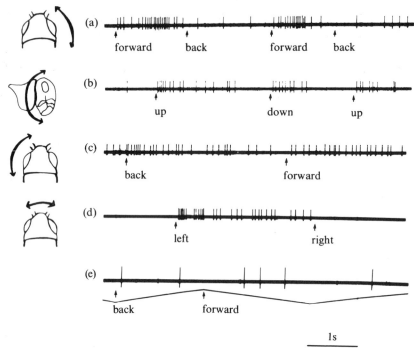

FIG. 18.2. A directionally sensitive fibre in the locust circumoesophageal connective. This neuron fires strongly to movements of a small lamp forward over the contralateral eye (a). It does not fire to movement in the opposite direction. There is little spontaneous firing. Dorsoventral movements over the contralateral eye (b) cause smaller responses equal for either direction of movement. The neuron responds to movement over the ipsilateral eye (c) but without directional sensitivity. The binocular zone (d) is dominated by the contralateral eye. Oscillating movements of a striped pattern (e) evoke no response, indicating that this type of fibre is not involved in the optomotor response.

directional selectivity. They have little or no spontaneous activity and their response is a sustained firing to movements in all directions except the null direction. Firing frequency is higher to movement in the preferred direction. Although there is no response in the null direction, no inhibitory mechanism can be deduced as the units have little or no spontaneous firing. They respond best to movements of a small object or small lamp and do not respond at all to large stripe patterns, indicating that they are not involved in the optomotor response. It is possible that neurons of this type respond to movement of objects, whereas the optomotor fibres respond to the animal's own movement.

18.4. Direction detecting behaviour: models for the optomotor response

As the optomotor reflex is so reliable it has been used in a large number of experiments to determine behaviourally the stimuli which can be detected as motion by the animal. A stimulus that evokes an optomotor response must satisfy the requirements of only the optomotor system. Other direction detecting systems may have quite different properties. The difference is emphasized by the presence of the type of directional unit in Fig. 18.2.

The use of optomotor responses for a black box analysis of motion perception in insects was pioneered by Hassenstein on the beetle *Chlorophanus* (see Reichardt 1969 for a review). Since then there has been a large body of work on the fly and locust (Thorson 1966*a*,*b*). Most of the experiments have not contradicted a model, shown in simplified form in Fig. 18.3 (Reichardt 1967; Reichardt and Varjú 1959; Reichardt 1961), which consists of a basic set of two facets whose receptors interact symmetrically through a delayed multiplication line. The algebraic sum of the multiplication products is the directional reaction. This sums with the similar reactions from other facet pairs to give the overall response. For fly and *Chlorophanus* various filters were added; these would found experimentally to be unnecessary in the locust. Some important but often neglected work on locust is that of Thorson (1966*a*,*b*). Using oscillating stimuli and small signal analysis Thorson showed that all the models in Fig. 18.3*c*–*g* could be used to fit his data.

All the mathematical models shown in Fig. 18.3*c*–*g* are similar in that they are symmetrical. It is possible that there is no useful distinction between them. For the insect to perceive constant velocity motion there must be either a general asymmetry or private symmetrical reactions between pairs of channels. Thus a major question in the choice of models becomes not the choice of symmetrical model, but a choice between a symmetrical or asymmetrical model. Local symmetric reactions have been thought to occur because insects can make optomotor responses to movements of $<1°$ (Thorson 1966*a*,*b*) or to movements viewed through two slits each 1° wide (McCann and MacGinitie 1965). In both cases the movements traverse a maximum distance from one facet to the next. Although at first sight this implies that only two facets are needed to code direction, in all cases the movement information was summed over many pairs of facets and this does not exclude the possibility of a general asymmetry. Results from recording in the optic lobe indicate that all direction sensitive cells have receptive fields much larger than would be expected if only two facets were involved. Other tests show that at least 50–100 facets must be utilized in the locust for selective directional motion detection to be possible (Kien 1974*a*).

An asymmetric model has been proposed for the inputs to the direction-sensitive neurons in the moth (Collett and Blest 1966). This was based on a model proposed for rabbit retina (Barlow and Levick 1965), which is illustrated in Fig. 18.3*a*,*b*. These models account for the large receptive fields of direction-detecting cells in that any number of non-directional channels can feed into the 'and' or 'and not' gates. The inhibitory version was used to account for the inhibition commonly observed for movement in

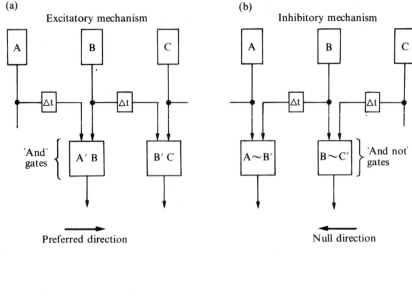

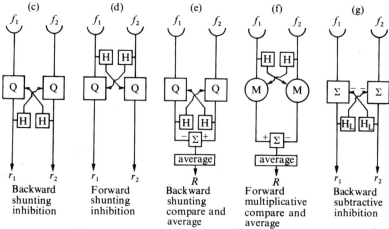

FIG. 18.3. Models proposed for inputs to direction-detecting neurons. Asymmetric models in the upper row as proposed by Barlow and Levick (1965) for the inputs to the directionally selective ganglion cells of the rabbit retina. The inhibitory model (b) is usually used to explain the decrease of firing below spontaneous rate for movement in the null direction. Combination of many 'and' or 'and not' gates produces a cell with a large receptive field. Symmetrical models in the lower row as proposed by Thorson (1966) for the inputs to the locust optomotor system. These are variations with symmetrical interaction between facet pairs. (f) is the cross correlation or multiplication model.

the null direction in the moth (Collett and Blest 1966). However, the facilitatory and the inhibitory mechanisms (in opposite directions) have both been observed in the fly (see Mimura, next Chapter).

Another method of behavioural analysis consists of using stimuli at threshold, not only in amplitude of movement but in complexity and pattern size. Locusts, which follow tiny movements of large patterns, may make consistent errors with small patterns. For the left eye, a single edge with black anterior will be followed by the head but the white anterior edge will cause the animal to move consistently against the direction of motion (Kien 1974a). Failure of this edge to cause inhibition in the null direction has been reported in medullary neurons (Northrop and Guignon 1970) and a response appropriate for the opposite direction of motion has been seen in the optomotor output fibres (Kien 1974b).

These responses have led to a new model for the input stage of the optomotor system (Fig. 18.4). It consists of two Barlow and Levick networks placed in opposition. The asymmetry is preserved as the sizes of the receptive fields are different for each network. Error responses to small stimuli may be explained as follows. A movement of a pattern in either direction excites either the 'on' or 'off' cells or both. After the first cell has been excited a movement in the null direction causes inhibition of the next cells which prevents any activity passing from them to the direction sensitive neuron. However, activity still passes from the first 'on' or 'off' cell which was excited. This happens, as mentioned above, even for movements in the null direction. When a simple stimulus, such as a single black stripe, moves past the eye the leading edge will excite 'off' cells and the trailing edge will excite 'on' cells. One edge must be moving in the null direction of the network it stimulates. If the movement is small enough only two of the larger field 'off' cells and in some cases only two of the small field 'on' cells will be excited. Even though the movement is in the null direction for one system the activity of one cell in this system will pass to the directional neuron. Thus a very similar discharge may reach both direction detecting neurons causing a near equal response in these neurons with opposite preferred directions. This opposition will cause the animal to make errors. In the case of the single stripe, as the amplitude of the movement increases the number of cells in whose preferred direction the stripe moves will increase thus removing any ambiguity.

A single black-white edge will excite either the 'on' or the off' system. As shown in Fig. 18.4 for the left eye the edge with white anterior always moves in the null direction of the system involved; a movement to posterior will stimulate only 'on' cells, an anterior movement will stimulate only 'off' cells. If the edge is the only moving stimulus in the visual field the only change in firing in the directional neurons will be caused by the firing of the first cell to be excited. This will always activate the directional neuron whose preferred direction is opposite that of the movement. In the absence of any other

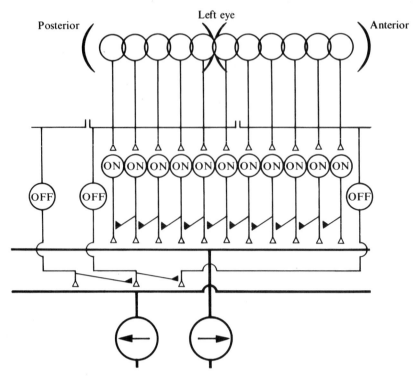

FIG. 18.4. Model of inputs to locust optomotor system proposed on the basis of error and anomalous responses to simple stimuli. The ON cells with small receptive fields (small circles) are connected asymmetrically so that the next order neuron is direction-detecting with preferred direction forward for the left eye. The type BC cell (Horridge *et al.* 1965) is suggested for this cell. The asymmetric connections form a modified Barlow and Levick network (see Fig. 18.3a). off cells which respond to a net dimming of their receptive fields (larger receptive field of off cells shown by large brackets) are connected similarly but in the opposite direction so that the optimum direction of the directional neuron is also in the opposite direction. The hypothesis is that the direction of the optomotor response depends on which of these two opposing networks is more strongly stimulated.

directional information the animal must interpret this as a movement in the preferred direction of that neuron and this will cause the observed behavioural response.

As the individual visual receptors have no directional specificity, the directional information must be built up from direction insensitive cells combined in such a way so as to create new information. Thus it is conceivable that all motion detecting systems draw their inputs from the same cells but have these connected to the motion-sensitive neuron in different ways. These inputs must be derived from every facet. The logical places where the facets are still represented individually and a variety of interactions can occur are the lamina and medulla where the facets are represented by cartridges

(Braitenberg 1970). Structural studies in the fly (Braitenberg 1970) support the lamina as the first site of the interaction between cartridges necessary for the response to motion. However, in the medulla there are a number of large tangential cells which could have the wide receptive fields of the direction detecting cells. Until the locust lamina is studied in detail functionally and structurally it cannot be known whether the inputs first interact here or if motion integration occurs only in the medulla.

Many questions remain to be answered. Only the motion sensitive neurons have been studied; orientation sensitive neurons found in the vertebrate visual cortex (Hubel and Wiesel 1959) have not been reported. How does the locust determine the position of a moving stimuli in its visual field? This information is lost in the optomotor system, the only system studied in detail. Also unknown is the output of the information gathered by the direction-insensitive motion fibres. These probably pass to numerous integrating areas of the brain. Answers to such questions may be found only after a concerted attack on the locust central nervous system using refined techniques such as microelectrode recording and intracellular marking.

Asymmetric networks as in Fig. 18.4 may exist in many insect optic lobes. For example, there is anatomical evidence for asymmetric collaterals in fly lamina (Braitenberg 1970). Also in the fly, units with receptive fields small enough for only two ommatidia to contribute (required by the multiplication theory for direction analyses) are all non-directional. All directional units have larger receptive fields (Mimura 1971). These studies support the existence of asymmetric nerve connections from small field non-directional cells to large field directional units. The details of these connections will vary in the different insect orders. The model shown in Fig. 18.4 is proposed for the locust which exhibits the anomalous edge responses and requires stimulation of a large number of facets for a response both in the optomotor output fibres and behaviourally.

18.5. Conclusion

The locust has variety of motion-detecting systems which range from novelty systems that alert the animal to the highly specific optomotor inputs which are responsible for head stabilization after body movement. Before trying to understand any mechanism of motion detection, one must know which system one is analysing. It is already clear that each system has its own optimum stimuli, intimately related to its function. This would not allow extrapolation from one system to another.

Functionally, all motion detecting systems may be related hierarchically. A non-directional motion system with an extremely low threshold responds to any new movement. Also a large command fibre responding to small movements primes the neuropile to respond to cerebral input rather than local or intersegmental reflexes. The command and the arousal type fibres prime the animal non-specifically for any subsequent response. A variety of more specific activities can now take place. A direction detecting network responding to small movements informs the animal of the direction of small external movements while another direction and velocity detecting network is responsive to large movements and allows stabilization of the head after body movements.

19. Units of the optic lobe, especially movement perception units of diptera

K. MIMURA

19.1. Introduction

IN this review, a general description of the types of neurons in the optic lobe will be followed by a summary of properties of neurons subserving movement perception in particular. The first part of this paper will be concerned with various features of the analytical processing of movement by the neural network of the optic lobe, as well as the question of integration with other sensory modalities which possibly contribute to the activity of optic lobe neurons.

19.2. Types of units encountered

19.2.1. *Responses of lamina units*

Excitation of the photoreceptor cells is a graded depolarization without spikes, well described by numerous authors (from Naka and Eguchi 1962, to Rehbronn 1972, on *Calliphora*). This depolarization is propagated to the first neuropile layer, the lamina, by electrotonic spread (Ioannides and Walcott 1971) along the projection of the retinula cell axons. In general, the projection is such that each lamina cartridge receives axons from receptors that look in one direction in the outside world, as shown in flies with open rhabdomeres (Braitenberg 1967; Kirschfeld 1967) and in bee, locust, and dragonfly with fused rhabdomeres (Horridge and Meinertzhagen 1970).

Responses of lamina units are of two kinds, those without spikes and those with. The lamina ganglion cells yield hyperpolarizing graded responses without spikes, similar in form to the retinula responses but inverted in sign (Shaw 1968; Autrum *et al.* 1970). These, and the units with spikes found by Arnett (1970, 1971), are reviewed by Laughlin in this volume, Chapter 15.

Recently, new units have been found in the lamina with spike discharge at light-on, light-off or at both light-on and -off (Mimura 1974). These three types of responses were, however, from two kinds of neurons. One type responded at both light-on and -off to a light stimulus given at any position

424	*Units of the optic lobe, especially units of diptera*

within the receptive field, and the other type responded to light-on *or* light-off to a spot of light in different positions within the receptive field. The area responding to light-on is arranged along both sides of the area in which light-off is effective (Arnett 1971). The neural organization and function of the on- and off-centre units remains to be explored in more detail in relation to visual perception. No neurons which respond specifically to movement have been found so far in the lamina, but only neurons responding to a brightness change (Mimura 1974).

19.2.2. *Responses of medulla and lobula units*

Several authors have described neuron types and activity patterns in the medulla and the lobula, with the following properties:

(1) Spontaneous units
   (a) Units not excited by any light stimulus (Bishop and Keehn 1967, fly)
   (b) Units which gave responses to light stimuli (Bishop and Keehn 1967, fly; Burtt and Catton 1960, locust)
(2) Units which do not show spontaneous discharges, but respond to light stimuli (Burtt and Catton 1960, locust; Bishop and Keehn 1967, fly; Swihart 1969, butterfly)
   (a) Units which respond to a stationary light stimulus
      (1) On units
         Some discharge phasically only at light-on (Bishop and Keehn 1967, fly; Burtt and Catton 1960, locust; Swihart 1969, butterfly) or tonically (Swihart 1969, butterfly). Other units discharge at both light-on and -off (Burtt and Catton 1956, in four species of insects; Burtt and Catton 1960, locust; Bishop and Keehn 1967, fly; Swihart 1969, butterfly)
      (2) Inhibitory units
         The spontaneous discharges are either phasically inhibited at light-on and -off, or tonically inhibited during light (Swihart 1969, butterfly)
   (b) Units responding to movement of an object
      Since Burtt and Catton (1956) described spikes elicited by movement of an object in the optic lobe of four species of insects. Many descriptions of these units have been made (Burtt and Catton 1959, 1960, locust; Horridge *et al.* 1965, locust; Bishop and Keehn 1967, honey bee). Details will be presented in the next section.
   (c) Novelty units quickly habituate but respond afresh to a novel stimulus, either stationary or moving (Horridge *et al.* 1965, locust; Swihart, 1969, butterfly). These units may have some 'arousal' function.

(d) Multi-modal units show convergence from two or more sensory inputs. Units of the locust medulla are excited by visual, auditory and mechanical stimuli (Horridge *et al.* 1965). Further details will be described in a later section.

19.3. Movement sensitive units

19.3.1. *Classification of movement units*

Movement units have been found in the medulla, lobula and protocerebrum, but not in the lamina. The following classification is based upon recordings with microelectrodes in the fly *Boettcherisca* (Mimura 1971):

(1) Units excited by a moving object

Units in this category discharge when an object is moved within the receptive field. Although the units respond also to a stationary light, such responses are usually only brief at light-on and -off. In contrast to this, the discharge elicited by a moving stimulus is higher in frequency than that caused by a stationary one, and continues while the motion continues. The following types can be distinguished:

(a) Non-directional type

Discharges are elicited by movement of an object in any direction within the receptive field. Units of this type are of two kinds, the first with a relatively small receptive field ($<90°$) and the other with a wide receptive field ($>90°$). The wide field includes up to half of the ipsilateral visual field. The unit with the small field corresponds to those described as 'jittery movement' fibres by Swihart (1969) and as class I by Bishop *et al.* (1968). Its discharge frequency hardly varies with change in velocity of the object. The unit with the wide field is equivalent to the non-directional units of flies described by Bishop and Keehn (1967). Units with small fields are mostly found in the medulla, but those with wide field are in the lobula.

(b) Uni-directional type

Units in this category respond best to movement of an object in one (preferred) direction and not (or less) to movement in the opposite (null) direction. Units of this type are also of two kinds with either small or wide receptive field. Although each kind is found in both medulla and lobula, small field units are most common in the region between the medulla and the lobula. The receptive field is either ipsilateral or contralateral to the recording site, and can be bilateral. Bishop *et al.* (1968) found four types of preferred direction in relation to the vertical midline axis of the head, namely, horizontally inward, horizontally outward, vertically downward, and upward, but Mimura (1972) finds the preferred direction not necessarily so exact (in a different species of fly). The units with wide field described here are considered equivalent to the directionally selective units with a receptive field of 180° as described by Bishop and Keehn (1967), and the complex sustaining-movement units of the butterfly reported by Swihart (1969), as well as the class II units described by Bishop *et al.* (1968). Directional motion perception in a unit with a field so wide as to extend over both hemispheres of the visual field is presumably accomplished by binocular interaction in the brain as described by McCann and Foster (1971). The speed of an object is coded in units of this type in that the frequency of impulses is approximately proportional to the log of the stimulus velocity.

(c) Semi-integrative type

The receptive fields of these units are complicated, as if two or more receptive fields of the unidirectional type, or those responding to movements in two antagonistic directions, are combined in adjacent or overlapping regions of the receptive field of a higher order unit, but there is not necessarily a smooth, continuous relation between the directions of movement in the two or more parts of the receptive field. Possibly fields of this type are on the way towards the more complex convergence found in the integrative type described below. Some neurons of this type discharge strongly when a spot is moved vertically in a narrow band on the midline in front of the fly, which may be functionally significant for the visual control of the position of the fly and interaction with the sense of equilibrium. Units of this type are distributed widely in the medulla and lobula. The coding of velocity by frequency is similar to that in the unidirectional type.

(d) Integrative type

These units respond to a pattern of different stimuli, which together make a biologically meaningful combination. As expected, the adequate stimuli also include modalities other than visual, as described in the following section. In this section will be considered only units upon which other units of the unidirectional or semi-integrative types converge to produce a sensitivity to a complex but continuous motion within the receptive field. The most characteristic type of movement is a continuous circular movement of an object with the insect in the centre. For different units the movements that are effective in eliciting discharges are about one or other of the major axes of the insect. Therefore they are meaningful as the mechanism by which the coordinates of surrounding movements are coded in separate units of the insects' nervous system. Units of this type are mainly found in the lobula and in the tract from lobula to protocerebrum. Units with similar properties have been demonstrated in the locust nerve cord (Burtt and Catton 1966, 1969). These units appear not to code the velocity of a movement because the response magnitude does not gradually increase with increase in the speed of movement; instead, it rapidly attains its maximum and then decreases as the speed increases.

(e) Alerting unit

This unit does not respond specifically to the direction of movement, but is excited by any motion of a small object. Because it rapidly adapts to any stimulus that is repeated, but responds at full strength to a novel stimulus, it serves as an alerting mechanism to new stimuli, and may correspond to the novelty unit of the preceding section.

(2) Units inhibited by a moving object

The inhibition of spontaneous discharges is of two types, one with and the other without directional selectivity towards a moving stimulus. In one sense, all units with directional selectivity could be included in this class, because the phenomena that no spike is elicited in the null direction may be considered to mean that the stimulus has positive inhibitory action rather than being indifferent. However, experimental results show that in some units a background discharge is actively inhibited by movement of a spot of light in the null direction (Fig. 19.1).

19.3.2. *The mechanism of sensitivity to direction of motion*

On the physiological side, one can infer from the locations of the above units that integration proceeds from simple responses such as light-on or -off at the most peripheral level, through an analytical level concerned with movement direction and speed, to the deepest level, where we find sensitivity to complex movement and summation with other inputs. On the histological side the pioneer work by Cajal and Sánchez (1915) has been confirmed and extended during recent years (Braitenberg 1967; Trujillo-Cenóz 1969; Horridge and Meinertzhagen 1970; Strausfeld 1971). However, only in exceptional cases do we know the physiological responses of the histological types, and it is as yet impossible to point to the neuronal mechanisms underlying motion detection in an anatomical framework. The following is relevant.

That the neurons which detect movement are not in the lamina but lie deeper has been suggested or assumed from the morphological evidence that there are no fibres with lateral spread between several optic cartridges in the appropriate directions in the lamina (Strausfeld 1971). In fact, no responses to movement have been recorded in the lamina although looked for (Horridge *et al.* 1965; Mimura 1974). Analysis of direction of movement

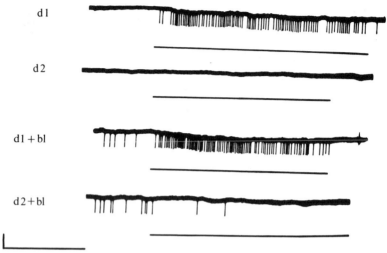

FIG. 19.1. The active inhibition of a background discharge by a spot moving in the null direction. Responses of a unit of the unidirectional type to movement of a spot of light in the preferred, d_1, and null direction, d_2. The two lower lines, $d_1 + bl$ and $d_2 + bl$, show responses to movement of the spot in the preferred and null directions while spikes are produced by a background light stimulus. Bars under each record show the durations of movement of the spot. Calibration, 1 mV and 0·5 s.

is necessarily performed below the lamina, which is not sufficiently complicated to provide the circuits or the numerous output fibres. However, it is certain that neurons of the lamina play a role in the analysis of movement if only because they are part of the pathway to the deeper regions.

By careful manipulation of the stimulus, it is possible to obtain further details of the physiological interactions which contribute to motion perception. When two or more stationary spots of light are arranged along the preferred axis of a unit of the unidirectional type, and flashed in succession, if they have the appropriate time and space interval, as well as being in the preferred direction, the spike discharges to the second spot are facilitated. On the other hand, when the spot sequence is in the null direction, discharges to the second spot are less than those to the first, or completely occluded. These experimental results show that movement perception is accomplished by a lateral interaction which actively exerts an excitatory process in the preferred direction and an inhibitory one towards the null side.

Furthermore, the lateral excitatory and inhibitory processes can be distinguished in other ways. In the preferred direction, the excitatory mechanism is increased at each 5° of visual angle, whereas in the null direction the closer the second spot to the first the greater the inhibition. The excitatory effect extends to 20° or more in visual angle, though values differ for each unit. The lateral extent of the inhibitory process is less than that of the excitatory one, with values from a minimum of 8° to a maximum of 20°. The increase in the excitatory process at each 5° may contribute to the continuous production of spikes by interrupting the adaption in some way, and may be the mechanism by which the movement unit discharges steadily during a continuous movement of a real object.

How these excitatory and inhibitory processes correspond to the neural elements in the optic lobe is quite unknown. As said above, sufficient lateral connections are available in the medulla but probably not in the lamina Strausfeld 1971).

19.4. The motion detecting system in insect behaviour

In this section the integrative action of the movement detecting units and the summation with inputs of other modalities will be discussed.

19.4.1. *Movement of an object not distinguished from movement of the insect*

Movement of an object around the resting insect, in a relative sense, corresponds to movement of the insect itself in a stationary environment. So, as a matter of course, several authors have taken interest in the interaction between these aspects and insect behaviour. One general conclusion is that directionally selective units code the information prerequisite to the opto-motor response (Bishop and Keehn 1967). In the nerve cord of a butterfly, Swihart (1969) found a directionally selective unit with a very wide receptive field sensitive to motion from behind to forward and suggested it to be part of the mechanism for flying on a straight course. Heide (1971) stimulated flies with a moving stimulus (the rotation of a cylindrical pattern), observed that some of the non-fibrillar flight muscles are active only when the cylinder rotates in one direction, and inferred that directionally sensitive motion detecting units are essential in the control of yawing. These conclusions imply that excitation of the directionally-selective units in a tethered insect is not different from excitation by movement of the insect itself in a stationary environment. This assumption has been tested as follows:

A microelectrode was inserted into a directionally-sensitive unit with a preferred direction from front to back in the left hemisphere of the visual field (Fig. 19.2a,b). A spot of light was set on the left side of the fly, which was rotated together with the base and electrode supports. Spikes were elicited by rotating to the right and were inhibited by rotating to the left (Fig. 19.2c,d). The responses caused by self-movement were indistinguishable from those caused by object-movement.

Quite a different type of response has been found in the DCMD neuron of the nerve cord in crickets. This neuron responds to object motion but not to the forced or voluntary movement of the whole eye (Palka 1969).

19.4.2. *Interaction between motion and air current stimuli*

It can readily be imagined that variations in the state of a living organism may influence the properties of the perception system so as to permit adaptation to the new conditions. Support for this idea comes from the study of

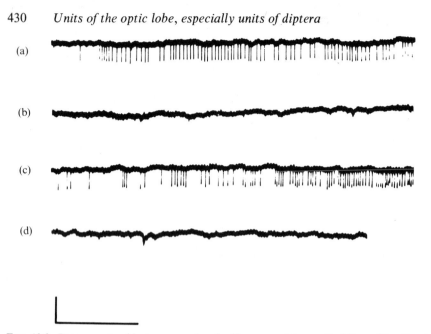

FIG. 19.2. Responses to object-movement and self-movement in a unit of the unidirectional type. (a) and (b) responses to a moving spot in the preferred and null directions, respectively. (c) and (d) responses to self-movement. With a stationary spot in the receptive field, spikes are produced by rotating the fly in the direction corresponding to the preferred one (c), while none are produced by rotating the fly in the opposite direction (d). Calibration, 1 mV; 0·5 s.

integration between flight behaviour and the visual response to a moving object. All investigations on movement perception have hitherto been made with a stationary animal. It is well known, however, that air currents on the head or body induce flight movement in many insects, including flies (Hollick 1940). The alterations in the responses of units to a moving object seen while the animal is also stimulated with an air current will now be described. Units were recorded from the medulla, lobula, and tracts to the protocerebrum of *Boettcherisca peregrina* with glass microelectrodes of tip diameter less than 0·3 μm. A spot of light, 2 mm in diameter and 1·23 W m^{-2} in intensity, was moved at a distance of 10 cm from the dark-adapted fly. An odourless air current of saturated humidity and velocity 0·3 m s^{-1} was directed on the front of the fly along the longitudinal axis.

Responses to both visual movement and air current were found in some of the movement sensitive units (Fig. 19.3), and the air current had either an excitatory or inhibitory effect. It is well-known that hairs on the frons, vertex, and antennae are receptors for air currents (Weis-Fogh 1949) but also that bending of the antenna by wind pressure is an important stimulus (Gewecke and Schlegel 1970). Records from the movement sensitive unit showed that mechanical bending of the antenna without an air current (Fig. 19.4) also

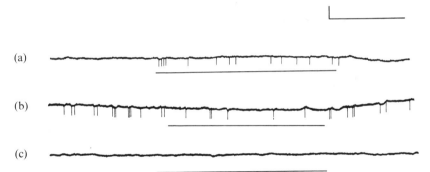

FIG. 19.3. Three kinds of responses of a movement unit in the optic lobe of the fly to the air current stimulus. (a) excitatory response to the air current; (b) inhibitory response; (c) no response. In many units, this indifference (c) to the air current has a subliminal effect upon the response to the moving spot. Bars under each record show the duration of the air current. Calibration, 1 mV; 0·5 s.

produced spikes, so possibly the responses to the air current had been caused by mechanoreceptors of the antenna joints. Some of the units responded to a stimulus on one antenna and others to bending of either. In Fig. 19.4 the response is to stimulation of only the antenna ipsilateral to the optic lobe where the spikes are recorded.

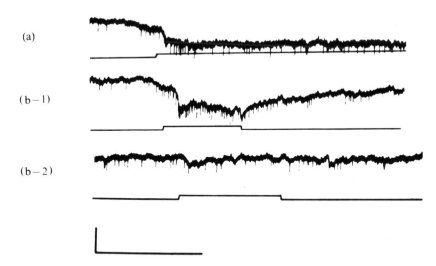

FIG. 19.4. Responses of a movement unit in the optic lobe of the fly to air current and to mechanical bending of the antenna. (a) discharges produced by the air current; (b) discharges induced by bending the antenna with a human hair. To prevent the hair being an effective visual stimulus, black hair was used in the dark. In the unit from the left optic lobe illustrated here, bending of the left antenna (b1) was effective but not bending of the right one (b2). Calibration, 1 mV; 0·5 s.

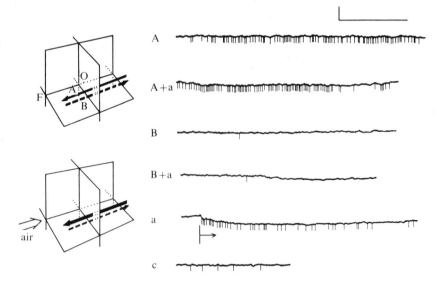

FIG. 19.5. Excitatory effects of air current upon visual responses to a moving spot. In the inset, 0 is the position of the fly with front at F, and F0 the head axis. A and B indicate directions of motion of the spot corresponding to the letters on the records. Solid arrows show the directions in which a spot excites the unit; of these, the thick arrow shows that the response is facilitated. Dashed arrows show that no responses, neither excitatory nor inhibitory, are induced by the spot moving in this direction. The upper inset indicates responses without the air current, and the lower with it. Records A and B show the spikes elicited by movement of the spot in directions A and B, respectively, and in *a* the spikes elicited by the air current alone, onset of which is denoted by an arrow below. Thus, A + *a* and B + *a* indicate that the movement stimulus was given during the air current stimulation. *c* is the background discharge in the dark. Calibration, 1 mV; 0·5 s.

The interactions between visual stimulus and air current are of four types:

(1) *Excitatory effects.* In this class, the unit responds to either air current or spot and the responses sum (Fig. 19.5). In the example illustrated in Fig. 19.6, the effect of the air current alone was subliminal, causing no spike, but it enhanced the response to movement of the spot when the two stimuli were presented together. The facilitatory effect of the air current is not necessarily equal in all directions in directional-selective units (Fig. 19.7, in direction E only).

(2) *Indifferent effects.* The units in this category responded to movement of the object, but gave only the spontaneous background discharge to the air current (Figs. 19.8 and 19.9). The unit in Fig. 19.8 responded to motion in two opposite directions until the air current was applied, and then stimulation only from front to back remained effective (A in Fig. 19.8). The unit in Fig. 19.9 failed to respond to movement of the spot while the air current stimulus was acting, and showed only a slight response to the air current alone.

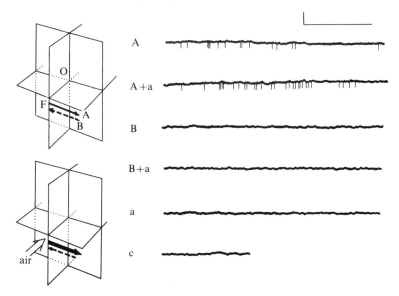

FIG. 19.6. Excitatory effects of the air current upon responses to the moving spot. For symbols see Fig. 19.5.

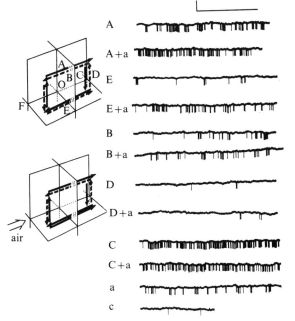

FIG. 19.7. Further excitatory effects of the air current upon responses to the moving spot. For symbols see Fig. 19.5. In this unit of the integrative type, the air stimulation was effective in direction E only.

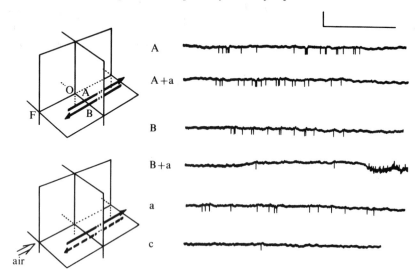

FIG. 19.8. Indifferent effect of the air current upon responses to the moving spot. For symbols see Fig. 19.5. The burst-like fluctuations at the end of record B+*a* are an artefact of muscular activity induced by the air current.

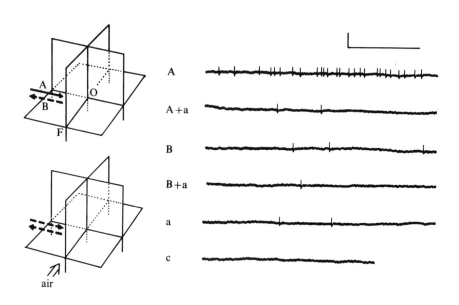

FIG. 19.9. Indifferent effects of the air current upon responses to the moving spot. For symbols see Fig. 19.5.

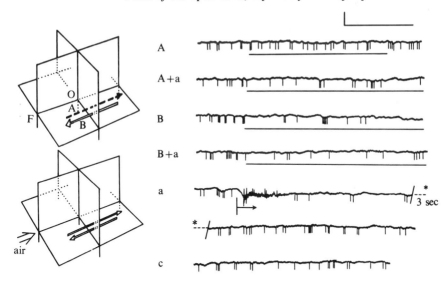

F IG. 19.10. Inhibitory effects of the air current upon responses to the moving spot. Open arrows indicate that spontaneous discharges are inhibited by a spot moving in this direction; for other symbols see Fig. 19.5. In record *a*, the air current started at the time indicated by the arrow, at first causing an artefact from movement of the wings. The record *a* continues in a second line, after the star, which marks a 3 s interval, after which the unit is adapted to the air current. In this unit a spot moving in direction A elicited no responses without the air current, whereas spontaneous discharges were suppressed by the air current.

(3) *Inhibitory effects.* In this category a definite inhibition was produced by the air current. In one example (Fig. 19.10) spontaneous discharges were inhibited by the onset of the air current, but this effect adapted in a few seconds. In the adapted state, the spontaneous discharges were inhibited by the movement stimulus whereas without the air current the unit had been indifferent to the movement stimulus. However, the air current had no effect upon the inhibition caused by movement of the spot in the opposite direction. In other words, by applying the air current, the effective stimulus causing inhibition of this unit was changed from one direction to two opposite directions.

(4) *Disinhibitory effects.* In this type, the inhibition caused by movement in the visual field is removed by the air current, and replaced by excitation. In the example (Fig. 19.11), the inhibition by movement in the opposite direction (a), however, was not affected by air current.

These changes in the properties of the directionally-selective units when the air current is turned on, vary in each unit and it is difficult to make sense of them. However, the inputs from the compound eye and the wind-sensitive mechanoreceptor certainly converge and therefore the pattern of visual

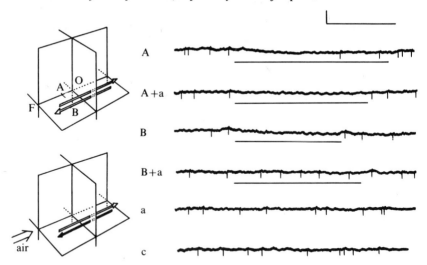

F IG. 19.11. Disinhibitory effects of the air current upon responses to the moving spot. For symbols, see the legend of Figs. 19.5 and 19.10. This unit did not respond to the air current alone, as in (a). However, the air current converted the inhibition by the spot moving in direction B to an excitation.

information in the insects' motion-sensitive system is not the same in a flying insect as in one seeing the same motion but not experiencing the wind velocity of flight. It is not possible to understand these differences as yet but we suppose them to be adaptive.

ACKNOWLEDGEMENT

Supported by a grant from the Education Ministry of Japan; I am grateful to Prof. M. Kuwabara and Dr. H. Tateda, Department of Biology, Faculty of Sciences, Kyushu University, for their support and encouragement in these experiments.

20. Vision during flight

T. COLLETT and A. J. KING

20.1. Introduction

An insect must be able to detect and identify objects in its environment during flight. It can then approach or follow some objects and flee from others, while certain features of the terrain may provide guiding landmarks. When an insect moves the image of the surroundings flows across the eyes. Although this retinal image movement is an important cue for maintaining stable flight, it also disrupts the detection and analysis of particular features in the environment.

The problem of analysing visual stimuli against a moving background is common to most animals and many have adopted the strategy of making eye or head movements to minimize the disturbing effects of locomotion. Thus some birds perform a series of rapid forward head movements while walking; between each such saccade they keep the head still relative to the ground and the body moves forward to catch up with the head. So for most of the time the head is stationary; vision is only disturbed during head saccades. However, complete stabilization is rarely a feasible solution and for many animals just one part of the retinal image remains stationary, while the rest of the visual field moves about this point. The dogfish bends its head from side to side as it swims; at the same time it counter-rotates its eyes so that the eye on the inside of the curve fixates a vertical strip three feet away, and only this region of the visual field is stable (Harris 1965). In mammals, saccades enable selected parts of the surroundings to be viewed by the region of retina with greatest acuity. Eye and head saccades are also used during locomotion to point the fovea in the direction in which the animal moves, which helps keep the image stationary on the fovea.

Some insects make head movements during flight for the same kinds of reasons. Land (1973) has shown that if a blowfly is suspended so that movement is restricted to rotation about a vertical axis, changes of direction occur as the result of head and body saccades. In the intervals, despite fluctuations in the direction in which the body points, compensatory movements keep the head clamped with respect to the visual surroundings. The visual consequences of turns made by means of saccades will be brief. For a fly able to move forwards, as well as turn, the situation will presumably be somewhat different. It will keep its head stationary relative to some point in front, and so only the images of very distant objects, or points directly in the line of flight, will be stationary. In this case the advantages of making saccades are twofold: it is easier to distinguish voluntary from involuntary turns, and at least for the front of the eye the retinal image will usually be almost stationary. Between saccades any movement of the image in this region can be attributed to unexpected perturbations, for instance a gust of wind. However, the moving background will still interfere with the detection of targets on the lateral retina.

In this chapter we consider some of the problems that flying insects must solve if they are to use their visual system effectively during flight. We present neurophysiological evidence from recent experiments on hoverflies for the existence of separate systems for feature detection and orientation. We then discuss some of the properties feature detectors must have if they are to be used during flight. Secondly, we consider the control of flight orientation by means of optomotor responses, and ways in which large-field motion detectors can generate optomotor torque to correct unintended deviation. Thirdly, we consider neurophysiological mechanisms enabling flying insects to track moving targets.

20.2. Feature detection in hoverflies

In the optic lobes of hoverflies we have recorded from two functionally distinct classes of movement detector. One class consists of directionally selective units with receptive fields (RFs) which cover most of one eye, with properties similar to those of large-field units in calliphorid flies described later. The other class are small-field units with properties which make it impossible for them to contribute to the responses of the large-field units and extremely unlikely that they play any part in a feedback system controlling orientation.

We have recorded the small-field units with electrodes placed in the medulla, external chiasm, and lobula. The units are not spontaneously active and give little or no discharge at on or off. Their RFs are approximately circular, c. 20° across. Strong responses are given to stimuli moving across the RF at more than c. 30° s^{-1}, if the stimulus conforms to a set of requirements, which varies in detail from unit to unit. Many, but not all, units are directionally selective. Most of those that are have similar preferred directions, responding best to approximately backward movement (Fig. 20.1). Some units show strong preferences for dark patterns, with almost no response to light ones (Fig. 20.2). Others have weaker preferences, or none.

The units described here responded to small spots moving through their RFs, but not to the movement of long bars extending beyond the RF. They also failed to respond to gratings or single edges. Not all stimuli smaller than

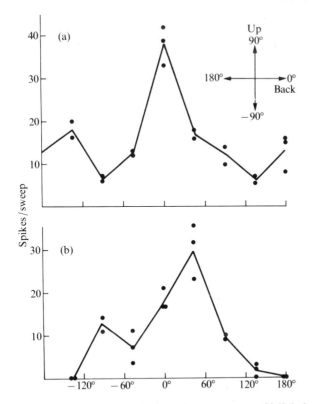

FIG. 20.1. Response of two directionally-selective units to movement of 2·5° dark spot through RF in various directions. Traverse starts and finishes outside RF. (a) Unit with weak directional selectivity. Spot moves 36°, at 86° s⁻¹. Unit 12; *Syrphus* sp. RF is 15° × 20°, with centre at (0°, +40°). (b) Unit with strong directional selectivity. Spot moves 37°, at 134° s⁻¹. Unit 16; *Syrphus* sp. RF is 15° × 20°, with centre at (0°, +55°). (RF dimensions are given throughout as horizontal and vertical angular extents, in that order. RF positions are given throughout as (azimuth, elevation) of the RF centre in the visual field. Azimuth is positive on the right of the mid-line, negative on the left. Elevation is positive above, negative below, the equator of the visual field. These are the co-ordinates used by Bishop, Keehn, and McCann (1968). 'Forwards' and 'backwards' refer to the visual field of one eye; so spots moving forwards are moving towards the mid-line. In Figs. 20.1–20.7, 20.9, 20.11, 20.12, there are two brightness levels: 'dark' is 8 cd m⁻², and 'bright' is 25 cd m⁻²).

the RF were equally effective. We studied the effects of varying the contrast, velocity and size of the pattern, altering its horizontal and vertical extent independently.

Many units responded strongly to spots between 2° and 4° across, but not to smaller or larger ones (Figs. 20.2 and 20.3). In order to analyse this selectivity we compared the responses to spots of different diameters with the responses to bars of different heights but of constant width. Some units behaved in the way illustrated in Fig. 20.4b. Larger bars elicited smaller

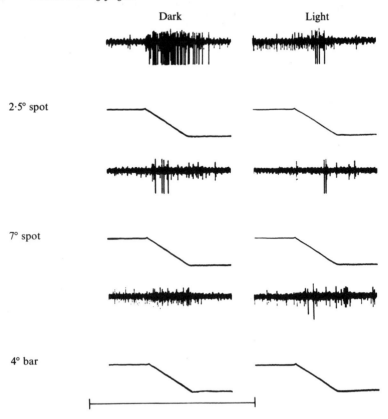

F IG. 20.2. Typical responses of unit which responds selectively to small dark spots. Unit has strong directional selectivity (same unit as Fig. 20.1b); no responses to any pattern for movement in anti-preferred direction. Responses shown for dark and bright patterns, moving backwards horizontally through RF. Patterns move 37°, at 134° s^{-1}. Time mark: 1 s.

responses, and the response was roughly the same for bars and spots of equal height. These units were thus only excited if the extent of the stimulus perpendicular to the direction of motion was much smaller than the RF. Extent along the direction of motion was less critical, and these units responded well to 3° tongues.

For reasons which are still uncertain, the size of the most effective spot is for some units (perhaps all) dependent on its velocity. One example is shown in Fig. 20.5. At 70° s^{-1} the unit responded strongly to a backward moving 2·5° spot, but not to a 7° spot, but at higher velocities the 7° spot was preferred. The extent of a stimulus perpendicular to the direction of motion determines the number of subunits excited simultaneously, whereas extent along the direction of movement determines the time interval between the passage of the leading and trailing edges. Therefore units which are affected

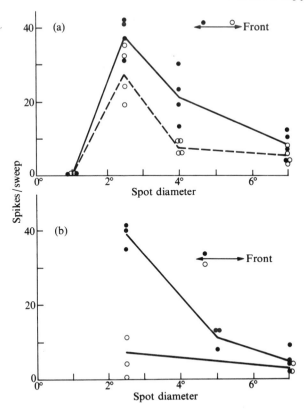

FIG. 20.3. Responses of two units to spots of various sizes. (a) Unit with weak directional selectivity. Spots move 35°, at 65° s⁻¹. Horizontal movement through RF. Lines join the means of several responses for each spot size and direction of movement. Solid line: backwards movement; broken line: forwards movement. Unit 9; *Eristalis* sp. (b) Unit with strong directional selectivity. Same unit as Fig. 20.1b. Response to backward movement of dark and bright spots of various sizes; lines connect means of several responses. Solid circles indicate dark spots; open circles indicate bright spots.

by both leading and trailing edges may show some reciprocity between size and velocity. Whatever the mechanism responsible for this change in preference, it has a possible functional role. It enables the unit to recognize at different distances objects of a fixed size moving at a characteristic speed (e.g. other insects). This property may be more valuable than specificity for absolute retinal image size.

Other kinds of unit responded best to different features, or combinations of features. The unit of Fig. 20.4a was excited best by 5° spots and less by smaller or larger spots. No response was given to 12° spots. However, when tested with 2·5° wide bars of different heights, the response failed only for bars higher than 20°. The selectivity of this unit appears to be a function of both the horizontal and vertical extents of the stimulus.

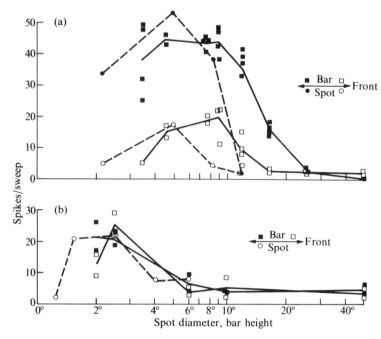

FIG. 20.4. Responses of two units to spots of different sizes and to 2·5° wide bars of different heights, moving horizontally through RF. Bars oriented vertically. (a) Unit with weak directional selectivity. Lines join the means of responses to several traverses. Solid lines indicate dark bar movement; broken lines indicate dark spot movement. Open symbols: forward movement; solid symbols: backward movement. For bars symbols represent responses to single traverses; for spots symbols represent the mean response of several traverses. All stimuli move 40° at 266° s⁻¹, starting and finishing outside RF. Unit 28; *Syrphus vitripennis*, ♂. RF is 25° × 12°, centred at (−12°, +26°). (b) Unit with weak directional selectivity. Same unit as Fig. 20.3a. Bright spots and bars. Lines join means of several responses. Broken line and open circles represent spot movement (only backward movement shown; symbols represent means). Solid line indicates bar movement; solid squares, individual responses to backward motion; open squares, responses to forward motion. All patterns move through 35° at 65° s⁻¹.

The units have inhibitory surrounds with properties different from those of the RF. The surround inhibition is not directionally or orientationally selective (Fig. 20.6) and it shows spatial summation: gratings have a more powerful inhibitory effect than a single edge or a bar. Rough plots of the size of the inhibitory region showed it was at least twice as large as the RF. The surround strength was graded, weakening with distance from the RF. Bars or edges moving through the RF at the same time as an effective spot abolish the response to the spot (Fig. 20.7). We do not know whether suppression under these circumstances can be attributed entirely to the non-directional surround mechanism.

The properties of the surround will prevent the unit from responding to effective features while large objects are moving close to the RF, or when the

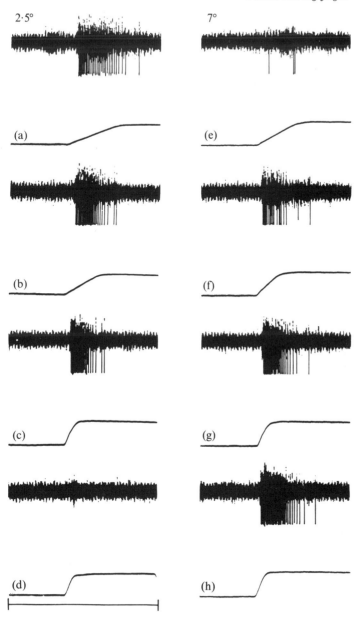

FIG. 20.5. Response to dark spots moving at different velocities. Directionally selective unit; spots move through RF in preferred direction (backwards, horizontal); no response to movement in anti-preferred direction. Unit did not respond to bright spots, or to bars of any size. (a)–(d): 2·5° spot; velocity changes from 72° s^{-1} in (a) to 430° s^{-1} in (d). (e)–(h): 7° spot; velocity changes from 92° s^{-1} in (e) to 370° s^{-1} in (h). Spots always move between same two points 25° apart. Unit 10; *Syrphus vitripennis*, ♂. RF: 20° × 20°, centred at (0°, +65°). Time mark: 1 s.

444   *Vision during flight*

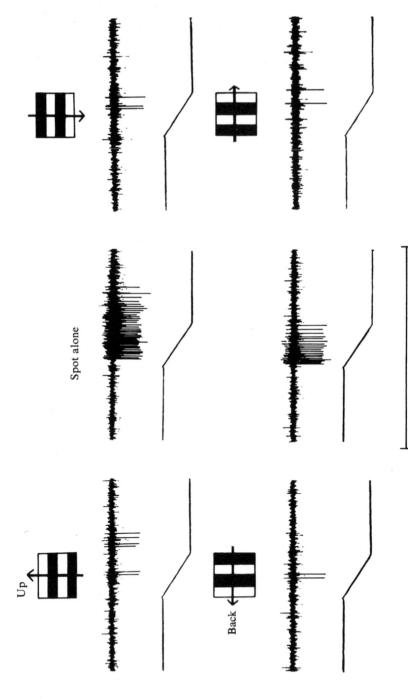

FIG. 20.6. Effect of grating moving outside RF on response to spot moving across RF. Spot always moves on same path: backwards, horizontal, between two points 35° apart, at 195° s^{-1}. Monitor indicates stimulus movement. Grating is 12° square wave; driven by same waveform as spot, so monitor indicates movement of spot or spot and grating. Directions of grating movement as indicated. Horizontal grating movement: grating moves in 70° × 30° region below RF. Vertical grating movement: grating moves in 30° × 70° region lateral to RF. Same unit as Fig. 20.1b.

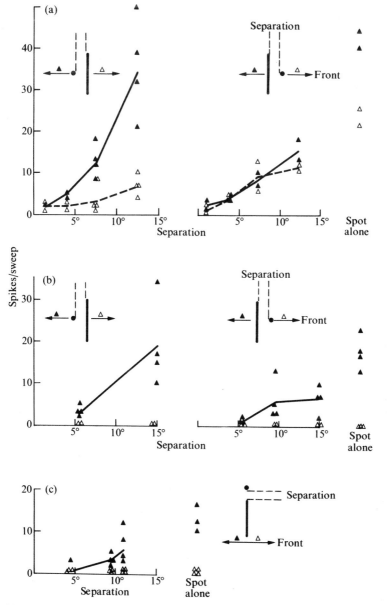

FIG. 20.7. Simultaneous movement of bar and spot at indicated separation in same horizontal direction between two points. Solid symbols: backward movement; open symbols: forward movement. Relative positions of bar and spot as indicated. 2·5° dark spot; 2·5° × 50° dark bar, with long axis oriented vertically. (a) Unit with weak directional selectivity. Same unit as Fig. 20.3a. Pattern moves through 35° at 65° s⁻¹. (b) Unit with strong directional selectivity. Same unit as Fig. 20.1b. Pattern moves 37° at 134° s⁻¹. (c) As (b) except that spot and bar separation varied vertically.

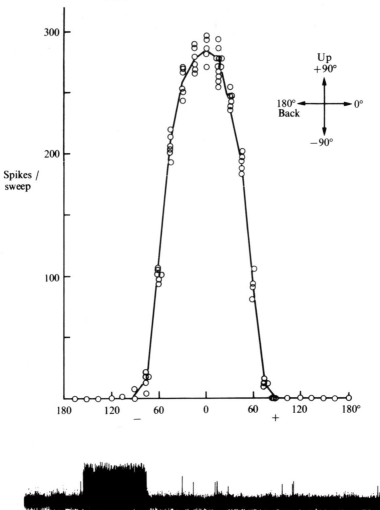

FIG. 20.8. Response of wide-field ipsilateral directionally selective movement detector, with no spontaneous activity to 9·5° square wave grating in 40° circular aperture centred at (+22°, 0°). Grating moves at 5·3° sec⁻¹. Bright bars are 136 cd m⁻², dark bars 22 cd m⁻², and surround 8 cd m⁻². Stimulus cycle composed of 1·7 s movement followed by 3·3 s no movement. Above: variation of response magnitude with direction of movement. Below: typical response to one cycle of horizontal movement. Time mark is 2 s. *Syrphus* sp, ♂. Unit 1.

whole background is moving. We have not yet measured the velocity characteristic of the surround. The relation between the effect of velocity on the RF and on the surround will determine the extent to which the units are active during flight. We suspect the surround is activated at somewhat lower velocities than the centre. If the surround is sensitive to low velocity then the units will be shut off during flight, except perhaps for a small group in the direction of flight, for which, in an open environment, movement of the background across the surround would be very slow.

Several properties of these units make it impossible for them to contribute to the large-field movement detectors. Inhibitory surrounds prevent these feature detectors from responding to gratings and large bars, which are powerful stimuli for the large-field units. The feature detectors have little or no response at on or off, while the large-field units give powerful on and off responses. The feature detectors have a higher velocity threshold than the large-field units, which give a rather uniform response during movement, starting when the movement starts, stopping when it stops (Fig. 20.8), with little habituation. The feature detectors respond with a triggered burst, which often outlasts the presence of the exciting pattern in the RF (Figs. 20.2 and 20.6). When an effective pattern is presented to feature detectors several times in rapid succession (e.g. at 2 s intervals), the response is likely to be weak the second time, and to disappear by the fourth time; intervals of 15–20 s are required if the response is to be maintained. Units whose response is unreliable, and not related in any simple way to the duration or velocity of movement, are not suitable elements for closed-loop servo-control of orientation; they may trigger open-loop responses to objects that they recognize.

20.3. Object recognition during flight and the problem of relative movement

If objects can be identified by features independent of movement, then background movement during flight will not impede recognition. Colour is such a feature. Morpho butterflies will turn to a blue lure at a distance of 15–20 m (Swihart 1972). This response simply requires blue sensitive neurons which are not inhibited by relative motion.

In some circumstances movement itself may be a cue. Male grayling butterflies will pursue many passing objects in the hope that one is a female grayling. Experiments with dummies (Tinbergen 1972) showed that colour and shape are not important; pursuit is excited and possibly in part maintained by a dark object moving erratically. The irregular flight of both male and female will ensure that from moment to moment the speed and direction of the image of the female across the male's retina will differ from the background. This raises the problem of the detection of relative motion. To be sensitive to a moving butterfly during flight a detector should respond to the moving target but not to the moving background.

Because movement in the surround inhibits their response, the units recorded in hoverflies can distinguish between target motion and image motion resulting from the insect's own movements. In this respect they are similar to visual cells in the ventral nerve cord of some Orthoptera (Palka 1969, 1972; Rowell 1971a). However, because the surround is activated by movement in any direction the hoverfly units do not detect relative motion; most will simply be shut off during flight.

Units which do detect the relative motion of a moving target against a moving background have been described in the anterior optic tract of the privet hawk-moth (Collett 1971a, 1972). These cells have directionally-selective excitatory centres and large directionally-selective inhibitory surrounds, so arranged that the centre response is only inhibited if surround motion is in the preferred direction of the centre. Consequently, the units will not respond to the motion of a large grating, or, presumably, to image motion of the environment when the insect moves. On the other hand they will respond if a target and grating move in opposite directions across the eye (see Section 20.6 on 'visual tracking' for a fuller description).

The anterior optic tract units are of little use if the background and target travel at different speeds in the same direction across the retina. Obviously, if the background moves very slowly and the target rapidly, units may respond to target motion, but not to the background, because the inhibition and excitation evoked by the moving background are both relatively weak. Indeed, Rowell (1971a) has shown for the orthopteran DCMD unit mentioned above that background motion at speeds of less than $5° \text{ s}^{-1}$ does not weaken the response to the target, and that above this lower threshold the strength of the inhibition increases with background speed. Whether there are neurons with more elaborate ways of detecting relative motion is not certain (but see Rowell 1971b).

The problem of recognizing the shape of a moving target has received little attention. In mammals the oculomotor system ensures that selected moving targets can be fixed on the retina. This may be true for insects as well, if saccades are used to bring selected stationary targets into the direction of flight. However, the images of objects not in the direction of flight will be distorted, as will all rapidly moving objects. If geometrical features are to be used for the recognition of moving images special tricks are needed. The behaviour of the hoverfly feature detectors suggests two possibilities. (1) Images of stationary objects and other moving insects will travel mainly horizontally across the retina. Thus vertical features will be less distorted than horizontal. Some of the hoverfly units fire to stimuli of any horizontal extent, but are only excited if the height of the image falls within a narrow range. Thus the specificity of these units is provided by vertical features. (2) Some units adjust their size preference according to the velocity of the image, and so are designed to recognise an object of a characteristic size and speed at different distances. Here the specificity is given by the ratio of speed to size. The lesson to be learnt from these examples is that geometric features of moving images can be used for shape recognition provided that the image distortions can be predicted. One can either use some parameter that remains undistorted, or one can combine properties of the image in such a way that the combination is invariant with image motion. See Fig. 22.2 this volume.

20.4. Orientation in flight: the role of the optomotor system

In addition to detecting targets during flight, an insect has to orient in its visual surroundings. In this section we discuss the role of the optomotor system in aerial navigation. A way in which it can act is seen clearly by considering the behaviour of hoverflies: they hang motionless in the air for seconds at a time. The optomotor system contributes to this hovering flight by providing an efficient velocity servo in which image motion in any direction across the retina produces a restoring force in the same direction, so minimizing the velocity of the image. The error signal for this servo could be provided by motion detecting neurons monitoring movement of the whole environment across the retina. Such neurons have been found in moths (Collett and Blest 1966), flies (Bishop and Keehn 1967), and bees (Kaiser and Bishop 1970). They are directionally selective to movement across the whole field of one or both eyes. Their preferred directions are either horizontal or vertical, so they are able to provide a combination of outputs signalling drift velocity of the eye. To give good stability the neurons providing the error signal should respond strongly at relatively low velocities (Fig. 20.9).

The optomotor system also helps to stabilize locomotion. Experiments by Wilson and Hoy (1968), for example, suggest strongly that the optomotor system can keep an insect on a straight course. In the dark, or in the light under open-loop conditions, the bug, *Oncopeltus*, tends to walk in circles, which for an individual bug are likely to be always in the same direction. However, when the bug walks freely in the light, it can maintain a straight course. In this case the optomotor system apparently corrects an inherent asymmetry in the locomotor system. For a hovering fly the optomotor reflex will tend to reduce the output of the movement detectors to zero, assuming for simplicity that the spontaneous rate of the movement detectors induces negligible torque. In the second example, where the system acts to keep the bug on a straight course while moving forwards, the situation is more complicated. It is this situation we wish to discuss here.

Consider an insect moving within an environment in which objects are uniformly distributed. When the insect moves forwards the images of objects will tend to travel backwards across the retina. The speed of retinal image motion will depend on both the distance of the objects from the fly and the

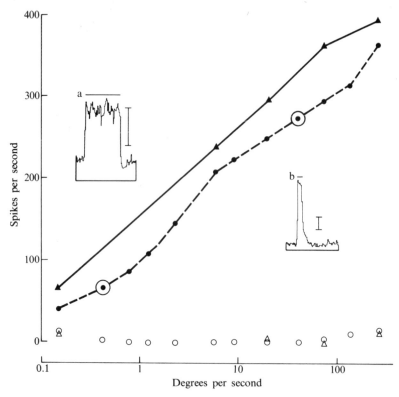

FIG. 20.9. Response of wide-field contralateral movement detector with forward preferred direction to stimuli moving at different velocities. Patterns projected on to a 70° × 70° tangent screen. Circles represent responses to 5° × 36° bar which moves through 25° between −30° and −5°. Bar extends vertically from +32° to −4°. Triangles indicate responses to 12° period square wave grating. Grating covers screen and moves 25°. Solid symbols indicate forward motion, open symbols backward movement. Data points represent average firing rate over five sweeps. Insets show computer print-outs of the response to bar movement at two velocities, averaged over five sweeps. The means of these responses are indicated by the encircled symbols. Time calibration for (a) is 59 s, for (b) 625 ms. Vertical calibration 50 spikes s^{-1}. *Calliphora erythrocephala*, ♀. Unit C31.

region of retina they stimulate. Thus the images of near objects will move faster than more distant ones, and images will move more rapidly across lateral than across anterior or posterior retina. However, if an insect is stationary, except for rotation about a vertical axis, the velocity of image movement will depend only on the angular velocity of the insect. These two situations are combined when an insect flying forwards deviates from a straight course and the image movement generated by the turn is super-imposed on that due to the forward component of movement. The eye on the outside of the turn usually sees images moving backwards, whereas the eye on the inside of the turn sees images moving both backwards and forwards. The exact distribution of image movement will depend on the translational and angular velocities of the insect with respect to the world, and on the distribution of objects in it. However, for the eye on the inside of the turn there will always be some forward motion in the front and the back, whereas lateral retina will predominantly be stimulated by backward motion. The velocity feedback signals are therefore derived from the sums and differences of forward and backward movement across the two eyes. The situation is slightly more complicated if translatory motion has a transient sideways component, caused for instance by a sudden cross-wind. However, in some Diptera at least, the present analysis is unaffected by this extra degree of freedom, as the resulting image movement over the front of the eye generates appropriate compensatory torque, whereas image movement over the back of the eye can be neglected because in this region the sensitivity of the optomotor system is weak (McCann and Foster 1971).

The feedback signals are separated in that they come from neurons with forward and backward preferred directions. There is little doubt that both classes of neurons can contribute to optomotor torque. A variety of flying insects when stimulated with backward (or forward) movement across one eye produce torque that makes the insect turn to the same (or opposite) side. Goetz (1972) has investigated the optomotor behaviour of walking *Drosophila* under open-loop conditions. If the two eyes are presented with similar stimuli moving backwards at different velocities, the insect turns to the side of the faster stimulus (but see Section 20.4.4 on anomalous torque). In this case torque is generated because of a difference in the speed of backward motion across the two eyes.

In some Diptera the neurons which probably mediate optomotor torque are most sensitive to movement in a frontal region of the eye, centred some 20° from the mid-line (Bishop, Keehn, and McCann 1968). Consequently, during forward flight the image of the background will travel relatively slowly across the region of highest sensitivity. Therefore, should the insect deviate from a straight course, the visual feedback resulting from the turn is, in the region of highest sensitivity, superimposed upon a relatively weak level of excitation resulting from the forward movement of the insect. It is thus likely that over part of the region of highest sensitivity there will be a net forward component to image movement for the eye on the *inside* of the turn.

Thus corrective torque during flight can be generated in two different ways: (1) the difference in the speed of backward motion across the two eyes; (2) the presence of forward motion across one eye. Our problem is to determine the relative contributions of the forward and backward motion detectors under normal conditions. To make accurate measurements of differences in the speed of backward motion across the two eyes requires more sophisticated neural equipment than a system which simply detects and eliminates forward motion. It is worth considering some of the difficulties that arise if flight is to be stabilized by equalizing the responses of backward sensitive neurons.

20.4.1. *Asymmetries in signal transmission*

If the neurons with backward preferred direction make a substantial contribution to stabilization during flight, it becomes essential to eliminate any inherent asymmetry that may exist in the transmission of signals from the two eyes. As Horridge (1966) emphasized, in feedback circuits which reduce the retinal slip speed to zero, good stabilization is achieved without any great accuracy in the open-loop gain, provided that it is high. But if optomotor torque is controlled by the difference in the speed of backward movement across the two eyes, to keep the insect on a straight course the forward gain must ideally be the same on both sides.

20.4.2. *Saturation*

Clearly, if velocity differences between the two eyes are to be compared, the responses of the wide-field units mediating the optomotor response should vary with image speed over a suitable range of speeds. Figure 20.9 shows the responses of a movement detector with forward direction (type 2al of Bishop, Keehn, and McCann 1968) to a black bar 36° high and 5° wide moving horizontally through 25° at different speeds. The bar swept through the most sensitive region of the unit's receptive field, and the cycle times used gave at least 15 s interval between each traverse. The firing rate varies linearly with the logarithm of the velocity over three orders of magnitude. Very much the same relation was obtained when the stimulus was a 70° × 70° vertical grating of 12° period. This evidence (see also Fig. 6 of Bishop and Keehn 1967) suggests that wide-field movement detectors are sensitive indicators of stimulus velocity both for large and relatively small stimuli over a wide range of speeds. On the other hand the data of Fermi and Reichardt (1963) using optomotor torque as an indicator of the velocity response shows that the torque generated by a fly (*Musca domestica*) placed in the centre of a rotating striped drum may (depending on the number and width of stripes) saturate at much lower velocities than Fig. 20.9 indicates.

20.4.3. *Movement detectors confound stimulus parameters*

If flight is to be stabilized accurately by neurons with backward preferred direction, then the neurons should act as reliable velocity sensors, that is, their responses should ideally vary with velocity and nothing else. However, it is known that the response of some movement detectors varies with the brightness and contrast of a stimulus (Bishop and Keehn 1967), as well as with its size and velocity. If movement detectors confound stimulus parameters, insects would automatically tend to turn towards objects which because of their contrast or size stimulate the neurons particularly strongly, and this may not be where the insect wishes to go.

20.4.4. *Anomalous torque*

Measurements of the optomotor torque generated by stationary flies placed inside a moving drum (e.g. Fermi and Reichardt 1963) show that as the velocity of a moving stimulus is increased the torque rises to a peak and then falls. If the units with backward preferred direction are the principal contributors to torque during normal flight, the system will behave anomalously for stimuli moving faster than the optimal velocity.

During flight the background passes backwards over the retina. If the insect veers to the left, the background will travel more rapidly across the right eye than the left. At relatively low speeds, backward-sensitive movement detectors with visual fields in the right eye will respond more strongly than those with left visual fields. As a consequence the torque induced by the backward units will tend to make the insect turn to the right, thus helping it keep to a straight course. Suppose now that the insect is blown by a strong wind or flies so fast that the image of the background travels across the eye at a greater speed than the peak velocity of the optomotor system. Now any tendency to turn to the left will mean that movement detectors with left visual fields will respond more strongly than those with right visual fields, so making the insect turn even further to the left. This unstable situation would lead to an insect flying upwind, or erratically, if its forward velocity is high. Kennedy (1939) gave a rather similar account of a mechanism which would make mosquitoes fly upwind, and more recently Goetz (1972) has considered the same problem.

None of the problems listed above arise if only the forward sensitive neurons are responsible for controlling torque during flight. If both sets of neurons are active, those with forward preferred direction will tend to reduce any deviation in flight path caused by the asymmetrical response of the backward sensitive ones. Although the relative contributions of the forward and backward motion detectors to flight stabilization have not been assessed under natural conditions, there are some indications that the forward-sensitive neurons may play the more important role.

(1) During flights which last more than a few seconds the response of neurons with backward-preferred direction will be partially adapted, whereas the forward-sensitive neurons which only respond during turns will be relatively fresh.

(2) In the privet hawk-moth, neurons with forward- and backward-preferred direction behave differently when presented with two moving stimuli, one travelling in the preferred direction, the other in the anti-preferred direction (Collett 1971b). Forward motion across the

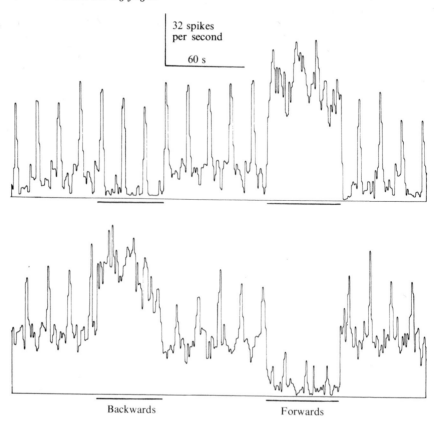

32 spikes
per second

60 s

Backwards   Forwards

FIG. 20.10. Responses of two wide-field movement detectors projecting from the optic lobe of the privet hawk moth recorded simultaneously. The preferred direction of the unit in the top trace was forwards, the bottom trace backwards. A 40° high, 3° wide light bar moves back and forth through 40° in 3 s in an anterior region of the field. The peaks in firing rate are caused by bar movement in the preferred direction, the troughs by movement in the anti-preferred direction. When indicated by the bars beneath the traces, a grating of 30° period moved at about 30° s^{-1} horizontally, backwards or forwards, in the posterior half of the field. Firing rate is averaged over 1 s bins. Reproduced from Collett (1971b).

receptive field of a neuron with backward-preferred direction suppresses the response to a target moving in the preferred direction elsewhere, but backward motion has no effect on the response of a neuron with forward-preferred direction to a second target moving in the preferred direction (Fig. 20.10). In blowflies, movement detectors with forward-preferred direction across one eye enhance the response of movement detectors with backward-preferred direction across the other eye, but the backward-sensitive neurons do not affect the forward-sensitive ones (McCann and Foster 1971). Both features will tend to amplify the effect of a forward-moving stimulus by suppressing the activity of units responding to backward movement on the same side, or enhancing the activity of the backward-sensitive neurons on the opposite side. However in *Drosophila* behavioural experiments suggest there is no such asymmetry (Goetz 1964).

(3) Most measurements of torque generated during flight have failed to separate the torque due to the optomotor system from that generated by the saccadic system. Since there is evidence that saccades may be visually triggered (Fig. 12 of McCann and Fender 1964; reproduced by Land in Chapter 21), statements about the relative strengths of optomotor torque resulting from forward and backward motion are difficult to interpret. It has recently been reported that, in some flies, backward-stimulus motion generates more torque than forward motion (Heisenberg 1972; Reichardt 1973). This difference could be the result of asymmetries in the saccadic or optomotor systems, or simply because during backward motion the optomotor and saccadic systems cause the fly to turn in the same direction, whereas during forward motion the torques generated by the two systems are of opposite sign. Land discusses this question in more detail in Chapter 21.

Thus the optomotor system provides some stability in flight, but it is supplemented by other visual and non-visual control mechanisms which may in addition determine the *direction* of flight. Bees and ants, for instance, are able to orient by maintaining a constant angle with respect to the sun. von Frisch's classical experiments (von Frisch 1967) have shown that foraging bees can use colour as a cue to a good feeding place, so in some way the colour of a target must be used to control the direction of flight. However, recent experiments by Kaiser (1972) and Kaiser and Liske (1972) argue compellingly that the bee optomotor system receives inputs principally from green receptors and is therefore colour blind. In these circumstances the optomotor system is unlikely to control the direction of flight, but merely stabilize a course determined in other ways. Land's (1973) studies of head saccades in blowflies suggest that flies may make rapid saccades towards particular features in the environment in order to change course, and between turns stabilize their flight course partly by means of the optomotor response. Furthermore, the saccadic system can prevent the fly from being pulled in the direction of irrelevant features which stimulate the optomotor system asymmetrically.

20.5. Directionally-selective units as velocity detectors

The questions raised in the previous section prompted us to examine how accurately the wide-field directionally-selective units of the blowfly serve as velocity sensors. Features in a natural environment will be of varying sizes and shapes, so to act as a reliable monitor of the speed of background movement the wide-field units should ignore the spatial properties of the stimulus. It was interesting, therefore, to find that the relation between firing rate and stimulus velocity is largely unaffected by the form of the moving stimulus, provided that the stimulus passes through the RF region of peak sensitivity.

Units sensitive to forward motion were the most convenient to work with. To minimize damage motion detectors with left visual field were usually recorded in the right optic lobe. Data from units were discarded if the spontaneous firing rate was more than *c*. 10 spikes per sec, since higher rates were associated with unreliable and weak responses. The number of tests that could be performed on a single unit was limited by the gradual deterioration of the preparation and because of the 15 s interval between each traverse of the stimulus necessary to make all the responses independent of each other.

In Figs. 20.9, 20.11, 20.12, and 20.13 the firing rate of a forward-sensitive unit is plotted against stimulus velocity for different patterns. In all cases the stimulus passed through the most sensitive region of the RF. Each point is the average firing rate during five sweeps of the stimulus. The slope of the relation does not vary greatly with the number of bars (Fig. 20.9), or with

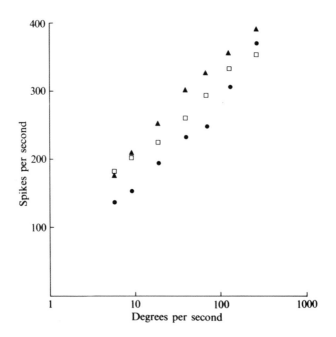

F1G. 20.11. Velocity characteristic of contralateral wide-field movement detector with forward preferred direction to single 36° high bars of different widths moving horizontally between −25° and 0°. Circles indicate bar width is 1·5°, squares 7°, and triangles 17°. Only responses to forward motion are shown. *Calliphora erythrocephala*, ♂, unit C32.

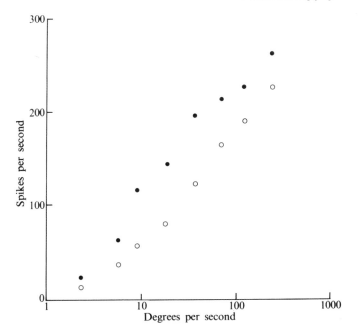

FIG. 20.12. Velocity characteristic of contralateral wide-field movement detector with forward preferred direction for bars 1·5° wide and of different heights. Open circles indicate bar height is 3°, solid circles a bar height of 39°. Bars move 25°, between −25° and 0°. *Calliphora erythrocephala*, ♀. Unit C33.

the width (Fig. 20.11), or height (Fig. 20.12) of individual bars. In the experiment of Fig. 20.13 the unit was stimulated by a single vertical bar, 16° wide by 36° high, on a light background, or by a light bar of the same dimensions on a dark background. It is clear that for these parameters contrast reversal has a negligible effect on the unit's velocity characteristic.

These findings suggest that the response of the unit is principally determined by stimulus velocity. However, more complicated results are obtained if stimuli do not pass through the most sensitive region of the RF. The velocity characteristic of the upper part of the eye differs considerably and consistently from that of the lower; the upper responds better than the lower to high speeds, whereas the converse holds for low speeds (Fig. 20.14). Despite these complications it is true to say that, for stimuli passing through the region of peak sensitivity, the unit acts as a remarkably faithful velocity sensor, ignoring other aspects of the stimulus.

In the previous section we argued that it is of little consequence if the velocity information gathered by motion detectors is distorted by other stimulus parameters, so long as the units do no more than participate in a velocity servo which drives their output to zero. However, the behaviour of

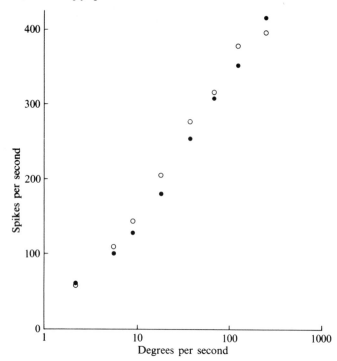

FIG. 20.13. Velocity characteristic of contralateral wide-field motion detector with forward preferred direction to a single light bar (open circles), or dark bar (solid circles). Bar moves 25° horizontally between −30° and −5°. *Calliphora erythrocephala* ♂. C35.

the units in the present experiments suggests they might contribute to a more sophisticated control system; one, for example, in which motor commands are generated to make the insect turn at a specified velocity. The optomotor system might then be part of a velocity servo which enables the commands to be executed accurately, in addition to its role in stabilizing locomotion.

However, this suggestion must be considered in the light of experiments on the velocity response of motion detectors when the stimulus is a rotating, striped drum. In this situation, unlike ours, the velocity characteristic of the optomotor response, measured behaviourally (Fermi and Reichardt 1963), and of motion detectors (Bishop and Keehn 1967) depends strongly on the spatial wavelength of the stimulating pattern. The response of the unit partially adapts to continuously moving stimuli (e.g. Bishop, Keehn, and McCann 1968) and the state of adaptation is very likely to be a function both of the velocity and of the spatial characteristic of the pattern. Whether the difference between our experiments and those of Bishop and Keehn can be explained in this way, and whether the effect of spatial wavelength on the

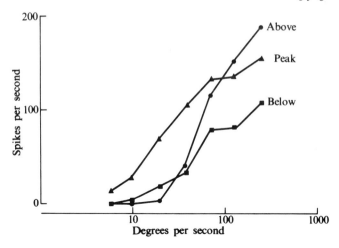

FIG. 20.14. Velocity characteristics of contralateral wide-field motion detector with forward preferred direction to light 4° diameter spot moving horizontally between −30° and −5° in three different vertical positions. The vertical coordinate of the centre of the stimulus for the response labelled peak is −3°, for that labelled above +13°, and for that below −19°. *Calliphora erythrocephala* ♂. C37.

velocity response is important under natural conditions are questions we cannot yet answer.

20.6. Visual tracking

Some insects are able to track moving targets during flight. In this section we will describe in more detail the behaviour of neurons in the anterior optic tract of the hawk-moth, since these neurons are well suited to play a role in a visually-guided tracking system.

To track accurately targets which subtend a small angle on the retina, animals presumably require a system additional to the optomotor system. Indeed, the optomotor system would be expected to work against any attempt to track with smooth pursuit movements a small moving target against a featured background, for the image movement of the background during any tendency to track will result in an opposing optomotor response. Thus rabbits will follow small moving targets with their eyes only if the background does not elicit an antagonistic optomotor response (Rademaker and ter Braak 1948). In flies, where the sensitivity of the optomotor system is highest towards the front of the eye, it may well be that so long as the moving target is large and stimulates large-field movement detectors strongly, a kind of transient tracing can be achieved by the optomotor system by itself, once the target is fixated by the front of the eye. In this case the opposing optomotor torque generated by the background would make the fly follow the target with a velocity error which would depend on the nature of the background and the position of the target on the retina. But for short periods, at least until the target was carried beyond the sensitive frontal region, the fly would tend to follow it. Preliminary experiments by Virsik (cited by Poggio and Reichardt 1973) suggest that this may be so for the house-fly, which follows a moving stripe against a featured background with the sort of lag expected, were the optomotor system responsible for the behaviour. Presumably under normal circumstances, head saccades could periodically abolish the position error, as do saccadic eye movements generated by the primate oculomotor system.

The behaviour of some neurons in the anterior tract of the privet hawk-moth suggests that possibly some insects may have a more sophisticated system for visual tracking. The receptive fields of these neurons are typically 30° or more across and located in the antero-ventral quadrant of the eye ipsilateral to the recording site. The neurons respond optimally to stimuli moving forwards over the eye, backward movement suppressing the resting discharge. They respond to light or dark targets of any shape or size so long as the target is not appreciably larger than the field. Forward motion over the rest of the eye, provided that the speed of the surround stimulus is greater than $3°$ s^{-1}, suppresses the centre response, but backward motion does not. Across the contralateral eye backward, but not forward motion inhibits the response to movement within the receptive field (Fig. 20.15). Directionally-selective inhibition of this kind allows the unit to respond to a target moving in the preferred direction, when the insect is still, or when the insect turns in the same direction, but more slowly than the target. And this is what happens during tracking.

When a grating in the surround moves backwards across the ipsilateral eye, or forwards across the contralateral eye, at speeds of more than $3°$ s^{-1}, the velocity characteristics of the centre region are changed. With no surround motion most of the units investigated only responded to target speeds above $2°$ s^{-1}, the response magnitude increasing with the velocity up to $20°$ s^{-1} (Fig. 20.16). However, with such facilitatory surround movement they are sensitive to targets moving as slowly as $0.06°$ s^{-1} within the centre region, while responses to higher speeds are enhanced (Fig. 20.17). Such behaviour would be appropriate for neurons which form part of a velocity servo for tracking a moving target, with the output of the neurons providing a command signal to torque generators.

Consider an insect starting to track an object moving at a constant speed. As the insect begins to follow the target, the slip speed of the target image across the eye decreases, and the image of the background moves in the reverse direction. When the target image is effectively stabilized on the retina, its slip speed approaches zero, while the background now moves across the retina at a rate equal to the eye speed. If the neurons are to respond

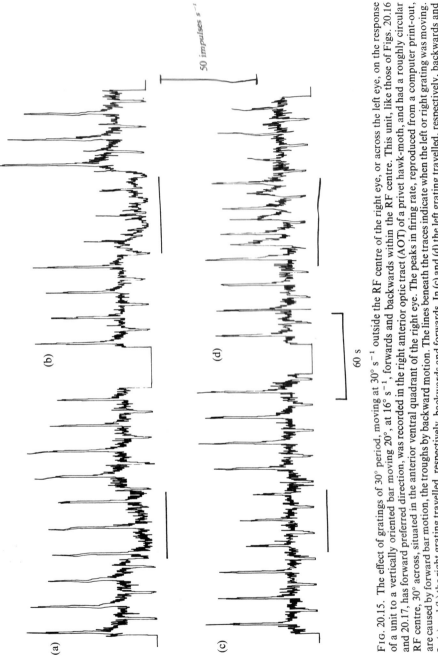

FIG. 20.15. The effect of gratings of 30° period, moving at 30° s⁻¹ outside the RF centre of the right eye, or across the left eye, on the response of a unit to a vertically oriented bar moving 20°, at 16° s⁻¹, forwards and backwards within the RF centre. This unit, like those of Figs. 20.16 and 20.17, has forward preferred direction, was recorded in the right anterior optic tract (AOT) of a privet hawk-moth, and had a roughly circular RF centre, 30° across, situated in the anterior ventral quadrant of the right eye. The peaks in firing rate, reproduced from a computer print-out, are caused by forward bar motion, the troughs by backward motion. The lines beneath the traces indicate when the left or right grating was moving. In (a) and (b) the right grating travelled, respectively, backwards and forwards. In (c) and (d) the left grating travelled, respectively, backwards and forwards. Firing rate is averaged over 250 ms bins. Reproduced from Collett, 1972.

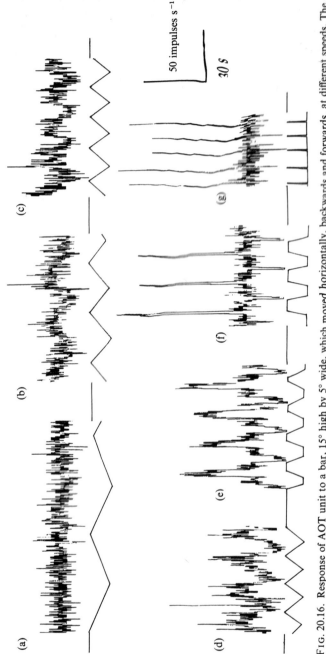

FIG. 20.16. Response of AOT unit to a bar, 15° high by 5° wide, which moved horizontally, backwards and forwards, at different speeds. The 25° sweep of the bar lay within the RF centre. The bar speeds were: (a) 0·8; (b) 1·7; (c) 2·8; (d) 3·6; (e) 6·2; (f) 20·8; (g) 100° s⁻¹. The stimulus monitor below the computer print-out of firing is a diagrammatic indication of the movements of the bar. Upward excursions signify forward movement. Firing rate is averaged over 250 ms bins. Reproduced from *Collett 1971b*.

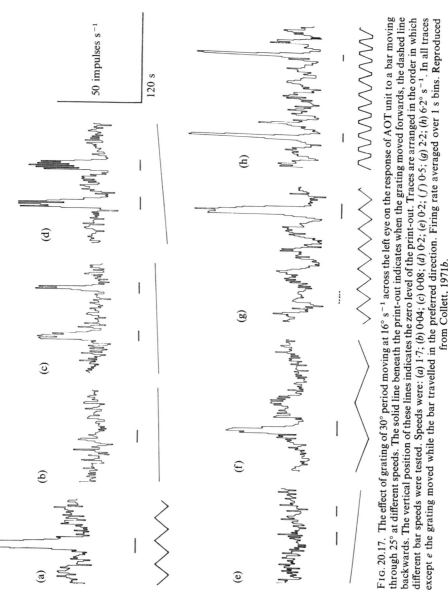

FIG. 20.17. The effect of grating of 30° period moving at 16° s^{-1} across the left eye on the response of AOT unit to a bar moving through 25° at different speeds. The solid line beneath the print-out indicates when the grating moved forwards, the dashed line backwards. The vertical position of these lines indicates the zero level of the print-out. Traces are arranged in the order in which different bar speeds were tested. Speeds were: (*a*) 1·7; (*b*) 0·04; (*c*) 0·08; (*d*) 0·2; (*e*) 0·2; (*f*) 0·5; (*g*) 2·2; (*h*) 6·2° s^{-1}. In all traces except *e* the grating moved while the bar travelled in the preferred direction. Firing rate averaged over 1 s bins. Reproduced from Collett, 1971*b*.

when the insect follows closely behind the target, they must be sensitive to low velocities, but for the initial detection and acquisition of the target only neurons capable of responding to higher velocities are required.

It is worth remarking on a possible advantage of making the neurons insensitive to low velocities unless there is appropriate movement through the surround. The main problem in making a simple velocity servo perform as well as the primate smooth pursuit system is that of stability. As the eye begins to follow a target, the slip speed of the target across the retina drops, so reducing the velocity error signal. This in turn lessens the signal to the effectors, whereupon the eye again lags behind the target and the velocity error signal rises. Oscillations generated in this way would occur if a high open-loop gain were coupled with appreciable delays round the feedback loop. Instability could, however, be avoided if some quantity related to eye speed provided a positive feedback signal to the system. In this case the fall in the velocity error signal when the eye catches up with the target would be balanced by an increase in the signal related to eye speed (Robinson 1971). The speed at which the background apparently travels across the eye is one measure of eye speed. Therefore, it is possible that the facilitation of low velocity responses by background movement helps increase the stability of a velocity error servo.

20.7. Centrifugal neurons to the optic lobe which monitor an insect's own movements

The hypothesis of the previous section suggested that a signal related to eye speed could be used to modify the responsiveness of movement detectors and so enhance the efficiency of a tracking system. Information of this kind relayed back to the optic lobe would also help the visual system distinguish target motion from background motion. In principle such a signal could be derived either from motor commands to the effectors, or from the visual or mechanical consequences of the movement. Both sources of information have their difficulties and inaccuracies. If signals obtained from motor commands are to be employed, they must be related accurately to the intended movement, rather than to the muscular forces needed to perform it. Too little has been discovered about the higher organization of motor behaviour in insects to know whether the appropriate signals are available at all. Visual feedback should provide information about the movement that actually occurred and should not be disturbed by any imprecision in the execution of a motor command. However, in practice, as we have already seen, the quantitative information provided by movement detectors is somewhat ambiguous. It is possibly because of these uncertainties that the behaviour of the anterior optic tract neurons is not a graded function of the speed of surround motion. The direction of surround movement determines whether the response is inhibited or facilitated but the amount of facilitation does not depend reliably on its speed.

Some of the many neurons which in arthropods project centrifugally to the optic lobe are directionally-selective movement detectors with large binocular or monocular receptive fields and response properties which make them suitable monitors of the visual consequences of the insect's movements (Collett 1970). Moreover, as befits neurons carrying such information, there is no sign that they receive anything but visual information: they do not respond to non-visual stimuli, and their response properties are not altered when the flight motor is active. Their behaviour in fact resembles rather

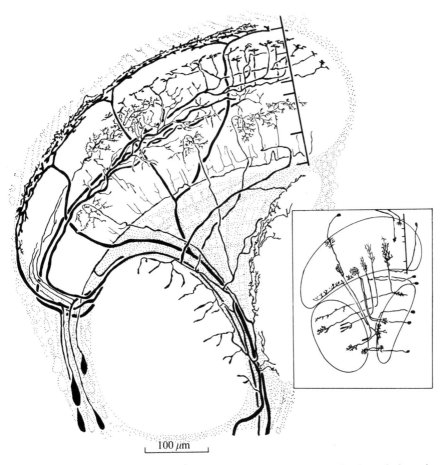

100 μm

FIG. 20.18. Drawing of some tangential cells in the medulla of the privet hawk-moth shown in horizontal section. Reconstructions made from wholly or partially-impregnated cells stained using the Golgi–Colonnier method. Inset shows perpendicular cells carrying the mosaic projection from the retina. Figure prepared by N. J. Strausfeld. Reproduced from Collett, 1970.

closely that of neurons recorded in the ventral nerve cord, which are possibly involved in generating optomotor torque, and there is some evidence that they and the thoracic neurons in part receive a common input (Collett 1971*b*). Thus the same visual information may be routed to the motor system to control orientation and back to both optic lobes to tell the visual system what movements were performed.

Centrifugal neurons have been shown to project to the medulla (Collett 1970) and to the lobula complex (Collett 1970; McCann and Foster 1971). There is evidence from the privet hawk-moth (Collett 1970) that the binocular directionally-selective movement detectors projecting to the medulla can be identified with an anatomically defined class of cell—the tangential cell (Strausfeld and Blest 1970) which has terminal branches that cover much of the retinal projection to the medulla (Fig. 20.18). Therefore one might expect the information carried by these cells to be distributed widely over the medulla.

There are probably two of the binocular neurons detecting horizontal movement projecting to each medulla with opposite preferred directions across the two eyes. These neurons respond optimally during yaw. However, because of the asymmetrical distribution of their inhibitory inputs, they are probably also active during forward flight (Collett 1971*b*). Consider, for example, a neuron with backward-preferred direction across the right eye and forward-preferred direction across the left eye. Backward motion across the left eye (i.e. in the anti-preferred direction) does not suppress the response to backward motion across the right eye, although forward motion across the right eye will suppress the response to forward motion across the left eye. In other words these neurons receive direct or indirect inhibitory inputs from movement detectors excited by forward motion, but not from ones with backward-preferred direction. This feature may well be important because centrifugal neurons which would respond optimally during forward flight (i.e. sensitive to backward motion across both eyes) have not been found in the medulla. Some information about the forward motion of the insect can thus be obtained from the summed activity of the neurons with clockwise and anticlockwise preferred direction, whereas the differences between the activities of these neurons will signal that the insect is turning.

ACKNOWLEDGEMENT

We are very grateful to M. F. Land for valuable discussion, to C. Atherton for photographic assistance, and to the S.R.C. for financial support.

Part V
Behavioural analysis

21. Head movements and fly vision

M. F. LAND

21.1. Introduction

ANIMALS whose eyes are movable relative to the rest of their bodies often use this capacity to keep their visual axes temporarily stationary with respect to the environment, in spite of movements of the body tending to disturb this relationship. In vertebrates this stability is achieved both by a visual feedback loop that opposes motion of the visual field across the retina, and by a feedforward pathway from the labyrinth that enables the eye muscles to compensate for head movements, independent of vision. It is interesting that in crabs, which also have movable eyes, a similar dual system of maintaining eye position is also employed (Horridge 1966; Sandeman and Okajima 1973).

Stabilizing eye movements are a common feature of the most highly-evolved visual systems and we might well seek their adaptive functions. We can speculate from our own visual experience that the emancipation of vision from ongoing body movements makes it possible to detect, localize, and recognize objects, especially small ones, unhindered by the blurring effect of retinal motion. Similarly it can be argued that the difficulties of navigation in a complex environment are likely to be reduced considerably if the visual world can be kept still temporarily, so that its contours can be seen and located without ambiguity. If the visual systems of insects are involved in any or all of these functions, as seems likely, one would expect them to possess mechanisms for stabilizing their heads, and hence their eyes.

Stabilizing eye movements, however, are no use on their own in a highly mobile animal. Changes of body direction must sooner or later necessitate changes in the direction of vision, and the visual system must have ways of periodically resetting the positions of the eyes, as locomotion causes the scene to change. One would expect such movements to be rapid, like human saccades, so that stabilized vision is suspended for as short a time as possible. It is interesting that in human vision this pattern of stabilization alternating with saccades is found not only in reflexly-induced optokinetic nystagmus, caused by rotating the visual world, but it is also the strategy employed during the voluntary scanning of the surroundings (Yarbus 1967; Noton and Stark 1971).

In this article I hope to show that insects—in this case blowflies (*Calliphora*)—use head movements in ways that are not very different from the use of eye movements in vertebrates. I shall show that flies during flight exhibit both stabilizing movements, tending to keep the direction of the head (and hence the eyes) constant, and also rapid saccadic movements which change the direction of the visual axis. Some of the consequences of such eye movements during vision have already been examined in the article by Collett and King (Chapter 20) but it is worth reiterating that most of our present information relating to fly vision has been obtained either from electrophysiological studies in which the eyes are necessarily stationary, or from optomotor studies in which the head is fixed to the thorax. Any attempt to relate the findings of these studies to vision during normal flight will have to take active head movements into consideration.

A preliminary account of some of the findings reported here has already been made (Land 1973).

21.2. Head movements

21.2.1. *Methods*

Blowflies (*Calliphora erythrocephala*) were obtained from stocks maintained by the Agricultural Research Council at Sussex University. Animals were temporarily anaesthetized with carbon dioxide and attached by their dorsal thorax to strong entomological pins, using insect wax. Relatively few (about 10 per cent) of flies showed strong consistent flight, and only these were subsequently used. The attachment pin was bent around the fly, and joined to a pivot (Fig. 21.1a). This consisted of a second pin revolving in a glass tube; this tube nearly fitted the pin, and was sealed by heat at the base to give a V-shaped bearing, and partially sealed at the top so as to permit free rotation of the pin, but a minimum of lateral play. The junction between the attachment pin and the pivot was carefully adjusted so that the axis of the pivot coincided with a vertical axis through the centre of the fly's thorax. This arrangement has very little rotational friction, and the flies rotated freely with no tendency to 'stick' at all, as judged from films of their movements. However, the pivot does substantially increase the fly's rotational inertia, and this is an important and inevitable weakness of this arrangement; the moment of inertia of the pivot was estimated to be approximately five times that of the fly itself (2000 mg mm^2 as opposed to 400 mg mm^2 for the fly). This certainly has two effects which should be considered in interpreting the results: one is to slow down the rate at which the fly's body is able to change direction, and the second is to increase the period and to a lesser extent the amplitude of the oscillatory body yaw seen in a number of records (e.g. Fig. 21.3). It is possible that in normal unrestrained flight these oscillations disappear. Attempts to reduce the mass of the pivot resulted in violent vibration during flight, and were abandoned.

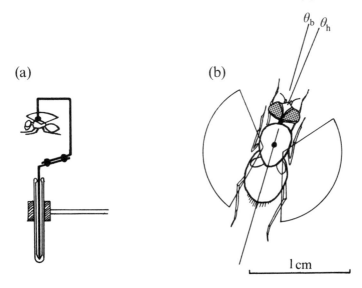

FIG. 21.1. (a) Attachment of the fly and construction of the pivot. (b) Appearance of a fly from above showing the two angles measured. The difference between these angles $(\theta_h - \theta_b)$ gives, over time, a measure of the activity of the neck muscles.

The fly and pivot were positioned upright in the centre of a white drum 12 cm in diameter. In most of the experiments this contained four vertical black stripes subtending 20° wide by 110° high at the fly, spaced at 90° intervals. This configuration was chosen because it could be used both to examine the extent to which the flies fixed individual stripes, as well as to provide a powerful optokinetic stimulus when rotated. The drum could be turned at known speeds by means of a synchronous motor fitted with a clutch and variable gear train. It was illuminated from inside and above using a fluorescent ring lamp, which gave the white parts of the drum a brightness of *ca.* 200 cd m^{-2} at the level of the fly. Films (16 mm) were taken from directly above using a Bolex camera with a macro-telephoto lens, focussed so that the fly filled most of the field of view. Most films were taken at 50 f.p.s. with a reduced shutter sector to minimize blur due to movement. They were analysed later frame by frame to obtain the direction of the body axis (θ_b) and the head axis (θ_h) as indicated in Fig. 21.1b. A small piece of celluloid of negligible mass was attached to the top of the head to facilitate the measurement of θ_h. Both angles could then be measured to an accuracy of *ca.* 1° relative to an external reference line, and *ca.* $\frac{1}{2}$° relative to the value in the preceding frame. In some cases the behaviour of flies was examined for long periods (Fig. 21.2) where only the approximate direction of flight was required; in this case the camera was removed and replaced by a half-silvered mirror behind which was a manually controlled pointer attached to

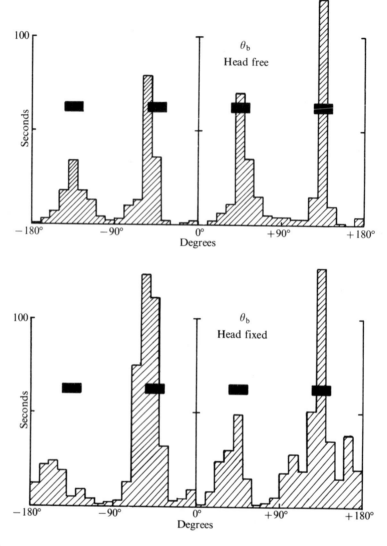

FIG. 21.2. Histograms of the angles taken up by the bodies of two flies in relation to the positions of the four stripes in the drum (black bars) measured with a matching pointer system (see methods) over a period of 15 min. The upper record was obtained from a normal fly, and the lower from a different fly whose head had been joined to the thorax with insect wax. In both records *c.* 300 s of observation time have been omitted, during which the flies were turning so fast that their positions could not be allotted to any one 10° sector.

Both histograms show a strong tendency for the flies to fly in the direction of one or another of the stripes.

a 360° potentiometer (see Land 1971). This pointer was visible together with the fly and was used to follow its direction, the output of the potentiometer going to a pen recorder. This method is only accurate to *c.* 5° and its speed is limited by the observer.

21.2.2. *Orientation in the drum*

In a stationary drum the flies show a strong tendency to fly in the direction of one of the four stripes, changing direction frequently from one to another (Fig. 21.2). The periods during which any one stripe was being fixated varied from about one second to over a minute. Stripe fixation, although the rule, seems to be by no means obligatory: flies would frequently position their axes between two stripes for several seconds before turning to one or the other. The overall impression obtained from watching several hours of this behaviour is that the fly is exhibiting a preference in flying towards a stripe, and is free to change direction at will. Movements between stripes are sudden and unexpected, they are not preceded by periods of particularly intense oscillation as one might expect if the fly needed to develop a large amount of rotational kinetic energy to escape from the visual 'grip' of a particular stripe. On the contrary, such changes of direction appear to be deliberate and effortless. Stripe fixation thus does not have the character of a binding reflex.

Interestingly, this overall preference for directing flight towards stripes seems little affected by whether or not the fly is free to move its head. There is very little difference in the two flight direction histograms in Fig. 21.2, although the second was made with a fly whose head was waxed to the thorax. This is in keeping with the finding of Reichardt (1973) using a totally different arrangement, that stripe fixation occurs in *Musca* under both conditions. Both results indicate that head movements are not essential for directed flight, and this is borne out by the observation that released fixed-head flies manage to avoid obstacles reasonably well.

21.2.3. *Head movements during stripe fixation*

A filmed record of the movements of head and body of a male *Calliphora* during a period of stripe fixation is shown in Fig. 21.3, on two different time scales. This is typical of a total of 15 min of film obtained from five different flies. The upper record shows the adjustments of head and body direction over a nine second period, and the lower record is an excerpt from this showing angles frame-by-frame. A short sequence from the film from which Fig. 21.3b was derived is shown in Fig. 21.3c.

The following conclusions can be drawn from Fig. 21.3.

(1) The head makes several abrupt changes in direction (fourteen in the whole record, three in the excerpt), fixating different points within and

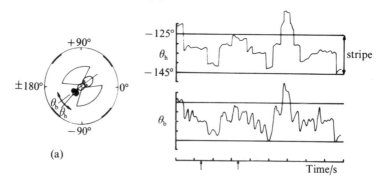

(a)

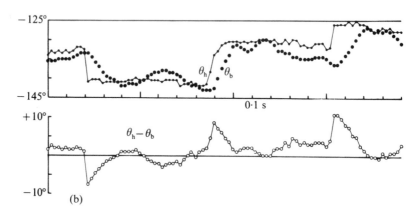

(b)

FIG. 21.3. Movements of the head and body during a period of stripe fixation. (a) left: relative position of the fly and stripes (not to scale). The convention used throughout this paper is that anti-clockwise rotations are defined as positive and clockwise as negative. right: head direction (θ_h) and body direction (θ_b) in relation to the position of a stripe over a 9 s period during which the head makes fourteen separate saccadic movements. (b) part of the record in (a) between arrows, plotted frame-by-frame on an expanded time scale. The lower trace shows the component of the head movement that is caused by the action of the neck. (c) Reproduction of a sequence from part of the same film as that analysed in Figs. 21.3(a) and 21.6. The sequence shows two saccadic head movements, between frames 3 and 5, and frames 12 and 14. Note that before, between, and after these movements the angles taken up by the head remain constant, and that following each saccade the body is slowly brought into line with the new direction of the head (e.g. between frames 4 and 9). The interval between frames is 20 ms, and the drum surrounding the animal is stationary and contains four 20° stripes situated diagonally as indicated in Fig. 21.2.

just outside the stripe. Between these movements its direction remains remarkably constant. Drifts of head position do occur, but these are unimpressive compared with the stability that is maintained in relation to the much more erratic behaviour of the body.

(2) Changes in head direction (θ_h) are the result of sudden movements of the neck, which will be termed *saccades* by analogy with the rapid eye

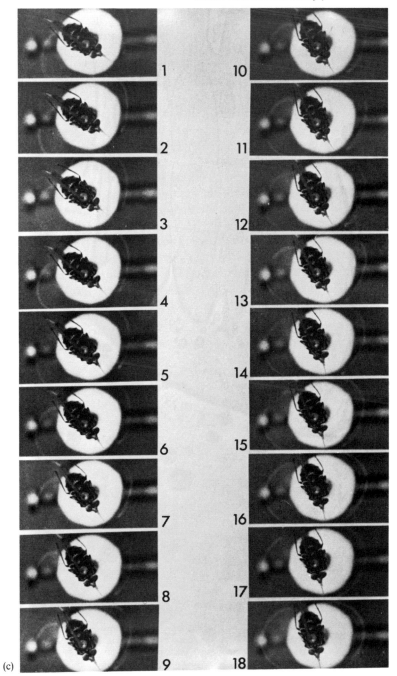

F IG. 21.3 (*Cont.*)

movements that shift the direction of fixation in human vision. These have amplitudes in the range 5–20° when measured as movements of the head relative to the body ($\theta_h - \theta_b$, Fig. 21.4), although their absolute size is rather greater, up to 25°, as the body is usually moving in the same direction as the head during a saccade. The maximum angle that the head has been seen to make with the body is *c.* 17° (Fig. 21.4). Saccades are usually so fast that they are almost completed between successive frames of film (20 ms) which means for a 10° movement that the angular velocity must be in excess of 500° s^{-1}.

(3) After a saccade, the direction of the body comes slowly into line with the new direction of the head, taking 100–200 ms to do so. The slowness of the body is certainly due in part to the extra inertial load of the pivot, and without this load the body could be expected to follow the head much more closely.

(4) The stability of the head in relation to the body, especially after each saccade, is achieved by active neck movements that are in the opposite direction to the body movements. This is clearly shown in Fig. 21.3b, where the neck movements ($\theta_h - \theta_b$) are almost the mirror image of the body movements (θ_b), except during the saccades themselves.

These conclusions establish that the muscles of the neck have two functions: (1) to cause rapid changes in head orientation (saccades) which are followed by reorientations of the body; (2) to compensate for imposed or inadvertent movements of the body, so that the direction of the axis of the head remains almost constant.

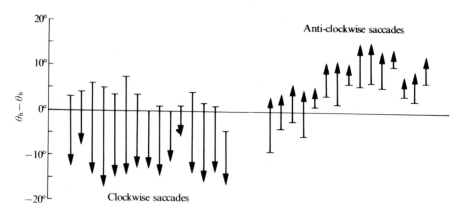

FIG. 21.4. Amplitudes of thirty saccades, fifteen in each direction. Each arrow shows the initial and final directions of the head relative to the body ($\theta_h - \theta_b$). The horizontal line represents the situation where the head and body are in line. There is considerable variation in the sizes of saccades, as well as in the initial and final neck deviations. The amplitudes of the saccades in relation to the surroundings are actually larger, on account of body movements in the same direction as the head.

21.2.4. *Timing of head and body movements*

Although, after a saccade, the body takes some time to catch up with the new direction of the head, this is not necessarily because it receives its 'instruction to turn' later. In a system with considerable inertia the most reliable guide to the timing of muscular activity must come from measurements of acceleration, rather than position or velocity. This is because, in this instance, the first indication of a change of body direction will be the production of yaw torque (T) in one or other direction. Then, since:

$$T = I\ddot{\theta}_b, \quad \text{where}$$

I is the moment of inertia of the fly plus the pivot (*ca.* 2400 mg mm^2), and $\ddot{\theta}_b$ is the angular acceleration of the body, changes in flight torque resulting from 'commands to turn' will show up immediately as changes in acceleration.

Some of the torque will also be taken up in overcoming aerodynamic friction, and this will lead to an underestimate of torque as derived from measurements of acceleration; with the high inertia conditions here, however, this effect is likely to be small and will be ignored.

Figure 21.5 gives an analysis of the same thirty saccades illustrated in Fig. 21.4, in which the timing of each head movement (θ_h—θ_b) is plotted on the same time axis as the body acceleration ($\ddot{\theta}_b$). It can be seen that for both anti-clockwise and clockwise saccades the body has begun to accelerate at the initiation of each head movement, and its acceleration is maximal as the head reaches its greatest deviation. After this the body acceleration ceases abruptly, from which it can be concluded that the turning torque has ceased.

Figure 21.5 thus shows that, although the body takes more than 100 ms to catch up with the head, its principal 'turning effort' is made at the same moment as the head saccade (or at least within 20 ms of it), which implies that the two systems, mediating yaw torque via the wings, and neck position via the neck muscles, probably receive the motor command to turn at the same time.

The magnitude of the maximum body torque can be calculated from the above equation. When I = 2400 mg mm^2 and $\ddot{\theta}_b$ = 3000° s^{-2}, T = 1·26 dyn cm. This is about the same torque as other workers have obtained from houseflies (*Musca*) in a moving optomotor drum which they are *not* free to follow (McCann and MacGinitie 1965; Fermi and Reichardt 1963).

21.2.5. *Major changes of flight direction*

In most instances where a fly changes direction from one stripe to another, it can be seen to do so via a cascade of individual saccades. This is illustrated in Fig. 21.6. Just as during stripe fixation (Fig. 21.3) the direction of the head (θ_h) is held constant between saccades. The essential difference between these

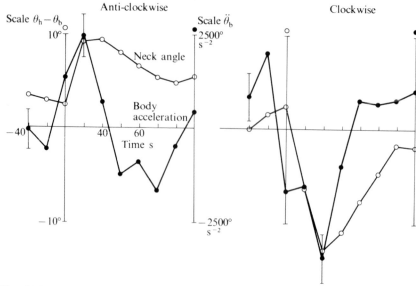

FIG. 21.5. The temporal relations of head movements (O) and body acceleration $\ddot{\theta}_b$ (●) during fifteen anti-clockwise and fifteen clockwise saccades.

For each saccade the angle of the head relative to the body $(\theta_h - \theta_b)$ was measured on the frame immediately preceding the saccade (0 on the time axis) and then on the two frames preceding this and seven following. The plotted values are the average angles for each set of fifteen saccades.

Body acceleration was measured as follows: consider the body direction (θ_b) on three successive frames of film, x, y, and z. The average velocity between frames x and y will be equal to

$$\frac{\theta_{b(y)} - \theta_{b(x)}}{0 \cdot 02} \text{ deg s}^{-1},$$

when 0·02 s is the interval between frames. Similarly, the acceleration at the instant (y) is equal to the difference between the velocities during the periods $(z - y)$ and $(y - x)$, again divided by the frame interval. This is

$$\frac{(\theta_{b(z)} - \theta_{b(y)}) - (\theta_{b(y)} - \theta_{b(x)})}{0 \cdot 02^2} \text{ deg s}^{-2}.$$

Accelerations are averaged and plotted in the same way as head positions. Vertical bars represent twice the standard errors of these measurements. An angular acceleration of 2500° s^{-2} represents a torque of c. 1 dyne cm.

saccades and those made during stripe fixation lies in their timing and direction: they tend to follow each other closely with stable intervals between lasting only 40–200 ms, and serial saccades like these tend to be always in the same direction. Otherwise they are no different from saccades made at other times. At first sight the movements of the body that accompany these head movements appear to follow a smooth curve, but closer inspection shows that each head saccade is accompanied by an acceleration of the body in the same direction, just as in the more widely spaced saccades (Figs. 21.3 and 21.5).

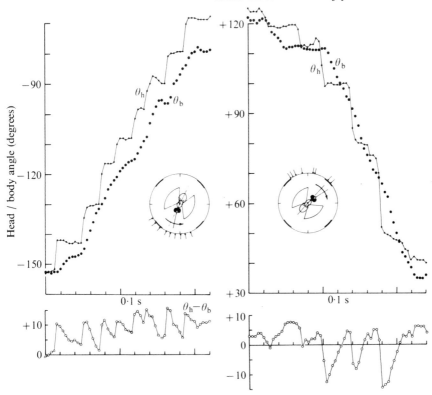

FIG. 21.6. Serial saccades made during turns from one stripe to another. Left: clockwise (eight saccades), right anticlockwise (six saccades).

These findings show that one method by which a fly steers itself is by the regulation of the direction and frequency, and possibly also the amplitude, of saccadic movements of the head and body. This is certainly in keeping with the appearance of free flight of flies, which seems to consist of periods of more or less straight flight with fairly abrupt turns of varying sizes. I would expect the larger turns to be composed of a number of smaller saccades, but it will be extremely difficult to determine whether this is so in free flight.

Occasionally turns appear to be made in a rather different manner. Instead of making a series of saccades, the fly tilts its head around the axis of the neck by *ca.* 30°, and then begins to turn, apparently smoothly, in the direction that the top of the head has been tilted. These movements usually result in the fly spinning past several stripes, with no attempts to stabilize the head, before a new stable direction is attained. Guessing, it seems possible that in this kind of turn the tilting of the head represents a way of deliberately minimizing the effectiveness of vertical contours that normally prevent rotational slip of the eyes (see the following section).

21.2.6. *Visual control of stabilization*

The accurate way in which the neck muscles maintain the direction of the head could be achieved in three ways; using information from the halteres about body movement (Pringle 1957), from the antennae which are reported to be sensitive to yaw-induced wind-direction changes on the head (Burkhardt and Gewecke 1966), or from vision which would involve detection and compensation of image slip across the eyes.

It is easy to demonstrate the importance of a visual mechanism controlling stabilization by examining the movements of head and body when the visual environment, represented by the drum, is rotated around the fly. Figure 21.7 shows a record of such behaviour, which includes a reorientation from one stripe to another and back again, while the drum is being turned at $20° \text{ s}^{-1}$. In general, the head follows the drum rotation with such accuracy that there is no detectable slip of the head relative to the drum. This is true not only in the periods of stripe fixation, but also during the brief intervals of stabilization between saccades. The conclusion must be that the mechanism that enables the neck muscles to maintain the directional stability of the head is mediated visually, and it is operative at all times except during the instant of each saccade. In the moving as well as the stationary drum the head behaves as if 'clamped' to the visual pattern. This does not rule out a role for the other sources of information about head or body movement (see the discussion later) but it does establish the overriding importance of visual feedback in maintaining a stationary relation between the environment and the head.

Another demonstration of the importance of vision in head stabilization is shown in Fig. 21.8, in which the striped drum is replaced by a plain white drum. Two things stand out, the first is that the four large saccades that are made in the plain drum do not terminate in a reinstatement of stabilization, as they do when stripes are present, and the second that neither the body nor the head is capable of maintaining a constant direction. There is nothing, it seems, to prevent either the head or the body 'slipping' relative to the surrounding space, as there clearly is in the upper figure. There is a small movement of the neck in the direction of compensation, even when stripes are absent; this might be attributable to extravisual information (halteres or antennae) or to residual visual cues like the pivot-support or asymmetries in illumination of the drum.

21.2.7. *Conclusions*

Flies make two kinds of head movement in flight: saccadic movements which alter the relation between the retina and the visual surroundings by

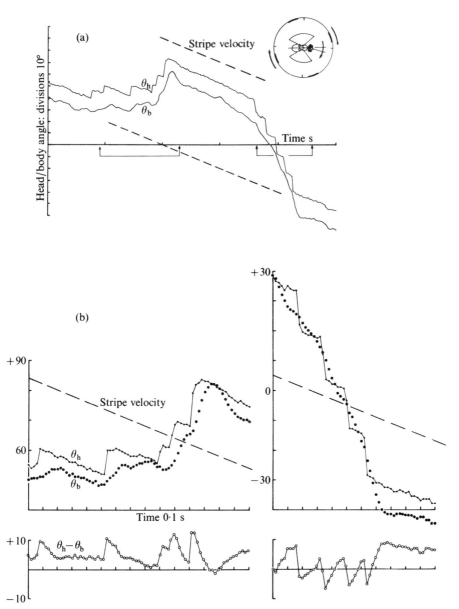

FIG. 21.7. (a) Movements of head and body during a 5 s period in which the fly changes direction from one stripe to another and then back again. The drum is rotating clockwise at $20° \, s^{-1}$ and the dashed lines indicate the velocity and approximate positions of two of the stripes. Body and head traces are displaced by $10°$ relative to each other to avoid confusion. This record shows that the head follows the movements of the drum closely at all times except during saccades. (b) Two excerpts from the record shown in (a) between the arrows which demonstrate the closeness with which the head follows the drum during periods of stabilization, when saccades are being made towards (right) and away from (left) the direction of movement of the drum. The dashed lines here only indicate the velocity of the pattern on the drum. Head and body traces are not displaced.

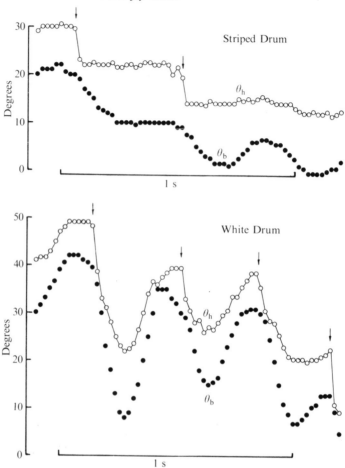

FIG. 21.8. Saccades made in a striped drum (above) and plain drum (below). The films were taken of the same fly within minutes of each other. Head movements (O) are stabilized after each saccade when stripes are present, but not in their absence. Body movements (●) show greater fluctuations in the absence of a visual pattern. (Body position has been displaced by 10° relative to head position in both records.) These brief records are fairly typical of each four minute film.

5–25°, and stabilization movements which tend to preserve the relation between the retina and surroundings as constant as possible. Stabilization is achieved by the visual counteraction of slip across the retina. Large turns are usually made by combining many small turns with short intervals between them.

21.3. Discussion

21.3.1. *Stabilization of head direction*

The observations described in the last section show that changes of direction are sudden discrete events and that direction of the head axis is held remarkably constant during flight. This constancy is achieved in spite of small variations in body axis orientation by active compensatory movements of the neck. Since both head and body follow the pattern in a moving drum (Fig. 21.7) it is clear that both the flight motor system controlling flight yaw, and the neck muscles, receive visual information about slip across the retina, and both systems act on it. With respect to head angle the two systems act co-operatively as coarse and fine adjustments: the flight motor system establishes the basic direction, or rate of turning in a moving drum, and the neck makes whatever minor compensatory movements are necessary to nullify irrelevant or accidental departures of the body from a straight course. This is well shown for example in Fig. 21.3b, where oscillations of the body of c. 5° are exactly counteracted by movements of the neck $(\theta_h - \theta_b)$ so that the head direction (θ_h) is accurately maintained.

Thus, although the overall control system for maintaining head direction can be represented by an optomotor feedback loop of the familiar kind (see e.g. Mittelstaedt 1964), this loop actually contains two 'motors', the flight motor and the neck muscles, whose operations are not identical. If we consider the situation in Fig. 21.7, where the head and body are following the moving drum, the differences between the actions of the flight motor and neck become clear. Basically, the body follows the drum, but with some degree of error, and we can write $B = D \pm e$, where B is the angular velocity of the body, D the drum velocity and e represents the velocity error of the body. The head, however, follows the drum almost perfectly, and the head velocity H is obviously the sum of the body velocity B and the neck velocity N, so that $H = D$ to a good approximation, and so $B + N = D$. Combining this with the other equation, $N = D - B$, or $N = D - (D + e) = -e$. This means that if the neck is to correct for body errors ($N = -e$), in the way that it appears to do, the part of the nervous system controlling the neck must have access not only to visual information about drum speed (D) but also to independent information about the angular velocity of the body (B), so that the final activity of the neck represents the difference $(D - B)$.

The only source of independent information about movements of the body, whether these are intended or accidental, is the halteres. These organs are known to be capable of detecting yaw (Pringle 1948, 1957). It seems likely, therefore, that the input to the neck muscles represents the difference between the optomotor signal which controls flight yaw torque, and a signal representing actual body yaw derived from the halteres. This tentative hypothesis is presented in more explicit form in Fig. 21.9b.

So far this hypothesis is quite conjectural; however, it should be reasonably easy to test and I have included it here because it seems *a priori* the only arrangement that can satisfactorily account for both the optomotor behaviour, and the stability of the head when the body motion includes unintended perturbations.

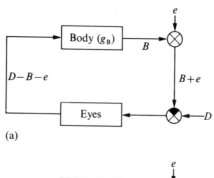

(a)

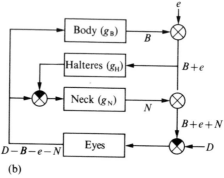

(b)

FIG. 21.9. (a) Basic optomotor loop responsible for maintaining body direction, which would apply to a fly with its head waxed to the thorax. All symbols represent angular velocities: body (B), drum (D) and imposed errors (e). If the 'gain' of the flight motor system is g_B then the body velocity B will be

$$\frac{g_B}{1+g_B}(D-e),$$

i.e. if the gain is large the fly will tend to follow the drum, and oppose errors. (b) Tentative hypothesis to account for the ability of the neck to compensate for body errors, within the overall loop given in (a). It assumes that actual body yaw ($B+e$) is measured by the halteres and this is subtracted from the optomotor signal reaching the neck, whose angular velocity is N. The final velocity of the head is the sum $B+e+N$. Then, for the body,

$$B = g_B(D-B-e-N), \quad \text{and for the neck,}$$

$$N = g_N\{D-B-e-N-g_H(B+e)\}.$$

Then if, for simplicity, g_B and g_N are assumed to be equal,

$$N = B-g_N g_H(B+e).$$

Arbitrarily setting g_H equal to $1/g_N$,

$$N = B-B-e,$$

i.e. $N = -e$, which is the observed condition.

It is likely that a second kind of mechanism also exists to prevent the body axis from straying too far from the head axis, since the latter can only make adjustments of up to *ca.* 17° in each direction before being inevitably 'dragged off course' by the body. The most likely basis for such a mechanism would be a feedback loop in which proprioceptive information about neck inclination, possibly from the prosternal hair plates, reflexly causes flight yaw tending to bring the body back into line with the head. A mechanism of this kind was shown by Mittelstaedt (1950) to be present in dragonflies, in relation to roll rather than yaw, and it is likely, though as yet unproven, that Diptera employ a similar mechanism as a 'back stop' to supplement the visual loop outlined above.

21.3.2. *Saccades and the optomotor loop*

In principle, the fly could change direction by injecting a 'command' into the visual loop, between the eyes and the flight motor. Such a command would act as a new 'reference input' to the loop, which would then operate to achieve the new commanded angular velocity. If this is represented by C, then the body angular velocity in the loop illustrated in Fig. 21.9a will become

$$\frac{g_B}{1+g_B}(C+D-e),$$

i.e. the flight motor system will execute the command, follow the drum and oppose errors as before. It would seem from this that the simplest way for a fly to change course would be via a series of commands to alter angular velocity. Such a system would, however, be relatively slow since the velocities involved would have to be within the normal range of working of the optomotor response, which in *Musca* becomes much less efficient at velocities greater than a few hundred degrees per second (Fermi and Reichardt 1963). It is a matter for speculation as to how long it would take for a fly to change direction in this way, with the loop coming back into equilibrium after each command, but it would probably be a matter of tens or possibly hundreds of milliseconds. In any case, this does not seem to be the way that direction changes are actually brought about.

The angular velocity of the head during saccades is typically at least $500°\ s^{-1}$, and since these are usually nearly completed between frames of film, the maximum velocity may be much higher. The data of Fermi and Reichardt (1963) obtained from fixed flies strongly suggests that at such speeds the optomotor response is considerably weaker than at lower speeds, presumably because the eyes are unable to generate responses to high velocity movement. Thus the saccadic turning commands, rather than working via the loop, as suggested in the previous argument, might in fact be circumventing it simply by going too fast. If so, the visual loop will be 'open' for the short duration of the saccade, as seems to be the case during human saccadic eye movements (Young and Stark 1963).

The analysis of the timing of body acceleration in relation to each head saccade presented in Fig. 21.5 shows that the body is turned by a sudden torque 'pulse' lasting *ca.* 60 ms and nearly coincident with the head movement. Like the head, it seems that the body receives a command to produce an acceleratory burst.

Because most previous records of optomotor behaviour have been made with flies that are not free to move, and whose heads are fixed to the thorax (Fermi and Reichardt 1963; McCann and MacGinitie 1965; McCann and Foster 1971) it is a matter of some importance to determine whether or not saccades, as well as sustained optomotor responses, occur in these situations as well as in the relatively less constrained situation discussed in this paper. Obviously head saccades cannot be made, but one would certainly expect in the measurements of body torque made by these authors to find sudden pulses of torque with a time course like that shown in Fig. 21.5. Unfortunately, very few records of yaw torque have been published that have not been either averaged, or fed through a low-pass filter. However, in an early paper McCann and Fender (1964) do publish 'unexpurgated' records of the torque produced by a fly in response to a 17° stripe moving clockwise around the animal. One such record is reproduced, with permission, in Fig. 21.10. This record shows very clearly that as the stripe moves away from the animal's front the fly produces rapid torque oscillations (5–10 s^{-1}) in the direction of the stripe which must represent attempts by the fly to 'catch up'. A smaller series in the opposite direction is made as the stripe approaches the front of the fly. It seems virtually certain that these high frequency components, which are superimposed on a steady optomotor torque in the direction of movement of the stripe, are identifiable with the saccades described here. The body torque occurring during a series of saccades like that shown in Fig. 21.6 would look exactly like McCann and Fender's record. In view of this it seems unhelpful to use averaged torque records as an index of optomotor behaviour, since they are likely to contain the combined torque produced by both the optomotor system and a saccadic system whose properties and preferences have yet to be explored.

21.3.3. *Saccades and fixation*

In the restricted situation of a drum containing four black stripes the flies usually fly in the direction of the stripes for most of the time (Fig. 21.2), changing direction at short but irregular intervals. This behaviour is not importantly different when the flies' heads are waxed to the thorax.

Behaviour that is essentially similar to this has been demonstrated by Reichardt (Reichardt and Wenking 1969; Reichardt 1970; Reichardt 1973) using an ingenious arrangement in which a fly (*Musca*) is suspended in a

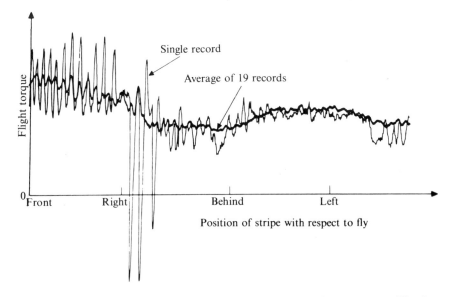

FIG. 21.10. Record of yaw torque made by a fixed flying fly (*Musca*) in response to a 17° stripe moved round it at a velocity of 29·3° s^{-1}. The total length of the record is *c.* 12 s. The record shows a series of rapid oscillations in the direction of motion of the stripe as it crosses the fly's axis, followed by a large torque in the opposite direction, and finally a series of small oscillations in the direction of the stripe as it approaches the front again. The record is taken from McCann and Fender (1964) and is reproduced with permission.

drum, and the yaw torque that it produces is transduced without the fly itself moving. An electrical signal proportional to the torque is then fed to a servomotor which drives the drum, so that if the fly produces clockwise torque (i.e. it is trying to turn to the right) the drum rotates anti-clockwise at a speed proportional to the torque, and in this way the fly without moving is able to alter the position of the drum and any visual structure it contains. The general result of these experiments was that if the drum contained a single stripe the fly positioned it in front—the equivalent result to that reported here (Fig. 21.2). If the drum contained a stripe wider than *ca.* 40° the fly fixed upon two positions inside but near each edge of the stripe, and when there were several stripes in the drum the fly spent part of the time fixing upon each one.

Reichardt (1973) and Poggio and Reichardt (1973) have suggested an explanation for fixation in the following terms. If there are asymmetries in the optomotor response, such that stimuli moving forwards across the retinae are relatively less effective in producing optomotor torque than stimuli which move backwards, then if the fly makes small spontaneous movements ('noise') in both directions the net result will be for a simple pattern like a stripe to move to the front. Or, if the fly is free to move, for the fly to turn towards the stripe. Poggio and Reichardt draw the analogy between a fly 'trapped' in this way by a stripe, and a particle in Brownian motion caught in a potential energy well.

While this explanation is elegant and self-consistent, it is doubtful that it represents the method by which fixation is usually brought about. As Figs. 21.3, 21.6, and 21.7 show, changes in direction are generally the result of saccadic movements of the head and body, with the optomotor response, tending to stabilize direction, operating apparently undisturbed in between. If this is so, then it seems likely that the optomotor asymmetries observed by Reichardt are not simply variations in the optomotor response itself, but include the tendency of the fly to make saccadic movements in the direction of prominent objects in the field of view. Certainly, when the fly is free to turn, the fixation of stripes and probably also the much more complicated sorts of navigation that flies show while flying in a structured environment are governed by the directions of saccades and the rates at which they are made. Whether the timing of saccades is determined solely by the visual content of the surroundings is an open question. However, I can state with some assurance that it is impossible to guess, while watching a film of a fly, when the next saccade will occur or what its direction will be. It does not seem inappropriate to use the word 'voluntary' about individual saccades, even though when enough are averaged as in Fig. 21.2 they do conform to a statistical pattern related to visible structures in the environment.

21.3.4. *Conclusions: head movements and vision*

I would like to be able to draw, from this study, the conclusion that the kinds of head movements flies make enable them to see better. One would think that a temporary stable visual image of the surroundings would make it easier for the fly's visual system to detect and localize structures of importance, whether these are merely obstacles to be avoided, or 'preferred' objects to be approached. It is true that in the world of a fly in motion only the field of view ahead will be properly stabilized, but this is a small handicap since this region, in the line of flight, is the most important for immediate navigation.

The difficulty in trying to extend this argument is that it is very hard to tell what flies do see, or need to see. Because domestic flies are vagrant scavengers they do not exhibit clear preferences for particular kinds of object. This is not true for all Diptera; some, like robber-flies (*Asilidae*), empid flies (*Empidae*), and shore flies (*Ephydridae*) catch other insects on the wing, implying at least the ability to detect and track small targets, and some hover-flies (*Syrphidae*) as well as empids also mate on the wing, often after complex dances, which suggests that they possess considerable powers of pattern recognition. However, although *Calliphora* and *Musca* get into occasional aerial fights, they otherwise give little indication that the fine structure of their visual environment is particularly important to them. Jander and Schweder (1971) showed that walking *Calliphora* show preferences for objects that are large, vertical, and have irregular outlines, but again this implied capacity for recognition seems hardly commensurate with either the anatomical complexity of the flies' visual system, or with the need for well-stabilized vision. It is possible that flies, like certain Hymenoptera, get to know the landmarks in their immediate surroundings, but there is certainly no clear evidence that this is so. Until future studies of the natural history of flies are more advanced, we are left with the rather lame conclusion that fly vision, including the head movements associated with it, is mainly concerned with not bumping into things.

21.4. Summary

While flying, but free only to rotate, flies (*Calliphora*) show two kinds of head movements.

(1) Rapid saccadic movements with amplitudes of up to 20° relative to the direction of the body axis and durations of *c*. 20 ms. These head movements, mediated by the neck muscles, are accompanied by body turns which slowly bring the body axis back into line with the head.

(2) Stabilization movements which tend to keep the axis of the head still with respect to the surroundings, in spite of fluctuations of direction of the body axis. These movements are dependent on visual feedback.

An analogy is drawn between these movements and eye movements made during human vision. The possible roles of the two kinds of head movement are discussed in relation to object fixation and other visual functions, and a possible control system to account for head stabilization is outlined.

ACKNOWLEDGEMENT

I thank T. Collett and S. King for helpful discussions and for reading the manuscript. This work was supported by a grant from the Science Research Council.

22. Landing and optomotor responses of the fly Musca

A. FERNANDEZ PEREZ DE TALENS and
C. TADDEI FERRETTI

22.1. Introduction

THE landing response of a fly is more or less an all-or-none reaction to a contrasting object that is approached in flight. The fly lifts its first pair of legs to the sides of the head and extends the third legs behind (Hyzer 1962). Expansion in the visual field is the principal effective component of the natural stimulus (Goodman 1960, 1964). While in flight the fly is also controlled in direction by its optomotor response, which depends in a different way upon motion in the visual field (Fermi and Reichardt 1963). A model that mimics the fly's ability to distinguish between contraction, expansion, and lateral displacement in one dimension in the visual field can be made from two motion perception systems A and B that are the inputs to four logical circuits (Fig. 22.1). This is simply a filter which detects the conjunction of two motions, so that, for example, expansion is detected because it provides movement away from the centre of the paper at two places.

The relevant parameter is the angular velocity ω relative to the centre of the eye. A point P moving at constant speed in a straight line does not move with constant ω, but its motion has to be projected upon the hemisphere around the eye. The angular velocity ω is then calculated as a function of the angle α between the fly's long axis and the point P, with $\alpha = 0$ when P is directly in front of the fly. The dependence of ω upon α for different moving stimuli is illustrated in Figs. 22.2, 22.3, and 22.4.

The value of ω can be calculated for any stimulus from the equation:

$$\omega = d\alpha(t)/dt = \frac{\pm v}{r}\cos^2\alpha(t) - \frac{\pm y}{s}\sin^2\alpha(t) =$$

$$= \left\{\left(\frac{\pm v}{r_0 \pm yt}\right)\cos^2 \arctan\left(\frac{s_0 \pm vt}{r_0 \pm yt}\right)\right\} - \left\{\left(\frac{\pm y}{s_0 \pm vt}\right)\sin^2 \arctan\left(\frac{s_0 \pm vt}{r_0 \pm yt}\right)\right\}$$

where r = distance from the projection of P on the anterior longitudinal

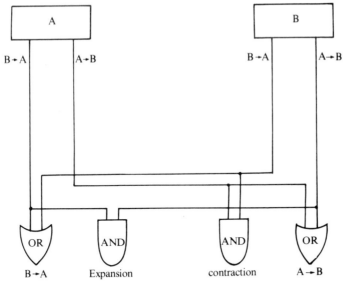

FIG. 22.1. A model which distinguishes between expansion, contraction, and lateral displacement.

semi-axis to the centre of the eye: $r_0 = r$ at time t_0; s = distance from P to the long axis; $s_0 = s$ at time t_0; v = velocity in a direction normal to the long axis; y = velocity in the direction of the longitudinal axis; α, defined as above, runs from $0°$ to $180°$ for the right eye and from $0°$ to $-180°$ for the left eye; ω is considered positive for a movement of P from anterior to posterior.

Values of ω are plotted in Figs. 22.2, 22.3, and 22.4 for a variety of motions including expansions and contractions as shown by the arrows (which should be considered as continuing in one or both directions). The value of ω clearly is different at different values of α, and the pattern of ω is different for each type of stimulus. If, instead of a point P, a succession of contrasts moves along in the same motion (as in moving stripes or other patterns), the curves of many different P values at the same time coincide with the curve for a single P at different times.

To characterize a visual stimulus one must consider all the various motions happening at a given time. As an example, consider in Fig. 22.2a a natural landing stimulus in which nineteen P values approach from in front of the fly until the $\pm 90°$ axis is reached. The advancing front of P values is a straight line at right angles to the line of motion, and placed symmetrically. Let $\omega = 100$ and $r_0 = 200$. The dotted curves refer to moving P values with $s = 0.5, 1, 2, 3.5, 5, 10, 20, 50,$ and 100 from the top (nearest ones) to the bottom (furthest away). These curves are as in Fig. 22.3i. Now look at the dashed curves, which are the isochrones, i.e. each dashed line is drawn

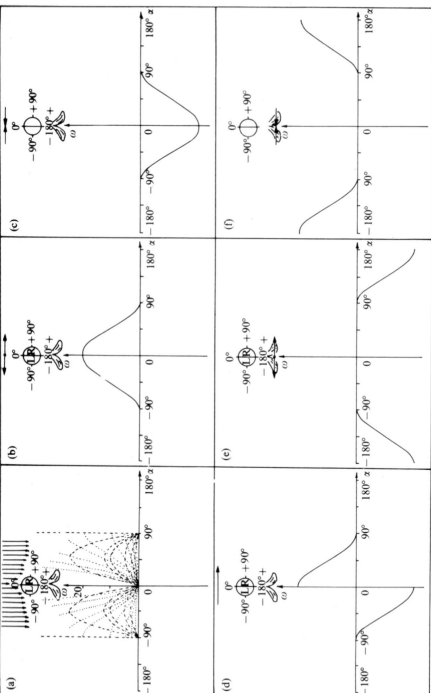

FIG. 22.2 The angular velocity ω of a point P, which moves in a straight line in various directions relative to the head of the fly, as a function of the angle α between P and the longitudinal axis. Blind zone on the back and binocular overlap are ignored. In the examples of different motions the arrows represent a movement that is continuous in the directions indicated. If, instead of points, one considers stripes, it is the contrast frequency ω/λ (where λ is the perceived spatial period of stripes), that determines the optomotor response (Götz 1964). The value of ω/λ does not then change with α. If one considers ω instead of ω/λ only the extreme values of the stripes and the size of the pupils to general are

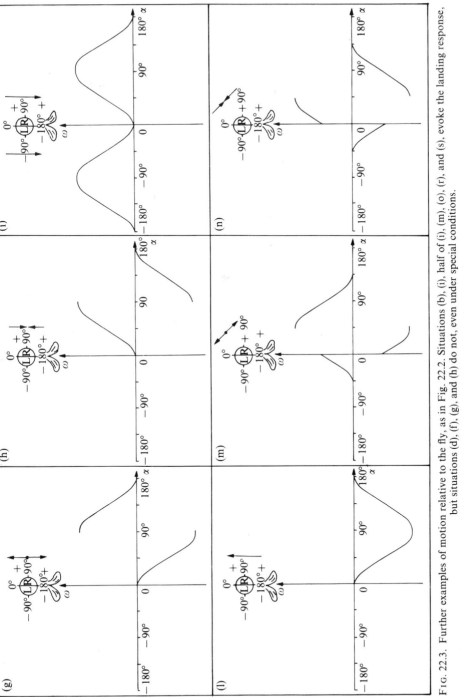

FIG. 22.3. Further examples of motion relative to the fly, as in Fig. 22.2. Situations (b), (i), half of (i), (m), (o), (r), and (s), evoke the landing response, but situations (d), (f), (g), and (h) do not, even under special conditions.

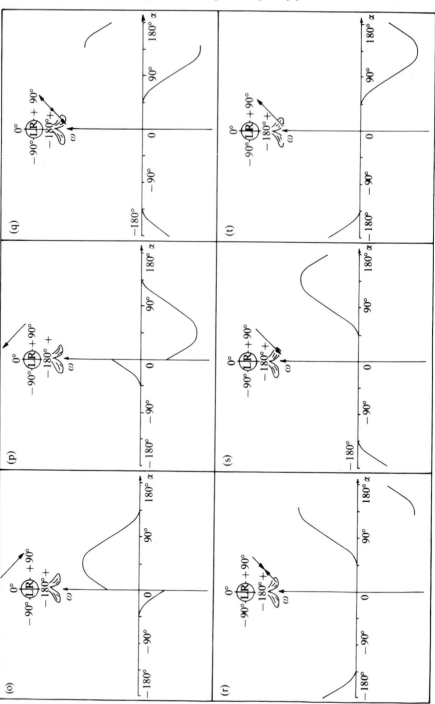

Fig. 22.4. Final examples of motion relative to the fly, as in Fig. 22.2. Situations (c), (e), (l), (n), (p), (q), and (t) do not evoke the landing response.

through the many perceived values of ω at a given time t. When $t = 200/100$ $= 2$ the isochrones are vertical and at times $t = 1\cdot98, 1\cdot96, 1\cdot95, 1\cdot9, 1\cdot8$, and $1\cdot5$ the isochrones are the dashed curves shown in order from top to bottom. This is a series of parabolas approximating to the curves of $\sin^2 2\alpha$.

The ommatidia that first perceive an ω of a given threshold value are those at $\pm45°$, and it should be noted that in this example the binocular overlap in the anterior of the visual field has been ignored. These relations between ω and α shown in Figs. 22.2, 22.3, and 22.4 are essential for the understanding of the various possible stimulus situations.

22.2. Methods

Musca domestica has been prepared in three ways: (1) Flying flies with both eyes intact were held stationary in space with the head glued to the thorax. (2) One-eyed flies in stationary flight with head fixed to thorax. Either the other eye was covered with black wax or a paper cone, or surgically removed some hours before the experiment; (3) Normal or one-eyed flies, head fixed to thorax, suspended by a 3·5 cm hair and free to fly in all directions.

Stimulus situations were as follows: (1) Relative motion between the fly and a contrasting object, so that the periphery of the object moves relative to the background, with an unavoidable change in total intensity and a value of ω for every contrast on the object. (2) Simulation of approach or lateral displacement of an object without change in brightness by rotation of black and white spirals on a disc with or without mask, to generate a relative radial motion of contrasting edges (Fig. 22.2b). (3) Reduction of intensity alone (Goodman 1960, 1964; Braitenberg and Taddei 1966).

22.3. Monocular and binocular vision

One-eyed flies give landing responses with all three of the above stimulus situations. When a small vertical stick is moved towards the flying binocular fly the stimulus is equally effective if moved from the front or from the side. On the other hand monocular flies react as follows: Movement from the side of the seeing eye causes a response of legs on both sides, but movement in the front segment between -20 and $+20°$, which is normally covered by binocular vision, causes leg responses only on the side of the seeing eye. From this we can infer connections of some kind that are of course not direct nervous pathways (Fig. 22.5).

Tests with an expansion in one direction show that in binocular flies a vertical expansion is the most effective stimulus in the region of binocular overlap (Fernandez and Taddei 1970) (Fig. 22.6a). In monocular flies, how-ever, horizontal expansion is most effective (Fig. 22.6b), showing that summation between the two eyes is important for vertical expansion in the

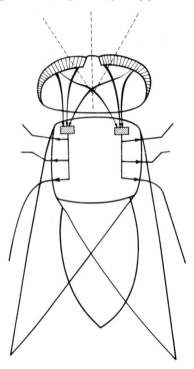

FɪɢG. 22.5. Scheme of pathways of excitation from different regions of the eyes to the various legs for the landing response.

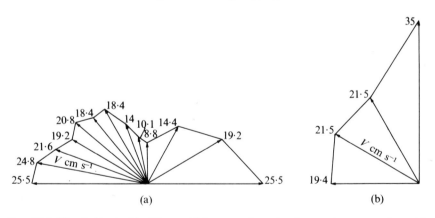

FɪɢG. 22.6. Threshold velocities V in cm s^{-1} for a one-dimensional expansion presented to a (a) two-eyed, and (b) one-eyed, fly. The stimulus was formed by expansion of black and white moving stripes in a direction at right angles to the edges in the pattern. The stimulus was presented symmetrically to the front of the fly and could be rotated to test the effect of direction of expansion. The pattern was 30 cm long, 3 mm wide, of spatial period 8 cm and was 9 cm from the eyes in (a), 5 cm in (b).

binocular region. The implications of this finding for normal vision are not known.

Although the response is all-or-none, the stimulus is said to be more effective when it evokes the response for a wider range of experimental conditions. A rotating spiral pattern, always at right angles to the line of sight, is most effective as a stimulus when in the zone of binocular overlap and least effective when at 90° to the body axis (Taddei and Fernandez Perez de Talens 1967). Data for a two-eyed fly, tested with expansion and contraction of a spiral stimulus are given in Fig. 22.7, and tested with a one-dimensional expanding line stimulus in Fig. 22.8. As shown, there is little effect of covering part of the stimulus or covering one eye, unless otherwise stated in the legend. The maximum response (maximum distance of the stimulus from the fly) in Figs. 22.7a and 22.8a and the minimum in Figs. 22.7b and 22.8c were obtained with the fly at the centre of the stimulating disc, confirming previous work. The results with one-dimensional stimuli are in agreement with Fig. 22.6.

Monocular flies were also tested by the approach of a vertical stick in the front part of the visual field with the head rotated 180° at the neck and fixed in the new position. With the head in this position most flies failed to fly but

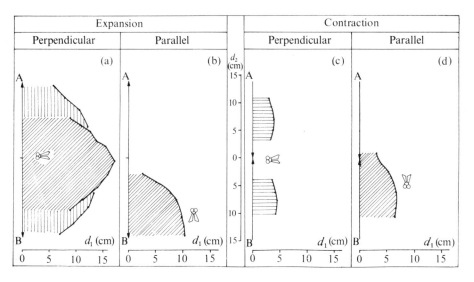

FIG. 22.7. Conditions in which the fly gives the landing response to a rotating disc with a spiral pattern to mimic expansion and contraction in the visual field. Diagonal cross hatch, zones of response with the fly near the horizontal diameter. Horizontal, or vertical cross hatch, zones where responses are more evident when the fly faces the lower or upper half of the disc (respectively). The line AB represents the horizontal section of the pattern, d_1 = distance in cm from the pattern surface; d_2 = distance from the pattern centre. The pattern has a period of 8 cm and velocity 61 cm s^{-1} measured along a radius.

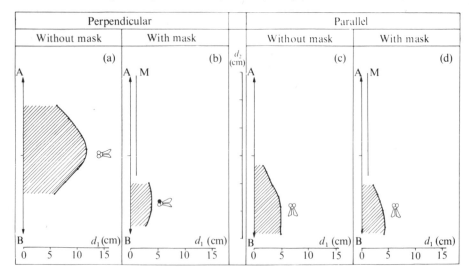

FIG. 22.8. Conditions in which the fly gives the landing response to a pattern of vertical stripes which expand in the horizontal direction, with the fly's axis perpendicular or parallel to the pattern. A mask M was available to cover half of the pattern. The line AB is the horizontal section of the pattern; d_1 = distance in cm from pattern surface; d_2 = distance from pattern centre. The stripes have a period of 8 cm, width of 2 cm in vertical direction, and velocity 61 cm s^{-1}.

some which flew and reacted did so with legs of both sides regardless of the stimulus. A few continued to react with the legs of only the same side as the seeing eye, and did so only temporarily. At successive presentations of the landing stimulus the response changed rapidly but persistently to one with six legs. Even after half an hour without stimulation, the three-leg followed by the persistent six-leg response was consistently evoked. Although few specimens showed this change, they did so repeatedly.

22.4. Landing and optomotor responses

A lateral displacement, for example when a screen obscures part of an expanding stimulus or when the fly is very close to the rotating spiral and able to see only part of it (Figs. 22.7 and 22.8), sometimes evokes the landing response. Two conditions are necessary for a stimulus moving in one direction. First, for a given speed the stimulus must be nearer to be effective, as shown by comparison of the different parts of Figs. 22.7 and 22.8. Secondly, the fly might be located well within the pattern, not near its edge. These results are in agreement with the finding that the effectiveness of a landing stimulus is proportional to the number of facets stimulated (Goodman 1960, 1964). There is an additional effect, however, because when the experiments of Fig. 22.8 are repeated with a constant length of stimulus in the visual field, the distance from fly to stimulus at threshold for the situation of Fig. 22.8d (see Fig. 22.3i, but monocular) is always less than that for the situation in Fig. 22.8a (see Fig. 22.2b). This result was checked by showing that one bar moved in translation is far less effective than two bars moved away from each other and simulating expansion.

In all the above experiments, and others to corroborate them, a translatory stimulus evokes the landing response only when the movement is from front to back across the eye. This is seen in the asymmetries of Figs. 22.7 and 22.8. This third condition was found valid also for movements on planes other than the horizontal one (in agreement with Coggshall 1971, 1972). A landing response caused by a decrease in total light flux can be counteracted by a simultaneous contracting pattern (Taddei and Fernandez 1967, 1971, 1972b, 1973c) and also by a linear movement forwards across the eye.

This directional effect of a linear motion shows that when the stimulus for landing is an expansion composed of two components in opposite directions (unidimensional expansion), one component acts on one eye and the other on the other, if the stimulus is to be maximally effective.

Another general condition for the landing response to be effective is that the stimulus must be presented to anterior ommatidia. In fact, a unidimensional contracting pattern is occasionally effective when presented to backward-looking facets at the side of the head (Fig. 22.4r), but less effective than the unidimensional expanding pattern on the antero-lateral part of the eye (Fig. 22.3m). Although all the vectors are in the right direction, a contracting two-dimensional pattern immediately behind the fly (Fig. 22.2f) does not evoke the landing response, showing that posterior facets are ineffective.

The stimuli used in these experiments can evoke optomotor responses and the escape response as well as landing, and it is of interest to inquire into the interaction between these actions. For this the fly must be free to turn and so is suspended on a hair. In this situation, a flying fly presented with a lateral displacement that is adequate to evoke a landing response in a fixed fly, first shows the landing response and then the optomotor reaction to the movement, but no longer shows the landing response. Conversely, the landing response is sometimes seen during classical optomotor reactions (Reichardt, personal communication). A fly suspended on a hair, flying straight into the centre of a contracting pattern, with head fixed to thorax, and not showing a landing response, oscillates as if following the motion first on one side and then on the other (Situation (o) in Fig. 22.4). A fly free to turn shows the landing response when presented with the centre of an expanding spiral pattern, and then quickly turns round and flies away from it. Similarly, a walking fly will follow a lateral displacement but takes off when presented with an expanding field. Therefore, fixed flies are not able to respond naturally.

When a fly flies towards an object on a background, in order to land on it, it perceives all the values of angular velocity ω for the object points, together with a different set of values for points at a different distance on the background. Therefore, there is a discontinuity in the angular velocities around the edge of the object. This discontinuity will increase as the fly approaches. If the object itself starts to move towards the fly there is an abrupt and unexpected increase in these discontinuities of velocity and this is the most obvious aspect of the stimulus, which is inferred to cause the escape response (Taddei and Fernandez 1972a,c, 1973b). Similarly, when a fly approaches freely towards a stationary disc painted with spirals, all the angular motions are in agreement with each other and with the discontinuity at the edge. On the other hand, when the disc is rotating, its angular velocities are not in agreement with those of the background and edge, and possibly this is the aspect which evokes an escape response. When a fly is fixed, many aspects of the pattern of angular velocities around it are not at all the expected consequences of its own efforts to fly, turn, or escape. Therefore care is necessary in the interpretation of responses of all kinds from a fixed preparation if the normal mechanics of visual discrimination is to be analysed.

22.5. Conclusion

Experiments on the landing response have yielded the following conclusions:

(1) The natural landing stimulus is represented as in Fig. 22.2a, in which the isochrones (dashed lines) have the form $\sin^2 2\alpha$ over a visual field 90° wide measured from the long axis of the fly, and the angular velocities ω are always positive.

(2) The main condition for a landing response is that all the angular velocities be positive. Examples (d), (g), (h), in Figs. 22.2 and 22.3 are at the lower limit.

(3) The velocities must be presented to the anterior part of the eye. Example F of Fig. 22.2 is at the lower limit.

(4) If condition (2) is satisfied, the $\sin^2 2\alpha$ curves may be modified to $\cos^2 \alpha$ (example (b) in Fig. 22.2) or even to $\sin^2 \alpha$ (example (i) in Fig. 22.3) or combinations and intermediates of these.

(5) The lowest threshold is attained when the positive velocities are distributed symmetrically in front of both eyes (Figs. 22.2b, 22.3i).

(6) As long as the velocities remain positive they can be unequal on the two eyes, although the threshold for the landing response is then raised.

(7) In contrast to these findings, the optomotor response is best evoked by opposite directions of motion across the two eyes, although it has long been known that motion across one eye can be effective (example l, Fig. 22.3).

(8) As a result of (6) and (7), stimulus situations can be found which cause both optomotor and landing responses.

(9) The landing response is the result of a filter, which abstracts a velocity pattern that is symmetrical with respect to the fly's axis, whereas the optomotor filter abstracts the average relative rotation around the fly.

23. The effect of illumination on the landing response

C. TADDEI FERRETTI and
A. FERNANDEZ PEREZ DE TALENS

23.1. Introduction

THE fly has several quite distinct visually-controlled behaviour patterns. The one studied in most detail is the optomotor turning response in walking and the optomotor torque response in flight (Fermi and Reichardt 1963; Götz 1968). Another response to a visual input is the approach which can be elicited by a stationary object on a distant background as the flying insect approaches it (Reichardt and Wenking 1969; Reichardt 1973). Yet another is the escape or turning to flee at the approach of an object (Goodman 1960; Taddei and Fernandez 1971, 1972a,c, 1973b). The landing response is quite distinct from the above and is elicited by either an expansion in the visual field or by a fall in illumination. Expanding visual stimuli generally involve a change in illumination but, by use of a rotating spiral pattern on a disc, the two effective features of the stimulus can be separated. This paper is concerned with the effect of illumination, which was excluded in the study of expansion (Braitenberg and Taddei 1966; Fernandez and Taddei 1970). This paper, with its companion in this volume (Chapter 22) represents a summary of the recent papers of the authors.

23.2. Methods

Expansion or contraction in the visual field is generated by rotation of a disc upon which are drawn $2n$ Archimedean spirals of equal increments i. Spaces between the spirals are alternately black and white. The fly, with head fixed to thorax, is stationary flying facing the centre of the disc. A 220 V, 60 W lamp powered by direct current, 40 cm from the disc, gives controlled illumination from 10 to 1500 lx.

23.3. Relations between pattern expansion and light level

Let the relevant measurements be defined as follows: I (lx) = illumination measured as illuminance on the disc; d (cm) = distance from fly to disc; R (cm s^{-1}) = radial velocity of stripes on the disc; $\alpha(t)$ = angle between the

anterior long semi-axis of the fly and a line from the centre of the head to a wavefront at any moment; L (cm) $= i/n =$ the spatial wavelength in radial direction; $\Delta\phi =$ interommatidial angle; $\Delta\rho =$ receptor acceptance angle.
    The perceived radial expansion speed of the wavefront is

$$P \text{ (rad s}^{-1}) = (R/d)\cos^2\alpha$$

(Taddei and Fernandez 1972c) and $P = R/d$ when $\alpha = 0$, i.e. at the centre of the disc; $T =$ threshold value of P just adequate to cause a landing response. Positive velocities imply expansion, negative velocities contraction.
    If the stimulus is below threshold the fly can be made to respond by suitable increase of P or by *fall* in illumination. Next we consider that each value of P implies a particular timing in the change in light flux that occurs in each receptor as the stripes pass over. Let $T_1 = \Delta\phi/P$ be the time required for a stimulus (a definite ΔI) to pass from an ommatidium to its immediate

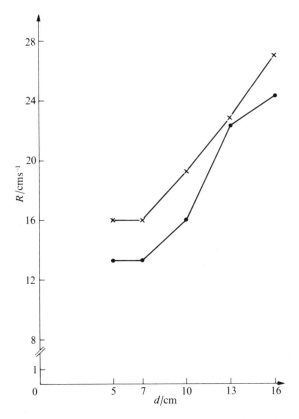

FIG. 23.1. The threshold radial velocity of expansion R (in cm s^{-1}) as a function of the screen distance of the fly, d, (in cm) at ($\bullet$) $I = 50$ lx and ($\times$) $I = 500$ lx.

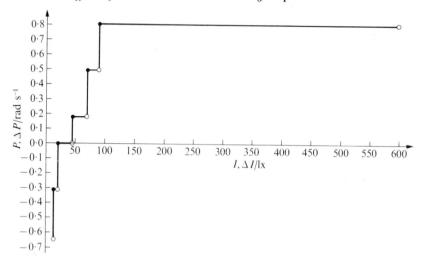

FIG. 23.2. Alternate yes (●) and no (○) states of the landing response obtained by alternate decrements of illumination level I (in lx) and perceived expansion speed P (in rad s^{-1}).

neighbour. Let $T_2 = (\arctan(L/2d))/P$ be the time required for two successive stimuli (two ΔI of opposite sign) to pass one receptor, where $P/\arctan(L/d)$ is the contrast frequency. Let T_3 be the time necessary for a stimulus to cover the receptor completely. For an expanding stimulus $T_3 = F(\Delta\rho/P)$. During a change of illumination T_3 is the time required to obtain a value of ΔI.

The threshold values of R are consistently lower at lower light levels when the fly has been adapted for 10 min over the range 50–500 lx (Fig. 23.1). Tests were next conducted by alternating decrements ΔP and ΔI in such a way that responses alternated with failures to respond (Fig. 23.2). Note that negative values of P arise because a contraction is necessary to balance a suitably large decrease in illumination (Taddei and Fernandez 1967, 1971, 1972b, 1973c). In a variety of experiments the values of ΔI and ΔP were measured with I and P as independent variables, keeping other parameters constant.

The relations between ΔI and P and between ΔI and I were then determined more directly. First, with P fixed in the range 0·5 to $-0·33$ rad s^{-1}, the threshold values of ΔI were measured after adaptation to various values of I. Then, with I fixed in the range 50–600 lx, the threshold values of ΔI were determined for various values of P (Figs. 23.3 and 23.4). The drop in illumination required is less in dimmer light, but at higher values of I higher values of $\Delta I/I$ are found. The dependence of ΔP on I and on P are without significance because the measured ΔP is only the increment required to bring P up to the threshold value T at the particular illumination used (Fig. 23.5). This curve shows that the threshold value T is higher in brighter light,

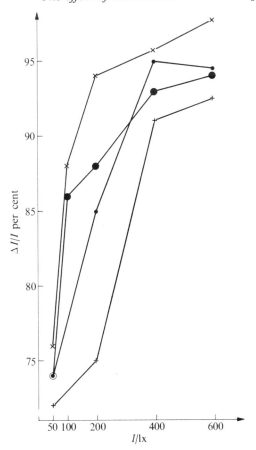

FIG. 23.3. Values of $\Delta I/I$ per cent at threshold values of ΔI as a function of I (in lx). The measurements were taken at the following values of P rad s^{-1}: ($+$) 0·5, ($\bullet$) 0·2, ($\bullet$) 0, ($\times$) $-$0·33.

which one would suppose to be contrary to the expectation that the response depends on contrast transfer in the ommatidia. This lowering of T in dim light is not attributable to acuity, as the spatial frequency is well above the lower limit.

The curve shows that T increases rapidly with I up to I^* (at about 60 lx) and more slowly at values over I^*. Our hypothesis is that the sudden change arises from the two types of receptors in the fly eye. A variety of work suggests that vision depends on retinula cells 1–6 in dim light and on cells 7–8 in bright light (Kirschfeld and Franceschini 1968, 1969). If this is so, we can argue that when the dim-light system is used the threshold is adjusted more rapidly to a reduction in I than when the bright light system is operating.

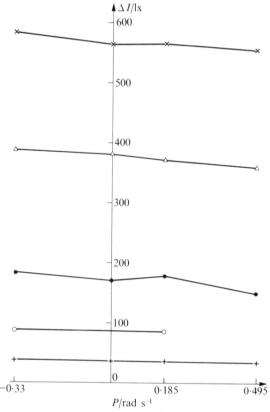

FIG. 23.4. Threshold decrements ΔI (in lx) as a function of the perceived expansion speed P (in rad s^{-1}) at the following levels of illumination I lx: ($+$) 50, ($\bigcirc$) 100, ($\bullet$) 200, ($\triangle$) 400, ($\times$) 600.

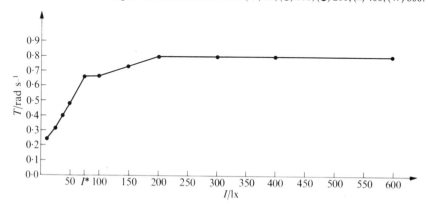

FIG. 23.5. The threshold perceived expansion speed. T (in rad s^{-1}) as a function of the illumination level I (in lx). At I^* there is an abrupt change attributed to a transition from one visual system to another.

23.4. Effect of the percentage of black in the pattern

A set of discs were made with various percentages of black (B per cent) by adjusting the width of the bands between the spirals. The light reflected from each disc, measured as (illuminance at the fly's eye position)/(illuminance on the disc) × 100, agrees to within 3 per cent with that expected from the geometry of each disc, showing that the discs were accurately made.

These discs have been used as an alternative method of testing the influence of variation of I. The value of T was measured for each pattern of the set (Fig. 23.6). A higher value of T implies that the pattern is less effective in evoking the landing response. The two general results are, first that the patterns with a greater percentage of black are more effective in evoking the

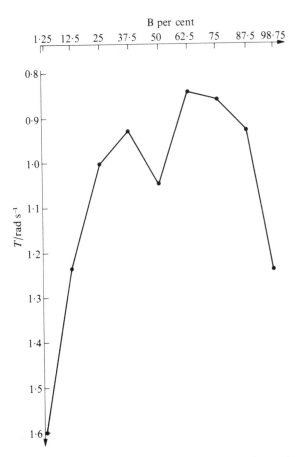

F IG. 23.6. The threshold perceived expansion speed T (in rad s^{-1}) as a function of the percentage of black B per cent in the spiral pattern. Pitch of spiral 8 cm; distance from pattern to fly 17 cm.

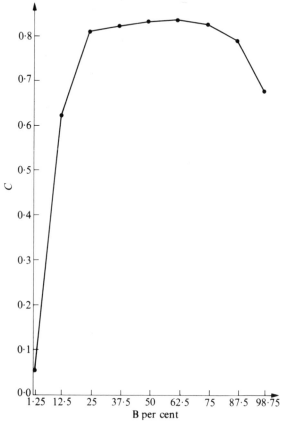

F IG. 23.7. Theoretical values of perceived contrast C of the pattern for various percentages of black B per cent in the spiral pattern. Pitch of spiral 8 cm, reflectance of black zones 5 per cent, of white zones 50 per cent, distance between fly and pattern 17 cm, acceptance angle of ommatidium 1·8°.

landing response than the same with contrast reversed, and secondly that the value of $1/T$ shows a maximum on each side of 50 per cent. The results are independent of the increments in the spirals.

One reason why there should be an increased tendency to land as the percentage of black is raised is that the *average* light flux in a receptor decreases. Therefore, the contrast transfer, $C = (I_{max} - I_{min})/2I_{av}$ increases as the average flux I_{av} is lowered. Tests to control for this show that when the illumination is increased to compensate for this effect, the asymmetry in Fig. 23.6 is reduced. Other examples are known where the optimum effect is obtained with more black than white in the pattern (Burtt and Catton 1969 on locust; Goodman 1960 on the approach of a simple object to a flying fly).

A contributory factor to this asymmetry can be calculated as follows: when the angular sensitivity of a retinula cell is considered as Gaussian in shape, the contrast transfer for any pattern in the visual field can be computed. This has been done (Taddei and Fernandez 1973*a*) for the spiral patterns used (Fig. 23.7) for various values of *B* per cent when the reflectance percentage of white zones is 50 per cent and of black zones 5 per cent, $d = 17$ cm and $\Delta\rho = 1\cdot8°$. The asymmetry, particularly at the two ends, is similar to that in Fig. 23.6.

Finally, the asymmetry can be further explained by calculation of the Fourier spectrum of the stimulus pattern, as follows;

A periodic function $f(x)$ with period L can be expanded into a Fourier series:

$$f(x) = \sum_{n=-\infty}^{+\infty} \beta_n \exp(in\mu_0 x)$$

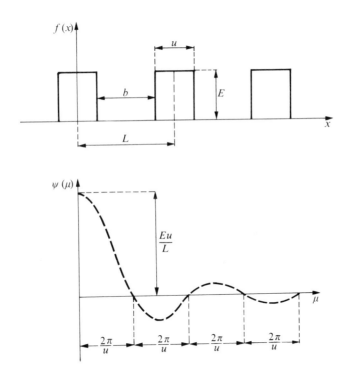

FIG. 23.8. (a) A periodically repeated pulse of intensity E, period L, on-period u, off-period b. (b) Envelope of the corresponding Fourier spectrum

$$\Psi(\mu) = \frac{Eu}{L} \frac{\sin \mu}{\mu}.$$

Discrete lines representing spectrum coefficients are defined by

$$\beta_n = \psi\left(\frac{nu}{L}\right).$$

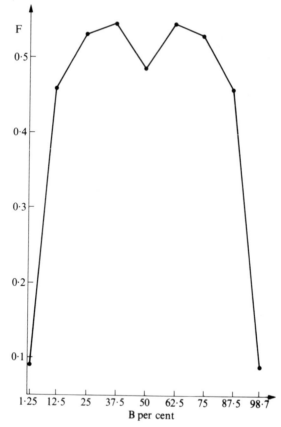

F

0·5

0·4

0·3

0·2

0·1

1·25 12·5   25   37·5   50   62·5   75   87·5 98·7

B per cent

FIG. 23.9. The computer evaluated sum F of the first 160 Fourier coefficients of the stimulating pattern as a function of the percentage of black B per cent in the pattern.

where $\mu_0 = 2\pi/L$ and the coefficients β_n are

$$\beta_n = 1/L \int_{-L/2}^{L/2} f(x)\exp(-in\mu_0 x)\,\mathrm{d}x$$

each representing the amplitude of the nth harmonic component of $f(x)$ having pulsation $n\mu_0$. In the case of a periodic succession of identical rectangular pulses of period L, amplitude E and duration u (Fig. 23.8a), the application of such formulae yields

$$\beta_n = Eu/L\{\sin(n\pi u/L)/(n\pi u/L)\}.$$

This Fourier spectrum shows an envelope of the form $(\sin z)/z$ (Fig. 23.8b) which is that of an harmonic oscillation of decreasing amplitude. β_0 corresponds to pattern average intensity. We will next multiply each β_n by the factor

$$\exp\{-\pi^2\,\Delta\rho^2/4\ln 2\,.\arctan^2(L/nd)\}$$

which represents the transductor properties of the ommatidium for each frequency in the stimulus (Götz, 1964).

In his original work on the beetle *Chlorophanus*, upon which much of the modelling of the insect optomotor system has been based, Hassenstein (1959) found that the optomotor response is proportional to the square of the perceived contrast. He also observed that when the optomotor stimulus is considered as the sum of its Fourier components, *Chlorophanus* is insensitive

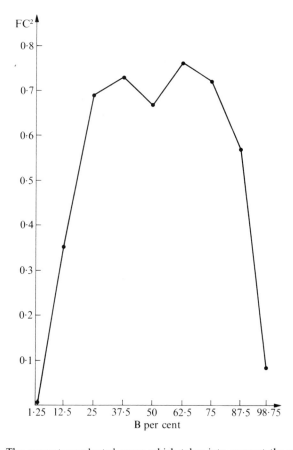

FIG. 23.10. The computer-evaluated curve which takes into account the square of the perceived contrast of the stimulus and the sum of the absolute values of the Fourier coefficients. For each value of *B* per cent this curve is obtained from the corresponding points on Figs. 23.7 and 23.9.

to the relative phase of these components. The optomotor response depends on the sums of the squares of the Fourier coefficients without reference to their phase or sign (Reichardt and Varjú 1959). Continuing this analysis with various flies, the model for movement perception, deduced from the analysis of *Chlorophanus* behaviour (Hassenstein and Reichardt 1956), was found to apply also to the optomotor response of *Musca* in flight (Fermi and Reichardt 1963) and of *Drosophila* walking (Götz 1968).

Following this line of thought with the landing response situation, we computed the sums of the first 160 Fourier coefficients of the spiral patterns for various values of B per cent (Fig. 23.9). The curve we obtained is a reasonable approximation to the experimental result with reference to the decrease at 50 per cent (Fig. 23.6) but with maxima of similar height at the sides of 50 per cent. We next took into account the asymmetry of the contrast transfer (Fig. 23.7), and multiplied the squares of the values in Fig. 23.7 by the corresponding values from Fig. 23.9. The result (Fig. 23.10) is the theoretical calculation of the influence of B per cent taking into account the perception of contrast and also the spatial characteristics of the stimulus. In fact, Fig. 23.10 agrees remarkably well with the measurements of Fig. 23.6. Thus it appears that the landing response could depend on the square of the perceived contrast and on the sum of the absolute values of the Fourier coefficients of the pattern, which would imply that it is insensitive to phase relations in the stimulus pattern. The mechanism of perception in the landing response therefore appears to be of the same nature as that of the optomotor response. At least, by this analysis we find no distinction between the two responses and therefore infer that they share one mechanism of motion perception.

Part VI
Ocellus

24. The neural organization and physiology of the insect dorsal ocellus

L. J. GOODMAN

24.1 Introduction

THE functions of the simple eyes of adult insects, the dorsal ocelli, have proved impossible to define with precision. Consisting of some hundreds of visual cells beneath a common cuticular lens with a relatively wide aperture and often backed with a tapetal layer, the ocelli appear to be well suited structurally for light gathering. The visual cells show a high degree of convergence upon wide field second-order units. In view of this and the fact that the principal focal plane of the lens appears not to fall within the retinal space in any ocellus yet examined, the ocelli seem to have no role in form perception. Recently, Rosser (1972) has stressed that the optical systems of some ocelli are capable of localizing light from different parts of the visual field upon the receptor layer, thus fitting them for a crude type of directional discrimination if the underlying neural organization can preserve this spatial distribution, and Zenkin and Pigarev (1971) have shown that dragonfly ocelli respond to moving-edge stimuli.

Ocellar implication in the entrainment of circadian rhythms has several times been proposed (Cloudsley-Thompson 1953; Harker 1956), but the work of Nishiitsutsugi-Uwo and Pittendrigh (1968) on the cockroach appears to preclude this function at least for that insect. Interest in a neurosecretory role for the ocelli has been revived by recent work: Schlein (1972) has shown that neurosecretory cells in the ocellar nerve of the fly, *Sarcophaga falculata* are responsible for initiating post emergence growth. Brousse-Gaury (1968, 1971) has reported third-order neurons running from the ocellar nerves through N.C.C.I. and N.C.C.II to the retrocerebral complex in *Periplaneta americana*, *Acheta domestica*, *Locusta migratoria*, and *Schistocerca gregaria*. The cell bodies of these neurons lie off the tracts of N.C.C.I and II. Cooter (personal communication) has confirmed the presence of an ocellar contribution to N.C.C.II in *P. americana* with cobalt dyeing techniques. Some of the larger neurons in the median and lateral ocellar nerves appear to give off short collaterals to the neurosecretory cells of the pars intercerebralis as they enter the brain in *S. gregaria* (Goodman and Hoare unpublished observations).

Behavioural studies give little guidance in the search for ocellar function. Many insects show a depression in their locomotor activity after ocellar occlusion and it has been suggested that the ocelli act as photokinetic stimulatory organs (Wolsky 1933; Médioni 1959) but it remains to be established that ocellar occlusion produces a general non-specific loss of sensitivity. Médioni and his co-workers (Médioni 1959*a*,*b*; 1964, 1967: Romaneli 1968: Vaysse 1973*a*,*b*) have reported some evidence in support of such a role in the Diptera, demonstrating reduced rates of response in the photopositive and geonegative reactions of *Drosophila* and increased latency in the commencement of precopulatory behaviour and of 'take-off' in response to light stimuli when the ocelli are occluded. Using anocellate mutants they have shown that possession of an ocellar apparatus, even if completely occluded, can be effective in reducing the latency of the 'take-off' response as well as of the courtship behaviour. In this context it is of interest to note that many Lepidoptera previously believed to be anocellate possess much reduced external ocelli (Dickens and Eaton 1973). In some moths a nerve runs from the reduced external ocellus to an internal ocellus which in turn sends an ocellar nerve to the brain. Some of the Apterygota also possess internal ocelli (Elofsson 1970). Light stimulation of the ocelli also increases the heart beat rhythm of *Calliphora* (Médioni, Campan, and Quéinnec 1971). This reaction is a tonic one with a long latent period. As the input relationships of the ocelli become better known (See Section 24.2) their modulating effects upon the activity of particular sensory interneurons and motoneurons may be examined and the neural mechanisms underlying some of these effects described in a more precise manner.

It has been noticed many times that the ocelli make some contribution to positive phototactic orientation mediated by the compound eyes (Goodman 1970), but Jander and Barry 1968, have been the only workers to formulate a specific role for them in the phototropotactic behaviour of *Gryllus bimaculatus*, showing that the lateral ocelli act synergistically or antagonistically with their neighbouring compound eye depending upon the light intensity and the integrity of the median ocellus. Since behavioural studies have never revealed any motor response which is directly initiated or controlled by stimulation of the ocelli, most physiological studies have been undertaken with the aim of analysing visual mechanisms in a simple photoreceptor rather than to elucidate the ocellar contribution to the direction and co-ordination of behaviour in the whole insect. Recent histological and physiological studies have revealed that the neural organization of this receptor and its relationships with the compound eyes and other sense organs are rather more complex than hitherto supposed and this may suggest new approaches towards solving the problem of the functional role of the ocellus.

24.2. The neural organization of the ocellus

24.2.1. *The retinula cells and their terminals*

In the retinula cells of many insects, particularly the Exopterygota, the retinula cells form small groups having a common rhabdom complex. In some species the number of cells contributing to the rhabdom is fairly constant over the greater part of the retina. In the dragonfly three cells contribute to form a three-limbed rhabdom with occasional complexes formed from a greater number of cells (Chappell and Dowling 1972). Other insects have much more irregularly shaped rhabdom complexes. In *Schisto-cerca gregaria*, for example, any number of cells between two and seven can contribute to the complex (Goodman 1970). In the higher insect orders, the roughly hexagonal retinula cells are closely packed together with the entire face of each cell modified to form a rhabdomere at its distal end. The micro-villi of the rhabdomeres on each face of the cell interdigitate with those of the neighbouring cells so that in transverse section the rhabdomeres appear to form a common meshwork across the retina although the integrity of each cell is, in fact, preserved (Goodman and Kirkham in preparation). The microvillar organization here would appear to preclude any polarized-light sensitivity. In the locust there are no preferred directions of orientation of the microvilli across the retina and polarized light discrimination seems unlikely on anatomical grounds. Even the dragonfly, with its regular three-limbed rhabdoms, appears not to possess any preferred direction or orien-tation of these limbs and the microvilli within them. Dark screening pigment is entirely absent in some ocelli as in *P. americana* and variously distributed in others. In the locust a thin ring of pigment cells surrounds the ocellus just above the distal ends of the retinula cells. Many dragonflies have pigment granules within the sheath cells surrounding the proximal ends of the retinula cells and pigment migration in response to changing light intensity has been reported (Lammert 1925). In the higher orders, pigment granules are more often found within the retinula cells themselves and in *Rhodnius*, at least, distal migration of the granules occurs in response to high light intensities (Goodman and Kirkham in preparation).

The proximal ends of the retinula cells narrow to form the retinula axons which descend between sheath cells or cells of the tapetal layer through a basement membrane into the first-order synaptic plexus. This synaptic area seen in section in the light microscope appears as a relatively disorganized region with many apparently identical retinula cell axons descending in an irregular manner upon a diffuse network of second-order dendrites formed from the multiple branching of the few large second-order neurons. Golgi preparations and E.M. studies have revealed that the structure of first- and second-order units and their synaptic organization is considerably more complex (Goodman, Kirkham, and Hoare 1974). Below the median ocellus of *S. gregaria* there are at least three types of retinula cell terminal and there are indications from preliminary studies on other ocelli that where the visual cells form discrete groups there is specialization amongst their axon terminals. Cajal (1918) has described two types of retinula cell ending in the dragonfly. In *S. gregaria* the commonest type of retinula cell terminal, hereafter called a Class I terminal, has a bulbous, irregular, club-like ending and bears brush-like arborizations at a higher level on the axon, Fig. 24.1A. A second class of retinula axon extends more deeply into the synaptic region, in many cases as far as the distal region of the ocellar nerve, giving off short, stubby collaterals before ending in fine branches each bearing small bulbous terminations, Fig. 24.1B. It has not proved possible to trace individual Class I and II terminals back to their respective retinula cells owing to the angled pathway taken by axons *en route* to the synaptic plexus. No obvious

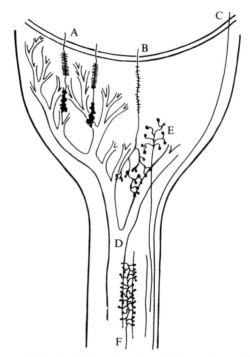

FIG. 24.1. A diagrammatic longitudinal section through the median ocellus of *Schistocerca gregaria* to show the structure and distribution of retinula cell terminals and second-order neurons. (A) Class I retinula cell terminal with a bulbous club-shaped ending and brush-like lateral extensions. (B) Class II retinula cell terminal with diffuse branching ending and short lateral projections. (C) retinula cell axons which pass ensheathed into the ocellar nerve and are believed to run through to the brain. (D) one of the six giant axons. (E) a small afferent second-order neuron. (F) a 'mossy' efferent neuron synapsing with the giant axons.

differences can be detected between individual retinula cells that might be correlated with these two types of terminal either in their fine structure or their manner of association with other members within a group. Although the cells within a group taper at different points there is no cell corresponding to an eccentric cell. The only indication that there may be some differentiation lies in the fact that the axon of one member of a group of retinulae will frequently separate from their descending bundle of axons and pass tangentially across the cup joining with another group of axons to pass through the basement membrane. Although an accurate estimate is not possible there appear to be many more Class I than Class II terminals. The two types of terminal are not clearly restricted to particular zones in the synaptic plexus although Class I axons tend to end higher in the ocellar cup and Class II axons to descend to the proximal part of the cup and into the distal region of the ocellar nerve. It is in this region that the second-order afferent neurons with small dendritic fields end and they appear to be associated exclusively with Class II terminals. The functional significance of two types of terminal is not yet known. Since at least two types of second-order neuron project into the synaptic plexus, and Class II terminals appear to be associated with the small field type, some of the photoreceptor cells could have differing spectral sensitivities with this information preserved to a higher level. Alternatively the association of Class II terminals with small field second-order neurons may be part of a movement-detecting mechanism.

Cajal (1918) mentioned a third type of retinula cell axon in the dragonfly whose ending he could not trace. In the locust, retinula cell axons are encountered which appear to bypass the synaptic plexus, enter the ocellar nerve and travel through to the brain, Fig. 24.1C. These retinula axons, hereafter called Class III axons, differ from the Class I and II axon terminals in retaining their glial sheaths after passing through the basement membrane. In fact, most of them are situated around the rim of the synaptic plexus and are separated from it by several layers of glial material. The Class III axons appear to arise from the small retinula cells located around the rim of the visual cell layer. A greater relative proportion of these cells is occupied by rhabdomere, and differentiation of the axon appears at a more distal point than in the majority of retinulae. It is not clear whether these differences have a functional significance or whether they are merely a consequence of the position of these cells packed around the distal-most edge of the receptor layer. Many synapses occur along the whole length of the ocellar nerve but these small outer axons appear to remain sheathed all the way to the brain. Visual cell axons which bypass the first-order synaptic region and run centrally are well known in the compound eye of the Lepidoptera, Diptera, and Hymenoptera (Strausfeld and Blest 1970; Strausfeld 1970) passing through the lamina and running on to the medulla. Amongst other possibilities the presence of such axons in the ocellus would mean that spatial distribution of intensity upon the retina could be preserved to a higher level in a rather crude manner.

24.2.2. *The neurons of the ocellar nerve*

In addition to the six large (10 μm diameter) second-order neurons found in the median ocellar nerve of *S. gregaria* there are seventy–eighty other axons present which have been shown by serial section to extend right through into the brain at least as far as the protocerebral bridge and further in the case of some individual axons. The number often appears higher in the distal part of the nerve since branching begins at this level. These neurons fall into two classes on the basis of diameter; twenty–thirty lie within the range 1·0–1·5 μm, and fifty–sixty are *ca.* 0·5 μm in diameter. The locust is not unique in this respect, all other species so far examined, including the bee, fly, *Rhodnius*, and the dragonfly, have been found to have a number of giant axons, several medium-sized axons and a larger number of smaller axons (Goodman 1970). The giant axons are second-order afferent neurons, some of them innervating more than one ocellus (See Section 24.2.5). They branch dichotomously in the distal part of the ocellar nerve and their branches extend as far as the basement membrane in the ocellar cup, Fig. 24.1D. Some of the smaller neurons present are second-order afferents with a restricted dendritic field. These small field afferents end mainly in the lower half of the ocellar cup and, as previously mentioned, are associated with Class II retinula axon terminals, Fig. 24.1E. Also present in the ocellar nerve are small efferent neurons with 'mossy' type endings. They do not appear to enter the ocellar cup, synapsing with the giant and medium–sized axons in the distal part of the ocellar nerve and possibly with some of the smaller afferent fibres as well, Fig. 24.1F. Kenyon (1896) mentions small efferent neurons in the ocellar nerves of the bee which arise in the brain at some level below the calyces and terminate in the lower part of the synaptic plexus. Third-order neurons in the ocellar pathway extend out into the proximal part of the ocellar nerve, their fine endings forming axo-axonic synapses with giant and medium second-order units. The through-going retinula cell axons also contribute to the nerve. In each ocellus examined one or two neurons can be found containing large vesicles (150–200 nm) with granular contents, Fig. 24.4a. They are first seen amongst the bundles of retinula axons piercing the basement membrane. There is never more than one in a bundle and they are only present in a few bundles. They can be found at all levels in the synaptic plexus and all the way along the ocellar nerve into the brain but synapses with other units have not been seen. They appear to contain neurosecretory material but their destination is not known nor is it clear whether they are first- or second-order units.

24.2.3. *First-order synaptic organization in the locust*

Unlike the lamina of the compound eye, the synaptic plexus of the ocellus which extends from the basement membrane into the distal half of the ocellar nerve does not present an ordered array of contacts between first- and second-order neurons. A low level of organization is imposed upon the distal synaptic region by the passage of semi-parallel bundles of axons through the basement membrane and the extension of some second-order dendrites upwards into these discrete bundles, but soon after the constraints of the basement membrane are removed the retinula axons in the bundles begin to diverge and an ordered arrangement is lost. Within the main synaptic area can be found retinula cell axons, arborizations, and terminals, large branches of the giant axons, medium-sized dendrites, and many very small dendrites. The larger branches of the giant axons may be readily identified by their size, the relative number and distribution of vesicles and mitochondria, and the relative concentrations of neurofilaments and microtubules present. The photoreceptor axons and terminals can be identified from similar criteria. The identification of the smaller dendrites is more difficult. Longitudinal sections have shown that quite large branches of the giant axons give off many very small dendrites at all levels in the synaptic plexus and a large proportion of those present must belong to the giant axons. The short branches or arborizations of the photoreceptor terminals have a characteristic vesicle distribution enabling them to be identified with some certainty. There remain some medium-sized and small dendrites with a characteristic disposition of organelles and of synaptic contacts which are believed to be the small field second-order neurons. No cells corresponding to horizontal cells have been found in the synaptic area. The following types of synapse have been found:

24.2.3.1. *Peg-type reciprocal synapses.* In these a peg-like projection extends from one cell into a neighbouring cell. Serial sections show that the projection is packed with vesicles and that vesicles within the second cell always cluster around the indentation containing the peg (Fig. 24.2d). Intersynaptic filaments can be resolved in the cleft between the peg and the recipient cell in some cases but there is little thickening of either cell membrane and no subsynaptic apparatus is present. On occasion vesicles from both the peg and its recipient cell have been resolved, apparently emptying their contents into the synaptic cleft. It is common for the recipient cell itself to contribute a peg to the donor cell. Several such pegs may project into a cell from neighbouring cells. These synapses are most commonly found between retinula cells. As these cells descend through the basement membrane they lose their glial sheaths and become irregular in shape, at this point peg-type synapses between them are extremely common. They are found between Class I terminals throughout the upper synaptic zone and also between Class I terminals and quite large branches of the giant axons.

24.2.3.2. *Ribbon-type synapses.* The great majority of the synapses found throughout the synaptic plexus are small ribbon-like synapses with an electron-dense organelle surrounded by a cluster of synaptic vesicles in the

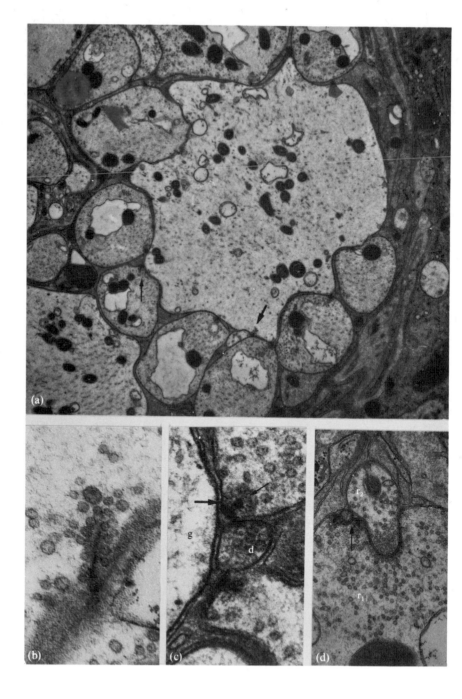

(a)

(b)

(c)

(d)

presynaptic cytoplasm, Fig. 24.2a,b. The presynaptic membrane appears thickened opposite this organelle; at high resolutions this is a separate electron-dense area lying immediately adjacent to the presynaptic membrane, Fig. 24.2c. This thickening bears a strong resemblance to the arciform density seen between the ribbon and the plasma membrane in the terminals of vertebrate rods and cones (Ladman 1958) and in the bipolar terminals of primates (Dowling and Boycott 1966). At low resolution, this structure frequently gives the synapse the appearance of possessing a T-shaped organelle. Opposite this area of the presynaptic plasma membrane is a widened synaptic cleft possessing intersynaptic filaments. There is usually a slight thickening of the postsynaptic plasma membrane at this point. The subsurface organelle of the presynaptic unit may appear completely round, oval, or elongated to varying degrees in section, the extreme conditions found are shown in Fig. 24.2b,c. It seems most probable that there is a narrow (30 nm), elongated, ribbon-type organelle present, lying roughly perpendicular to the plasma membrane which is seldom cut along its long axis, as in Fig. 24.2b, although it is possible that there are two distinct types of organelle, one ribbon-like, the other spherical or oval. It is also possible that the subsurface organelle shows different degrees of development depending upon its position within the cell. The rounded or oval forms tend to be found in the corners of the cells, the elongated form has generally been found in a less constricted part of the cell. Dowling and Chappell (1972) have found somewhat similar synapses in the dragonfly ocellus, where the pre-synaptic organelle is always a round, electron-opaque body, 40–50 nm in diameter, in which some substructure may be distinguished, separated from the plasma membrane by a clear space of 5 nm. There is no sign of the arciform-dense area of the locust adjacent to the plasma membrane. A survey of the ocellar ribbon-type synapses of several insects shows a general resemblance between these synapses and the ribbon synapses found in the plexiform layers of the vertebrate retina (Dowling and Boycott 1966). Ribbon-type synapses appear to be general in the ocelli having been seen in *Rhodnius*, the

FIG. 24.2. (a) A transverse section through the synaptic region of the median ocellus of *S. gregaria* × 8000. A ring of nine retinula cell terminals are making ribbon-type synaptic contact with a branch of a giant axon. Dyad synapses can be seen in which the retinula cell terminals are presynaptic to the giant axon and to a smaller second-order unit, thin arrow, and in which the giant axon branch is the presynaptic unit, thick arrow. (b) A ribbon-type synapse in a retinula cell. The subsurface organelle has apparently been cut through the long axis and is surrounded by a cluster of vesicles. (c) A ribbon-type synapse in a retinula cell. The subsurface organelle appears round and is surrounded by a cluster of vesicles (thin arrow). The arciform density lying between the ribbon and the plasma membrane can be seen (thick arrow). This ribbon-type synapse is forming a dyad synapse with a small dendrite, d, and a branch of a giant axon, g. (d) A peg-type reciprocal synapse is shown between two neighbouring retinula cell terminals, r_1 and r_2. Vesicles can be seen adjacent to the plasma membrane in both retinula cells. Two ribbon-type synapses can be seen in r_1, one of which is forming a dyad synapse with a small dendrite and r_2 (thin arrow).

honey bee, the fly (Goodman, Kirkham, and Hoare 1973) and the cockroach (Cooter unpublished observations). The development of the subsurface organelle appears to vary between species being relatively elongated in *Rhodnius*. Large numbers of these ribbon-type synapses are found in the photoreceptors situated in the corners of the descending axons, in their short branches and terminal bulbs in both Class I and Class II terminals. They are also found in the giant axons, both in their dendrites and along the axons themselves as far as the brain, and in the dendrites of the smaller second-order neurons. Many of the ribbon-type synapses are dyad synapses with the presynaptic ribbon organelle pointing towards two postsynaptic units both of which show specializations indicating that they are actively associated with the presynaptic unit. The commonest arrangement, in 41 per cent of all locust first-order synapses examined, is for a retinula cell axon or terminal bulb to be presynaptic to a large branch of a giant axon, to a very small second-order dendrite or to another retinula cell, Fig. 24.2a,c,d. Combinations of any two of these three postsynaptic possibilities may be seen but the most usual one is a giant branch and a small dendrite. Both large and small second-order dendrites may synapse back upon the photoreceptor terminals or upon adjacent second-order terminals, Fig. 24.2a. Ribbon-type synapses may also form reciprocal synapses between adjacent retinula cells and, less commonly, between giant axon branches or smaller dendrites and retinula terminals, Fig. 24.2a. These reciprocal synapses are quite distinct from the peg-type reciprocal synapses.

24.2.3.3. *Conventional synapses.* Synapses lacking any subsurface organelles but having an increased electron-density of both the pre- and postsynaptic plasma membrane are found in increasing numbers as one descends through the ocellar cup into the distal region of the ocellar nerve. A cluster of vesicles lies adjacent to the presynaptic membrane and intersynaptic filaments can be resolved at high magnifications. These conventional synapses (Gray and Guillery 1966) are found occasionally in the side walls of second-order neurons within the ocellar cup or much more frequently within small rounded terminals which are seen indenting the surface of the giant axons and their branches in the distal regions of the ocellar nerve, Fig. 24.4a. The latter endings are believed to be the synaptic terminals of the efferent fibres, Fig. 24.1F.

24.2.3.4. *Conventional synapses within giant axon bulbs.* In the lower part of the ocellar cup the giant axons show a new feature giving off stem-like projections which balloon out into bulbous endings, Fig. 24.3. The thin neck of this projection is packed with microtubules and the smaller bulbous endings with mitochondria and vesicles. Class II terminals make contact with these bulbs with ribbon-type synapses but the bulbs themselves appear to make conventional synaptic contacts with the retinula cell terminals. In some instances the vesicles are associated with the microtubules in the neck of the bulb in the manner described for the lamprey nervous system (Smith, Järlfors, and Beránek 1970) where it has been suggested that the microtubules have a role in transporting synaptic vesicles to the synaptic points.

24.2.3.5. *Vesicles.*   No conspicuous differences have been found between the vesicles in all the types of synapse described above, after using a variety of fixation techniques and tilting the sections upon a goniometer stage. The great majority are roughly spherical, lucent vesicles, of the order of 30–35 nm, with occasional larger vesicles of the same type associated with them. Very rarely single dense-cored vesicles of the order of 60–70 nm are seen in the second-order units. Small dense-cored vesicles are established as a component of adrenergic nerves but rather larger dense-cored vesicles are often reported in small numbers from other neurons not generally associated with adrenergic systems. De Iraldi and de Robertis (1968) suggest they may be present in small numbers in all types of neuron.

Having identified the types of neuron and synapse in the synaptic plexus it is possible to characterize different zones. The upper synaptic zone comprises the greater part of the ocellar cup extending from the basement membrane to the point at which the cup begins to narrow to form the nerve. Here there is a short transitional zone followed by the lower synaptic zone which comprises the distal end of the ocellar nerve. The upper zone is characterized by many peg-type reciprocal synapses between retinula cell terminals, particularly as they pass through the basement membrane. The majority of the Class I terminals, i.e. the majority of the photoreceptors, end in this region and the corners of their axons, branches, and terminal bulbs contain many ribbon-type synapses making contact with giant axon branches, small dendrites, and other retinula cells, mostly in the form of dyad contacts as described earlier. There are some ribbon-type reciprocal synapses present in this region mostly between photoreceptor terminals and giant axon branches but few conventional synapses, and Class II terminals make synaptic contacts with giant branches and small dendrites as they pass through. Thus the main feature of the major synaptic zone of the ocellus is the presence of reciprocal and ribbon-type dyad synapses forming a system of extensive lateral and feedback connections. Dowling and Chappell (1972) have pointed out that the dyads of the ocellus have a striking resemblance to those of the inner plexiform layer of the vertebrate retina and somewhat similar systems are reported from other invertebrates, notably the neural plexus of the *Limulus* lateral eye (Whitehead and Purple 1970) and the spider eye (Trujillo-Cenóz 1965). The implications of this system for transmission at the first-order synapse are discussed in 24.3. In the transitional zone, near the base of the cup and extending into the nerve, are an increasing number of smaller-diameter retinula axon terminals which contain prominent

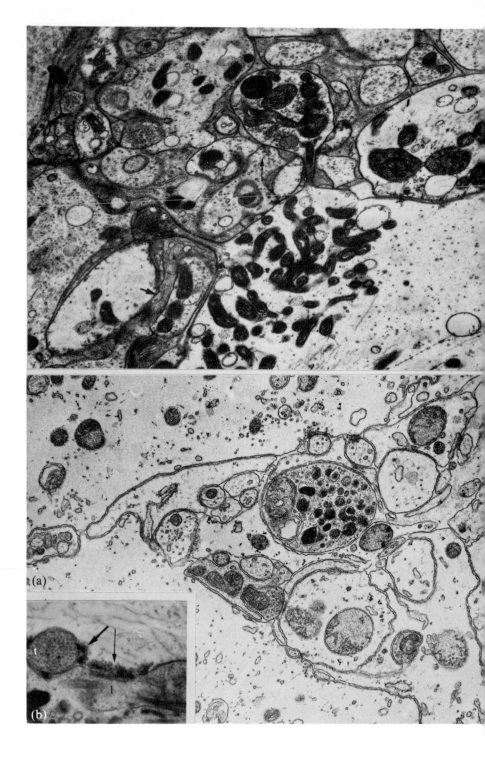

(a)

(b)

neurotubules. These are the Class II terminals, the majority of which descend to this level of the cup. Their endings are smaller and more regular in outline than those found in the upper zone but they contain similar ribbon-type synapses and make dyad contacts with large and small second-order neurons and other receptor terminals. Class I terminals are still present but become relatively fewer as the ocellar nerve is approached. It is in this region that the giant axon branches give off their bulb-like projections, Fig. 24.3. Class II terminals and small second-order dendrites synapse with these projections and they participate in dyad formations, acting as the presynaptic unit in some cases. Class II terminals in this region make contact with a third type of dendrite whose endings are quite distinct being completely packed with mitochondria and vesicles of varied sizes. Only Class II photoreceptor terminals appear to synapse with them, they never form part of a dyad array or act as the presynaptic unit. Golgi studies show small-field second-order neurons ending at this point in the cup and possibly these are the terminals of such neurons.

The lower synaptic zone, lying completely within the distal ocellar nerve, contains Class II retinula axon terminals only, synapsing with the very large branches and stems of the giant axons and with small-field second-order afferents. Class II terminals extend at least a third of the way down the ocellar nerve. The giant axons show a new feature at this stage repeatedly synapsing with each other by means of ribbon-type synapses.

24.2.4. Second-order synapses within the ocellar nerve

A distinct type of conventional synaptic contact is found in the lower synaptic zone. Very small rounded terminals are seen on the surface of the giant axons and their branches, Fig. 24.4a, having an increased electron density of their membranes where they are opposed to the giant axons with a corresponding increase in the density of the giant axon membrane. The highest concentration of vesicles appears generally to lie within the small

FIG. 24.3. A longitudinal section through the base of the ocellar cup of *S. gregaria* × 10 000. A branch of a giant axon is seen giving off bulbous projections, borne on a thin stem containing many neurotubules. Small photoreceptor terminals make synaptic contact with these projections, thin arrow. A retinula cell terminal, probably a Class II terminal, is presynaptic to the giant bulb and to another photoreceptor terminal. Note one of the terminal branches of a Class II retinula axon cut in longitudinal section, thick arrow.

FIG. 24.4. (a) A transverse section through the distal region of the ocellar nerve of *S. gregaria* showing the small terminals of the 'mossy' efferent neurons making synaptic contact with the giant axons and their branches, arrowed, × 15 000. An axon containing membrane bound granular material, possibly neurosecretory material, as well as many small vesicles is seen centrally. (b) A longitudinal section of the median ocellar nerve close to its entry into the brain. A giant axon is seen synapsing with a smaller neighbouring longitudinally-running axon l (thin arrow). It also makes synaptic contact with small rounded terminals, t, which indent the surface (thick arrow).

terminals. These are believed to be the synaptic terminals of the efferent fibres, Fig. 24.1F, and they are never seen within the ocellar cup.

Other arrangements resembling synapses occur along the whole length of the ocellar nerve until the neurons diverge at the protocerebral bridge. The giant axons continue to make synaptic contact with each other at all levels and in the proximal half of the nerve they make axo-axonic contacts with the fine extensions of third-order neurons which run out into the ocellar nerve. Some of these third-order axons terminate in a spray of fine bulbs, packed with vesicles, which indent the surface of the giant axons and some of the medium-sized axons as well. The giant axons make ribbon-type synaptic contacts with them, Fig. 24.4b.

The synaptic plexus of the ocellus thus shows a much greater degree of complexity and of organization than supposed previously. The giant axons have a very wide dendritic field making contacts with very many receptor terminals in all parts of the cup in most cases and making many contacts with each individual terminal, but second-order units with more restricted dendritic fields exist and there is some indication that they synapse with receptor terminals of Class II. The possibility of through pathways to the brain from the retinula cells also exists. The functional significance of the different types of receptor cell ending and of the small-field pathways remains to be determined. The wide-field pathways of the giant axons appear to be associated with the rapid signalling of 'dimming' to the level of the thoracic ganglia amongst other functions (see Section 24.3). It seems likely on anatomical grounds that giant axon activity at least can be modulated by efferent activity but as yet there is no evidence that efferent activity ascends to the photoreceptors.

24.2.5. Ocellar pathways within the brain

In spite of the conspicuous tracts formed by the ocellar nerves with their giant axons as they enter the brain and pass postero-ventrally towards the protocerebral bridge their pathways have been described only in the most general terms (Viallanes 1886; Kenyon 1896; Cajal 1918; Müller 1931; Hanström 1940; Bierbrodt 1942). Information about individual giant axon pathways, the destinations of the smaller ocellar neurons, ocellar association centres and pathways of higher-order neurons is scarce, indeed the smaller ocellar neurons are omitted entirely from some accounts. A combination of light microscopy, serial electron-microscopy sectioning and cobalt dyeing techniques has made it possible to describe the ocellar pathways of *S. gregaria* in more detail. The nerves enter the brain in the median antero-dorsal region of the protocerebrum. They retain a thin glial sheath around them as far as the protocerebral bridge which makes it easier to count the total number of axons present at different levels and to follow the course of some of the larger ones in serial sections. The median nerve passes posteriorly between the

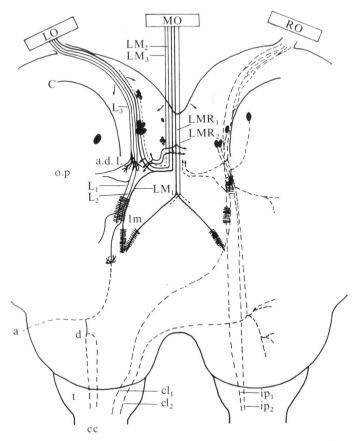

FIG. 24.5. The pathways of the larger ocellar neurons are shown diagrammatically in a transverse section through the brain of *S. gregaria*. The giant second-order neurons of the left lateral ocellus are shown together with those giant median neurons which innervate the left half of the brain. Neurons innervating the right ocellus are omitted. Neurons filled with dye from the circumoesophageal connectives and presumed to be third-order ocellar neurons are shown on the right side of the brain only. Second-order neuron cell bodies are shown on the left, third-order neuron cell bodies on the right. Small arrows indicate the points at which the giant axons send branches ramifying between the neurosecretory cells of the pars intercerebralis. a, antennal nerve; a.d.l, anterior dorso-lateral ocellar complex; C, calyx of protocerebrum; cc, circumoesophageal commissures; cl_1 and cl_2, contralateral third-order ocellar neurons; collaterals from cl_2 run to the antennal centres. d, deutocerebrum; ip_1 and ip_2, ipsilateral third-order ocellar neurons; L_1, L_2, and L_3, left lateral ocellar giant axons; LM_1, median neuron contributing to the left latero-medial ocellar complex, LM_2, LM_3, giant axons which run from the median ocellar cup and enter the left lateral ocellar tracts; recent studies have shown that LM_1 has only a short branch running in the left lateral ocellar tract but that LM_2 and LM_3 extend as far as the left lateral ocellar synaptic plexus. LMR_1, LMR_2, giant axons which run in the median and both left and right lateral ocellar tracts; lm, latero-medial ocellar complex; LO, MO, and RO, left lateral, median, and right lateral ocelli; o.p., region of posterior optic tract; t, tritocerebrum.

neurosecretory cells of the pars intercerebralis towards the protocerebral bridge. The two lateral nerves pass postero-ventrally around the edges of the corpora pedunculata towards the bridge. Division of the median tract into two begins just above the protocerebral bridge and is complete when the tracts have passed below the bridge. As each lateral tract reaches the edge of the protocerebral bridge it begins to divide into two, forming an inner and an outer tract as it passes through the edge of the bridge. The two median tracts pass laterally each one to join with the lateral tract on its side. Some median fibres in each tract join the inner lateral tract and together pass posteriorly to form the latero-medial ocellar complex, others join the outer lateral tract and just below the level of the protocerebral bridge form a second complex which will be referred to as the anterior dorso-lateral ocellar complex, Fig. 24.5.

24.2.5.1. The latero-medial ocellar complex. Each lateral ocellar nerve contains two giant axons, L1/R1 and L2/R2 whose dendrites supply all parts of the synaptic plexus of the ocellus and which run posteriorly along the sides of the corpora pedunculata to end with a clearly demarkated area of terminal arborizations within the latero-medial complex, Fig. 24.5. The cell bodies of these two axons lie just dorsal to the ocellar tract as it runs down towards the protocerebral bridge as shown in Fig. 24.5. Cobalt dyeing has shown that a third giant axon on each side contributes to the complex and this one innervates the median ocellus. This neuron, LM1/RM1 sends a branch out in the median ocellar nerve which sends dendrites only to the ipsilateral half of the median ocellar cup. This neuron runs back into the latero-medial complex and produces a well-defined area of terminal arborizations at a more posterior point than those of the two lateral giants, Figs. 24.5 and 24.6a,c. The cell body of this neuron is a very large cell lying lateral to the lateral ocellar tract, Figs. 24.5 and 24.6a. A further contribution to the complex is made by a neuron from the median ocellar nerve which is somewhat smaller than the giant axons when it enters the brain, of the order of 4–5 μm, but which rapidly increases in diameter as it approaches the bridge. When the other median ocellar neurons start to descend through the protocerebral bridge this one remains above the bridge and sends branches into both lateral ocellar tracts, these branches subdivide within the lateral tracts

FIG. 24.6. (a) A whole brain preparation seen from above in which the median nerve has been filled with cobalt dye. The dendritic fields of LM_1 and RM_1 can be seen on the left and right, and the cell bodies CLM_1 and CRM_1 can be seen lateral to the left and right ocellar tracts. Three median axons can be seen running out in the right lateral ocellar tract and their cell bodies can be seen lying above the right ocellar tract. (b) L_2 from the left latero-medial complex is seen extending to a dendritic field d belonging to a sensory neuron running in the antennal nerve. The contributions of R_1 and R_2 to the latero-medial ocellar complex on the right side of the brain can also be seen. (c) The dendritic field produced by LM_1 which lies posteriorly to that produced by L_1 and L_2. LMR_1 can be partly seen deeper in the brain.

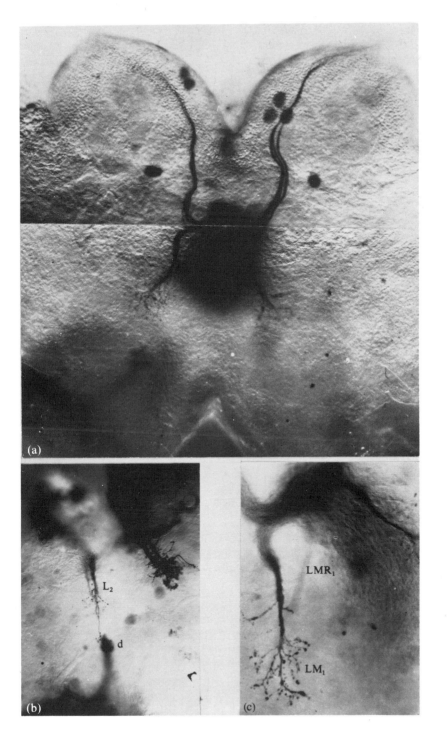

(a)

(b)

L₂

d

(c)

LMR₁

LM₁

and one branch at least continues outwards towards the ocellus although it has not been traced as far as the ocellar synaptic plexus. This neuron, LMR1, probably has an input from all three ocelli and it continues posteriorly when the majority of median ocellar axons have moved laterally, finally branching dichotomously when each branch passes laterally towards the region of the latero-medial complex but at a more ventral level in the brain. LMR1 produces a well defined area of terminal arborizations parallel to but deeper than those of LM1 or RM1, Figs. 24.5 and 24.6c. The latero-medial complex on each side of the brain thus contains an input from two of the wide-field giant axons of the ipsilateral ocellus (L1/R1 and L2/R2), from a giant axon innervating the ipsilateral half of the median ocellus (LM1/RM1) and from an axon which innervates the whole of the median ocellus and possibly the ipsi- and contra-lateral ocelli (LMR1) and the arborizations of these units, whilst gathered locally in one complex, are each at a slightly deeper level within the brain. The neurons associated with this complex send collaterals to other areas of the brain, to the anterior and posterior optic tracts running to the optic lobes and, notably, to the sensory and motor antennal centres of the deutocerebrum. In Fig. 24.6b one of the lateral giant axons of this complex can be seen sending a branch posteriorly to meet a collateral of what is probably an antennal sensory interneuron descending into the suboesophageal commissure. They also appear to link with the DCMD neuron.

24.2.5.2. *The anterior dorso-lateral ocellar complex.* This complex is formed from the short stout branches of other median and lateral giant axons at the level of the posterior optic tract just below, and lateral to, the proto-cerebral bridge. At least one giant axon in each lateral ocellar nerve L3/R3 makes a major contribution producing an array of branches at this point. One major branch passes into the ipsilateral protocerebral lobe, others enter the posterior optic tract and many appear to terminate at this level. Four more of the giant neurons which can be traced all the way into the median ocellar cup appear to innervate both the median and one lateral ocellus and to be involved with this complex. Two axons on the left side of the median ocellar nerve turn outwards to the left lateral tract, LM2 and LM3, and two on the right side turn outwards to the right lateral ocellar tract, RM2 and RM3. At this point they branch; one arm appears to produce terminal arborizations in the complex, the other turns outward in the lateral ocellar tract towards the ocellus and has been traced as far as the ocellar cup. The cell bodies of these four neurons lie just anterior to the lateral ocellar tracts, two on each side of the brain, Figs. 24.5 and 24.6a. The median ocellar nerve contains six giant axons and two large medium-sized ones ($5 \mu m$). The pathways of the six giant axons and of one of the $5 \mu m$ ones (LMR1) have been described. The remaining smaller axon, LMR2, appears to contribute to this complex. Its pathway parallels that of LMR1, sending branches into both lateral ocellar tracts which then divide, one branch running outwards in the lateral ocellar tract traced as far as the proximal end of the nerve, the other apparently terminating in the complex. This neuron remains above the protocerebral bridge with LMR1 later descending and branching dichotomously with LMR1. Its connections have not been traced beyond that point. Each anterior dorso-lateral complex thus receives an input from its own ipsilateral ocellus (L3/R3), from two neurons which certainly contain an input from the median ocellus and possibly the ipsilateral ocellus (LM2 and 3 and RM2 and 3) and from LMR2 which has an input from the median ocellus and possibly the ipsi- and contra-lateral ocelli. Alternatively, if LM2 and 3/RM2 and 3 and LMR2 do not receive an input from the lateral ocellus, the complex will only receive an input from the ipsilateral and median ocelli and the median input to the ocellar tract presumably modifies the input of that tract in some way. The anterior dorso-lateral complex appears from its position to be associated with the optic lobes and the anterior optic pathway which runs between the optic lobes. Branches run towards the ipsilateral medulla from this point and they are joined by a contribution from L1/R1 in the latero-median complex. Light microscopy suggest that branches from the anterior dorso-lateral complex may cross to the other optic lobe in the posterior optic tract but this has not so far been confirmed by dye techniques. A preliminary examination of the ocellar tracts in the bee suggests that there is basically the same arrangement of the giant axons at least.

24.2.5.3. *The medium-sized and small neurons.* These neurons have been examined principally in the median ocellar nerve. The number of axons enclosed within the sheath as this nerve enters the brain remains approximately constant until the level of the protocerebral bridge. Above the central body the sheath is disrupted on the ventral surface and some of the giant and medium-sized axons send branches into the anterior optic tract which runs between the optic lobes at this point. Some of the smaller axons also pass into this tract. The sheath around the nerve disappears at the level of the protocerebral bridge and the smallest axons present pass upwards into the bridge where they make synaptic contact with fibres of unknown origin. The smallest axons in the lateral tracts behave in a like manner. In view of the possibility that some of these fibres are extended retinula cell axons it is interesting to note that some of them contain ribbon-type synapses very similar to those found in Class I and II terminals. The axon connections of the protocerebral bridge are varied, including connections with the dorsal protocerebrum, optic tubercle, and possibly the antennal centres (Bullock and Horridge 1965). The bridge was once suggested as a centre for optic correlation, but this view later seemed untenable as few optic fibres were found there and ocellar fibres were reported lacking. Since the ocellus does appear to make a contribution perhaps this idea should be re-examined. The protocerebral bridge is also a possible point of correlation between ocellar and antennal inputs which may be of as much importance to the insect as that between the ocelli and the compound eyes (See Section 24.3).

Little is known of the fate of the medium-sized axons. Some of their cell bodies lie either side of the median nerve as it runs through the pars inter-cerebralis, Fig. 24.5. Most of the medium-sized axons and some smaller ones can still be seen when the median nerve divides into two tracts and passes laterally but beyond this point they are difficult to trace. There is some indication that second-order medium-sized neurons are entering N.C.C.I. on each side but this has yet to be confirmed by cobalt studies for the locust. It is not clear whether the small-field second-order afferent neurons are amongst those which terminate in the protocerebral bridge or whether they enter one of the two optic tracts.

24.2.5.4. *Contacts with the median and lateral neurosecretory cells.*   Electron-microscope studies have shown that soon after the ocellar nerves enter the brain some of the giant axons present give off fairly fine branches which penetrate the sheath and ramify amongst the smaller median and lateral neurosecretory cells, respectively. They appear to be making tight-junctions with these cells. The presence of these fine branches has been confirmed by cobalt studies. Four types of neurosecretory cell have been identified in this region (Highnam 1961) but it is not known if the ocellar connections are restricted to any particular type. They certainly do not appear to extend as far as the four very large neurosecretory cells situated just below the median ocellar nerve as it reaches the protocerebral bridge but rather to be confined to cells quite near the surface of the brain.

24.2.5.5. *Higher-order ocellar neurons.*   Older histologists were not agreed as to whether the giant second-order neurons continued through the brain into the circumoesophageal connectives. Viallanes (1886) could not follow ocellar neurons through into the connectives in the bee but Kenyon (1896) traced fibres back from the connectives into the ocellar tracts and thought that they were probably second-order neurons. Cajal (1918) was unable to resolve this point in his studies on the bee and the dragonfly since he could not stain the giant axons by the Golgi method. More recently Power (1942) and Satija (1957) have reported large second-order fibres entering both the ipsi- and contra-lateral connectives in the fly and locust, respectively. In cobalt studies (Goodman and Patterson 1974), second-order neurons have so far never been traced from the distal end of the ocellar nerve into the circumoesophageal connectives. When the circumoesophageal connectives are cut and dye introduced ipsi- and contra-lateral neurons can be mapped up into the ocellar tracts to the point at which the ocellar nerves enter the brain, Fig. 24.5, but not beyond. Units filled from this direction appear to branch several times within the proximal region of the tract and their cell bodies do not appear to coincide precisely with those of the second-order neurons. These observations suggest that there are large third-order ocellar interneurons which run up through the ocellar complexes and into the three ocellar tracts and, if so, this could account for the confusion of earlier workers and would explain the presence of fine fibres and small terminals synapsing so extensively with the giant axons in the proximal part of the ocellar tracts. The latencies of the large ocellar 'off' responses found in the ventral connectives also suggest that they are third-order units (Goodman 1970). There are at least eight of these large third-order neurons in the locust; two ipsilateral and two contralateral entering each connective, Fig. 24.5. Initial observations suggest that one ipsilateral and one contralateral third-order neuron run up through the latero-median ocellar complex, with dendritic fields at appropriate points sending branches into the median and lateral ocellus of that side. The other pair of ipsi- and contra-lateral units run in a slightly more lateral position, one of them at least having a dendritic field in the region of the anterior dorso-lateral ocellar complex but this is probably an over-simplification. The contralateral units in particular show collaterals running to the optic lobes and to the antennal centres. Information from all three ocelli must reach each connective and this is borne out by electrophysiological evidence (See Section 24.3).

24.3. The physiology of the ocellus

24.3.1. *The retinula cells and events at the first-order synapse*

How do these observations upon the neural organization of the ocellus relate to its known physiology? The earliest extracellular recordings were made by Parry (1947) who failed to find spike activity in the ocellar nerve of *Locusta* and supposed that light falling upon the ocellus produced a depolarization which was conducted electrotonically along the ocellar nerve. Hoyle (1955) reported an arrhythmic dark discharge in the ocellar nerve of *Locusta* during darkness or dim light which was silenced by illumination. Ruck (1961*a*,*b*,*c*) found a similar effect in the dragonfly ocellar nerve and in a series of electroretinogram (ERG) studies on different species showed that activity in the photoreceptors exerted an inhibiting effect upon the second-order neurons, a finding which has since been repeated in other arthropod photoreceptors examined in this way (Shaw 1968).

Figure 24.7 shows the averaged evoked ERG of *S. gregaria* at different stimulus intensities (Goodman and Patterson in preparation). The complete waveform of the steady state intact preparation, column 1, is very similar to that obtained by Ruck (1961*a*) from the dragonfly: it consists of an initial positive peak with the onset of illumination, except at very low intensities, followed by a sustained positive plateau phase which at high intensities may

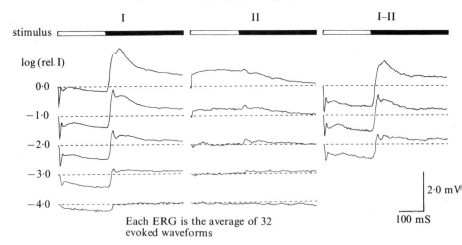

Each ERG is the average of 32
evoked waveforms

FIG. 24.7. The averaged electroretinogram (ERG) of the median ocellus of *S. gregaria* at
different light intensities. Each ERG is the average of thirty-two consecutive responses in the
steady-state animal. Light flashes of 250 ms duration were presented at 1/2·5 s, upward deflection negative, d.c. recordings. Column I shows the response of the intact animal. Column II
the response on the same preparation one hour after the ocellar nerve had been cut. The waveform in Column II has been subtracted from that in Column I in Column III, (I − II).

sometimes become slightly negative. With the onset of darkness there is a
marked negative peak followed by a slow decay which may occasionally
overshoot the resting level. Ruck supposed this waveform to be formed from
the algebraic addition of four separate components. Column 2 shows the
averaged response of the presynaptic units only in the same preparation. This
comprises the slow depolarizing potential of the retinula cells, Ruck's
Component I, and consists of a rapid corneal negative phase which declines
to form a sustained negative plateau. At light 'off' there is a further small
rapid depolarization before a slow decline to the resting level. More than
one such negative 'off' transient may be observed and at low intensities the
'off' transients become the most conspicuous feature of the response. The
rapid initial negative phase is preceded by a small positive transient which
may correspond to Ruck's Component 2 which he believed to be a secondary
depolarization produced in the retinula cell axons as a result of the slow
depolarization of the retinula cells. Figure 24.7, column 3, shows the contribution of the postsynaptic units to the normal intact ERG and this consists
of a positive transient at the beginning of illumination followed by a sustained
positive plateau corresponding to Ruck's Component 3. At the end of
illumination there is a pronounced rapid depolarization with a large overshoot of the resting potential followed by a slow decay to the resting state.
This hyperpolarizing potential in the second-order neurons inhibits their
spontaneous afferent impulses (Ruck's Component 4). Various questions

were left open in Ruck's analysis including the manner in which the photo-receptor potential was transmitted to the first-order synapse and the nature of the activity in the majority of axons in the ocellar nerve. Intracellular studies have extended our knowledge of electrical events at the first-order synapse, and of the response characteristics of the receptor cells and post-synaptic units, and similarities in electrical activity between the ocelli and the compound eyes have emerged (Goodman and Patterson in preparation). Intracellular recordings from the presynaptic units in the dragonfly (Chappell and Dowling 1972) and the locust (Chappell, Goodman, and Patterson unpublished observations) confirm that the photoreceptors respond to light with a sustained depolarization, Fig. 24.8a,b. There are certain differences between the two insects. In the dragonfly a single 'on' spike is found at higher intensities resembling that found in the bee drone eye (Baumann 1968) but

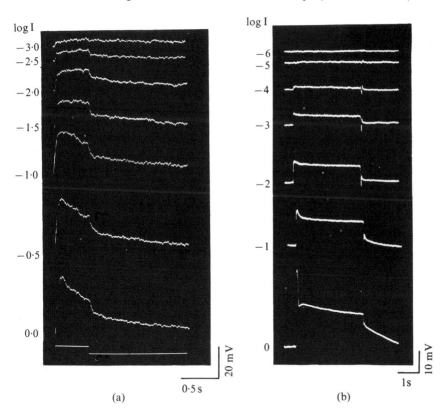

(a)   (b)

FIG. 24.8. Receptor cell intracellular responses in the dragonfly (from Chappell and Dowling 1972) and the locust showing a sustained depolarization at light 'on'. (b) In the dragonfly response note the 'on' spike at high intensities and the 'off' oscillation. In dim light this is often the most conspicuous feature of the response. (a) The locust shows no 'on' spike, less con-spicuous 'off' effects, and a slower return to the baseline.

this is absent in the locust. The locust ocellar retinula cell response bears a marked resemblance to that of the locust compound eye retinula cell (Shaw 1968) having a less well developed 'off' effect and a slower recovery rate than the dragonfly. No repetitive transients which might represent spike activity in the retinula cell axons have ever been observed in the locust ocellus. Ruck never observed such activity in the dragonfly but accounted for its absence by suggesting that the desynchronized firing of retinula cell axons would be masked in extracellular recordings. With one exception, attributable to cell membrane damage, Chappell and Dowling (1972) have found no evidence of repetitive firing in dragonfly retinula cell axons. Further, they go on to show that application of tetrodotoxin, although removing the initial spike of the retinula cell response, does not block the light-evoked slow potential responses of postsynaptic units nor change their characteristics. It appears that in the ocellus the slow receptor potential is responsible for transmission to the synapse and the initiation of the synaptic activity which produces the postsynaptic hyperpolarization as in the locust compound eye (Shaw 1968).

Some features of the postsynaptic response are of particular interest. In intracellular recordings from post-synaptic units, Chappell and Dowling (1972) rarely recorded 'spikes' but when impulse spontaneous activity was found in the dark it was inhibited by illumination. More usually they recorded small discrete hyperpolarizing potentials at low intensities, presumably IPSPs, which, at slightly higher intensities, summed to give a sustained but fluctuating hyperpolarization, Fig. 24.9. At higher intensities an initial large transient hyperpolarization is superimposed upon the sustained hyperpolarization at light 'on'. At the end of the illumination the potential in the cells rebounds above the original dark membrane potential. The amplitudes of the transient and sustained phases of the postsynaptic response reach a maximum at lower intensities and start to decline at high intensities whereas in the receptor cells these parameters show no signs of saturation even at high intensities. There is also less sustained potential relative to the initial transient wave in the postsynaptic response compared with that of the presynaptic units. This is particularly marked at higher intensities where the initial hyperpolarization and the 'off' depolarization are very large compared to the sustained component. The intracellular postsynaptic response of the locust shows similar features and again bears resemblances to intracellular recordings from lamina cells in the compound eye (Shaw 1968); the sustained phase of the 'on' response is virtually ungraded with stimulus intensity and the 'on' transient only weakly graded, the peak amplitude and latency reaching their maximum value at lower intensities. Here also the amplitude may start to decline at higher intensities. In the locust the relationship of the initial rate of hyperpolarization of the 'on' effect to stimulus intensity is by far the most markedly graded of all the parameters studied (see Fig. 24.7) the response of the postsynaptic units

log I

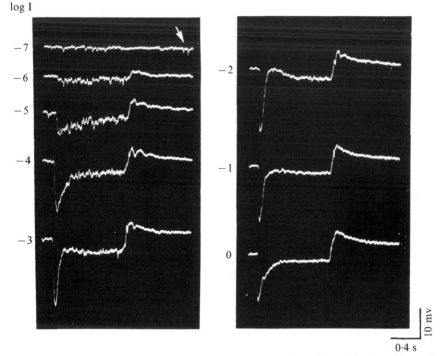

0·4 s

Fig. 24.9. Intracellular postsynaptic activity from the dragonfly median ocellus from Chappell and Dowling (1972). Intracellular slow potentials alone are seen; impulse activity is absent. Small discrete hyperpolarizing potentials seen in response to dim lights are believed to be inhibitory postsynaptic potentials (IPSPs). These responses are also seen in the dark (arrow). At higher intensities these discrete events apparently sum to produce a sustained hyperpolarization. At high intensities the 'on' and 'off' transients are the most marked feature of the response.

being more rapid than that of the presynaptic units. Post-synaptic activity also appears more sensitive to light since the slow depolarizing potential of the retinula cells only contributes significantly to the ERG of the intact animal at higher intensities; below log (relative intensity) = −1·0 the ERG is produced predominantly by component 3. Chappell and Dowling have shown that postsynaptic activity in the dragonfly appears more sensitive to light by 1 or 2 log units, Fig. 24.10, as occurs in compound eyes (see Laughlin, Chapter 15).

The most striking feature of the postsynaptic response of the ocellus is the transient 'on' and 'off' response, Fig. 24.9. Chappell and Dowling have shown that in the dragonfly when incremental stimuli are imposed upon a steady background illumination the postsynaptic activity becomes entirely transient in nature. The background illumination normally silences most spontaneous spike activity and a flash of the same intensity as the background illumination is just sufficient to elicit an 'off' response in the ocellar nerve. Cooter (1973) reports similar effects in the cockroach. Chappell and

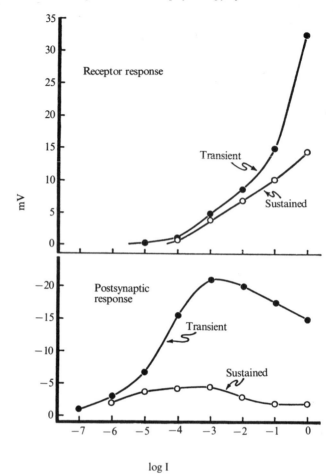

FIG. 24.10. Intensity response relationships for photoreceptor and postsynaptic units, from Chappell and Dowling (1972). The amplitude of the peak of the transient wave and the sustained component of the photoreceptor response and of the postsynaptic response have been plotted as a function of intensity. The sustained component of the photoreceptor response was measured 3 s after the onset of light.

Dowling find that $\Delta I/I$ for the 'off' threshold is nearly constant over five decades of stimulus intensity. In diurnal insects subject to constantly fluctuating background intensities the normal response of the giant axons is thus likely to be a phasic one signalling dimming with a short burst of impulses and, in spite of their great sensitivity, having the same $\Delta I/I$ over a wide range of intensities. This wide range is perhaps in part due to the operation of feedback at the first-order synapse. The decline in the peak amplitude of the postsynaptic 'on' effect in the locust and dragonfly at higher intensities is a sign of feedback operating at this point cutting off the post-synaptic response. Dowling and Chappell (1972) suggest that the extensive lateral and feedback connections demonstrated between both photoreceptor terminals and postsynaptic dendrites function to enhance the transients in the postsynaptic response and so produce a phasic response in giant axons, and they refer to a similar function for the lateral inhibitory mechanisms of *Limulus* which enhance phasic activity in the eccentric cell (Ratliff *et al.* 1963). In other visual systems extensive lateral and reciprocal innervation between receptor cells or higher-order units occurs (Sjostrand 1961; Dowling and Boycott 1966). The high degree of convergence upon the giant axons, from which the postsynaptic recordings are probably taken, makes it unlikely that feedback mechanisms aiding image discrimination are of importance. On the other hand the development of a feedback system which would allow the highly-photosensitive giant second-order units to signal small changes of intensity over a very wide operational range could be of great importance in the development of a visual alarm or alerting system. Ocellar units firing briefly in response to small decreases of intensity over a similar operational range have been found running in the ventral connectives to the thoracic ganglia (Goodman 1970). The photoreceptors synapsing upon the small-field second-order neurons appear to lack these feedback connections (Goodman, Kirkham, and Hoare 1974). The underlying mechanism for enhancement of phasic activity in the ocellar postsynaptic units is unknown, Dowling and Chappell, however, draw attention to the fast oscillatory 'off' responses which are a prominent feature of the dragonfly photoreceptor response, particularly at low intensities, which they suggest may indicate local feedback activity in the retinula axons. Such activity can just be detected in the averaged, extracellular recording of the photoreceptor response of *S. gregaria*, Fig. 24.7, but it is much more conspicuous in the averaged ERG of the cockroach (Cooter 1973, see Fig. 24.11) where the hyperpolarizing component recorded in the cup is very small, the resulting waveform largely comprising the response of the presynaptic units. Cooter has shown that the 'off' oscillations are not linked with the 'off' spikes generated in the post-synaptic units since, amongst other parameters, the latency–intensity relationship is different in the two cases, the latency of the 'off' oscillations increasing with stimulus intensity. Chappell and Dowling (1972) have shown that oscillatory activity in the dragonfly ocellus is not affected by any form of spike activity, and thus is presumably not the result of an efferent discharge, a point of some interest in view of the speculation as to the role of efferent units in the nerve (See Section 24.2). These efferent units, however, do not enter the cup, but synapse with second-order neurons within the length of the ocellar nerve. Little is known of the response characteristics of the majority of ocellar second-order units.

Spike activity in the ocellar nerve has been recorded extracellularly in

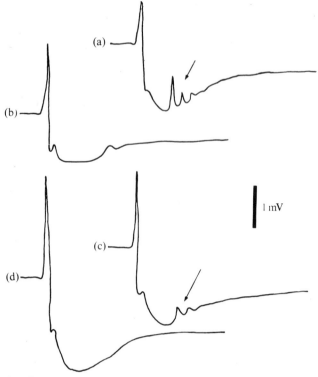

FIG. 24.11. The effect of stimulus intensity and level of background illumination on the off-components of the ocellar electroretinogram. Each trace shows the averaged response to sixty-four consecutive stimuli delivered at a rate of one per sec. Positivity of corneal electrode upwards, d.c. coupled. All traces start at the stimulus and last for 160 ms. Stimulus intensity in (a) and (b) was low, and (c) and (d) high. The top trace in each pair was recorded with a light background; the lower traces in the dark. The off-oscillations are arrowed. From Cooter (1973).

Locusta (Hoyle 1955), the dragonfly (Ruck 1958; Chappell and Dowling 1972; Rosser 1974) *Calliphora* (Metschl 1963) *Schistocerca* (Goodman 1968) *Boettcherisca* (Mimura *et al.* 1969) but all these authors record only two or three or at the most five units. From their size, they are usually assumed to be the responses of the giant axons. No spike activity that could convincingly represent the many medium-sized and smaller ocellar neurons has ever been reported. Possibly this is due to the difficulty of recording from such small units but Shaw (1968) was unable to find convincing spike discharges in the locust lamina units and concluded that the lamina response reached the next relay station in the medulla by passive transmission of a slow potential over a distance of *ca.* 400 μm. The summed hyperpolarizing potential of all the locust ocellar postsynaptic units as recorded in the ERG (Fig. 24.7) has many features in common with the locust compound eye lamina response (see earlier) and it may be that in many of the smaller neurons at least transmission is by the spread of slow graded potentials. Since the contribution each axon makes to the waveform is a function of axonal surface area the smaller axons present must contribute significantly to the total waveform of component 3. The distances involved in the case of the ocellar second-order afferents are much greater than those of the lamina units, being of the order of 1–2 mm. Shaw (1972) has recently examined the implications of transmission by slow graded potentials over long distances in small axons in the visual cell axons of the barnacle lateral ocellus which are only a few microns wide but up to 11 mm long. He concluded that decremental conduction was effective even over such long distances in narrow axons. Transmission by slow graded potentials would seem feasible for the smaller ocellar neurons over a considerably shorter distance and this may be why spikes are recorded in so few units. Zettler and Järvilheto (1971a,b) have recently proposed decrement-free spread of a graded hyperpolarizing potential in the Type 1 monopolar neuron of the lamina of *Calliphora*. In this neuron the relationship of the initial rate of hyperpolarization of the 'on' effect to stimulus intensity is by far the most markedly graded of all the stimulus parameters studied and the same is true for the locust ocellar hyperpolarizing postsynaptic response, Fig. 24.12 (Goodman and Patterson 1973). Zettler and Järvilheto have shown that these postsynaptic units of the lamina in *Calliphora* have a greater capacity for the temporal resolution of stimuli than the presynaptic units and suggest that this explains the observations of Autrum and Gallwitz (1951) that the integrity of the optic lobes is necessary to achieve a high rate of flicker-fusion frequency in the fly. Ruck (1958b) showed that the flicker-fusion frequency in ocelli matched that of their respective compound eyes, an observation which led him to suggest that it must be characteristics of the retinula cells which determined whether or not an eye could resolve high rates of flicker since the neural correlate of the lamina was apparently lacking in the ocellus, but some of the second-order neurons of ocelli seem to have similar temporal characteristics to these lamina neurons of the compound eye and this may account for the high flicker-fusion frequency seen in some ocelli.

Transmission in any visual cell axons present in the ocellar nerve is presumably by means of a slow graded depolarization in line with the visual cell axons ending within the cup. No signs of such activity have been found but it would be difficult to detect against the activity of the postsynaptic units and it has never been looked for specifically. Examination of the response characteristics of pre- and post-synaptic units in the ocellus has revealed that they resemble in many ways those of compound eye photoreceptor cells and postsynaptic units with some specializations, for example, in the giant axons, for the ocellar function of signalling 'dimming' rapidly to the thorax and other centres. Many aspects of ocellar anatomy remain to be

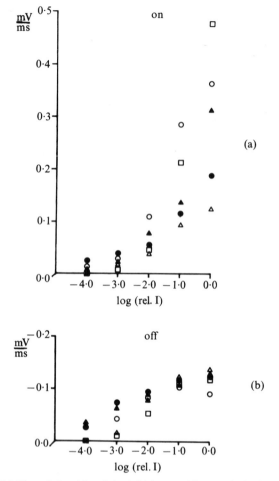

FIG. 24.12. (a) The relationship of the initial rate of hyperpolarization of the postsynaptic response to the stimulus intensity in the median ocellus of *S. gregaria*. The responses of five different insects are shown. (b) The relationship of the rate of depolarization of the postsynaptic response to the stimulus intensity in the same five insects.

explained in terms of physiological activity, for example, no physiological correlate of the anatomical specializations of the photoreceptor cells has yet been sought. Efferent activity is reported only once, in the dragonfly, where Rosser (1972) describes spike activity in four units in each of the median and lateral ocellar nerves.

24.3.2. Ocellar units in the brain and ventral nerve cord

24.3.2.1. *The brain.* As yet few ocellar units have been recorded from within the brain and their sites have not been precisely located. 'On-off' units can be recorded in the vicinity of the ocellar tracts as they pass back towards the protocerebral bridge and ocellar units have also been reported from the protocerebral lobes (Horridge *et al.* 1965). An ocellar input to the optic lobes has been demonstrated (see Section 24.2), all three ocelli contributing branches of the giant, medium and small axons to the anterior optic tract which runs between the lobes. Some of the protocerebral ocellar units show signs of lateral ocellar interaction, the 'off' response to a contralateral light being abolished by an ipsilateral stimulus. The danger of light spread within the head capsule, however, makes it difficult to be certain of the stimulus parameters. Although there is histological evidence of an ocellar input to the optic lobes no ocellar units have ever been encountered in extracellular recordings from the optic commissures of the lobes. Now that the ocellar pathways are becoming better known it may be feasible to examine the possibility that transmission by ocellar units in this region is by means of graded potentials.

Ocellar fibres have been shown to send collaterals to the antennal centres in the deutocerebrum (Goodman and Patterson in preparation) and a further possible point of correlation of these two types of input occurs in the protocerebral bridge (see Section 24.2). Some electrophysiological evidence exists of ocellar interaction with antennal input at various sites within the brain. In *Boettcherisca peregrina*, Mimura *et al.* (1969) have shown that ocellar illumination can influence the level of spontaneous discharge in antennal units responding to either mechanical or chemical stimulation. Antennal units whose response to stimulation was enhanced by ocellar illumination were encountered in the median region of the brain, possibly these authors were recording in the vicinity of the protocerebral bridge where antennal and ocellar inputs are brought together (see Section 24.2), or in the central body, although the position of the electrode tips was not localized. In the dorsal protocerebral lobes ocellar illumination reduced responses in antennal units and this was also true for antennal units in the ventral region of the deuto- and trito-cerebrum. Compound eye units in the ventral cord of *S. gregaria* show the influence of ocellar illumination although the effects vary depending, amongst other factors, upon the particular unit examined, which individual ocellus is stimulated and the relative intensity of illumination of the compound eye and ocellus. This interaction is believed to take place at brain level although the site has not been identified (Goodman 1970).

24.3.2.2. *The ventral nerve cord.* The largest and most readily recordable units in the ventral connectives of many insects, including the cockroach, dragonfly, locust, bee, and fly, are pure ocellar units, firing with one or two spikes when the ocelli are dimmed (Goodman 1970). Short trains of spikes can be produced by stepwise dimming. Two large ipsilateral and two contralateral units, believed to be ocellar third-order sensory interneurons enter each ventral connective (see 24.2.5.5) and it is probable that these units are the source of the large 'off' spikes. One each of these ipsilateral and contra-lateral units seem to pass through the lateromedial ocellar complex and presumably thus have the possibility of receiving an ipsi, median and possibly contralateral ocellar input at this point as well as ipsi and median inputs on their respective sides at a higher level in the tract (see Fig. 24.5). The other pair pass through the anterior dorso-lateral complex where they receive an input from the ipsi and median ocellus on their respective sides. In the locust 'off' spike activity is symmetrically distributed in each connective down to the level of the metathoracic ganglion. In whole connective recordings a pair of spikes are seen in each connective after brief stimulation of either of the lateral ocelli. They have the same latency in each connective (see Fig. 24.13). Stimulation of the median ocellus produces a pair of compound-action potentials presumably reflecting their contribution to both ipsi- and contra-lateral third-order units in each connective. No sign of interaction between any of the three ocelli or between the ocelli and the compound eyes has been found in whole connective recordings of 'off' spikes (Goodman 1970) but single unit studies have shown some signs of interaction between ocellar and compound eye input. These higher-order ocellar units have a large receptive field, a low threshold, and a wide operational range. They do not habituate with repetitive stimulation, a feature which is perhaps related to the develop-ment of an extremely phasic response in the giant axons, and have a relatively shorter latency than any of the larger compound eye units which respond to 'dimming' (Goodman 1970; Patterson and Goodman in preparation).

Other ocellar units have been reported from the ventral connectives. Tonic units that fire in the dark, showing long (up to 30 s) post-stimulus inhibition and disposed symmetrically within the connectives with respect to each ocellus, have been found in both the cockroach (Cooter 1974) and the locust. Small 'on/off' and sustaining 'on' units, which can reflect interaction between the three ocelli and between the ocelli and the compound eyes, are found, although much less frequently, in *S. gregaria*. These are presumably third- or higher-order units. No bimodal units, fired by either compound eye or ocellar stimulation, or antennal and ocellar stimulation have ever been encountered (Goodman 1970).

The large ocellar off-units obviously function to signal the onset of 'dimming' rapidly to the thoracic ganglia but it is not clear what effect this has on subsequent motor activity. No escape reactions are mediated solely by the ocelli and they are certainly not command fibres directly initiating motor activity. The state of ocellar illumination can influence thoracic motorneuron activity, briefly speeding up or slowing down some of the small, tonically firing, slow motor units controlling posture and upon occasion affecting their post-stimulus firing rate over periods of up to 30 min. Short bursts of firing can follow changes in ocellar illumination in some of the

Five sweeps superimposed   Single sweep

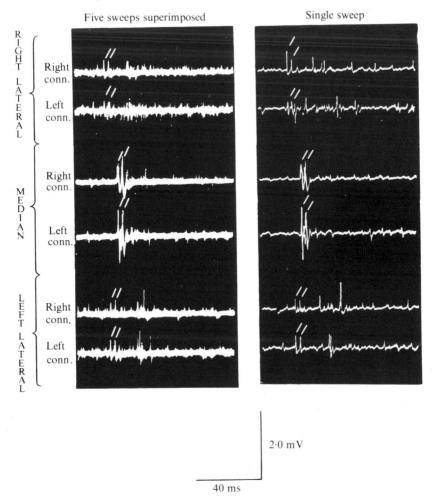

R
I
G
H
T
L
A
T
E
R
A
L

Right
conn.

Left
conn.

M
E
D
I
A
N

Right
conn.

Left
conn.

L
E
F
T
L
A
T
E
R
A
L

Right
conn.

Left
conn.

2·0 mV

40 ms

FIG. 24.13. Ocellar 'off' responses in the left and right ventral connectives of the locust, *S. gregaria* in response to stimulation of the right lateral, median, and left lateral ocelli. Records taken from whole connectives between the pro- and meso-thoracic ganglia. A pair of spikes are produced in each connective upon stimulation of the right and left lateral ocelli. These spikes have the same latency in each connective. Stimulation of the median ocellus produces a compound action potential in each connective with a longer latency. Ocellar off-spikes indicated by white arrows.

larger fast motorneurons and this activity can reflect ocellar and compound eye interaction. None of these responses can be directly correlated with the activity of the large 'off' units in the cord and they are possibly due to activity in the smaller higher-order ocellar units. No system has been found with an output specifically driven by ocellar stimulation, indeed this would not really be expected, and no simple or consistent relationship with ocellar stimulation has ever been found in any of the many motorneurons examined in the thorax. Ocellar stimulation can produce slowing down or speeding up of the firing rate of the same motorneurons at different times. This is of course also true for the input from the compound eye and other sense organs and it would be surprising if there were any simple relationship between ocellar activity and motorneuron output in view of the many factors which can influence activity at each stage in the pathway (Frazer Rowell 1971*a,b*). Examination of a large number of thoracic leg motorneurons suggests that direct effects of ocellar stimulation are more readily seen in the slow tonically firing units than in the larger, fast units where compound eye stimulation is more effective (Patterson and Goodman in preparation). Loss of tonus has frequently been reported as a consequence of ocellar blinding and older authors suggested that stimulation of the ocelli was necessary for the maintenance of muscular tone (Wolsky 1933). However, a similar apparent loss of tonus can often be produced by covering the compound eyes and the effects seen may simply be due to a sudden drop in the level of sensory input rather than a specific reaction to loss of ocellar input. The entire stance and pattern of leg movements of some insects, for example the bee, is altered when one set of photoreceptors is covered but these effects are rarely maintained over very long periods (Goodman 1970).

The second-order ocellar neurons have an input to the optic lobes and antennal centres and possibly to the DCMD visual interneuron, and the ocellar interneurons which enter the circumoesophageal connectives and are presumed to be the site of the large ocellar 'off' spikes in the cord send collaterals to the same sites. Ocellar 'dimming', whilst not producing any overt behavioural response, often results in small antennal movements, changes of posture and increased muscle tonus similar to the signs of arousal described by Frazer Rowell (1971*a*). On occasion ocellar 'dimming' may also result in suspension of on-going activity, even if only briefly. This is particularly noticeable in the cockroach (Cooter 1974). Suspension of on-going activity is presumably a necessary prelude to any change to new behavioural activity such as the initiation of an escape reaction. It seems reasonable to suppose that the major function of the giant ocellar pathways is to promote arousal in response to sudden shadowing presumably modulating to some degree activity of certain sensory interneurons and possibly thoracic motorneurons.

References

AGEE, H. R. *and* ELDER, H. W. (1970). Histology of the compound eye of the Boll Weevil. *Ann. ent. Soc. Amer.* **63**, 1654.

ALAWI, A. A. (1971). On transient of insect electroretinogram: its cellular origin. *Science* **172**, 1055.

ALBRECHT, F. O. (1953). *The anatomy of the migratory locust.* Univ. of London, Athlone Press, London.

ALLEN, J. L. *and* BERNARD, G. D. (1967). Superposition optics—a new theory, *M.I.T. Res. Lab. Electron. quart. Progr. Rep.*, **86**, 113.

AMBRONN, H. *and* FREY, A. (1926). *Das polarisationsmikroskop*, Akad. Verlags. Ges., Leipzig.

ANDERSON, A. (1972). The ability of honeybees to generalize visual stimuli. In *Information processing in the visual systems of arthropods* (ed. R. Wehner), Berlin, Heidelberg, New York: Springer, p. 207.

AOKI, Y. (1966). Light rays in lens-like media, *J. opt. Soc. Amer.*, **56**, 1658.

—— *and* SUZUKI, M. (1967). Imaging properties of a gas lens, *IEEE Trans. Microwave Techniques*, **MTT-15**, 2.

ARNETT, D. W. (1971). Receptive field organization of units in the first optic ganglion of Diptera. *Science* **173**, 929.

—— (1972). Spatial and temporal integration properties of units in the first optic ganglion of Dipterans. *J. Neurophysiol.* **35**, 429.

ARNOLD, M. T., BLOMQUIST, G. L. *and* JACKSON, L. L. (1969). Cuticular lipids of insects—The surface lipids of the aquatic and terrestrial life forms of the big stonefly *Pteronarcys californica Newport. Comp. Biochem. Physiol.* **31**, 685.

ATTNEAVE, F. *and* OLSON, R. K. (1967). Discriminability of stimuli varying in physical and retinal orientation. *J. exp. Psychol.* **74**, 149.

AUTRUM, H. (1955). Die spektrale Empfindlichkeit der Augenmutante white-apricot von *Calliphora erythrocephala. Biol. Zbl.* **74**, 515.

—— (1958). Electrophysiological analysis of the visual system in insects. *Exp. Cell. Res. (Suppl.)* **5**, 426.

—— (1961). Physiologie des Sehens. *Fortschr. Zool.* **13**, 257.

—— (1968). Colour vision in man and animals. *Naturw.* **55**, 10.

—— *and* BURKHARDT, D. (1960). Die spektrale Empfindlichkeit einzelner Sehzellen. *Naturw.* **47**, 527.

—— *and* —— (1961). Spectral sensitivity of single visual cells. *Nature Lond* **190**, 639.

—— *and* GALLWITZ, U. (1951). Zur Analyse der Belichtungspotentiale der Insectenaugen. *Z. vergl. Physiol.* **33**, 407.

—— *and* KOLB, G. (1968). Spektrale Empfindlichkeit einzelner Sehzellen der Aeschniden. *Z. vergl. Physiol.* **60**, 450.

AUTRUM, H. *and* KOLB, G. (1972). The dark adaptation in single visual cells of the compound eye of Aeschna cyanea. *J. comp. Physiol.* **79**, 213.

—— *and* WIEDEMANN, I. (1962). Versuche über den Strahlengang im Insektenauge. *Z. Naturforsch.* **17b**, 480.

——, ZETTLER, F. *and* JÄRVILEHTO, M. (1970). Postsynaptic potentials from a single monopolar neuron of the ganglion opticum I of the blowfly *Calliphora*. *Z. vergl. Physiol.* **70**, 414.

—— *and* —— (1962). Die Sehzellen der Insekten als analyzatoren für polarisierter Licht. *Z. vergl. Physiol.* **46**, 1.

—— *and* ZWEHL, V. VON (1964). Die spektrale Empfindlichkeit einzelner Sehzellen des Beinenauges. *Z. vergl. Physiol.* **48**, 357.

BADGER, G. M. (1949). The relative reactivity of aromatic double bonds. *J. chem. Soc.* **50**, 456.

—— (1951). The aromatic bond. *Quart. Rev.* **5**, 147.

BARLOW, H. B. (1952). The size of ommatidia in apposition eyes. *J. exp. Biol.* **29**, 667.

—— (1965). Visual resolution and the diffraction limit. *Science* **149**, 553.

—— *and* LEVICK, W. R. (1965). The mechanism of directionally selective units in rabbit's retina. *J. Physiol.*, **178**, 477.

BARRA, J. A. (1971*a*). Les photorécepteurs des Collemboles. Etude histochimique du cristallin. *Rev. ecol. biol. Sol (Paris)*, **8**, 49.

—— (1971*b*). Les photorécepteurs des Collemboles, étude ultrastructurale. I. L'appareil dioptrique.—Z. Zellforsch. **117**, 322.

BARROS-PITA, J. C. *and* MALDONADO, H. (1970). A fovea in the praying mantis eye. II. Some morphological characteristics. *Z. vergl. Physiol.* **67**, 79.

BAUMANN, F. (1966). Stimulation lumineuse de différents secteurs d'une cellule rétinienne de l'Abeille. *J. Physiol. (Paris)* **58**, 458.

—— (1968). Slow and spike potentials recorded from retinula cells of the honeybee drone in response to light. *J. gen. Physiol.* **52**, 855.

—— (1972). Influence of light adaptation and intracellular injection of sodium on the receptor potential of drone retinula cells. *J. Physiol. (Lond.)* **226**, 114.

—— *and* HADJILAZARO, B. (1971). Afterpotentials in retinula cells of the drone. *Vision Res.* **11**, 1198.

—— *and* —— (1972). A depolarizing aftereffect of intense light in the drone visual receptor. *Vision Res.* **12**, 17.

—— *and* MAURO, A. (1973). Effect of anoxia on the change in membrane conductance evoked by illumination in arthropod photoreceptors. *Nature (Lond.)* in the press.

BAUMGAERTNER, H. (1928). Der Formensinn und die Sehschaerfe der Bienen. *Z. vergl. Physiol.* **7**, 56.

BAYLOR, D. A. *and* O'BRYAN, P. M. (1971). Electrical signaling in vertebrate photo-receptors. *Fed. Proc. (Fed. Amer. Soc. exp. Biol.)* **30**, 79.

BEDAU, K. (1911). Das Facettenauge der Wasserwanzen, *Z. wiss. Zool.* **97**, 417.

BEDINI, C. *and* TONGIORGI, P. (1971). The fine structure of the pseudoculus of Acentero-mid Protura (Insecta, Apterygota). *Monitore zool. ital. (n.s.)*, **5**, 25.

BEHRENS, M. E. *and* WULFF, V. J. (1965). Light-initiated responses of retinula and eccentric cells in the *Limulus* lateral eye. *J. gen. Physiol.* **48**, 1081.

—— *and* —— (1967). Functional autonomy in the lateral eye of the horseshoe crab, *Limulus polyphemus*. *Vision Res.* **7**, 191.

BEIER, W. *and* MENZEL, R. (1972). Untersuchungen über der Farbensinn der deutschen Wespe. *Zool. Jb. Physiol.* **76**, 441.

—— *and* —— (1973). Das Farbensehvermögen der deutschen Wespe, *Paravespula germanica*. *Zool. Anz.* In the Press.

BENNETT, R. (1967). Spectral sensitivity studies on the whirligig beetle *Dineutes ciliatus*. *J. Insect Physiol.* **13**, 621.

—— *and* RUCK, P. (1970). Spectral sensitivities of dark- and light-adapted Notonecta compound eyes. *J. Insect Physiol.* **16**, 83.

——, TUNSTALL, J. *and* HORRIDGE, G. A. (1967). Spectral sensitivity of single retinula cells of the locust. *Z. vergl. Physiol.* **55**, 195.

BENOLKEN, R. M. (1961). Reversal of photoreceptor polarity recorded during the graded receptor potential response to light in the eye of *Limulus*. *Biophys. J.* **1**, 551.

—— *and* RUSSELL, C. J. (1966). Dissection of a graded visual response with tetrodotoxin. In *The functional organization of the compound eye* (ed. C. G. Bernhard), pp. 231–250, Pergamon Press, Oxford.

BERLESE, A. (1909. *Monografia dei Myrientomata*. *Redia* **6**, 1.

BERNARD, G. D. (1971). Evidence for visual function of corneal interference filters. *Insect Physiol.* **17**, 2287.

—— *and* MILLER, W. H. (1968). Interference filters in the corneas of Diptera. *Invest. Ophthalmology*, **7**, 416.

BERNHARD, C. G. (1967). Light transmission and its regulation in the compound eye. *Med. biol. J.* **17**, 100.

——, BOETHIUS, J., GEMME, G. *and* STRUWE, G. (1970). Eye ultrastructure, colour reception and behaviour. *Nature (Lond.)* **226**, 865.

——, GEMME, G. *and* MØLLER, A. R. (1968). Modification of specular reflexion and light transmission by biological surface structures. *Quart. Rev. Biophys.* **1**, 89.

——, —— *and* SALLSTRÖM, J. (1970). Comparative ultrastructure of corneal surface topography in insects with aspects on phylogenesis and function. *Z. vergl. Physiol.* **67**, 1.

—— *and* MILLER, W. H. (1962). A corneal nipple pattern in insect compound eyes. *Acta physiol. Scand.* **56**, 385.

—— *and* —— (1968). Interference filters in the corneas of Diptera. *Invest. Ophthalmology* **7**, 416.

——, —— *and* MØLLER, A. R. (1963). Function of the corneal nipples in the compound eyes of insects. *Acta. physiol. Scand.* **58**, 381.

——, ——, *and* —— (1965). The insect corneal nipple array—a biological broad-band impedance transformer that acts as an antireflection coating. *Acta physiol. Scand.* **63**, Suppl. 243, 1.

—— *and* OTTOSON, D. (1960a). Comparative studies on dark adaptation in the compound eyes of insects, *J. gen. Physiol.* **44**, 195.

—— *and* —— (1960b). Studies on the relation between the pigment migration and the sensitivity changes during dark adaptation in diurnal and nocturnal Lepidoptera, *J. gen. Physiol.* **44**, 205.

—— *and* —— (1961). Further studies on pigment migration and the sensitivity changes during dark adaptation in the compound eye of nocturnal insects, *Acta physiol. Scand.* **52**, 99.

BERNHARD, C. G. *and* OTTOSON, D. (1962). Pigment position and light sensitivity in the compound eye of noctuid moths, *Acta physiol. Scand.* **54**, 95.

BERNHARDS, H. (1916). Der Bau des Komplexauges von *Astacus fluviatilis* (*Potamobius astacus* L.) *Z. wiss. Zool.* **116**, 649.

BERNSTEIN, S. *and* FINN, C. (1971). Ant compound eye: size-related ommatidium differences within a single wood ant nest. *Experientia* **27**, 708.

BERTHOLF, L. M. (1931). The distribution of stimulative efficiency in the ultraviolet spectrum for the honeybee. *J. agric. Res.* **43**, 703.

——(1932). The extent of the spectrum for *Drosophila* and the distribution of stimulative efficiency in it. *Z. vergl. Physiol.* **98**, 32.

BERTRAND, D. PERRELET, A., *and* BAUMANN, F. (1972). Propriétés physiologiques des cellules pigmentaires de l'oeil du fauxbourdon. *J. Physiol.* (*Paris*) **65**, 102A.

BIERBRODT, E. (1942). Der Larvenkopf von *Panorpa communis* L. und seine Verwandlung, mit besonderer Berücksichtigung des Gehirns und der Augen. *Zool. Jb.* (*Anat*) **68**, 49.

BIRUKOW, G. (1950). Vergleichende Untersuchungen ueber das Helligkeits- und Farbensehen bei Amphibien. *Z. vergl. Physiol.* **32**, 348.

BISHOP, L. G. (1969). A search for color encoding in the responses of a class of fly interneurones. *Z. vergl. Physiol.* **64**, 355.

——(1970). The spectral sensitivity of motion detector units recorded in the optic lobe of the honeybee. *Z. vergl. Physiol.* **70**, 374.

—— *and* KEEHN, D. G. (1967). Neural correlates of the optomotor response in the fly. *Kybernetik*. **3**, 288.

——, ——, *and* McCANN, G. D. (1968). Motion detection by interneurons of optic lobes and brain of the flies *Calliphora phaenicia* and *Musca domestica. J. Neurophysiol.* **31**, 509.

BLEST, A. D. *and* COLLETT, T. S. (1965a). Microelectrode studies of the medial protocerebrum of some Lepidoptera—I. Responses to simple, binocular visual stimulation. *J. Insect. Physiol.* **11**, 1079.

—— *and* —— (1965b). Microelectrode studies of the medial protocerebrum of some Lepidoptera. II. Responses to visual flicker. *J. Insect. Physiol.* **11**, 1289.

BORN, M. *and* WOLF, E. (1965). *Principles of optics*, p. 161, 435. Pergamon Press, New York.

BORSELLINO, A., FUORTES, M. G. F., *and* SMITH, T. G. (1965). Visual responses in *Limulus. Cold Spring Harbor Symp. quant. Biol.* **30**, 429.

BOSCHEK, C. B. (1970). On the structure and synaptic organization of the first optic ganglion in the fly. *Z. Naturforsch.* **25B**, 560.

——(1971). On the fine structure of the peripheral retina and lamina ganglionaris of the fly, *Musca domestica. Z. Zellforsch.* **118**, 369.

BOULET, P. C. (1968). Le fonctionnement de l'oeil composé et le comportement prédateur des Mantides, expérimentation sur des modèles, *Psychol. fr.* **13**, 345.

BOULIGAND, Y. (1965). Sur une architecture torsadée repondue dans de nombreuses cuticles d'arthropodes. *C.r. Acad. Sci.* (*Paris*) **261**, 3665.

BRAITENBERG, V. B. (1967). Patterns of projection in the visual system of the fly. I. Retina–lamina projections. *Exp. Brain Res.* **3**, 271.

——(1970). Ordnung und Orientierung der Elemente im Sehsystem der Fliege. *Kybernetik* **7**, 235.

BRAITENBERG, V. B. (1971*a*). What sort of computer we expect to find associated with the compound eye of flying insects?—*Biokybernetik*, Bd. III, pp. 215–220. VEB Fischer Verlag. Jena.

—— (1971*b*). The structures of the visual ganglia in relation to studies on movement perception in the fly. *Atti Congresso Nazionale Cibernetica, Casciano Terme*, pp. 42–53.

—— (1972). Periodic structures and structural gradients in the visual ganglia of the fly. In *Information Processing in the Visual Systems of Arthropods* (ed. R. Wehner), pp. 3–15 Berlin, Heidelberg, New York: Springer.

—— and TADDEI FERRETTI, C. (1966). Landing reaction of *Musca domestica*. *Naturw.* **53**, 155.

BRAMMER, J. D. (1970). The ultrastructure of the compound eye of a mosquito *Aedes aegypti* L. *J. exp. Zool.* **175**, 181.

BRANDENBURG, J. (1960). Die Feinstruktur des Seitenauges von *Lepisma saccharina*. *Zool. Beitr. (n.F.)* **5**, 291.

BRIGGS, M. H. (1961). Retinene-I in insect tissues. *Nature (Lond.)* **192**, 874.

BRITTON, E. B. (1970). Coleoptera. In *The Insects of Australia*. (Ed. CSIRO. Staff). University Press, Melbourne.

BROUSSE-GAURY, P. (1968). Les organes para-ocelliares des Blattes: point de départ de réflexes neuro-endocriniens. *C.r. Acad. Sci. (Paris)* **267**, D. 649–650.

—— (1971). Modification de la neurosécrétion au niveau de la pars intercerebralis de *Periplaneta americana* L. en l'absence de stimuli ocellaires. *Bull. biol. Fr. Belg.* **105**, 84.

BROWN, H. M., MEECH, R. M., KOIKE, H., *and* HAGIWARA, S. (1969). Current–voltage relations during illumination: photoreceptor membrane of a barnacle. *Science* **166**, 240.

——, HAGIWARA, S., KOIKE, H., *and* MEECH, R. M. (1970). Membrane properties of a barnacle photoreceptor examined by the voltage clamp technique. *J. Physiol.* **208**, 385.

BROWN, J. E. *and* LISMAN, J. E. (1972). An electrogenic sodium pump in *Limulus* ventral photoreceptor cells. *J. gen. Physiol.* **59**, 720.

BROWN, P. L. *and* WHITE, R. H. (1972). Rhodopsin of the larval mosquito. *J. gen. Physiol.* **59**, 401.

BRUCKMOSER, P. (1968). Die spektrale Empfindlichkeit einzelner Sehzellen des Rückenschwimmers *Notonecta glauca* L. (Heteroptera). *Z. vergl. Physiol.* **59**, 187.

DE BRUIN, G. H. P. *and* CRISP, D. J. (1957). The influence of pigment migration on vision of higher Crustacea. *J. exp. Biol.* **34**, 447.

BUDDENBROCK, W. VON (1929). Einige Bemerkungen zum augenblicklichen Stand der Frage nach dem Farbensinn der Tiere: *Zool. Anz.* **84**, 189.

—— (1952). *Vergleichende Physiologie, Vol. I: Sinnesphysiologie*. Verlag Birkhaeuser, Basel.

—— and FRIEDRICH, H. (1933). Neue Beobachtungen ueber die kompensatorischen Augenbewegungen und den Farbensinn der Taschenkrabbe (*Carcinus maenas*). *Z. vergl. Physiol.* **19**, 747.

—— and MOLLER-RACKE, I. (1952). Neue Beobachtungen ueber den Farbensinn der Insekten. *Experientia* **8**, 62.

—— and —— (1953). Ueber das Wesen der optomotorischen Reaktionen. *Experientia* **9**, 191.

BULLOCK, T. H. *and* HORRIDGE, G. A. (1965). *Structure and function in the nervous systems of invertebrates*. Freeman: San Francisco.

BURCHER, J. (1967). Les combinaisons optiques, *Rev. optique, Paris*, 713.

BURKHARDT, D. (1962). Spectral sensitivity and other response characteristics of single visual cells in the arthropod eye. *Symp. Soc. exp. Biol.* **16**, 86.

—— (1964). Colour discrimination in insects. *Adv. Insect Physiol.* **2**, 131.

—— and GEWECKE, M. (1966). Mechanoreception in Arthropoda: the chain from stimulus to behavioral pattern. *Cold Spring Harbor Symp. quant. Biol.* **30**, 601.

——, DE LA MOTTE, I., *and* SEITZ, G. (1966). Physiological optics of the compound eye of the blowfly. *Wenner-Gren Center Int. Symp.* Ser. 7, 51.

—— and STRECK, P. (1965). Das Sehfeld einzelner Sehzellen—eine Richtigstellung. *Z. vergl. Physiol.* **51**, 151.

—— and —— (1972). Electrophysiological studies on the eyes of Diptera, Mecoptera and Hymenoptera. In *Information processing in the visual systems of arthropods.* (ed. R. Wehner) 147, Springer Verlag, Heidelberg, New York.

BURTT, E. T. *and* CATTON, W. T. (1954). Visual perception of movement in the locust. *J. Physiol.* **125**, 566.

—— and —— (1956). Electrical responses to visual stimulation in the optic lobes of the locust and certain other insects. *J. Physiol.* **133**, 68.

—— and —— (1959). Transmission of visual responses in the nervous system of the locust. *J. Physiol.* **146**, 492.

—— and —— (1960). The properties of single-unit discharges in the optic lobe of the locust. *J. Physiol.* **154**, 479.

—— and —— (1962). A diffraction theory of insect vision. I. An experimental study of visual acuity in certain insects. *Proc. R. Soc.* B **157**, 53.

—— and —— (1966). Perception by locusts of rotated patterns. *Science* **151**, 224.

—— and —— (1966). Image formation and sensory transmission in the compound eye. In *The functional organization of the compound eye.* (ed. C. G. Bernhard). Pergamon Press, Oxford.

—— and —— (1966). The role of diffraction in compound eye vision. In *The functional organization of the compound eye.* (ed. C. G. Bernhard), Pergamon Press, New York, pp. 63–76.

—— and —— (1969). Resolution of the locust eye measured by rotation of radial striped patterns. *Proc. R. Soc.* B **173**, 513.

BUSH, B. M. H., WIERSMA, C. A. G., *and* WATERMAN, T. H. (1964). Efferent mechano-receptive responses in the optic nerve of the crab, *Podophthalmus. J. cell. comp. Physiol.* **64**, 327.

BUTLER, L., ROPPEL, R., *and* ZEIGLER, J. (1970). Post-emergence maturation of the eye of the adult Black Carpet Beetle *Attagenus megatoma* (Fab.)—an EM study. *J. Morph.* **130**, 103.

BUTLER, R. (1971). The identification and mapping of spectral cell types in the retina of *Periplaneta americana. Z. vergyl. Physiol.* **72**, 67.

—— (1972). The anatomy of the compound eye of *Periplaneta americana*, I, II. *J. comp. Physiol.* **83**, 223.

—— and HORRIDGE, G. A. (1973). The electrophysiology of the retina of *Periplaneta americana*. I. Changes in receptor acuity upon light/dark adaptation. *J. comp. Physiol.* **83**, 263.

CAJAL, S. R. (1909). Nota sobre la estructura de la retina de la mosca. *Trab. Lab. de Invest. Biol. Madrid* 7, 217.

CAJAL, S. R. (1918). Observaciones sobre la estructura de los ocelos y vias nerviosas ocelares de algunos insectos. *Trab. Lab. Invest. Biol. Univ. Madrid* **16**, 109.
—— and SÁNCHEZ, D. (1915). Contribucion al conocimiento de los centros nerviosos de los insectos. *Trab. Lab. Invest. Biol. Univ. Madrid* **13**, 1.
CAMPAN, R., GALLO, A., and QUEINNES, Y. (1965). Determination electroretinographique de la frequence critique de fusionnement visuel: étude comparative portant sur les yeux composés de dix-sept espèces d'insectes. *C.r. Soc. biol.* **159**, 2521.
CAMPBELL, F. W. (1968). The human eye as an optical filter. *Proc. IEEE* **56**, 1009.
—— (1969). Trends in physiological optics. In *Processing of optical data by organisms and by machines*, pp. 137–143. (ed. W. Reichardt), Academic Press, New York.
CAMPOS-ORTEGA, J. A. and STRAUSFELD, N. J. (1972). Columns and layers in the second synapic region of the fly's visual system: the case for two superimposed neuronal architectures. In *Information processing in the visual systems of arthropods* (ed. R. Wehner), 31–36. Springer: Berlin, Heidelberg, New York.
CARLSON, S. D. (1972). Microspectrophotometry of visual pigments. *Quart. Rev. Biophys.* **5**, 349.
—— and PHILIPSON, B. (1972). Microspectrophotometry of the dioptric apparatus and compound rhabdom of the moth *Manduca sexta* eye. *J. Insect. Physiol.* **18**, 1721.
CARRICABURU, P. (1965). Essai de détermination des constantes optiques de l'ommatidie d'*Eupagurus bernhardus*. *Bull. Soc. Hist. nat. Afr. Nord* **56**, 51.
—— (1966a). Dioptrique de l'ommatidie de la Crevette *Aristeus antennatus* Risso. *Vision Res.* **6**, 597.
—— (1966b). Etude interférométrique des cônes cristallins du Crabe *Callinectes sapidus*. *C.r. Acad. sci. (Paris)*, D **263**, 1408.
—— (1967a). Structure optique de l'ommatidie de la Limule *Xyphosura polyphemus*. *C.r. Acad. sci. Paris*, D **264**, 1476.
—— (1967b). L'acuité visuelle de l'oeil composé. Expérimentation sur un modèle. *Vision Res.* **7**, 909.
—— (1968). Contribution à la dioptrique oculaire des Arthropodes: détermination des indices des milieux transparents de l'ommatidie. Thèse. Paris 1967. publié. *Mém. Soc. Hist. nat. Afr. Nord* **9**, 1.
—— (1969). Catadioptrique de l'oeil composé. *Vision Res.* **9**, 1523.
—— (1972). L'oeil du Crabe *Carcinus maenas* en tant qu'optique de fibres. *C.r. Acad. sci. Paris* D. **274**, 2348.
—— and CHARDENOT, P. (1967). Spectres d'absorption de la cornée de quelques arthropodes. *Vision Res.* **7**, 43.
CARTHY, J. D. (1951). The orientation of two allied species of British ants. *Behaviour* **3**, 153.
CASTLE, E. S. (1936). The double refraction of chitin. *J. gen. Physiol.* **19**, 797.
CATTON, W. T. and CHAKRABORTY, A. (1969). Single neuron response to visual mechanical stimuli in thoracic nerve cord of locust. *J. Insect Physiol.* **15**, 245.
CAVENEY, S. (1971). Cuticle reflectivity and optical activity in scarab beetles: the rôle of uric acid. *Proc. R. Soc. Lond.* B **178**, 205.
CHAUVIN, R. (1964). Experiences sur l'apprentissage par equipe du labyrinthe chez Formica polyctena. *Insect. Soc.* **11**, 1.
CHAPPELL, R. L. and DOWLING, J. E. (1972). Neural organisation of the median ocellus of the dragonfly. I. Intracellular electrical activity. *J. gen. Physiol.* **60**, 121.

CHMURZYNSKI, J. A. (1967). On the orientation of the housefly, *Musca domestica*, towards white light of various intensities. *Bull. Acad. pol. Sci., Sér. sci. biol.* **15**, 415.
—— (1969). Orientation of blowflies, Calliphoridae, towards white light of various intensities. *Bull. Acad. pol. Sci.*, Cl. II, **17**, 321.
CHRETIEN, H. (1959). *Calcul des combinaisons optiques*, 862. Librairie du Bac, Paris.
CLAUS, C. (1879). Der Organismus der Phronimiden. *Arb. Zool. Inst. Univ. Wien.* **2**, 59
CLOUDSLEY-THOMPSON, J. L. (1953). Studies in diurnal rhythms. III. Photo-periodism in the cockroach, *Periplaneta americana*. *Ann. Mag. nat. Hist.* **6**, 705.
COGGSHALL, J. C. (1971). Sufficient stimuli for the landing response in *Oncopeltus fasciatus*. *Naturw.* **2**, 100.
——(1972). The landing response and visual processing in the milkweed bug, *Oncopeltus fasciatus*. *J. exp. Biol.* **57**, 401.
COLLETT, T. (1970). Centripetal and centrifugal visual cells in medulla of the insect optic lobe. *J. Neurophysiol.* **33**, 239.
—— (1971a). Visual neurones for tracking moving targets. *Nature Lond.* **232**, 127.
—— (1971b). Connections between wide-field monocular and binocular movement detectors in the brain of a hawk-moth. *Z. vergl. Physiol.* **75**, 1.
—— (1972). Visual neurones in the anterior optic tract of the privet hawk-moth. *J. comp. Physiol.* **78**, 396.
—— and BLEST, A. D. (1966). Binocular directionally-selective neurones, possibly involved in the optomotor response of insects. *Nature Lond.* **212**, 1330.
COLLEWIJN, H. (1969). Optokinetic eye movements in the rabbit. Input–output relations. *Vis. Res.* **9**, 117.
COLLINS, D. L. (1945). Iris-pigment migration and its relation to behavior in the codling Moth. *J. exp. Zool.* **69**, 164.
COLLINS, F. D. (1953). Rhodopsin and indicator yellow. *Nature Lond.* **171**, 469.
COOK, J. W. and SCHOENTAL R. (1948). Oxidation of anthracene by osmium tetroxide. *Nature Lond.* **161**, 237.
COOTER, R. J. (1974). Visual and multimodal interneurons in the ventral connectives of the cockroach *Periplaneta americana*. *J. exp. Biol.* (in the press).
—— (1974). Ocellus and ocellar nerves of *Periplaneta americana*. In preparation.
CORNSWEET, T. N. (1970). *Visual perception*. Academic Press, New York.
COWLEY, J. M. and MOODIE, A. F. (1957a). Fourier images, I. The point source. *Proc. phys. Soc. Lond.* B **70**, 486.
—— and —— (1957b). Fourier images, II. The out-of-focus patterns. *Proc. phys. Soc. Lond.* B **70**, 497.
—— and —— (1957c). Fourier images, III. Finite sources. *Proc. phys. Soc. Lond.* B **70**, 505.
COX, A. (1965). *A system of optical design*. Focal Press, London.
CREUTZFELDT, O. D., POEPPL, E., and SINGER, W. (1971). Quantitativer Ansatz zur Analyse der funktionellen Organisation des visuellen Cortex (Untersuchungen an Primaten). In *Pattern Recognition in Biological and Technical Systems*, (ed. O. J. Gruesser and R. Klinke.) pp. 81–96. Springer, Berlin, Heidelberg, New York.
——, SAKMANN, B., SCHAICH, H., and KORN, A. (1970). Sensitivity distribution and spatial summation within the receptive field center of retinal on-center ganglion cells and transfer function of the retina. *J. Neurophysiol.* **33**, 654.
CRIEGEE, R. (1936). Osmiumsäure-ester als Zwischenprodukte bei Oxydationen. *J. Liebigs Ann. Chem.* **522**, 75.

CRUSE, H. (1972*a*). Versuch einer quantitativen Beschreibung des Formensehens der Honigbiene. *Kybernetik* **11**, 185.

—— (1972*b*). A qualitative model for pattern discrimination in the honey bee. In *Information Processing in the Visual Systems of Arthropods*, pp. 201–206 (ed. R. Wehner). Springer, Berlin, Heidelberg, New York.

CUTRONA, L. J., LEITH, E. N., PALERMO, C. J., *and* PORCELLO, L. J. (1960). Optical data processing and filtering systems. *IRE Trans. on Information Theory* **6**, 386.

DARTNALL, H. J. A. (1953). The interpretation of spectral sensitivity curves. *Brit. med. Bull.* **9**, 24.

—— (1962). *The Eye* (ed. H. Davson), Vol. 2, Chapt. 17, Academic Press, London.

—— (1972). Photosensitivity. In *Photochemistry in vision*. Vol. VII/1 of Handbook of Sensory Physiology, (ed. H. Autrum, *et al.*) Springer Verlag, Heidelberg, New York.

DAUMER, K. (1956). Reizmetrische Untersuchung des Farbensehens der Biene. *Z. vergl. Physiol.* **38**, 413.

DAY, M. F. (1951). Pigment migration in the eye of the Moth *Ephestia kühniella* Zeller. *Biol. Bull.* **80**, 275.

DAYET, J. *and* VINCENT-GEISSE, J. (1968). Etude des formules de réflexion à la surface d'un milieu absorbant dans le cas des faibles absorptions. Application à la méthode de réflexion totale atténuée. *J. Phys.* **29**, 1005.

DEANE, C. (1932). New species of Corylophidae. *Proc. Linn. Soc. N.S. Wales* **57**, 336.

DEBAISIEUX, P. (1944). Les yeux des Crustacés. *La Cellule*, **50**, 5.

DINGLE, H. *and* FOX, S. S. (1966). Microelectrode analysis of light responses in the brain of the cricket (*Gryllus domesticus*). *J. cell. Physiol.* **68**, 45.

DLUZKY, G. M. (1967). *Ants of the genus* Formica, "*Nauka*", *Moscow* (in Russian).

DØVING, K. B. *and* MILLER, W. H. (1969). Function of insect compound eyes containing crystalline tracts. *J. gen. Physiol.* **54**, 250.

DOWLING, J. E. (1963). Neural and photochemical mechanisms of visual adaptation in the rat. *J. gen. Physiol.* **46**, 1287.

—— (1968). Discrete potentials in the dark-adapted eye of *Limulus*. *Nature Lond.* **217**, 28.

—— *and* BOYCOTT, B. B. (1966). Organisation of the primate retina: electron microscopy. *Proc. R. Soc. Lond.* B **166**, 80.

—— *and* CHAPPELL, R. C. (1972). Neural organisation of the median ocellus of the dragonfly. II. Synaptic structure. *J. gen. Physiol.* **60**, 148.

—— *and* RIPPS, H. (1972). Adaptation in skate photoreceptors. *J. gen. Physiol.* **60**, 698.

DURUZ, C. *and* BAUMANN, F. (1968). Influence de la température sur le potentiel de repos et le potential récepteur d'une cellule photoréceptrice. *Helv. physiol. Acta* **26**, CR341.

DE IRALDI, A. P. *and* DE ROBERTIS, E. (1968). The neurotubular system of the axon and the origin of granulated and non-granulated vesicles in regenerating nerves. *Z. Zellforsch.* **87**, 330.

DELACHAMBRE, J. (1971). Etudes sur l'épicuticule des insectes—Modifications de l'épiderme au cours de la sécrétion de l'épicuticule imaginale chez *Tenebrio molitor* L. *Z. Zellforsch.* **112**, 97.

DEMOLL, R. (1914). Die Augen von *Limulus. Zool. Jahrb. Abt.f. Anat. u. Ontog.* **38**, 443.

DENTON, E. J. (1959). The contributions of the oriented photo-sensitive and other molecules to the absorption of whole retina. *Proc. R. Soc.* B **150**, 78.

558   *References*

DETHIER, V. G. (1963). The physiology of insect senses, Methuen and Wiley, London.
DE VALOIS, R. L. (1965). Behavioural and electrophysiological studies of primate vision. In *Contributions to sensory physiology* (ed. W. D. Neff) **1**, 137.
DE VOE, R. D. (1972). Dual sensitivities of cells in wolf spider eyes at ultraviolet and visible wavelengths of light. *J. gen. Physiol.* **59**, 247.
DICKENS, J. C. *and* EATON, J. L. (1973). External ocelli in Lepidoptera previously considered to be anocellate. *Nature, Lond.* **242**, 205.
DIESENDORF, M. *and* HORRIDGE, G. A. (1973). Theory of the partially-focussed clear-zone compound eyes. *Proc. R. Soc. Lond.* B. **183**, 141.
DIJKGRAAF, S. (1953). Über das Wesen der optomotorischen Reaktionen. I. *Experientia* **9**, 112.
EAKIN, R. *and* BRANDENBURGER, J. L. (1970). Osmic staining of amphibian and gastropod photoreceptors. *J. Ultrastruct. Res.* **30**, 619.
ECKERT, M. (1968). Hell-Dunkel-Adaptation in aconen Appositionsaugen der Insekten. *Zool. Jb. Physiol.* **74**, 102.
—— (1970). Verhaltensphysiologische Untersuchungen am visuellen System der Stubenfliege *Musca domestica* L.—*Doctural dissertation*, University of Berlin.
—— (1971). Die spektrale Empfindlichkeit des Komplexauges von Musca (Bestimmung aus Messungen der optomotorischen Reaktion). *Kybernetik* **9**, 145.
EGUCHI, E. (1971). Fine structure and spectral sensitivities of retinula cells in the dorsal sector of compound eyes in the dragonfly *Aeschna*. *Z. vergl. Physiol.* **71**, 201.
—— *and* WATERMAN, T. H. (1966). Fine structure patterns in crustacea rhabdom. In *The functional organisation of the compound eye.* (ed. C. G. Bernhard) Pergamon Press, Oxford.
—— *and* —— (1967). Changes in retinal fine structure induced in the crab *Libinia* by light and dark adaptation. *Z. Zellforsch.* **79**, 209.
—— *and* —— (1968). Cellular basis for polarized light perception in the spider crab, *Libinia*. *Z. Zellforsch.* **84**, 87.
EHEIM, W. P. *and* WEHNER, R. (1972). Die Sehfelder der zentralen Ommatidien in den Appositionsaugen von *Apis mellifera* und *Cataglyphis bicolor* (Apidae, Formicidae; Hymenoptera). *Kybernetik* **10**, 168.
ELOFSSON, R. (1970). Brain and eyes of Zygentoma (Thysanura). *Ent. scand.* **1**, 1.
ELTRINGHAM, H. (1919). Butterfly vision. *Trans R. ent. Soc. Lond.* **79**, 1.
ENOCH, J. M. (1963). Optical properties of retinal photoreceptors. *J. opt. Soc. Am.* **53**, 57.
—— (1967). Comments on 'Excitation of Waveguide Modes in Retinal Receptors'. *J. opt. Soc. Am.* **57**, 548.
ETIENNE, A. S. (1969). Analyse der schlagausloesenden Bewegungsparameter einer punktfoermigen Beuteattrappe bei der Aeschna-Larve. *Z. vergl. Physiol.* **64**, 71.
EXNER, S. (1876). Ueber das Sehen von Bewegungen und die Theorie des zusammengesetzten Auges. *Sitz. -ber. Akad. Wiss. Wien, math. -nat. Kl., 3. Abt.* **72**, 156.
—— (1885). Ein Mikro-Refraktometer. *Archiv. mikr. Anat.* **25**, 97.
—— (1886). Cylinder welche optische Bilder entwerfen. *Pflüger's Arch. ges. Physiol.* **38**, 274.
—— (1889). Das Netzhautbild des Insectenauges. Sitzungsber. math. -wissen. -Cl. Kaiserl. *Akad. Wiss. Abt.* 3, **98**, 13.
—— (1891). *Die Physiologie der facettirten Augen von Krebsen und Insecten.* Leipsiz und Wien, Franz Deuticke.

EYSEL, U. T. *and* GRUESSER, O. J. (1971). Neurophysiological basis of pattern recognition in the cat's visual system. In *Pattern Recognition in Biological and Technical Systems*, (ed. O. J. Gruesser and R. Klinke) pp. 60–80. Springer, Berlin, Heidelberg, New York.

EYZAGUIRRE, C. *and* KUFFLER, S. W. (1955). Process of excitation in the dendrites and in the soma of single isolated sensory nerve cells of the lobster and crayfish. *J. gen. Physiol.* **39**, 87.

FAHRENBACH, W. H. (1969). The morphology of the eyes of *Limulus*. II. Ommatidia of the compound eye. *Z. Zellforsch.* **93**, 451.

FEDER, D. (1971). Fourier optics analysis of a radial pattern used to test locust eye resolution. *M.S.E.E. Thesis*, University of Washington, Seattle.

FERMI, G. *and* REICHARDT, W. (1963). Optomotorische Reaktionen der Fliege *Musca domestica*. Abhaengigkeit der Reaktion von der Wellenlaenge, der Geschwindigkeit, dem Kontrast und der mittleren Leuchtdichte bewegter periodischer Muster. *Kybernetik* **2**, 15.

FERNANDEZ PEREZ DE TALENS, A. *and* TADDEI FERRETTI, C. (1970). Landing reaction of *Musca domestica*: dependence on dimension and position of the stimulus. *J. exp. Biol.* **52**, 233.

FISCHER, A. *and* HORSTMANN, G. (1971). Der Feinbau des Auges der Mehlmotte *Ephestia kuehniella* (Lepidoptera, Pyralididae). *Z. Zellforsch.* **116**, 275.

FILSHIE, B. K. (1970). The resistance of epicuticular components of an insect to extraction with lipid solvents. *Tissue and Cell* **2**, 181.

FLETCHER, A., MURPHY, T. *and* YOUNG, A. (1954). Solution of two optical problems. *Proc. R. Soc. Lond.* A **223**, 216.

FOSTER, A. B. *and* WEBBER, J. M. (1960). Chitin. *Adv. Carbohydr. Chem.* **15**, 371.

FRAENKEL, G. *and* RUDALL, K. M. (1947). The structure of insect cuticles. *Proc. R. Soc. Lond.* B **134**, 111.

FRANCESCHINI, N. (1972). Sur le traitement optique de l'information visuelle dans l'oeil à facettes de la Drosophile. *Thèse Doc. Sci. phys.*, Grenoble.

—— *and* KIRSCHFELD, K. (1971*a*). Etude optique *in vivo* des éléments photorécepteurs dans l'oeil composé de *Drosophila*. *Kybernetik* **8**, 1.

—— *and* —— (1971*b*). Les phénomènes de pseudopupille dans l'oeil composé de *Drosophila*. *Kybernetik* **9**, 159.

FRANTSEVICH, L. I. (1970). Some observations on directionally-sensitive neurons in scarabaeid beetles. (In Russian). *J. Evolutionary Biochem. Physiol.* **6**, 446.

—— *and* MOKRUSHOV, P. A. (1970). The dynamic properties of directionally-selective visual neurons in scarabaeid beetles (*Coleoptera, Scarabaeidae*). *Ukrain. Acad. Sci.* (in Russian).

FRAZER ROWELL, C. H. (1971*a*). The orthopteran descending movement detector (DMD) neurones: A characterisation and a review. *Z. vergl. Physiol.* **73**, 167.

—— (1971*b*). Antennal cleaning, arousal and visual interneuron responsiveness in a locust. *J. exp. Biol.* **55**, 749.

FREY-WYSSLING, A. *and* BLANK, F. (1948). Ermittlung submikroskopischer Strukturen. *Tabulae Biol.* **19**, 30.

FRIEDERICHS, H. F. (1931). Beiträge zur Morphologie und Physiologie der Sehorgane der Cicindelliden (Col.). *Z. Morph. Ökol. d. Tiere* **21**, 1.

FRITSCH, K. VON (1914). Der Farbsinn und Formensinn der Biene. *Zool. Jb., Abt. allg. Zool. u. Physiol.* **35**, 1.

—— (1914). Demonstration von Versuchen zum Nachweis des Farbensinnes bei angeblich total farbenblinden Tieren. *Verh. Dtsch. Zool. Ges. Freiburg,* 50.

—— (1914). Der Farbensinn und Formensinn der Biene. *Zool. Jb. Physiol.* **37**, 1.

—— (1949). Die Polarisation des Himmelslichtes als orientierender Faktor bei den Tänzen der Bienen. *Experientia* **5**, 142.

—— (1965). *Tanzsprache und Orientierung der Bienen.* Springer-Verlag, Berlin, Heidelberg, New York.

—— (1967). *The dance language and orientation of bees.* Harvard University Press, Cambridge, Mass.

FULPIUS, B. *and* BAUMANN, F. (1969). Effects of sodium, potassium and calcium ions on slow and spike potentials in single photoreceptor cells. *J. gen. Physiol.* **53**, 541.

FUORTES, M. G. F. (1958). Electrical activity of cells in the eye of *Limulus. Am. J. Opthalmol.* **46**, 210.

—— (1959). Initiation of impulses in visual cells of *Limulus. J. Physiol.* **148**, 14.

—— *and* HODGKIN, A. L. (1964). Changes in time scale and sensitivity in the ommatidia of *Limulus. J. Physiol. (Lond.)* **172**, 239.

—— *and* MANTEGAZZINI, F. (1962). Interpretation of repetitive firing in nerve cells. *J. gen. Physiol.* **45**, 1163.

FUGE, H. (1967). Die Pigmentbildung im Auge von *Drosophila melanogaster* und ihre Beeinflussung durch den white-Locus. *Z. Zellforsch.* **83**, 468.

FUKUSHIMA, K. (1969). Visual feature extraction by a multi-layered network of analog threshold elements. *IEEE Trans. Systems Sci. Cybernetics,* **5**, 322.

FUSTER, J. M. *and* UYEDA, A. A. (1962). Facilitation of tachistoscopic performance by stimulation of midbrain tegmental points in the monkey. *Exp. Neurol.* **6**, 384.

GARNER, W. R. (1970). Good patterns have few alternatives. *Am. Scient.* **58**, 34.

GAVEL, L. VON (1939). Die kritische Steifenbreite als Mass der Sehschaerfe bei *Drosophila melanogaster, Z. vergl. Physiol.* **27**, 80.

GEMPERLEIN, R. *and* SMOLA, U. (1972a). Uebertragungseigenschaften der Sehzelle der Schmeissfliege *Calliphora erythrocephala.* 1. Abhängigkeit vom Ruhepotential. *J. comp. Physiol.* **78**, 30.

—— *and* —— (1972b). Transfer characteristics of the visual cell of *Calliphora erythrocephala.* 3. Improvement of the signal-to-noise ratio by presynaptic summation in the lamina ganglionaris. *J. comp Physiol.* **79**, 393.

GEORGE, M. (1963). Studies on *Campodea* (Diplura). the anatomy of the glands and sense-organs of the head. *Quart. J. micr. Sci.* **104**, 1.

GERSTEIN, G. L. *and* PERKEL, D. H. (1972). Mutual temporal relationships among neuronal spike trains. *Biophys. J.* **12**, 453.

GEWECKE, M. (1970). Antennae: Another wind-sensitive receptor in locust. *Nature, Lond.* **225**, 1263.

—— (1971). Der unterschiedliche Einfluss der Antennen und Stirnhaare von *Locusta migratoria* auf die Fluggeschwindigkeit. *Naturw.* **58**, 101.

—— *and* SCHLEGEL, P. (1970). Die Schwingungen der Antenne und ihre Bedeutung für die Flugsteuerung bei *Calliphora erythrocephala. Z. vergl. Physiol.* **67**, 325.

GOGALA, M. (1967). Die spektrale Empfindlichkeit der Doppelaugen von *Ascalaphus macaronius* Scop. (Neuroptera, Ascalaphidae). *Z. vergl. Physiol.* **57**, 232.

GOGALA, M., HAMDORF, K., *and* SCHWEMER, J. (1970). UV-Sehfarbstoff bei Insekten. *Z. vergl. Physiol.* **70**, 410.

GOLDSMITH, T. H. (1958*a*). The visual system of the honeybee. *Proc. Nat. Acad. Sci.* **44**, 123.

—— (1958*b*). On the visual system of the bee (*Apis mellifera*). *Ann. N.Y. Acad. Sci.* **74**, 223.

—— (1960). The nature of the retinal action potential and the spectral sensitivities of ultraviolet and green receptor systems of the compound eye of the worker honeybee. *J. gen. Physiol.* **43**, 775.

—— (1961). The physiological basis of wavelength discrimination in the eye of the honeybee. In *Sensory communication*. M.I.T. Press and John Wiley and Sons, Inc. (ed. W. A. Rosenblith).

—— (1963). The course of light and dark adaptation in the compound eye of the honeybee. *Comp. biochem. Physiol.* **10**, 227.

—— (1964). The visual system of insects. In *The physiology of Insecta* (ed. M. Rockstein), 394–462. Academic Press, New York, London.

—— (1965). Do flies have a red receptor? *J. gen. Physiol.* **49**, 265.

—— (1970). Retinaldehyde of honeybee. *Opthalm. Res.* **I**, 292.

—— (1972). The natural history of invertebrate visual pigments. In *Handbook of Sensory Physiology*, Vol. VII/1 Photochemistry of vision (ed. H. J. A. Dartnall), 685–715. Springer Verlag, Berlin, Heidelberg, New York.

——, DIZON, A. E., *and* FERNANDEZ, H. R. (1968). Microspectrophotometry of photoreceptor organelles from eyes of the prawn *Palaemonetes*. *Science* **161**, 468.

—— *and* FERNANDEZ, H. R. (1966). Some photochemical and physiological aspects of visual excitation in compound eyes. In *The functional organization of the compound eye* (ed. C. G. Bernhard), Pergamon Press, London.

—— *and* —— (1968). The sensitivity of housefly photoreceptors in the mid-ultraviolet and the limits of the visible spectrum. *J. exp. Biol.* **49**, 669.

—— *and* PHILPOTT (1962). Fine structure of the retinulae in the compound eye of the honey bee. *J. cell Biol.* **14**, 489.

—— *and* WARNER, L. (1962). The role of vitamin A in the visual cycle of insects. *J. gen. Physiol.* **46**, 360A.

—— *and* —— (1964). Vitamin A in the vision of insects. *J. gen. Physiol.* **47**, 433.

GOODMAN, J. W. (1968). *Introduction to Fourier optics*, p. 120. McGraw-Hill, New York.

GOODMAN, L. J. (1960). The landing responses of insects. I. The landing response of the fly, *Lucilia sericata*, and other Calliphorinae. *J. exp. Biol.* **37**, 854.

—— (1964). The landing responses of insects. II. The electrical response of the compound eye of the fly, *Lucilia sericata*, upon stimulation by moving objects and slow changes of light intensity. *J. exp. Biol.* **41**, 403.

—— (1970). The structure and function of the insect dorsal ocellus. In *Advances in insect physiology*, Vol. 7. (ed. J. E. Treherne, J. W. L. Beament, and V. B. Wigglesworth) Academic Press, London and New York.

—— *and* PATTERSON, J. A. (1974). Electrical activity of pre- and post-synaptic units in the median dorsal ocellus of *Schistocerca gregaria* (in preparation).

—— *and* KIRKHAM, J. B. (1973). The fine structure of the dorsal ocelli of *Rhodnius prolixus* (in preparation).

562　References

GOODMAN, L. J., PATTERSON, J. A., and HOARE, L. (1974). The fine structure of the ocelli of *Schistocerca gregaria*. II. The neural organisation of the synaptic plexus (in preparation).

GÖTZ, K. G. (1964). Optomotorische Untersuchung des visuellen Systems einiger Augenmutanten der Fruchtfliege *Drosophila*. *Kybernetik* **2**, 77.

—— (1965). Die optischen Uebertragungseigenschaften der Komplexaugen von *Drosophila*. *Kybernetik* **2**, 215.

—— (1968). Flight control in *Drosophila* by visual perception of motion. *Kybernetik* **4**, 199.

—— (1971). Spontaneous preferences of visual objects in *Drosophila*. *Drosophila Inf. Service* **46**, 62.

—— (1972). Processing of cues from the moving environment in the *Drosophila* navigation system. In *Information Processing in the Visual Systems of Arthropods* pp. 255–263. (ed. R. Wehner). Springer, Berlin, Heidelberg, New York.

—— (1972). Visual control of orientation patterns. In *Information processing in the visual system of Arthropods* (ed. R. Wehner), pp. 255–263. Springer Verlag, Berlin.

—— and GAMBKE, C. (1968). Zum Bewegungssehen des Mehlkäfers *Tenebrio molitor*. *Kybernetik* **4**, 225.

GOVARDOVSKII, V. I. (1971). Sodium and potassium concentration gradient in the retinal rod outer segment. *Nature New Biol.* **234**. 53.

GRAFFON, M. (1934). Untersuchungen über das Bewegungssehen bei Libellenlarven, Fliegen und Fischen. *Z. vergl. Physiol.* **20**, 299.

GRAY, E. G. and GUILLERY, R. W. (1966). Synaptic morphology in the normal and degenerating nervous system. *Int. Rev. Cytol.* **19**, 111.

GRENACHER, H. (1879). Untersuchungen über das Sehorgan der Arthropoden insbesondere der Spinnen, Insekten und Crustaceen. 188. Göttingen: Vandenhoek and Ruprecht.

GRIBAKIN, F. G. (1967a). Ultrastructural organization of photoreceptor cells of compound eye of the honeybee *Apis mellifera* (Russian). *Zh. evolutsionnoi biochemii i fiziologii* **3**, 66.

—— (1967b). The types of photoreceptor cells of the compound eye of the honeybee worker as revealed by electron microscopy. (Russian). *Tsitologia* **9**, 1276.

—— (1969a). The ultrastructural organization of photoreceptors of insects. (Russian). *Trudy Vsesoyuznogo Entomologicheskogo Obshchestva*, **53**, 238.

—— (1969b). The types of photoreceptor cells of the compound eye of the honeybee relative to their spectral sensitivities. (Russian). *Tsitologia* **11**, 308.

—— (1969c). Cellular basis of colour vision in the honey bee. *Nature, Lond.* **223**, 639.

—— (1972). The distribution of the long wave photoreceptors in the compound eye of the honeybee as revealed by selective osmic staining. *Vis. Res.* **12**, 1125.

GRUESSER, O. J. and GRUESSER-CORNEHLS, U. (1970). Die Neurophysiologie visuell gesteuerter Verhaltensweisen bei Anuren. *Verh. dtsch. Zool. Ges.* **64**, 201.

—— and KLINKE, R. (1971). *Pattern recognition in biological and technical systems.* Springer, Berlin, Heidelberg, New York.

——, VIERKANT, J., and WUTTKE, W. (1968). Die räumliche Verteilung und die Ausbreitungsgeschwindigkeit der lateralen Hemmung in den rezeptiven Feldern der Katzenretina. *Biokybernetik* **2**, 175.

GUIGNON, E. F. (1973). Information processing in the insect compound eye. *Ph.D. dissertation*. University of Connecticut, Storrs, Conn.

GÜNTHER, K. (1912). Die Sehorgane der Larve und Imago von *Dytiscus marginalis*. *Z. Wiss Zool.* **100**, 60.

HAGINS, W. A. (1965). Electrical signs of information flow in photoreceptors. *Cold Spring Harbor Symp. quant. Biol.* **30**, 403.

——, ZONANA, H. V., *and* ADAMS, R. G. (1962). Local membrane current in the outer segments of squid photoreceptors. *Nature, Lond.* **194**, 844.

HAMDORF, K. (1970). Korrelationen zwischen Sehfarbstoffgehalt und Empflindichkeit bei Photorezeptoren. *Verhandl. Deutsch. Zool. Ges.* **64**, 148.

——, GOGALA, M., *and* SCHWEMER, J. (1971). Beschleunigung der Dunkeladaption eines UV-Rezeptors durch sichtbare Strahlung. *Z. vergl. Physiol.* **75**, 189.

——, HÖGLUND, G., *and* LANGER, H. (1972). Mikrophotometrische Untersuchungen an der Retinula des Nachtschmetterlings *Deilephila elpenor. Verh. Dt. Zool. Ges.* **65**, 276.

——, PAULSEN, R., SCHWEMER, J., *and* TAEUBER, U. (1972). Photoreconversion of Invertebrate Visual Pigments. In *Information processing in the visual systems of arthropods.* (ed. R. Wehner), 97–208, Springer Verlag, Heidelberg, New York.

HANSEN, W. N. (1965). Extended formulas for attenuated total reflection and the derivation of absorption rules for single and multiple ATR spectrometer cells. *Spectrochim. Acta* **21**, 815.

HANSTRÖM, B. (1940). Inkretorische Organe, Sinnesorgane und Nervensystem des Kopfes einiger niederen Insektenordnungen. *K. svenska Vetensk. Akad. Handl.* (3) **18**, 1,

HARKER, J. E. (1956). Factors controlling the diurnal rhythm of activity in *Periplaneta americana. J. exp. Biol.* **33**, 224.

HARMON, L. D. (1968). Modelling studies of neural inhibition. In *Structure and function of inhibitory neural mechanisms.* (ed. C. von Euler *et al.*), pp. 537–563. Pergamon Press.

HARTH, E. *and* PERTILE, G. (1971). The role of inhibition and adaptation in sensory information processing. *Kybernetik* **10**, 32.

HARTLINE, H. K. (1928). A quantitative and descriptive study of the electric response to illumination of the arthropod eye. *Am. J. Physiol.* **83**, 466.

—— *and* GRAHAM, C. H. (1932). Nerve impulses from single receptors in the eye. *J. cell. comp. Physiol.* **1**, 277.

—— *and* RATLIFF, F. (1972). Inhibitory interaction in the retina of *Limulus.* In *Handbook of sensory physiology*, Vol. VII/2, *Physiology of Photoreceptor Organs* (ed. M. G. F. Fuortes), pp. 381–447. Springer, Berlin, Heidelberg, New York.

——, WAGNER, H. G., *and* MACNICHOL, E. F. (1952). The peripheral origin of nervous activity in the visual system. *Cold Spring Harbor Symp. quant. Biol.* **17**, 125.

HARRICK, N. J. *and* DU PRÉ, F. K. (1966). Effective thickness of bulk materials and of thin films of internal reflection spectroscopy. *Appl. Opt.* **5**, 1739.

HARRIS, A. J. (1965). Eye movements of the dogfish *Squalus acanthias. J. exp. Biol.* **43**, 107.

HASSENSTEIN, B. (1951). Ommatidienraster und afferente Bewegungsintegration. *Z. vergl. Physiol.* **33**, 301.

—— (1954). Über die Sehschärfe von Superpositionsaugen (Versuche an *Lysmata seticaudata* und *Leander serratus*). *Pubbl. Staz. zool. Napoli* **25**, 168.

HASSENSTEIN, B. (1958). Ueber die Wahrnehmung der Bewegung von Figuren und unregelmaessigen Helligkeitsmustern (Nach verhaltensphysiologischen Untersuchungen an dem Ruesselkaefer *Chlorophanus viridis*): *Z. vergl. Physiol.* **40**, 556.

—— (1959). Optokinetische Wirksamkeit bewegter periodischer Muster (Nach Messungen am Rüsselkäfer *Chlorophanus viridus*) *Z. Naturforsch.* **14b**, 659.

—— and REICHARDT, W. (1953). Der Schluss von Reiz-Reaktions-Funcktionen auf System-Strukturen. *Z. Naturforsch.* **8b**, 518.

—— and —— (1956). Systemtheoretische Analyse der Zeit-, Reihenfolgen- und Vorzeichenauswertung bei der Bewegungsperzeption des Rüsselkäfers *Chlorophanus*. *Z. Naturforsch.* **11b**, 513.

HAUPT, J. (1972). Ultrastruktur des Pseudoculus von *Eosentomon* (Protura). *Z. Zellforsch.* **135**, 539.

HAYS, D. *and* GOLDSMITH, T. H. (1969). Microspectrophotometry of the visual pigment of the spider crab, *Libinia emarginata*. *Z. vergl. Physiol.* **65**, 218.

HECHT, S. *and* WALD, G. (1934). The visual acuity and intensity discrimination of *Drosophila*, *J. gen. Physiol.* **17**, 517.

HEIDE, G. VON (1971). Die Funktion der nicht-fibrillären Flugmuskeln von *Calliphora*. Teil II. Muskuläre Mechanismen der Flugsteuerung und ihre nervöse Kontrolle. *Zool. Jb. Physiol.* **76**, 99.

HEINTZ, E. (1959). La question de la sensibilité des abeilles à l'ultraviolet: *Insectes Sociaux* **6**, 223.

HEISENBERG, M. (1972). Comparative behavioral studies on two visual mutants of *Drosophila*. *J. comp. Physiol.* **80**, 119.

HELLER, J. (1968a). Structure of visual pigments: I. Purification, molecular weight and composition of bovine visual pigments$_{500}$. *Biochemistry* **7**, 2906.

—— (1968b). Structure of visual pigments: II. Binding of retinal and conformational changes of light exposure in bovine visual pigment. *Biochemistry* **7**, 2914.

HELVERSEN, O. VON (1972). Zur spektralen Unterschiedsempfindlichkeit der Honigbiene. *J. comp. Physiol.* **80**, 439.

HENGSTENBERG, R. (1971). Das Augenmuskelsystem der Stubenfliege *Musca domestica*. I. Analyse der "clock spikes" und ihrer Quellen. *Kybernetik* **9**, 56.

—— (1972). Eye movements in the housefly *Musca domestica*. In *Information processing in the visual systems of arthropods* (ed. R. Wehner), pp. 93–96. Springer, Berlin, Heidelberg, New York.

—— and GÖTZ, K. G. (1967). Der Einfluss des Schirmpigmentgehalts auf die Helligkeits- und Kontrastwahrnehmung bei Drosophila-Augenmutanten. *Kybernetik* **3**, 276.

HENNIG, W. (1966). *Phylogenetic systematics*. Urbana-Chicago. Univ. Illinois Press.

HERTZ, M. (1934a). Zur Physiologie des Formen-und Bewegungssehen. I. Optomotorische Versuche an Fliegen. *Z. vergl. Physiol.* **20**, 430.

—— (1934b). Zur Physiologie des Formen-und Bewegungssehen. II. Auflösungsvermögen des Bienenauges und optomotorische Reaktion. *Z. vergl. Physiol.* **21**, 579.

—— (1934c). Zur Physiologie des Formen- und Bewegungssehens. III. *Z. vergl. Physiol.* **21**, 604.

—— (1935). Die Untersuchungen ueber das Formensehen der Honigbiene. *Naturw.* **23**, 618.

Hesse, R. (1901). Untersuchungen über die Organe der Lichtempfindung bei niederen Tieren. VII. Von den Arthropodenaugen. *Z. wiss. Zool.* **70**, 347.

Hesse, R. (1908). *Das Sehen der niederen Tiere.* Jena: G. Fischer.

Highnam, H. .C. (1961). The histology of the neurosecretory system of the adult female desert locust, *Schistocerca gregaria. Quart. J. microsc. Sci.* **102**, 27.

Himstedt, W. (1972). Untersuchungen zum Farbensehen von Urodelen: *J. comp. Physiol.* **81**, 229.

Höglund, G. (1966a). Pigment migration, light screening and receptor sensitivity in the compound eye of nocturnal Lepidoptera. *Acta. physiol. scand.* **69**, Suppl. 282.

—— (1966b). Pigment migration and retinula sensitivity. In *The functional organization of the compound eye* (ed. S. Bernhard), p. 77–88. Pergamon Press, Oxford and New York.

—— and Struwe, G. (1970). Pigment migration and spectral sensitivity in the compound eye of moths. *Z. vergl. Physiol.* **67**, 229.

—— and —— (1971). Pigment migration and illumination of single photoreceptors in a moth. *Z. vergl. Physiol.* **74**, 336.

——, ——, and Thorell, B. (1970). Spectral absorption by screening pigment granules in the compound eyes of a moth and a wasp. *Z. vergl. Physiol.* **67**, 238.

Hollick, F. S. J. (1940). The flight of the dipterous fly *Musca stabulans* Fallén. *Phil. Trans. R. Soc. Lond.* B **230**, 357.

Horn, G. and Hill, R. M. (1969). Modifications of receptive fields of cells in the visual cortex occurring spontaneously and associated with body tilt. *Nature, Lond.* **221**, 186.

—— and Rowell, C. H. F. (1968). Medium and long-term changes in the behaviour of visual neurones in the tritocerebrum of locusts. *J. exp. Biol.* **49**, 143.

Horridge, G. A. (1964). Multimodal interneurones of locust optic lobe. *Nature, Lond.* **204**, 499.

—— (1965). Arthropoda: receptors for light, and optic lobe. In *Structure and function in the nervous systems of invertebrates* (T. H. Bullock and G. A. Horridge), Freeman, San Francisco.

—— (1966a). Optokinetic responses of the crab *Carcinus*, and of the locust. *J. exp. Biol.* **44**, 233, 247, 255, 263, 275.

—— (1966b). The retina of the locust. In *The functional organization of the compound eye* (ed. S. Bernhard), p. 53. Pergamon Press, Oxford and New York.

—— (1966c). Study of a system, as illustrated by the optokinetic response. *Symp. Soc. exp. Biol.* **20**, 179.

—— (1968a). *Interneurons.* W. H. Freeman and Co., San Francisco.

—— (1968b). Pigment movement and the crystalline threads of the firefly eye. *Nature, Lond.* **218**, 778.

—— (1969a). The eye of *Dytiscus* (Coleoptera). *Tissue and Cell* **1**, 425.

—— (1969b). The eye of the firefly (*Photuris*). *Proc. R. Soc. Lond.* B **171**, 445.

—— (1969c). Unit studies on the retina of dragonflies. *Z. vergl. Physiol.* **62**, 1.

—— (1971). Alternatives to superposition images in clear-zone compound eyes. *Proc. R. Soc.* B **179**, 97.

—— (1972). Further observations on the clear zone eye of *Ephestia. Proc. R. Soc. Lond.* B **181**, 157.

—— and Barnard, P. B. T. (1965). Movement of palisade in locust retinula cells when illuminated. *Q. J. microsc. Sci.* **106**, 131.

HORRIDGE, G. A. *and* GIDDINGS, C. (1971a). Movement on dark and light-adaptation in beetle eyes of the neuropteran type. *Proc. R. Soc. Lond.* B **179**, 73.

—— *and* —— (1971b). The retina of *Ephestia* (Lepidoptera). *Proc. R. Soc. Lond.* B **179**, 87.

——, ——, *and* STANGE, G. (1973). The superposition eye of Skipper butterflies. *Proc. R. Soc. Lond.* B **182**, 457.

—— *and* MEINERTZHAGEN, I. A. (1970a). The exact neural projection of the visual fields upon the first and second ganglia of the insect eye. *Z. vergl. Physiol.* **66**, 369.

—— *and* —— (1970b). The accuracy of the patterns of connexions of the first- and second-order neurons of the visual system of *Calliphora*. *Proc. R. Soc. Lond.* B **175**, 69.

——, NINHAM, B. W., *and* DIESENDORF, M. O. (1972). Theory of the summation of scattered light in clear zone compound eyes. *Proc. R. Soc. Lond.* B **181**, 137.

—— *and* SANDEMAN, D. C. (1964). Nervous control of optokinetic responses in the crab *Carcinus. Proc. R. Soc.* B **161**, 216.

——, SCHOLES, J. H., SHAW, S., *and* TUNSTALL, J. (1965). Extracellular recordings from single neurones in the optic lobe and brain of the locust. In *The physiology of the insect central nervous system.* (ed. J. E. Treherne and J. W. C. Beament). Academic Press, London and New York.

——, WALCOTT, B. *and* IOANNIDES, A. C. (1970). The tiered retina of *Dytiscus;* a new type of compound eye. *Proc. R. Soc. Lond.* B **175**, 83.

HORSTMANN, E. (1935). Die tagesperiodische Pigmentwanderung in Facettenaugen von Nachtschmetterlingen. *Biol. Zentralbl.* **55**, 93.

HOYLE, G. (1955). Functioning of the insect ocellar nerve. *J. exp. Biol.* **32**, 397.

HUBEL, D. H. *and* WIESEL, T. N. (1959). Receptive fields of single neurons in the cat's striate cortex. *J. Physiol.* **148**, 574.

HYZER, W. G. (1962). Flight behaviour of a fly alighting on a ceiling. *Science* **137**, 609.

ILSE, D. (1932). Eine neue Methode zur Bestimmung der subjektiven Helligkeitswerte von Pigmenten. *Biol. Zbl.* **52**, 660.

IOANNIDES, A. C. *and* WALCOTT, B. (1971). Graded illumination potentials from retinula cell axons in the bug *Lethocerus. Z. vergl. Physiol.* **71**, 315.

ISHIKAWA, S. (1962). Visual response patterns of single ganglion cells in the optic lobe of the silkworm moth, *Bombyx mori* L. *J. Insect Physiol.* **8**, 485.

JACKSON, L. L. *and* BAKER, G. L. (1970). Cuticular lipids of insects. *Lipids* **5**, 239.

JACOBS, G. H. *and* YOLTON, R. L. (1971). Visual sensitivity and colour vision in ground squirrels. *Vision Res.* **11**, 511.

JAHN, T. L. *and* WULFF, V. J. (1941). Retinal pigment distribution in relation to diurnal rhythm in the compound eye of *Dytiscus. Proc. Soc. exp. Biol. Med.* **48**, 656.

JANDER, R. (1957). Die optische Richtungsorientierung der Roten Waldameise. *Z. vergl. Physiol.* **40**, 162.

—— (1964). Die Detektortheorie optischer Ausloesemechanismen von Insekten. *Z. Tierpsychol.* **21**, 302.

—— (1971). Visual pattern recognition and directional orientation in insects. *Ann. N.Y. Acad. Sci.* **188**, 5.

—— *and* BARRY, C. K. (1968). Die phototaktische Gegenkopplung von Stirnocellen und Facettenaugen in der Phototropotaxis der Heuschrecken und Grillen (Saltatoptera: *Locusta migratoria* und *Gryllus bimaculatus*). *Z. vergl. Physiol.* **57**, 432.

JANDER, R. *and* SCHWEDER, M. (1971). Ueber das Formunterscheidungsvermoegen der Schmeissfliege *Calliphora erythrocephala*. *Z. vergl. Physiol.* **72**, 186.

—— *and* VOLK-HEINRICHS, I. (1970). Das strauchspezifische visuelle Perceptorsystem der Stabheuschrecke (*Carausius morosus*). *Z. vergl. Physiol.* **70**, 425.

JÖRSCHKE, H. (1914). Die Facettenaugen der Orthopteren und Termiten. *Z. wiss. Zool.* **61**, 153.

JÄRVILEHTO, M. *and* ZETTLER, R. (1970). Micro-localisation of lamina located visual cell activities in the compound eye of the blowfly *Calliphora*. *Z. vergl. Physiol.* **69**, 134.

—— *and* —— (1971). Localized intracellular potentials from pre- and postsynpatic components in the external plexiform layer of an insect retina. *Z. vergl. Physiol.* **75**, 422.

—— *and* —— (1973). Electrophysiological-histological studies on some functional properties of visual cells and second order neurons of an insect lamina. *Z. Zellforsch.* **136**, 291–306.

KABUTA, H., TOMINAGA, Y., *and* KUWABARA, M. (1968). The rhabdomeric microvilli of several arthropod compound eyes kept in darkness. *Z. Zellforsch.* **85**, 78.

KAHMANN, H. (1947). Das Auge der Wirbellosen. *Tab. Biol.* **21**, 1.

KAISER, W. (1968). Zur Frage des Unterscheidungsvermoegens fuer Spektralfarben: Eine Untersuchung der Optomotorik der koeniglichen Glanzfliege *Phormia regina* Meig. *Z. vergl. Physiol.* **61**, 71.

—— (1972*a*). A preliminary report on the analysis of the optomotor system of the honey-bee—single unit recordings during stimulation with spectral lights. In *Information processing in the visual system of arthropods* (ed. R. Wehner) pp. 167–170, Springer Verlag, Berlin.

—— (1972*b*). Optomotor response studies with spectral lights—behavioural evaluation and single unit recording in honeybees. In *Abstracts of the 14th International Congress of Entomology*, Canberra, 22–30 August. p. 151 (1972).

—— *and* BISHOP, L. G. (1970). Directionally selective motion detecting units in the optic lobe of the honeybee. *Z. vergl. Physiol.* **67**, 403.

—— *and* LISKE, E. (1974). Die optomotorischen Reaktionen von fixiert fliegenden Bienen bei Reizung mit Spektrallichtern: *J. comp. Physiol.*, in press.

KAMMER, A. (1970). A comparative study of motor patterns during preflight warm-up in hawk moths. *Z. vergl. Physiol.* **70**, 45.

KAMPA, E. M., ABBOTT, B. C., *and* BODEN, B. P. (1963). Some aspects of vision in the Lobster *Homarus vulgaris* in relation to the structure of its eye. *J. mar. biol. Ass. U.K.* **43**, 683.

KAPANY, N. S. (1967). *Fiber optics.* Academic Press, London.

KAPRON, F. P. (1970). Geometrical optics of parabolic index-gradient cylindrical lenses. *J. opt. Soc. Am.* **60**, 1433.

KATZ, B. *and* MILEDI, R. (1965). Propagation of electric activity in motor nerve terminals. *Proc. R. Soc.* B **161**, 453.

KAY, R. E. (1969). Fluorescent materials in the insect eyes and their possible relationship to ultra-violet sensitivity. *J. Insect Physiol.* **15**, 2021.

KENNEDY, J. S. (1939). The visual responses of flying mosquitoes. *Proc. Zool. Soc. Lond.* ser. A **109**, 221.

KENYON, F. C. (1896). The brain of the bee. A preliminary contribution to the morphology of the nervous system of the Arthropoda. *J. comp. Neurol.* **6**, 133.

KIEN, J. (1974a). Sensory integration in the locust optomotor system: I Behavioural analysis. *Vision Res.* (in press).

—— (1974b). Sensory integration in the locust optomotor system: II Direction selective neurons in the circumoesophageal connectives and the optic lobe. *Vision Res.* (in press).

KIEPENHEUER, J. (1968). Farbunterscheidungsvermögen bei der roten Waldameise *Formica polyctena* Forster. *Z. vergl. Physiol.* **57**, 409.

KIESEL, A. (1894). Untersuchungen zur Physiologie des facettirten Auges. *Sitz. -ber. Akad. Wiss. Wien. math. -nat. Kl., 3. Abt.* **103**, 97.

KIKUCHI, R., NAITO, K., *and* TANAKA, I. (1962). Effect of sodium and potassium ions on the electrical activity of single cells in the lateral eye of the horseshoe crab. *J. Physiol.* **161**, 319.

—— *and* TAZAWA, M. (1960). Effect of intensity, duration and interval of stimulus on retinal slow potential. In *Electrical Activity of Single Cells.* pp. 25–38. Igakushoin, Tokyo, Hongo.

KIM, C. W. (1964). Formation and histochemical analysis of the crystalline cone of compound eye in *pieris rapae* L. (*Lepidoptera*). *Kor. J. Zool.* **7**, 19.

KIRCHHOFFER, O. (1908). Untersuchungen über die Augen pentamerer Käfer. *Arch. Biontol.* **2**, 237.

KIRMSE, W. *and* LAESSIG, P. (1971a). Strukturanalogie zwischen dem System der horizontalen Blickbewegungen der Augen beim Menschen und dem System der Blickbewegungen des Kopfes bei Insekten mit Fixationstreaktionen. *Biol. Zbl.* **90**, 175.

—— *and* —— (1971b). The characteristics of visually triggered head movements of the dragon-fly. In *Biokybernetik*, vol. 3 (ed. H. Drischel and N. Tiedt), pp. 261–264, G. Fischer. Jena.

KIRSCHFELD, K. (1965). Das anatomische und das physiologische Sehfeld der Ommatidien im Komplexauge von *Musca. Kybernetik* **2**, 249.

—— (1967). Die Projektion der optischen Umwelt auf das Raster der Rhabdomere im Komplexauge von *Musca. Exp. Brain Res.* **3**, 248.

—— (1968). The *Musca* compound eye. *Symp. zool. Soc. Lond.* **23**, 165.

—— (1969a). Absorption properties of photopigments in single rods, cones and rhabdomes. In *Processing of optical data by organisms and machines.* Rendiconti S. I.F. **43**, 116.

—— (1969b). Optics of the compound eye. Rendiconti delle Scuela internazionale di Fisica "E. Fermi", 43° Corso 144.

—— (1971). Aufnahme und Verarbeitung optischer Daten im Komplexauge von Insekten. *Naturw.* **58**, 201.

—— (1972). The visual system of *Musca*: Studies on optics, structure and function. In *Information Processing in the Visual Systems of Arthropods* (ed. R. Wehner), 61–74. Springer, Berlin, Heidelberg, New York.

—— (1972). Die notwendige Anzahl von Rezeptoren zur Bestimmung der Richtung des elektrischen Vektors linear polarisierten Lichtes. *Z. Naturforsch.* **27b**, 578.

—— (1973). Optomotorische Reaktionen der Biene auf bewegte "*Polarisations-Muster*". *Z. Naturforsch.* **28c**, 329–338.

—— *and* FRANCESCHINI, N. (1968). Optische Eigenschaften der Ommatidien im Komplexauge von *Musca. Kybernetik* **5**, 47.

KIRSCHFELD, K. *and* FRANCESCHINI, N. (1969). Ein Mechanismus zur Steuerung des Lichtflusses in den Rhabdomeren des Komplexauges von *Musca*. *Kybernetik* **6**, 13.

—— *and* REICHARDT, W. (1964). Die Verarbeitung stationaerer optischer Nachrichten im Komplexauge von *Limulus*. (Ommatidien-Sehfeld und raeumliche Verteilung der Inhibition). *Kybernetik* **2**, 43.

—— *and* —— (1970). Optomotorische Versuche an *Musca* mit linear polarisiertem Licht. *Z. Naturforsch.* **25b**.

KOIKE, H., BROWN, H. M., *and* HAGIWARA, S. (1971). Hyperpolarization of a barnacle photoreceptor membrane following illumination. *J. gen. Physiol.* **57**, 723.

KOLB, G. *and* AUTRUM, H. (1972). Die Feinstruktur im Auge der Biene bei Hell- und Dunkeladaptation. *J. comp. Physiol.* **77**, 113.

——, ——, *and* EGUCHI, E. (1969). Die spektrale Transmission des dioptrischen Apparates von *Aeschna cyanae* Müll. *Z. vergl. Physiol.* **63**, 434.

KOPAL, Z. (1955). *Numerical analysis*. John Wiley and Sons, New York.

KORN, A. *and* VON SEELEN, W. (1972). Dynamische Eigenschaften von Nervennetzen in visuellen System. *Kybernetik.* **10**, 64.

KRINSKY, N. J. (1958). The enzymatic esterification of vitamin A. *J. biol. Chem.* **232**, 881.

KUEHN, A. (1924). Zum Nachweis des Farbenunterscheidungsvermoegens der Bienen. *Naturwiss.* **12**, 116.

KÜHN, A. (1927). Über den Farbensinn der Bienen. *Z. vergl. Physiol.* **5**, 762.

KUIPER, J. W. (1964). On the optics of the superposition eye. *Arch. néérl. Zool.* **16**, 171.

—— (1962). The optics of the compound eye. In *Biological Receptor Mechanisms.* (ed. J. W. L. Beament). pp. 58–71, Cambridge University Press, London.

—— (1966). On the image formation in a single ommatidium of the compound eye in Diptera. In *The functional organization of the compound eye* (ed. C. G. Bernhard), pp. 35–50. London, Pergamon Press.

KUNZE, P. (1961). Untersuchung des Bewegungssehens fixiert fliegender Bienen. *Z. vergl. Physiol.* **44**, 656.

—— (1969). Eye glow in the moth and superposition theory. *Nature, Lond.* **223,** 1172.

—— (1970). Verhaltensphysiologische und optische Experimente zur Superpositions-theorie der Bildentstehung in Komplexaugen. *Verhandl. Deutsch. zool. Ges.* **64**, 234.

—— (1972). Comparative studies of Arthropod superposition eyes. *Z. vergl. Physiol.* **76**, 347.

—— *and* HAUSEN, K. (1971). Inhomogeneous refractive index in the crystalline cone of a moth eye. *Nature, Lond.* **231**, 392.

LABHART, T. (1972). The discrimination of light intensities in the honey bee. In *Information Processing in the Visual Systems of Arthropods*, pp. 115–119. (ed. R. Wehner). Springer, Berlin, Heidelberg, New York.

—— *and* WEHNER, R. (1972). Die Unterschiedsempfindlichkeit für Lichtintensitaeten bei *Apis mellifera, Rev. Suisse Zool.* **79**, 1068.

LADMAN, A. J. (1958). The fine structure of the rod bipolar synapse in the retina of the albino rat. *J. biophys. biochem. Cytol.* **4**, 459.

LAMMERT, A. (1925). Pigment movements in ocelli. *Z. vergl. Physiol.* **3**, 225.

LAND, M. F. (1969). Structure of the retinae of the principal eyes of jumping spiders (Salticidae: Dendryphantinae) in relation to visual optics. *J. exp. Biol.* **51**, 443.

—— (1971). Orientation by jumping spiders in the absence of visual feedback. *J. exp. Biol.* **54**, 119.

570   References

LAND, M. F. (1973). Head movements of flies during visually guided flight. *Nature, Lond.*, **243**, 299.

LANGE, H. *and* STRUWE, G. (1972). Spectral absorption by screening pigment granules in the compound eye of butterflies (*Heliconius*). *J. comp. Physiol.* **79**, 203.

LANGER, H. (1965). Spektrophotometrische Untersuchungen der Absorptionseigenschaften einzelner Rhabdomere im Facettenauge. *Zool. Anz., (Suppl.)* **29**, 329.

—— (1967). Grundlagen der Wahrnehmung von Wellenlaenge und Schwingungsebene des Lichtes: *Verh. Dtsch. zool. Ges.* Goettingen, (1966), 195.

—— (1967). Über die Pigmentgranula im Facettenauge von *Calliphora erythrocephala*. *Z. vergl. Physiol.* **55**, 354.

—— (1972). Metarhodopsin in single rhabdomeres of the fly, *Calliphora erythrocephala*. In *Information processing in the visual systems of arthropods*. (ed. R. Wehner), 109, Springer Verlag, Heidelberg, New York.

—— *and* HOFFMAN, C. (1966). Elektro- und stoffwechselphysiologische Untersuchungen über den Einfluss von Ommochromen und Pteridinen auf die Funktion des Facettenauges von *Calliphora erythrocephala*. *J. Insect Physiol.* **12**, 357.

—— *and* PATAT, U. (1962). Über die Bedeutung einer neuen Augenfarbenmutante von *Calliphora erythrocephala*, für die Untersuchung der Funktion der Facettenaugen. *Zool. Anz. Suppl.* **25**, 174.

—— *and* SCHNEIDER, L. (1970). Zur Struktur und Funktion offener Rhabdome in Facettenaugen. *Zool. Anz. Suppl.* **33**, 494.

—— *and* —— (1972). Lichtsinneszellen: In *Lehrbuch der Cytologie*. (ed. H. Ruska). VEB Fischer, Jena.

—— *and* THORELL, B. (1966a). Microspectrophotometric assay of visual pigments in single rhabdomeres of the insect eye. In *Functional organization of the compound eye*. pp. 145–149. (ed. C. G. Bernhard). Pergamon Press, Oxford.

—— *and* —— (1966b). Microspectrophotometry of single rhabdomeres in the insect eye. *Exp. cell Res.* **41**, 673.

LASANSKY, A. (1967). Cell junctions in ommatidia of *Limulus*. *J. cell. Biol.* **33**, 365.

—— *and* FUORTES, M. G. F. (1969). The site of origin of electrical responses in visual cells of the leech, *Hirudo medicinalis*. *J. cell. Biol.* **42**, 241.

LAUER, J. *and* LINDAUER, M. (1971). Genetisch fixierte Lerndisposition bei der Honigbiene. *Akad. Wiss. Lit. Mainz. Inf. Org.* **1**, 1.

LAUGHLIN, S. B. (1973). Neural integration in the first optic neuropile of dragonflies. I. Signal amplification in dark-adapted second order neurons. *J. comp. Physiol.* **84**, 335–355.

—— *and* HORRIDGE, G. A. (1971). Angular sensitivity of the retinula cells of dark-adapted worker bee. *Z. vergl. Physiol.* **74**, 329.

LEHNINGER, A. (1964). *The mitochondrion. Molecular basis of structure and function.* Benjamin, New York, Amsterdam.

LETTVIN, J. Y., MATURANA, H. R., McCULLOCH, W. S., *and* PITTS, W. H. (1959). *What the frog's eye tells the frog's brain.* Proc. IRE. **47**, 1940.

LEUTSCHER-HAZELHOFF, J. T. *and* KUIPER, J. W. (1964). Responses of the blowfly *Calliphora erythrocephala* to light flashes and to sinusoidally modulated light. *Documenta Opthalmologica* **18**, 275.

LEVIN, L. *and* MALDONADO, H. (1970). A fovea in the praying mantis eye. III. The centring of the prey. *Z. vergl. Physiol.* **67**, 93.

LIEBMAN, P. A. (1972). Microspectrophotometry of photoreceptors. *Handbook of Sensory Physiology.* vol. 7 (ed. H. J. Dartnall) Springer Verlag, Berlin.

LIPSON, S. G. *and* LIPSON, H. (1969). *Optical Physics,* Chapt. 10. Cambridge University Press, Cambridge.

LISMAN, J. E. *and* BROWN, J. E. (1972). The effects of intracellular iontophoretic injection of calcium and sodium ions on the light response of *Limulus* ventral photoreceptors. *J. gen. Physiol.* **59**, 701.

LOCKE, M. (1965). Permeability of insect cuticle to water and lipids. *Science* **147**, 295.

—— (1966). The structure and formation of the cuticulin layer in the epicuticle of an insect, *Calpodes ethlius* (Lepidoptera, Hesperiidae). *J. Morph.* **118**, 461.

LONGHURST, R. S. (1967). *Geometrical and physical optics.* Longman's, Green & Co., London.

LOTSCH, H. K. V. (1968). Reflection and refraction of a beam of light at a plane interface. *J. opt. Soc. Am.* **58**, 551.

LUBBOCK, J. (1885). *Ants, Bees, and Wasps.* Kegan Paul, Trench Trübner and Co., London.

LÜDTKE, H. (1953). Retinomotorik und Adaptionsvorgänge in Auge des Rücken-Schwimmers (*Notonecta glauca* L.) *Z. vergl. Physiol.* **35**, 129.

McCANN, G. D. *and* ARNETT, D. W. (1972). Spectral and polarization sensitivity of the Dipteran visual system. *J. gen. Physiol.* **59**, 534.

—— *and* DILL, J. C. (1969). Fundamental properties of intensity, form, and motion perception in the visual nervous systems of *Calliphora phaenicia* and *Musca domestica. J. gen. Physiol.* **53**, 385.

—— *and* FENDER, D. H. (1964). Computer data processing and systems analysis applied to research on visual perception. In *Neural theory and modeling* (ed. R. F. Reiss). pp. 232–252. Stanford University Press, Stanford.

—— *and* FOSTER, S. F. (1971). Binocular interactions of motion detection fibers in the optic lobes of flies. *Kybernetik* **8**, 193.

—— *and* McGINITIE, G. (1965). Optomotor response studies of insect vision. *Proc. R. Soc.* B **163**, 369.

McINTYRE, P. D. *and* SNYDER, A. W. (1974). Power transfer between optical fibres. *J. opt. Soc. Am.* **63**, 1518.

—— *and* —— (1974). Power transfer between non-parallel and tapered optical fibers. *J. opt. Soc. Am.* **64**, 285.

McLEOD, J. H. (1954). The axicon: a new type of optical element. *J. opt. Soc. Am.* **44**, 592.

MacNICHOL, E. F. JR. (1956). Visual receptors as biological transducers. In *Molecular structure and functional activity of nerve cells.* American Institute of Biological Sciences, Washington, D.C.

MAHAN, A. D., BITTERLI, C. V., *and* UNGER, H. J. (1972). Reflection and transmission properties of cylindrically guided electromagnetic waves. *J. op. Soc. Am.* **62**, 361.

MALDONADO, H. *and* BARROS-PITA, J. C. (1970). A fovea in the praying mantis eye. I. Estimation of the catching distance. *Z. vergl. Physiol.* **67**, 58.

MARAK, G. E., GALLIK, G. J., *and* CORNESKY, R. A. (1970). Light-sensitive pigment in insect heads. *Ophthal. Res.* **1**, 65.

MARCHAND, E. W. (1970). Ray tracing in gradient-index media. *J. opt. Soc. Am.* **60**, 1.

MARECHAL, A. (1952). Imagerie géométrique, aberrations. *Rev. d'Optique, Paris* **244**.

MASSOUD, Z. (1969). Étude de l'ornementation épicuticulaire du tégument des Collemboles en microscope éléctronique à balayage. *C.r. Acad. Sci.* (*Paris*) **268**, 1407.

MATTHIESSEN, L. (1886a). Über eine neue Etagenloupe. Centr. Z. Opt. Mech. 7, 109.

—— (1886b). Über den Strahlenduchgang durch coaxial continuirlich geschichtete Cylinder mit Beziehung auf den physikalisch-optischen Bau der Augen verschiedener Insekten, Repert. Phys. 22, 333.

—— (1879). Die Differnetialgleichungen der Dioptrik der geschichteten Krystallinse. Pflüger's Arch. ges. Physiol. 19, 480.

MAYER, H. (1957). Zur biologie und Ethologie einheimischer Collembolen. Zool. Jb. Syst. 85, 501.

MAZOCHIN-PORSHNYAKOV, G. A. (1956). On the colour vision in insects. Biofizika 1, 98. (Russian).

—— (1959). Discrimination between green, yellow and orange colours in bees. (Russian). Biofizika 4, 48.

—— (1960). Colour-vision in Calliphora. Biophys. 5, 295, 697, 790.

—— (1962). Colorimetric evidence of trichromatic vision in bumble-bees. (Russian). Biofizika 7, 211.

—— (1968). Die Insekten und ihre Faehigkeit, die visuellen Sehobjekte zu generalisieren. Rev. Ent. CCCP 47, 362 (Russian, German summary).

—— (1969a). Generalization of visual stimuli as an example of solution of abstract problems by bees. Zool. Jb. 48, 1125 (Russian, English summary).

—— (1969b). Die Faehigkeit der Bienen, visuelle Reize zu generalisieren. Z. vergl. Physiol. 65, 15.

—— (1969c). Insect vision. Plenum Press, New York.

—— and TRENN, W. (1972). Electrophysiological study of eye in ants. Zool. J. 52, 1007 (in Russian).

——, VISHNEVSKAY, T. M., GOLUBZOV, K. V. and BOTSHAROV, O. I. (1967). Vision of Anacanthotermes ahngerianus according to the data of electrophysiological experiments. Zool. J. 46, 1668 (in Russian).

MÉDIONI, J. (1959a). Mise en évidence et évaluation d'un "effet de stimulation" du aux ocelles frontaux dans le phototropisme de Drosophila melanogaster. C.R. Soc. Biol., 153, p. 1587.

—— (1959b). Les organes de stimulation photique chez Drosophila melanogaster. Non-spécificité de cette stimulation rélativement aux réactions à la lumière. C.R. Soc. Biol., 153, p. 1845.

—— (1964). Sur le rôle dynamogénique des organes sensoriels dans l'orientation des Insectes. Proc. 17th Int. Congr. Entomology, London, 279–81.

—— (1967). Données nouvelles sur les déplacements géotactiques de Drosophila melanogaster Meigen (Diptères, Drosophilides). Annales des Epiphyties, 18, 107–8.

——, CAMPAN, R. and QUÉINNEC, Y. (1971). Sur le rôle activateur des ocelles frontaux chez les Insectes: tachycardie induite par la stimulation ocellaire chez Calliphora vomitoria. Proc. 96th Congrès des Sociétés Savantes, Toulouse, in press.

METSCHL, N. (1963). Electrophysiologische Untersuchungen an des Ocellen von Calliphora. Z. vergl. Physiol. 47, 230.

MELAMED, J. and TRUJILLO-CENÓZ, O. (1968). The fine structure of the central cells in the ommatidia of dipterans. J. Ultr. Res. 21, 313.

MENZEL, R. (1967). Untersuchungen zum Erlernen von Spektralfarben durch die Honigbiene (Apis mellifica). Z. vergl. Physiol. 56, 22.

—— (1971). Über den Farbensinn von Paravespula germanica F. (Hymenoptera) ERG und selektive Adaptation. Z. vergl. Physiol. 75, 86.

MENZEL, R. (1972a). Feinstruktur des Komplexauges der Roten Waldameise, *Formica polyctena* (Hymenoptera, Formicidae). *Z. Zellforsch.* **127**, 356.

MENZEL, R. (1972b). The fine structure of the compound eye of *Formica polyctena*— Functional morphology of a hymenopteran eye. In *Information processing in the visual systems of arthropods.* (ed. R. Wehner), 37–49, Springer Verlag, Heidelberg, New York.

—— (1973). Spectral response of moving detecting and 'sustaining' fibres in the optic lobe of the bee. *J. comp. Physiol.* **82**, 135.

—— and LANGE, G. (1971). Änderungen der Feinstruktur im Komplexauge von *Formica polyctena* bei Helladaptation. *Z. Naturforsch.* **26B**, 357.

——, MOCH, K., WLADARZ, G., and LINDAUER, M. (1969). Tagesperiodische Ablagerungen in der Endokutikula der Honigbiene. *Biol. Zbl.* **88**, 61.

—— and SNYDER, A. W. (1974). Polarized light detection in the Bee, *Apis mellifera.* *J. comp. Physiol.* **88**, 247.

—— and WEHNER, R. (1970). Augenstrukturen bei verschidenen großen Arbeiterinnen von *Cataglyphis bicolor.* *Z. vergl. Physiol.* **68**, 446.

MEYER, H. and MARK, K. H. (1930). Der Aufbau der hochpolymeren organischen Naturstoffe. *Akad. Verlags. Ges. Leipzig.*

MEYER, H. W. (1971a). Visuelle Schlüsselreize für die Ausloesung der Beutfanghandlung beim Bachwasserlaeufer *Velia caprai* (Hemiptera, Heteroptera). I. Untersuchung der raeumlichen und zeitlichen Reizparameter mit formverschiedenen Attrappen. *Z. vergl. Physiol.* **72**, 260.

—— (1971b). Visuelle Schluesselreize fuer die Ausloesung der Beutefanghandlung beim Bachwasserlaeufer *Velia caprai* (Hemiptera, Heteroptera). 2. Untersuchung der Wirkung zeitlicher Reizmuster mit Flimmerlicht. *Z. vergl. Physiol.* **72**, 298.

—— (1972). Ethometrical investigations into the spatial interaction within the visual system of *Velia caprai* (Hemiptera, Heteroptera). In *Information Processing in the Visual Systems of Arthropods,* (ed. R. Wehner). pp. 223–229. Springer, Berlin, Heidelberg, New York.

MEYER-ROCHOW, V. B. (1971). Fixierung von Insektenorganen mit Hilfe eines Netzmittels—Das Dorsalauge der Eintagsfliege *Atalophlebia costalis. Mikrokosmos* **60**, 348.

—— (1972a). The eyes of *Creophilus erythrocephalus* F. and *Sartallus signatus* Sharp (Staphylinidae: Coleoptera)—Light-, interference-, scanning electron-, and transmission electron microscope examinations. *Z. Zellforsch.* **133**, 59.

—— (1972b). A crustacean-like organization of insect-rhabdoms. *Cytobiologie* **4**, 241.

—— (1973). The dioptric system of the eye of *Cybister* (Dytiscidae: Coleoptera). *Proc. R. Soc. Lond. B.* **183**, 159.

MICHAEL, C. R. (1968). Receptive fields of single optic nerve fibers in a mammal with an all-cone retina. II. Directionally selective units: *J. Neurophysiol.* **31**, 257.

MILAS, N. A., TREPAGNIER, J. H., NOLAN, J. T. and ILIOPULOS, M. I. (1959). A study of hydroxylation of olefins and the reaction of osmium tetroxide with 1,2-glycols. *J. Am. Chem. Soc.* **81**, 4730.

MILLECCHIA, R. and MAURO, A. (1969a). The ventral photoreceptor cells of *Limulus.* II. The basic photoresponse. *J. gen. Physiol.* **54**, 310.

—— and —— (1969b). The ventral photoreceptor cells of *Limulus.* III. A voltage clamp study. *J. gen. Physiol.* **54**, 331.

MILLER, W. H. (1957). Morphology of the compound eye of *Limulus. J. biophys. biochem. Cytol.* **3**, 421.

—— and BERNARD, G. D. (1968a). Butterfly glow. *J. Ultrastruct. Res.* **24**, 286.

MILLER, W. H. *and* BERNARD, G. D. (1968). Skipper glow. *MIT Res. Lab. Electron. qu. Progr. Rep.* **88**, 114.

——, ——, *and* ALLEN, J. L. (1968). The optics of insect compound eyes. *Science* **762**, 760.

——, MØLLER, A. R., *and* BERNHARD, C. G. (1966). The corneal nipple array. In *The functional organization of the compound eye* (ed. C. G. Bernhard) 21–33. Pergamon Press, London.

MIMURA, K. (1970). Integration and analysis of movement information by the visual system of flies. *Nature, Lond.* **226**, 964.

—— (1971). Movement discrimination by the visual system of flies. *Z. vergl. Physiol.* **73**, 105.

—— (1972). Neural mechanisms subserving directional selectivity of movement in the optic lobe of the fly. *J. comp. Physiol.* **80**, 409.

—— (1974). Analysis of visual information in lamina neurones of the fly. *J. comp. Physiol.* (in press).

——, TATEDA, H., MORITA, H., *and* KUWABARA, M. (1969). Regulation of insect brain excitability by ocellus. *Z. vergl. Physiol.* **62**, 382.

MITTELSTAEDT, H. (1949). Telotaxis und Optomotorik von *Eristalis* bei Augeninversion. *Naturw.* **3**, 90.

—— (1950). Physiologie des Gleichgewichtssinnes bei fliegenden Libellen. *Z. vergl. Physiol.* **32**, 422.

—— (1954). Regelung und Steuerung bei der Orientierung der Lebewesen. *Regelungstechnik* **2**, 226.

—— (1964). Basic control patterns of orientational homeostatis. *Symp. Soc. exp. Biol.* **18**, 365.

MOLLER-RACKE, I. (1952). Farbensinn und Farbenblindheit bei Insekten: *Zool. Jb., Abt. allg. Zool. u. Physiol.* **63**, 237.

MONDADORI, C. *and* WEHNER, R. (1972). Interferenz- und polarisationsmikroskopische Untersuchungen an der Cornea von *Cataglyphis bicolor* (Formicidae, Hymenoptera). *Rev. Suisse Zool.* **79**, 1106.

MOODY, M. F. (1964). Photoreceptor organelles in animals. *Biol. Rev.* **39**, 43.

—— *and* PARRISS, J. K. (1961). The discrimination of polarized light by Octopus: a behavioural and morphological study. *Z. vergl. Physiol.* **44**, 268.

MOTE, M. I. (1970a). Focal recording of responses evoked by light in the lamina ganglionaris of the fly *Sarcophaga bullata*. *J. exp. Zool.* **175**, 149.

—— (1970b). Electrical correlates of neural superposition in the eye of the fly *Sarcophaga bullata*. *J. Zool.* **175**, 159.

—— *and* GOLDSMITH, T. H. (1970). Spectral sensitivities of colour receptors in the compound eye of the cockroach *Periplaneta*. *J. exp. Zool.* **173**, 137.

—— *and* —— (1971). Compound eyes: Localization of two color receptors in the same ommatidium. *Science* **171**, 1254.

MOTOKAWA, K. (1970). *Physiology of colour and pattern vision*. Springer, Tokyo, Shoin, Berlin, Heidelberg, New York.

MUELLER, J. (1826). *Zur vergleichenden Physiologie des Gesichtssinnes*. C. Cnobloch, Leipzig.

MÜLLER, E. (1931). Experimentelle Untersuchungen an Bienen und Ameisen über die Funktionsweise der Stirnocellen. *Z. vergl. Physiol.* **14**, 348.

MÜLLER, J. (1970). Feinbau und Dunkelanpassung der Komplexaugen von *Rhodnius prolixus*. *Zool. Jb. Abt. allg. Zool. u. Physiol.* **75**, 111.

MUSOLFF, W. (1955). Untersuchungen ueber Farbensinn und Purkinjesches Phaenomen bei drei oekologisch verschiedenen Typen der Echsen (Lacertilia) mit Hilfe der optomotorischen Reaktionen. *Zool. Beitr. N. F.* **1**, 399.

NAKA, K. I. *and* EGUCHI, E. (1962). Spike potentials recorded from the insect photoreceptor. *J. gen. Physiol.* **45**, 663.

—— *and* KISHIDA, K. (1969). Retinal action potentials during dark and light adaptation. In *The functional organization of the compound eye*, (ed. C. G. Bernhard), 251–266, Pergamon Press, Oxford.

NEVILLE, A. C. (1967). Chitin orientation in cuticle and its control. In *Advances in Insect Physiology*, (ed. J. W. L. Beament, J. E. Treherne, and V. B. Wigglesworth) Vol. **4**, 213–284, Academic Press, London.

—— *and* CAVENEY, L. (1969). Scarabaeid beetle exocuticle as an optical analogue of cholesteric liquid crystals. *Biol. Rev.* **44**, 531.

—— *and* LUKE, B. M. (1971). Form optical activity in Crustacean cuticle. *J. Insect. Physiol.* **17**, 519.

——, THOMAS, M. G., *and* ZELAZNY, B. (1969). Pore canal shape related to molecular architecture of arthropod cuticle. *Tissue and Cell* **1**, 183.

NIEMANN, H. (1970). Eigenschaften diskreter Ortsfilter. *Kybernetik.* 7(2); 78–88.

NINOMIYA, N., TOMINAGA, Y., *and* KUWABARA, M. (1969). The fine structure of the compound eye of a damsel-fly. *Z. Zellforsch.* **98**, 17.

NISHIITSUTSUJI-UWO, J. *and* PITTENDRIGH, C. S. (1968a). Central nervous system control of circadian rhythmicity in the cockroach. II. The pathway of light signals that entrain the rhythm. *Z. vergl. Physiol.* **58**, 1.

NOWOTNY, H. *and* ZAHN, H. (1942). Über die Feinstruktur von Wollkeratin *Z. physik. Chem.* **51**, 265.

NOLTE, J. D. *and* BROWN, J. E. (1969). The spectral sensitivities of single cells in the median ocellus of *Limulus*. *J. gen. Physiol.* **54**, 636.

—— *and* —— (1972a). Electrophysiological properties of cells in the median ocellus of *Limulus*. *J. gen. Physiol.* **59**, 167.

—— *and* —— (1972b). Ultraviolet-induced sensitivity to visible light in ultraviolet receptors of *Limulus*. *J. gen. Physiol.* **59**, 186.

NORTHROP, R. B. (1971). Report AFOSR-68-1539 on the neurophysiology of data processing in the optic ganglia of insect compound eyes.

NORTHROP, R. B. *and* GUIGNON, E. F. (1970). Information processing in the optic lobes of the lubber grasshopper. *J. Insect Physiol.* **16**, 691.

NOSAKI, H. (1969). Electrophysiological study of color encoding in the compound eye of crayfish, *Procamburus clarkii*. *Z. vergl. Physiol.* **64**, 318.

NOTON, D. *and* STARK, L. (1971). Scan paths in eye-movements during pattern perception. *Science* **171**, 308.

NOWIKOFF, M. (1931). Untersuchungen über die Komplexaugen von Lepidopteren nebst einigen Bemerkungen über die Rhabdome der Arthropoden im allgemeinen. *Z. wiss. Zool.* **138**. 1.

—— (1932). Über den Bau der Komplexaugen von *Periplaneta (Stylopiga) orientalis* L. Jena. *Z. Naturw.* **67**, 58–69.

NUNNEMACHER, R. F. (1959). The retinal image of Arthropod eyes. *Anat. Rec.* **134**, 618.

OHLY, K. P. (1968). Das Gesichtsfeld des Komplexauges der Honigbiene *Apis mellifica*. Diplomarbeit, Dept. of Zoology, University of Frankfurt-M.

ÖHMAN, P. (1971). The photoreceptor outer segments of the river lamprey (*Lampetra fluviatilis*). An electron-, fluorescence- and light-microscopic study. *Acta Zool.* **52**, 287.

ORKAND, R. K., NICHOLLS, J. G., *and* KUFFLER, S. W. (1966). Effect of nerve impulses on the membrane potential of glial cells in the central nervous system of amphibia. *J. Neurophysiol.* **29**, 788.

OUDEMANS, J. T. (1887). *Bijdrage tot de Kennis der Thysanura en Collembola*. Amsterdam 1887.

PAGE, L. *and* ADAMS, N. I. (1937). Electromagnetic waves in conducting tubes. *Phys. Rev.* **52**, 647.

PALKA, J. (1965). Diffraction and visual acuity of insects. *Science* **149**, 551.

—— (1967). An inhibitory process influencing visual responses in a fibre of the central nerve cord of locusts. *J. Insect Physiol.* **13**, 235.

—— (1969). Discrimination between movements of eye and object by visual interneurones of crickets. *J. exp. Biol.* **50**, 723.

—— (1972). Moving movement detectors. *Am. Zoologist* **12**, 497.

PANDAZIS, G. (1930). Über die relative Ausbildung der Gehirnzentren bei biologisch verschiedenen Ameisenarten. *Z. Morph. Ökol. Tiere*, **18**, 114.

PAPOULIS, A. (1968). *Systems and transforms with applications in optics*. McGraw-Hill, New York.

PARKER, G. H. (1932). The movement of the retinal pigment. *Ergeb. Biol.* **9**, 239.

PARRY, D. A. (1947). The function of the insect ocellus. *J. exp. Biol.* **24**, 211.

PASK, C. *and* SNYDER, A. W. (1973). Image detection by the bee ommatidium. *J. opt. Soc. Am.* submitted for publication.

PATTERSON, J. A. *and* GOODMAN, L. J. (1973). Ocellar units in the ventral nerve cord of *Schistocerca gregaria* (in preparation).

PAULUS, H. F. (1970a). Zur Feinstruktur des zusammengesetzten Auges von *Orchesella*. *Z. Naturforsch.* **25b**, 380, 2 Taf.

—— (1970b). Das Komplexauge von *Podura aquatica*, ein primitives Doppelauge. *Naturwiss.* **57**, 502.

—— (1971). Einiges zur Cuticulastruktur der Collembolen mit Bemerkungen zur Oberflächenskulptur der Cornea. *Rev. Ecol. Sol.* (*Paris*) **8**, 37.

—— (1972a). Zum Feinbau der Komplexaugen einiger Collembolen, eine vergleichend-anatomische Untersuchung (Insecta, Apterygota). *Zool. Jb. Anat.* **89**, 1.

—— (1972b). The ultrastructure of the photosensible elements in the eyes of Collembola and their orientation. *Information processing in the visual systems of arthropods*. Symposium Zürich (ed. R. Wehner), Springer Verlag.

—— (1972c). Die Feinstruktur der Stirnaugen einiger Collembola (Insecta, Entognatha). *Z. zool. Syst. Evolutionsforsch.* **10**, 81.

—— (1973a). Die Feinstruktur des Seitenauges von *Dicyrtomina ornata* (Insecta, Collembola). I. Die Ommatidien (in preparation).

—— (1973b). Die Feinstruktur des Doppelauge von *Entomobrya muscorum*. (in preparation).

—— (1974). Die phylogenetische Bedeutung der Ommatidien der apterygoten Insekten (Collembola, Archaeognatha, Zygentoma). *Pedobiologica* (in press).

PERRELET, A. (1970). The fine structure of the retina of the honey bee drone. An electron microscopical study. *Z. Zellforsch.* **108**, 530.

—— and BAUMANN, F. (1969*a*). Evidence for extracellular space in the rhabdome of the honeybee drone eye. *J. cell Biol.* **40**, 825.

—— and —— (1969*b*). Presence of three small retinula cells in the ommatidium of the honeybee drone eye. *J. Microscopie* **8**, 49.

PHILLIPS, E. F. (1905). Structure and development of the compound eye of the honeybee. *Proc. Acad. Nat. Sci. Philad.* **57**, 123.

PICHT, J. (1929). Beitrag zur Theorie der Totalreflexion. *Ann. Phys.* **3**, 433.

PINTER, R. B. (1972). Frequency and time domain properties of retinular cells of the desert locust, *Schistocerca gregaria*, and the house cricket, *Achaeta domesticus*. *J. comp. Physiol.* **77**, 383.

POLOTAREV, V. M. *and* KISLOVSKII, L. D. (1965). On the choice of the optimum conditions for the production of spectra by the frustrated total internal reflection method. *Opt. Spectrosc.* **19**, 346.

PITTMAN, R. M. *et al.* (1972). Branching of central neurons: intracellular cobalt injection for light and electron microscopy. *Science* **176**, 412.

POGGIO, T. *and* REICHARDT, W. (1973). A theory of pattern-induced flight orientation in the fly *Musca domestica*. *Kybernetik* **12**, 185.

PORTER, K. *and* KALLMAN, F. (1953). The properties and effect of osmium tetroxide as a tissue fixative with special reference to its use for electron microscopy. *Exp. Cell Res.* **4**, 127.

POTTER, R. J. (1961). Transmission properties of optical fibers. *J. opt. Soc. Amer.* **51**, 1079.

PORTILLO, J. DEL (1936). Beziehungen zwischen den Oeffnungswinkeln der Ommatidien, Kruemmung und Gestalt der Insektenaugen und ihrer funktionellen Aufgabe. *Z. vergl. Physiol.* **23**, 100.

POST, C. T. *and* GOLDSMITH, T. H. (1965). Pigment migration and light-adaptation in the eye of the moth. *Biol. Bull.* **128**, 473.

—— and —— (1969). Physiological evidence for colour receptors in the eye of a butter-fly. *Ann. ent. Soc. Am.* **62**, 1497.

POWER, M. E. (1942). The thoracicoabdominal nervous system of an adult insect, *Drosophila melanogaster*. *J. comp. Neurol.* **88**, 347.

PRAAGH, J. P. VAN *and* VELTHUIS, H. H. W. (1971). Is the lux an appropriate measure for brightness in a bee flight room? *Bee World* **52**, 25.

PRINGLE, J. W. S. (1948). The gyroscopic action of the halteres of Diptera. *Phil. Trans. R. Soc.* B **233**, 347.

—— (1957). *Insect flight*. Cambridge Univ. Press, London.

RABAUD, E. *and* VERRIER, M. L. (1940). Vision et comportement des Pagures et théorie d'Exner. *C.r. Acad. sci. Paris* **211**, 300.

RADL, E. (1903). Untersuchungen ueber den Phototropismus der Tiere. Leipzig.

RATLIFF, F., HARTLINE, H. K., *and* MILLER, W. H. (1963). Spacial and temporal aspects of retinal inhibitory interaction. *J. Opt. Soc. Am.* **53**, 110.

RADEMAKER, C. G. J. *and* TER BRAAK J. W. G. (1948). On the central mechanism of some optic reactions. *Brain* **71**, 48.

REHBONN, W. (1972). Gleichzeitige intrazelluläre Doppelableitungen aus dem Komplexauge von *Calliphora erythrocephala*. *Z. vergl. Physiol.* **76**, 285.

REICHARDT, W. (1957). Autokorrelationsauswertung als Funktionprinzip des Zentralnervensystems. *Z. Naturforsch.* **12b**, 447.

578   *References*

REICHARDT, W. (1961). Nervous integration in the facet eye. *Biophys. J.* **2**, 121.

—— (1961). Ueber das optische Aufloesungsvermoegen der Facettenaugen von *Limulus. Kybernetik* **1**, 57.

—— (1969*a*). Autocorrelation, a principle for the evaluation of sensory information by the central nervous system. In '*Sensory Communication*', (ed. W. Rosenblith), pp. 303–17, M.I.T. and John Wiley, New York.

——(1969*b*). Movement perception in insects. In *Processing of Optical Data by Organisms and by Machines*, (ed. W. Reichardt), pp. 465–93, Academic Press, New York.

—— (1970). The insect eye as a model for analysis of uptake, transduction, and processing of optical data in the nervous system. In *The Neurosciences, 2nd study program*, (ed. F. O. Schmitt), pp. 494–511. Rockefeller Univ. Press, New York.

—— (1971). Visual detection and fixation of objects by fixed flying flies. In *Pattern Recognition in Biological and Technical systems*, (ed. O. J. Gruesser and R. Klinke), pp. 55–59. Springer, Berlin, Heidelberg, New York.

—— (1972). First steps in a behavioral analysis of pattern discrimination in Diptera. In *Information processing in the visual systems of arthropods*, (ed. R. Wehner), pp. 213–15. Springer, Berlin, Heidelberg, New York.

—— (1973*a*). Musterinduzierte Flugorientierung der Fliege *Musca domestica. Naturw.*, **60**, 122.

—— (1973*b*). Verhaltens-Analyse der Fixierung elementarer Muster durch das visuelle System der Fliege Musca domestica. *Naturw.* in the press.

—— and VARJÚ, D. (1959). Übertragungseigenschaften im Auswerte System für das Bewegungssehen. *Z. Naturf.* **14b**, 724.

—— and WENKING, H. (1969). Optical detection and fixation of objects by fixed flying flies. *Naturw.* **56**, 424.

RESCH, B. (1954). Untersuchungen ueber das Farbensehen von *Notonecta glauca* L. *Z. vergl. Physiol.* **36**, 27.

REUTER, T. *and* VIRTANEN, K. (1972). Border and colour coding in the retina of the frog. *Nature, Lond.* **239**, 260.

RITCHIE, J. M. (1971). Electrogenic ion pumping in nervous tissue. *Curr. Topics Bioenerget.* **4**, 327.

ROBINSON, D. A. (1972). Models of oculomotor organisation. In *The control of eye movements* (eds. P. Bach-y-Rita and C. C. Collins), pp. 519–38. Academic Press, New York.

ROGERS, G. L. (1962). A diffraction theory of insect vision. II. Theory and experiments with a simple model eye. *Proc. R. Soc.* **B 157**, 83.

RÖHLICH, P. (1967). Fine structural changes of photoreceptors induced by light and prolonged darkness. *Symp. on Neurobiology of Invertebrates*, Budapest.

—— and TÖRÖK, L. (1962). The effect of light and darkness on the fine structure of the retinal club in *Dendrocoelum lacteum. Q. J. micr. Sci.* **104**, 543.

—— and —— (1965). Fine structure of the compound eye of *Daphnia* in normal, dark- and strongly light-adapted state. *Eye Structure*, 2nd Symp. (ed. J. W. Rohen), Schattauer-Verlag, Stuttgart.

ROKOHL, R. (1942). Ueber die regionale Verschiedenheit der Farbentuechtigkeit im zusammengesetzten Auge von *Notonecta glauca: Z. vergl. Physiol.* **29**, 638.

ROMANELI, E. (1968). Etude expérimentale de l'influence de quelques facteurs sur les réactions d'envol chez un Insecte: *Drosophila melanogaster* Meigen. *Unpublished Doctor Thesis*, Univ. of Toulouse, France, p. 104.

ROSENGREN, R. (1971). Route fidelity, vision memory and recruitment behaviour foraging wood ants of the genus Formica. *Acta zool. Fenn.* **133**, 1.

ROSSER, B. (1974). A study of the afferent pathways of the dragonfly lateral ocellus from extracellularly recorded spike discharges. *J. exp. Biol.* **60**, 135.

ROSSER, R. (1972). *Ph.D. Thesis*, University of Bristol.

ROTH, H. *and* MENZEL, R. (1972). ERG of *Formica polyctena* and selective adaptation. In *Information processing in the visual systems of Arthropods*. (ed. R. Wehner), pp. 177–182, Springer Verlag, Heidelberg, New York.

ROWELL, C. H. F. (1971*a*). The orthopteran descending movement detector (DMD) neurones: a characterisation and review. *Z. vergl. Physiol.* **73**, 167.

—— (1971*b*). Variable responsiveness of a visual interneurone in the free-moving locust, and its relation to behaviour and arousal. *J. exp. Biol.* **55**, 727.

—— (1971*c*). Antennal cleaning, arousal and visual interneurone responsiveness in a locust. *J. exp. Biol.* **55**, 749.

—— *and* HORN, G. (1968). Dishabitation and arousal in the response of single nerve cells in an insect brain. *J. exp. Biol.* **49**, 171–183.

RUCK, P. (1958*a*). Dark adaptation of the ocellus in *Periplaneta americana*: a study of the electrical response to illumination. *J. Insect Physiol.* **2**, 189.

—— (1958*b*). A comparison of the electrical responses of compound eyes and dorsal ocelli in four insect species. *J. Insect Physiol.* **2**, 261.

—— (1961*a*). Electrophysiology of the insect dorsal ocellus. I. Origin of the components of the electroretinogram. *J. gen. Physiol.* **44**, 605.

—— (1961*b*). Electrophysiology of the insect dorsal ocellus. II. Mechanisms of generation and inhibition of impulses in the ocellar nerve of dragonflies. *J. gen. Physiol.* **44**, 629.

—— (1961*c*). Electrophysiology of the insect dorsal ocellus. III. Responses to flickering light of the dragonfly ocellus. *J. gen. Physiol.* **44**, 641.

—— (1965). The components of the visual system of a dragonfly. *J. gen. Physiol.* **49**, 289.

RUSHTON, W. A. H. (1959). A theoretical treatment of Fuortes' observations upon the eccentric cell activity in *Limulus*. *J. Physiol.* **148**, 29.

SANDEMAN, D. C. *and* OKAJIMA, A. (1973). Statocyst-induced eye movements in the crab *Scylla serrata* II. The responses of the eye muscles. *J. exp. Biol.* **58**, 197.

SANDER, W. (1933). Phototaktische Reaktionen der Bienen auf Lichter verschiedener Wellenlaenge: *Z. vergl. Physiol.* **20**, 267.

SANNASI, A. (1970). Resilin in the lens cuticle of the firefly, *Photinus pyralis* Linnaeus. *Experientia* **26**, 154.

SATIJA, R. C. (1957). Studies on the course and functions of the giant nerve fibres from the eye and the tegumentary nerve to the ventral nerve cord in *Locusta migratoria*. *Res. Bull. Panjab. Univ. (Zool.)* **132**, 511.

—— (1958). A hystological and experimental study of nervous pathways in the brain and thoracic nerve cord of *Locusta migratoria migratoriodes* (R. & F.). *Res. Bull. Panjab Univ.* **137**, 13.

SATO, S., KATO, M., *and* TORIUMI, M. (1957). Structural changes of the compound eye of *Culex pipiens* var. pallens Coquillet in the process of dark adaptation. *Sci. Rep. Res. Inst. Tôhoku Univ.* **29**, 91.

SCHALLER, F. (1969). Zur Frage des Formsehens bei Collembolen. *Verh. dt. Zool. Ges.* 1968, *Zool. Anz. Suppl.* **32**, 368.

SCHIFF, H. (1963). Dim light vision of *Squilla mantis* L. *Am. J. Physiol.* **205**, 927.

SCHLEIN, H. (1972). Postemergence growth in the fly *Sarcophaga falculata* initiated by neurosecretion from the ocellar nerve. *Nature, Lond.* **236**, 217.

SCHLEGTENDAL, A. (1934). Beitrag zum Farbensinn der Arthropoden. *Z. vergl. Physiol.* **20**, 545.

SCHLIEPER, C. (1926). Der Farbensinn von *Hippolyte*, zugleich ein Beitrag zum Bewegungssehen der Krebse. *Verh. Dtsch. Zool. Ges. Kiel* **31**, 188.

—— (1927). Farbensinn der Tiere und optomotorische Reaktionen der Tiere. *Z. vergl. Physiol.* **6**, 453.

—— (1928). Ueber die Helligkeitsverteilung im Spektrum bei verschiedenen Insekten. *Z. vergl. Physiol.* **8**, 281.

SCHMIDT, W. J. (1938). Polarisationsoptische Analyse eines Eiweiss-Lipoid Systems, erläutert am Aussenglied der Sehzellen. *Kolloidzschr.* **85**, 137.

SCHNEIDER, G. (1956). Zur spektralen Empfindlichkeit des Komplexauges von *Calliphora. Z. vergl. Physiol.* **39**, 1.

SCHNERLA, T. C. (1960). L'appentissage et la question du conflict chez la foutme. Comparison avec le rat. *J. Physiol. norm. pathol.* **57**, 11.

SCHNETTER, B. (1968). Visuelle Formunterscheidung der Honigbiene im Bereich von Vier- und Sechsstrahlsternen. *Z. vergl. Physiol.* **59**, 90.

—— (1972). Experiments on pattern discrimination in honey bees. In *Information Processing in the Visual Systems of Arthropods*, (ed. R. Wehner), pp. 195–200. Springer, Berlin, Heidelberg, New York.

SCHOENE, H. (1953). Farbhelligkeit und Farbunterscheidung bei den Wasserkaefern *Dytiscus marginalis, Acilius sulcatus* und ihren Larven. *Z. vergl. Physiol.* **35**, 27.

SCHOLES, J. (1969). The electrical responses of the retinal receptors and the lamina in the visual system of the fly *Musca. Kybernetik* **6**, 149.

—— and REICHARDT, W. (1969). The quantal content of optomotor stimuli and the electrical responses of receptors in the compound eye of the fly *Musca. Kybernetik* **6**, 74.

SCHULTZE, M. S. (1868). *Untersuchungen über die zusammengesetzten Augen der Krebse und Insekten* pp. 1–32. *Anat. Inst. Bonn.*

SCHULZE-SCHENKING, M. (1970). Untersuchungen zur visuellen Lerngeschwindigkeit und Lernkapazität bei Bienne, Hummeln und Ameisen. *Z. Tierpsychol.* **27**, 513.

SCHUEMPERLI, R. (1972). Wavelength-specific behavioural reactions in *Drosophila melanogaster*. In *Information processing in the visual systems of Arthropods.* (ed. R. Wehner), p. 155, Springer Verlag, Heidelberg, New York.

SCHWARZ, A. (1885). Ueber das Gesetz der Quellung von Leimcylindern. *Exner's Repert. Phys.* **21**, 702.

SCHWEMER, J., GOGALA, J. and HAMDORF, K. (1971). Der UV-Sehfarbstoff der Insekten. *Z. vergl. Physiol.* **75**, 174.

SEELEN, W. VON (1970). Zur Informationsverarbeitung in visuellen System der Wirbetiere. I. *Kybernetik* **7**, 44.

SEIBT, U. (1967). Der Einfluss der Temperatur auf die Dunkeladaptation von *Apis mellifica. Z. vergl. Physiol.* **57**, 77.

SEITZ, G. (1968). Der dioptrische Apparat im Insektenauge. Sonderdruck aus Verhandlungen der Deutschen Zoologischen Gesellschaft in Innsbruck, Akademische Verlagsgesellschaft. Geest and Portig K.-G., Leipzig, 361.

—— (1968). Der Strahlengang im Appositionsauge von *Calliphora erythrocephala* (Meig.) *Z. vergl. Physiol.* **59**, 205.

SEITZ, G. (1969a). Untersuchungen am dioptrischen Apparat des Leuchkäferauges. *Z. vergl. Physiol.* **62**, 61.

—— (1969b). Polarisationsoptische Untersuchungen am Auge von *Calliphora erythrocephala* (Meig.). *Z. Zellforsch.* **93**, 525.

—— (1970). Evidence for a longitudinal pupil in the blowfly eye from studies with light and electron microscopes. *Z. vergl. Physiol.* **69**, 169.

—— (1971). Bau und Funktion des Komplexauges der Schmeissfliege. *Naturw.* **58**, 258.

SELDIN, E. B., WHITE, R. H., and BROWN, P. K. (1972). Spectral sensitivity of larval mosquito ocelli. *J. gen. Physiol.* **59**, 415.

SHAW, S. R. (1967). Simultaneous recording from two cells in locust retina. *Z. vergl. Physiol.* **55**, 183.

—— (1968). Organization of the locust retina. *Symp. zool. soc. Lond.* **23**, 135.

—— (1969a). Optics of arthropod compound eye. *Science* **164**, 88.

—— (1969b). Interreceptor coupling in ommatidia of drone honeybee and locust compound eyes. *Vision. Res.* **9**, 999.

—— (1969). Sense cell structure and interspecies comparisons of polarized light absorption in arthropod compound eye. *Vis. Res.* **9**, 1031.

—— (1972). Decremental conduction of the visual signal in the barnacle lateral eye. *J. Physiol.* **220**, 145.

SHAW, T. I. (1972). The circular dichroism and optical rotatory dispersion of visual pigments. In *Photochemistry of vision*, (ed. M. J. A. Dartnall), Vol. VII/1 *Handbook of sensory physiology*, 180, Springer Verlag, Heidelberg, New York.

SCHNEIDER, L. and LANGER, H. (1969). Die Struktur des Rhabdoms im "Doppelauge" des Wasserläufers *Gerris lacustris*. *Z. Zellforsch.* **99**, 538.

SJOSTRAND, F. S. (1961). Electron microscopy of the retina. In *The structure of the eye*. (ed. G. K. Smelser), Academic Press, London and New York.

SKRZIPEK, K.-H. and SKRZIPEK, H. (1971). Die Morphologie der Bienenretina (*Apis mellifica* L.) in elektronenmikroskopischer und lichtmikroskopischer Sicht. *Z. Zellforsch.* **119**, 552.

—— and —— (1971). Zur funktionellen Bedeutung der räumlichen Anordnung des Kristallkegels zum Rhabdom im Auge der Trachtbiene (*Apis mellifica* L.). *Experientia* **27**, 409.

SLIFER, E. H. and SEKHON, S. S. (1970). Sense organs of a thysanuran, *Ctenolepisma lineata*, with special reference to those on the antennal flagellum (Thysanura, Lepismatidae). *J. Morph.* **132**, 1.

SMITH, D. S., JÄRLFORS, U., and BERÁNEK, R. (1970). The organisation of synaptic axoplasm in the lamprey (*Petromyzon marinus*) central nervous system. *J. cell. Biol.* **46**, 199.

SMITH, T. G. and BAUMANN, F. (1969). The functional organization within the ommatidium of the lateral eye of *Limulus*. *Prog. Brain Res.* **31**, 313.

——, STELL, W. K., and BROWN, J. E. (1968). Conductance changes associated with receptor potentials in *Limulus* photoreceptors. *Science* **162**, 454.

——, ——, ——, FREEMAN, J. A. and MURRAY, G. C. (1968). A role for the sodium pump in photoreception in *Limulus*. *Science* **162**, 456.

SMOLA, U. and GEMPERLEIN, R. (1972). Transfer characteristics of the visual cell of *Calliphora erythrocephala*. 2. Dependence on the location of measurement: Retina—lamina ganglionaris. *J. comp. Physiol.* **79**, 363.

SNITZER, E. (1961). Cylindrical dielectric waveguide modes. *J. opt. Soc. Am.* **51**, 491.

582   References

SNYDER, A. W. (1966). Surface waveguide modes along a semi-infinite dielectric fibre excited by a plane wave. *J. opt. Soc. Am.* **56**, 601.

—— (1969a). Asymptotic expressions for eigenfunctions and eigenvalues of a dielectric or optical waveguide. *IEEE Trans. Microwave Theor. Techn.* **17**, 1130.

—— (1969b). Excitation and scattering of modes on a dielectric or optical fiber. *IEEE Trans. Microwave theor. techn.* **17**, 1138.

—— (1970). Coupling of modes on a tapered dielectric cylinder. *IEEE Trans. Microwave theor. techn.* **18**, 383.

—— (1972a). Angular sensitivity of the bee ommatidium. *Z. vergl. Physiol.* **76**, 438.

—— (1972b). Power loss on optical fibers. *Proc. IEEE*, **60**, 757.

—— (1972c). Coupled mode theory for optical fibres. *J. opt. Soc. Am.* **62**, 1267.

—— (1973). Polarization sensitivity of individual retinula cells. *J. comp. Physiol.* submitted for publication.

—— (1974). Light absorption in visual photoreceptors. *J. opt. Soc. Am.* **64**, 216.

—— and HAMER, M. (1972). The light capture area of a photoreceptor. *Vis. Res.* **12**, 1749.

—— and HORRIDGE, G. A. (1972). The optical function of changes in the medium surrounding the cockroach rhabdom. *J. comp. Physiol.* **81**, 1.

—— and KAMER, M. (1972). The light capture area of a photoreceptor. *Vis. Res.* **12**, 1749.

——, MENZEL, R. and LAUGHLIN, S. B. (1973). Structure and function of the fused rhabdom, *J. comp. Physiol.* **87**, 99–135.

—— and MILLER, W. H. (1972). Fly colour vision. *Vis. Res.* **12**, 1389.

—— and PASK, C. (1972a). A theory for changes in spectral sensitivity induced by off-axis light. *J. comp. Physiol.* **79**, 423.

—— and —— (1972b). Light absorption in the bee photoreceptor. *J. Opt. Soc. Am.* **62**, 999.

—— and —— (1972c). Detection of polarization and direction by the bee rhabdom. *J. comp. Physiol.* **78**, 346.

—— and —— (1972d). Can an individual bee ommatidium detect an image? *J. comp. Physiol.* **80**, 51.

—— and —— (1972e). How bees navigate. *Nature, Lond.* **239**, 48.

—— and —— (1973a). Failure of geometric optics for optical waveguides. *J. opt. Soc. Am.* **64**, 608.

—— and —— (1973b). Diffraction properties of a photoreceptor compensate for lens diffraction. *J. comp. Physiol.* To be submitted.

—— and —— (1973c). Spectral and polarization sensitivity of *Diptera* retinula cells. *J. comp. Physiol.* **84**, 59.

—— and —— (1973d). Waveguide modes and light absorption in visual photoreceptors, *Vision Res.* **13**, 2605.

——, ——, and MITCHELL, D. J. (1973d). Light acceptance property of an optical fibre. *J. opt. Soc. Am.* **63**, 59.

—— and RICHMOND, P. (1972). Effect of anomalous dispersion on visual photopigments. *J. opt. Soc. Am.* **62**, 1278.

—— and —— (1973). Anomalous dispersion in visual photoreceptors. *Vis. Res.* **13**, 511.

—— and SAMMUT, R. (1973). Direction of E for maximum response of a retinula cell. *J. comp. Physiol.* **85**, 37–45.

SOTAVALTA, O., TUURALA, O., and OURA, A. (1962). On the structure and photo-mechanical reactions of the compound eyes of craneflies (Tipulidae, Limnobiidae). *Ann. Acad. Sci. Fenn.* A IV **62**, 1.

STEVENS, S. S. (1972). A neural quantum in sensory discrimination. *Science* **177**, 749.

STEYSKAL, G. C. (1957). Notes on colour pattern of eye in *Diptera. Bull. Brooklyn Entomolog. Soc.* **52**, 89.

STIEVE, H. (1964). Das Belichtungspotential der Isolierten Retina des Einsiedlerkrebses (*Eupagurus bernhardus* L.) in Abhängigkeit von den Extrazellulären Ionenkonzentrationen. *Z. vergl. Physiol.* **47**, 457.

—— (1965). Interpretation of the generator potential in terms of ionic processes. *Cold Spring Harbor Symp. quant. Biol.* **30**, 451.

——, BOLLMAN-FISCHER, H. *and* BRAUN, B. (1971). The significance of metabolic energy and the ion pump for the receptor potential of the crayfish photoreceptor cell. *Z. Naturforsch.* **26**, 1311.

—— *and* WIRTH, C. (1971). Über die Ionen-Abhängigkeit des Rezeptorpotentials der Retina von *Astacus leptodactylus. Z. Naturforsch.* **26**, 457.

STOCKHAMMER, K. (1956). Zur Wahrnehmung der Schwingungsrichtung linear polarisierten Lichtes bei Insekten. *Z. vergl. Physiol.* **38**, 30.

STRAUSFELD, N. J. (1970a). Golgi studies on insects. Part II. The optic lobes of Diptera. *Phil. Trans.* B **258**, 135.

—— (1970b). Variations and invariants of cell arrangements in the nervous system of insects. (A review of neuronal arrangements in the visual system and corpora pedunculata). *Verh. d. Dtsch. Zool. Ges.* **64**, 97.

—— (1971a). The organization of the insect visual system (light microscopy). I. Projections and arrangements of neurons in the lamina ganglionaris of Diptera. *Z. Zellforsch.* **121**, 377.

—— (1971b). The organisation of the insect visual system (light microscopy). II. Projection of fibres across the first optic chiasma. *Z. Zellforsch.* **121**, 442.

—— *and* BLEST, A. D. (1970). Golgi studies on insects. Part I. The optic lobes of Lepidoptera. *Phil. Trans.* B **258**, 81.

—— *and* BRAITENBERG, V. (1970). The compound eye of the fly (*Musca domestica*): connections between the cartridges of the lamina ganglionaris. *Z. vergl. Physiol.* **70**, 95.

—— *and* CAMPOS-ORTEGA, J. A. (1972). Some interrelationships between the first and second synaptic regions of the fly's (*Musca domestica*) visual system. In: *Information Processing in the Visual Systems of Arthropods* (ed. R. Wehner), Springer, Berlin.

STRECK, P. (1972a). Der Einfluss des Schirmpigmentes auf das Sehfeld einzelner Sehzellen der Fliege *Calliphora erythrocephala. Z. vergl. Physiol.* **76**, 372.

—— (1972b). Screening pigment and visual field of single retinula cells of *Calliphora*. In *Information Processing in the Visual Systems of Arthropods* (ed. R. Wehner), pp. 127–131. Springer, Berlin, Heidelberg, New York.

STROTHER, G. K. (1966). Absorption of *Musca domestica* screening pigment. *J. gen. Physiol.* **49**, 1087.

—— *and* CASELLA, A. J. (1972). Microspectrophotometry of arthropod visual screening pigments. *J. gen. Physiol.* **59**, 616.

—— *and* SUPERDOCK, D. A. (1972). In situ absorption of *Drosophila melanogaster* visual screening pigments. *Vis. Res.* **12**, 1545.

STRUWE, G. (1972). Spectral sensitivity of single photoreceptors in the compound eye of a tropical butterfly (*Heliconius numata*). *J. comp. Physiol.* **79**, 197.

SUGA, N. *and* KATSUKI, Y. (1962). Vision in Insects in terms of the electrical activities of the descending nerve fibres. *Nature, Lond.* **194**, 658.

584   *References*

S<small>UTHERLAND</small>, N. S. (1968). Outlines of a theory of visual pattern recognition in animals and man. *Proc. R. Soc.* B **171**, 297.

S<small>UTRO</small>, L. L. *and* K<small>ILMER</small>, W. L. (1969). Assembly of computers to command and control a robot. *A.F.I.P.S. Conf. Proc.* **34**, 113.

S<small>WAMMERDAM</small>, J. (1737–1738). Bybel der Natuure. Vols. I–II. Published by H. Boer-haave, Severinus, B. Vander and P. Vander, Leyden.

S<small>WIHART</small>, S. L. (1968). Single unit activity in the visual pathway of the butterfly *Heliconius erato*. *J. Insect Physiol.* **14**, 1589.

—— (1969). Colour vision and the physiology of the superposition eye of a butterfly (Hesperiidae). *J. Insect. Physiol.* **15**, 1347.

—— (1970). The neural basis of colour vision in the butterfly, *Papilio troilus*. *J. Insect Physiol.* **16**, 1623.

—— (1972). The neural basis of colour vision in the butterfly, *Heliconius erato*, *J. Insect Physiol.* **18**, 1015.

—— *and* G<small>ORDON</small>, W. C. (1971). Red photoreceptor in butterflies. *Nature, Lond.* **231**, 126.

T<small>ADDEI</small> F<small>ERRETTI</small>, C. *and* F<small>ERNANDEZ</small> P<small>EREZ</small> <small>DE</small> T<small>ALENS</small>, A. (1967). La reazione d'atter-raggio della *Musca domestica*. *Atti Convegno Gruppo Nazionale Cibernetica, Pisa*, 24.

—— *and* —— (1971). Neurosistemi: esperimenti sulla reazione d'atterraggio della *Musca domestica*. *Atti I Congresso Nazionale Cibernetica, Casciana Terme*, 268.

—— *and* —— (1972). Conditions for an object moving in a fly's visual field to be stimulating for landing. *Atti. II Congresso Nazionale Cibernetica, Casciana Terme*, 203.

—— *and* —— (1973a). Filter characteristics of Insect ommatidium—Automatica, in the press.

—— *and* —— (1973b). Landing reaction of *Musca domestica*, IV: A. monocular and binocular vision; B. relationships between landing and optomotor reactions. *Z. Naturf.* B, in the press.

—— *and* —— (1973c). Landing reaction of *Musca domestica*, III: dependence on the luminous characteristics of the stimulus. *Z. Natuf.* **28c**, 568–578.

T<small>AMIR</small>, T. *and* O<small>LINER</small>, A. A. (1969). Role of the lateral wave in total reflection of light. *J. opt. Soc. Am.* **59**, 942.

T<small>ASAKI</small>, K., T<small>SUKAHARA</small>, Y., I<small>TO</small>, S., W<small>AYNER</small>, M. J., *and* Y<small>U</small>, W. Y. (1968). A simple, direct, and rapid method for filling microelectrodes. *Physiol. Behav.* 3, 1009.

T<small>HIBAUD</small>, J. M. (1967a). Structure et régression de l'appareil visuel chez les Hypogastruridae (Collemboles) épigés et cavernicoles. *Ann. Spéléol.* **22**, 407.

—— (1967b). Étude de l'appareil oculaire chez *Typhlogastrura balazuci*, Hypogastru-ridae (Collembole) cavernicole. *Ann. Spéléol.* **22**, 797.

T<small>HOMAS</small>, E. (1955). Untersuchungen ueber den Helligkeits- und Farbensinn der Anuren: *Zool. Jb., Abt. allg. Zool. u. Physiol.* **66**, 129.

T<small>HOMAS</small>, I. *and* A<small>UTRUM</small>, H. (1965). Die Empfindlichkeit der dunkel- und hell-adaptier-ten Biene (*Apis mellifica*) fuer spektrale Farben: Zum Purkinje-Phaenomen der Insekten. *Z. vergl. Physiol.* **51**, 204.

T<small>HORSON</small>, J. (1964). Dynamics of motion perception in the desert locust. *Science* **145**, 69.

—— (1966a). Small signal analysis of a visual reflex in the locust. I. Input parameters. *Kybernetik* 3, 41.

—— (1966b). Small signal analysis of a visual reflex in the locust. II. Frequency dependence. *Kybernetik* 3, 53.

TINBERGEN, N. (1966). *Animal Behavior*. Time-Life International, New York.

—— (1972). *The animal in its world*. George Allen and Unwin, London.

——, MEEUSE, B. J. D., BOREMA, L. K., *and* VAROSSIEAU, W. W. (1942). Die Balz des Samtfalters *Eumenis* (= Satyrus) *semele* (L.): *Z. Tierpsychol.* **5**, 182.

TOMITA, T. (1956). The nature of action potentials in the lateral eye of the horseshoe crab as revealed by simultaneous intra- and extra-cellular recording. *Jap. J. Physiol.* **6**, 327.

—— (1957). Peripheral mechanism of nervous activity in lateral eye of horseshoe crab. *J. Neurophysiol.* **20**, 245.

——, KIKUCHI, R., *and* TANAKA, I. (1960). Excitation and inhibition in lateral eye of horseshoe crab. In *Electrical Activity of Single Cells*. pp. 11–23. Tokyo, Igakushoin, Hongo.

TRINCKER, D. *and* BERNDT, P. (1957). Optomotorische Reaktionen und Farbensinn beim Meerschweinchen. *Z. vergl. Physiol.* **39**, 607.

TRUJILLO-CENÓZ, O. (1966). Some aspects of the structural organization of the arthropod eye. *Cold Spring Harbor Symp. quant. Biol.* **30**, 371.

—— (1965). Some aspects of the structural organisation of the intermediate retina of Dipterans. *J. Ultrastruct. Res.* **13**, 1.

—— (1969). Some aspects of the structural organization of the medulla in muscoid flies. *J. Ultrastruct. Res.* **27**, 533.

—— *and* BERNARD, G. D. (1972). Some aspects of the retinal organization of *Sympycnus lineatus* Loew (Diptera, Dolichopodidae). *J. Ultrastruct. Res.* **38**, 149.

—— *and* MELAMED, J. (1966*a*). Compound eye of Dipterans: anatomical basis for integration. An electron microscope study. *J. Ultrastruct. Res.* **16**, 395.

—— *and* —— (1966*b*). *The Functional Organisation of the Compound Eye* (ed. C. G. Bernhard), Pergamon Press, London.

—— *and* —— (1970). Light and electron microscopical study of one of the systems of centrifugal fibers found in the lamina of Muscoid flies. *Z. Zellforsch.* **110**, 336.

TSIRULIS, T. P. (1966). An electronmicroscopic study of mitochondria of ellipsoid of photoreceptors adapted to the light and to the dark. (Russian). *Tsitologia* **8**, 90.

TSUNEKI, K. (1950). Some experiments on the colour vision in ants. *J. Fac. Sci. Hokkaido Univ., Ser.* 6 *Zool.* **10**, 77.

TUNSTALL, J. *and* HORRIDGE, G. A. (1967). Electrophysiological investigation of the optics of the locust retina. *Z. vergl. Physiol.* **55**, 167.

TUURALA, O. (1963). Bau und Photomechanische Erscheinungen in Auge einiger Chironomiden (Dipt.). *Ann. Ent. fenn* **29**, 209.

—— (1954). Histologische und Physiologische Untersuchungen über die Photomechanischen Erscheinungen in den Augen der Lepidopteran. *Ann. Acad. Sci. fenn.* A IV **24**, 5.

UMBACH, W. (1934). Entwicklung und Bau des Komplexauges der Mehlmotte *Ephestia kuhniella* Zeller, nebst einigen Bemerkungen über die Enstehung der optischen Ganglien. *Z. Morph. Ökol. Tiere* **28**, 561.

VAN DER LUGT, A. (1964). Signal detection by complex spatial filtering. *IEEE Trans. Information Theory* **10**, 139.

VARELA, F. G. (1970). Fine structure of the visual system of the honeybee (*Apis mellifera*). II. The lamina. *J. Ultrastruct. Res.* **31**, 178.

586   *References*

VARELA, F. G. *and* PORTER, K. R. (1969). Fine structure of the visual system of the honeybee (*Apis mellifera*). I. The retina. *J. Ultrastruct. Res.* **29**, 236.

—— *and* WIITANEN, W. (1970). The optics of the compound eye of the honeybee (*Apis mellifera*). *J. gen. Physiol.* **55**, 336.

VARJU D. (1959). Optomotorische Reaktionen auf die Bewegung periodischer Helligkeitsmuster. *Z. Naturforsch.* **14b**, 724.

VAYSSE, G. (1973*a*). Sur le rôle dynamogène éventual des ocelles frontaux dans le comportement sexuel du mâme chez *Drosophila melanogaster*. Unpublished Doctor Thesis, University of Toulouse, France, p. 156.

—— (1973*b*). Sur le rôle activateur des ocelles frontaux dans la compétition sexuelle entre les mâles chez *Drosophila melanogaster*. *C.R. Soc. Biol.* (in press).

VIALLANES, H. (1886). La structure du cerveau des Hyménoptères. *Bull. Soc. Philom. Paris* **10**, 82.

VIGIER, P. (1908). Mecanisme de la synthèse des impressions lumineuses recuellies par les yeux composés des Diptères. *C.R. Soc. Biol. Paris* **64**, 1221.

VINNIKOV, YA. A. (1969). The ultrastructural and cytochemical bases of the mechanism of function of the sense organ receptors. In *The structure and function of nervous tissue* (ed. G. H. Bourne), Academic Press, New York, London.

VOGEL, G. (1954). Das optische Weibehenschema bei *Musca domestica*. *Naturw.* **41**, 482.

—— (1957). Verhaltensphysiologische Untersuchungen über die den eibehenbesprung des Stubenfliegen—Männchens (*Musca domestica*) auslösenden optischen Faktoren. *Z. Tierpsychol.* **14**, 309.

VOGT, P. (1966). Bienen unterscheiden Flimmerfrequenzen im Verhaltensexperiment. *Naturw.* **53**, 536.

—— (1969). Dressur von Sammelbienen auf sinusfoermig moduliertes Flimmerlicht. *Z. vergl. Physiol.* **63**, 182.

VOGT, W. (1964). Ueber die optischen Schluesselreize beim Beuteerwerb der Larven der Libelle *Aeschna cyanea*. *Zool. Jb. Physiol.* **71**, 171.

VOSS, CH. (1955). Die Bedeutung von Streifenmustern für die Orientier ung der Roten Waldameise. *Verh. Deutsch. Zool. Ges. Kiel*, 540.

—— (1967). Das Formensehen der Roten Waldameise, *Formica rufa*. *Z. vergl. Physiol.* **55**, 225.

VOWLES, D. M. (1950). Sensitivity of ants to polarized light. *Nature, Lond.* **165**, 282.

—— (1954). The orientation of ants. *J. exp. Biol.* **31**, 314.

—— (1955). The structure and connections of the corpora pedunculata of bees and ants. *Quart. J. Microscop. Sci.* **96**, 239.

—— (1965). Maze learning and visual discrimination in the wood ant. *Brit. Psych.* **56**, 15.

—— (1966). The receptive fields of cells in the retina of the housefly (*Musca domestica*). *Proc. R. Soc.* B **164**, 552.

VRIES, H. DE (1956). Physical aspects of the sense organs. *Progr. Biophysics* **6**, 208.

—— *and* KUIPER, J. W. (1958). Optics of the Insect eye. *Ann. N.Y. Acad. Sci.* **74**, 196.

WACHMANN, E. (1972*a*). Zum Feinbau des Komplexauges von *Stylops* sp. (Insecta, Strepsiptera). *Z. Zellforsch.* **123**, 411.

—— (1972*b*). Das Auge des Hühnerflohs. *Z. Morph. Tiere* **73**, 315.

WADDINGTON, C. H. *and* PERRY, M. M. (1960). The ultra-structure of the developing eye of *Drosophila*. *Proc. R. Soc. Lond.* B **153**, 155.

—— *and* —— (1963). Inter-retinular fibers in the eyes of *Drosophila*. *J. Insect Physiol.* **9**, 475.

WALD, G., BROWN, P. K., *and* GIBBONS, I. R. (1962). Visual excitation: a chemo-anatomical study. *Symp. Soc. exp. Biol.*

—— *and* KRAININ, J. M. (1963). The median eye of *Limulus*, an ultraviolet receptor. *Proc. Nat. Acad. Sci.* **50**, 1011–1017.

WALLACE. G. K. (1959). Visual scanning in the desert locust *Schistocerca gregaria* Forskäl. *J. exp. Biol.* **36**, 512.

WALLS, G. L. (1963). *The vertebrate eye*. Hafner, New York.

WALTHER, J. B. (1958a). Changes induced in the spectral sensitivity and form of retinal action potential of the cockroach eye by selective adaptation. *J. Insect Physiol.* **2**, 142.

—— *and* DODT, E. (1959). Die Spektralsensitivität von Insektenkomplexaugen im Ultraviolett bis 290 mµ. Elektrophysiologische Untersuchungen an *Calliphora* und *Periplaneta*. *Z. Naturf.* **14b**, 273.

WALCOTT, B. (1969). Movement of retinula cells in insect eyes on light adaptation. *Nature, Lond.* **223**, 971.

—— (1971a). Cell movement on light adaptation in the retina of *Lethocerus* (Belostomatidae Hemiptera). *Z. vergl. Physiol.* **74**, 1.

—— (1971b). Unit studies on receptor movement in the retina of *Lethocerus* (Belostomitidae, Hemiptera). *Z. vergl. Physiol.* **74**, 17.

—— *and* HORRIDGE, G. A. (1971). The compound eye of *Archichauliodes* (Megaloptera). *Proc. R. Soc. Lond.* B **179**, 65.

WASHIZU, Y., BURKHARDT, D., *and* STRECK, P. (1964). Visual field of single retinula cells and interommatidial inclination in the compound eye of the blowfly *Calliphora erythrocephala*. *Z. vergl. Physiol.* **48**, 413.

WASSERMAN, G. S. (1970). Membrane excitability changes associated with the hyper-polarizing response. *Physiol. Behav.* **5**, 601.

WATERMAN, T. H. (1954). Directional sensitivity of single ommatidia in the compound eye of *Limulus*. *Proc. natn. Acad. Sci. U.S.A.*, **40**, 252.

—— (1954). Polarized light and angle of stimulus incidence in the compound eye of *Limulus*. *Proc. natn. Acad. Sci. U.S.A.* **40**, 258.

——, FERNANDEZ, H. R., *and* GOLDSMITH, T. H. (1969). Dichroism of photosensitive pigment in rhabdoms of the crayfish *Orconectes*. *J. gen. Physiol.* **54**, 415.

—— *and* WIERSMA, C. A. G. (1954). The functional relation between retinal cells and optic nerve in *Limulus*. *J. exp. Zool.* **126**, 59.

——, ——, *and* BUSH, B. M. H. (1964). Afferent visual responses in the optic nerve of the brac, *Podophthalmus*. *J. cell. and comp. Physiol.* **63**, 135.

WEHNER, R. (1967a). Pattern recognition in bees. *Nature, Lond.* **215**, 1244.

—— (1967b). Zur Physiologie des Formensehens bei der Honigbiene. II. Winkel-unterscheidung an Streifenmustern bei variabler Lage der Musterebene im Schwerefed. *Z. vergl. Physiol.* **48**, 413.

—— (1968). Die Bedeutung der Streifenbreite fuer die optische Winkelmessung der Biene, *Apis mellifera*. *Z. vergl. Physiol.* **58**, 322.

—— (1968). Optische Orientierungsmechanismen im Heimkehr-Verhalten von *Cataglyphis bicolor* Fab. *Rev. suisse zool.* **75**, 1076.

—— (1969). Die optische Orientierung nach Größenklassen von *Cataglyphis bicolor* F. *Rev. suisse zool.* **76**, 371.

—— (1969). Der Mechanismus der optischen Winkelmessung bei der Biene, *Apis mellifera*. *Zool. Anz. Suppl.* **33**, 586.

588   References

WEHNER, R. (1971). The generalization of directional visual stimuli in the honey bee, *Apis mellifera*. *J. Insect Physiol.* **17**, 1579.

—— (1972*a*). Dorsoventral asymmetry in the visual field of the bee, *Apis mellifera*. *J. comp. Physiol.* **77**, 256.

—— (1972*b*). Pattern modulation and pattern detection in the visual system of Hymenoptera. In *Information processing in the visual systems of arthropods*, (ed. R. Wehner), pp. 183–194. Springer, Berlin, Heidelberg, New York.

—— (1972*c*). Visual orientation performances of desert ants, *Cataglyphis bicolor*, towards astromenotactic directions and horizon landmarks. *Proc. AIBS Symp. Animal Orientation and Navigation. Wallops Station, Virginia*, pp. 421–436.

—— (1972*d*). Spontaneous pattern preferences of *Drosophila melanogaster* to black areas in various parts of the visual field. *J. Insect Physiol.* **18**, 1531.

—— (1973). Das Koordinatensystem des Sehfeldes bei Arthropoden. *Fortschr. Zool.*, in press.

——, EHEIM, W. P., *and* HERRLING, P. L. (1971). Die Rastereigenschaften des Komplexauges von *Cataglyphis bicolor* (Formicidae, Hymenoptera). *Rev. suisse Zool.* **78**, 722.

—— *and* LINDAUER, M. (1966). Zur Physiologie des Formensehens bei der Hongbiene. I. Winkelunterscheidung an vertikal orientierten Steifenmustern. *Z. vergl. Physiol.* **52**, 290.

—— *and* MENZEL, R. (1969). Homing in the ant *Cataglyphis bicolor*. *Science* **164**, 192.

—— *and* SCHUEMPERLI, R. (1969). Das Aktionsspektrum der phototaktischen Spontantendenz bei *Drosophila melanogaster*. *Rev. suisse Zool.* **76**, 1087.

—— *and* TOGGWEILER, F. (1972). Verhaltensphysiologischer Nachweis des Farbensehens bei *Cataglyphis bicolor* (Formicidae, Hymenoptera). *J. comp. Physiol.* **77**, 239.

—— *and* WEHNER-VON SEGESSER, S. (1973). Calculation of visual receptor spacing in *Drosophila melanogaster* by pattern recognition experiments. *J. comp. Physiol.*, in press.

WEIS-FOGH, T. (1949). An aerodynamic sense organ stimulating and regulating flight in locusts. *Nature, Lond.* **164**, 873.

WEIZSAECKER, E. VON (1970). Dressurversuche zum Formensehen der Bienen, insbesondere unter wechselnden Helligkeitsbedingungen. *Z. vergl. Physiol.* **69**, 296.

WENK, P. (1953). Der Kopf von *Ctenocephalus canis* (Aphaniptera). *Zool. Jb. Anat.* **73**, 103.

WERBLIN, F. S. (1971). Adaptation in a vertebrate retina: intracellular recordings in *Necturus*. *J. Neurophysiol.* **34**, 228.

WERRINGLOER, A. (1932). Die Sehorgane und Sehzentren der Dorylinen nebst Untersuchungen über die Facettenaugen der Formiciden. *Z. wiss. Zool.* **141**, 432.

WHITEHEAD, R. *and* PURPLE, R. L. (1970). Synaptic organisation in the neuropile of the lateral eye of *Limulus*. *Vis. Res.* **10**, 129.

WIEDEMANN, I. (1965). Versuche ueber den Strahlengang im Insektenauge (Appositionsauge). *Z. vergl. Physiol.* **49**, 526.

WIERSMA, C. A. G., BUSH, B. M. H., *and* WATERMAN, T. H. (1964). Efferent visual responses of contralateral origin in the optic nerve of the crab, *Podophthalmus*. *J. cell. and comp. Physiol.* **64**, 309.

—— *and* FIORE, L. (1971). Factors regulating discharge frequency in optomotor fibres of *Carcinus Maenus*. *J. exp. Biol.* **54**, 497.

WIERSMA, C. A. G. *and* OBERJAT, T. (1968). The selective responsiveness of various crayfish occulomotor fibres to sensory stimuli. *Comp. Biochem. Physiol.* **26**, 1.

—— *and* YAMAGUCHI, T. (1967). Integration of visual stimuli by the crayfish central nervous system. *J. exp. Biol.* **47**, 409.

—— *and* YANAGISAWA, K. (1971). On types of interneurons responding to visual stimulation present in the optic nerve of the rock lobster, *Panulirus interruptus*. *J. Neurobiol.* **2**, 291.

WIESE, K. (1972). Das mechanorezeptorische Beuteortungssystem von *Notonecta*. I. Die Funktion des tarsalen Scolopidialorgans. *J. comp. Physiol.* **78**, 83.

WIITANEN, W. *and* VARELA, F. G. (1971). Analysis of the organization and overlap of the visual fields in the compound eye of the honey bee (*Apis mellifera*: Hymenoptera, Apidae). *J. gen. Physiol.* **57**, 303.

WILLEM, V. (1897). Les yeux et les organes post-antennaires des Collemboles. *Ann. Soc. Eng. Belg.* **41**, 225.

—— (1900). Recherches sur les Collemboles et les Thysanoures. *Mém. cour. mém. sav. étrang. Acad. roy. sci. Belg.* **58**, 1.

WILSON, D. M. *and* HOY, R. R. (1968). Optomotor reaction, locomotory bias, and reactive inhibition in the milkweed bug *Oncopeltus* and the beetle *Zophobas*. *Z. vergl. Physiol.* **58**, 135.

WINTHROP, J. T. *and* WORTHINGTON, C. R. (1966). Superposition image formation in Insect eyes. *Biophys. J.* **6**, 124P.

WOLFE, D. E. *and* NICHOLLS, J. G. (1967). Uptake of radioactive glucose and its conversion to glycogen by neurons and glial cells in the leech central nervous system. *J. Neurophysiol.* **30**, 1593.

WOLF, E. (1935). Der Einfluss von intermittierender Reizung auf die optische Reaktion von Insekten. *Naturw.* **23**, 369.

WOLF, K. L. (1968). *Tropfen, Blasen und Lamellen oder von den Formen flüssiger Körper.* Springer Verlag, Berlin, Heidelberg, New York.

WOLKEN, J. J. *and* SCHEER, J. J. (1963). An eye pigment of the cockroach. *Exp. eye Res.* **2**, 182.

WOLSKY, A. (1933). Stimulationsorgane. *Biol. Rev.* **8**, 370.

WULFF, V. J. (1971a). Modification of the receptor potential of the *Limulus* lateral eye by current and light. *Physiol. Behav.* **6**, 513.

—— (1971b). The effect of cyclic AMP on *Limulus* lateral eye retinular cells. *Vis. Res.* **11**, 1493.

—— *and* MENDEZ, C. (1970). Visual receptor potential: modification by injected current in the *Limulus* lateral eye. *Science* **168**, 1351.

WURTZ, R. H. *and* GOLDBERG, M. E. (1971). Superior colliculus cell responses related to eye movements in awake monkeys. *Science* **171**, 82.

YAGI, N. *and* KOYAMA, N. (1963). *The compound eye of Lepidoptera.* Shinkyo-Press & Co. Ltd., Tokyo, Japan.

YARBUS, A. L. (1967). *Eye movements and vision.* Plenum, New York.

YEANDLE, S. S. (1967). Some properties of the components of the *Limulus* ommatidial potential. *Kybernetik* **3**, 250.

YOUNG, L. R. *and* STARK, L. (1963). Variable feedback experiments testing a sampled data model for eye tracking movements. *IEEE Trans. Human Factors in Electronics* HFE—**4**, 38.

ZAENKERT, A. (1939). Vergleichend-morphologische und physiologisch-funktionelle Untersuchungen an Augen beutefangender Insekten. *Sitzungsber. Ges. Naturforsch. Berlin*, 1–3, 82–169.

ZENKIN, G. M. *and* PIGAREV, I. N. (1971). Visually conditioned activity in the dragonfly. *Biophysics.* **16**, 299.

ZETTLER, F. (1969). Die Abhaengigkeit des Uebertragungsverhaltens von Frequenz und Adaptationszustand; gemessen am einzelnen Lichtrezeptor von *Calliphora erythrocephala. Z. vergl. Physiol.* **64**, 432.

—— *and* JAERVILEHTO, M. (1970). Histologische Lokalisation der Ableiteelektrode. Belichtungspotentiale aus Retina und Lamina bei *Calliphora. Z. vergl. Physiol.* **68**, 202.

—— *and* —— (1971). Decrement-free conduction of graded potentials along the axon of a monopolar neuron. *Z. vergl. Physiol.* **75**, 402.

—— *and* —— (1972a). Lateral inhibition in an insect eye. *Z. vergl. Physiol.* **76**, 233.

—— *and* —— (1972b). Intraaxonal visual responses from visual cells and second order neurons of an insect retina. In *Information processing in the visual systems of arthropods*, (ed R. Wehner), pp. 217–222. Springer, Berlin, Heidelberg, New York.

ZOLOTOV, V. (1972). Orientation by polarized light in the honeybee. (Russian). *Pchelovodstvo* **8**, 16.

ZOLOTOV, V. *and* FRAUTSEVICH, L. (1973). Orientation of bees by the polarized light of a limited area of the sky. *J. comp. Physiol.* **85**, 25.

ZORKOCZY, P. I. (1966). Cybernetic models of pattern sensitive units in the visual system. *Kybernetik.* **3**, 143.

ZSAGAR, H. J. (1972). Eigenschaften von Ortsfiltern mit zeitveränderlichen Koppelfunktionem. *Kybernetik.* **10**, 16.

ZYZNAR, E. S. (1970). The eyes of white Shrimp, *Penaeus setiferus* (Linnaeus) with a note on the rock shrimp *Sicyonia brevirostris* Stimpson. *Contr. mar. Sci.*, **15**, 87.

Index